To Sam Hallmark,

My life-long friend and colleague. Best wishes for tight lines and big fish.

Jerald S. R.

Biology and Management *of the* World Tarpon and Bonefish Fisheries

MARINE BIOLOGY
SERIES
Peter L. Lutz, Editor

PUBLISHED TITLES

Biology of Marine Birds
E. A. Schreiber and Joanna Burger

Biology of the Spotted Seatrout
Stephen A. Bortone

The Biology of Sea Turtles, Volume II
Peter L. Lutz, John A. Musick, and Jeanette Wyneken

Biology of Sharks and Their Relatives
Jeffrey C. Carrier, John A. Musick, and Michael R. Heithaus

Early Stages of Atlantic Fishes: An Identification Guide for the Western Central North Atlantic
William Richards

The Physiology of Fishes, Third Edition
David H. Evans

Biology of the Southern Ocean, Second Edition
George A. Knox

Biology of the Three-Spined Stickleback
Sara Östlund-Nilsson, Ian Mayer, and Felicity Anne Huntingford

Biology Management *of the* World Tarpon and Bonefish Fisheries

Edited by
Jerald S. Ault

CRC Press
Taylor & Francis Group
Boca Raton London New York

CRC Press is an imprint of the
Taylor & Francis Group, an **informa** business

The cover painting is courtesy of Jorge Martinez.

The logo on the book spine and title page is courtesy of Guy Harvey.

CRC Press
Taylor & Francis Group
6000 Broken Sound Parkway NW, Suite 300
Boca Raton, FL 33487-2742

© 2008 by Taylor & Francis Group, LLC
CRC Press is an imprint of Taylor & Francis Group, an Informa business

No claim to original U.S. Government works
Printed in the United States of America on acid-free paper
10 9 8 7 6 5 4 3 2 1

International Standard Book Number-13: 978-0-8493-2792-6 (Hardcover)

This book contains information obtained from authentic and highly regarded sources. Reprinted material is quoted with permission, and sources are indicated. A wide variety of references are listed. Reasonable efforts have been made to publish reliable data and information, but the author and the publisher cannot assume responsibility for the validity of all materials or for the consequences of their use.

No part of this book may be reprinted, reproduced, transmitted, or utilized in any form by any electronic, mechanical, or other means, now known or hereafter invented, including photocopying, microfilming, and recording, or in any information storage or retrieval system, without written permission from the publishers.

For permission to photocopy or use material electronically from this work, please access www. copyright.com (http://www.copyright.com/) or contact the Copyright Clearance Center, Inc. (CCC) 222 Rosewood Drive, Danvers, MA 01923, 978-750-8400. CCC is a not-for-profit organization that provides licenses and registration for a variety of users. For organizations that have been granted a photocopy license by the CCC, a separate system of payment has been arranged.

Trademark Notice: Product or corporate names may be trademarks or registered trademarks, and are used only for identification and explanation without intent to infringe.

Visit the Taylor & Francis Web site at
http://www.taylorandfrancis.com

and the CRC Press Web site at
http://www.crcpress.com

Dedication

*To my Dad,
Captain Frank W. Ault, USN (Ret.),
the father of "Top Gun,"
August 26, 1921–August 20, 2006,
who taught me everything that I know about fishing
and the sea,
but not everything that he knew*

Contents

Foreword xi
Preface xiii
Introduction xv
The Editor xix
Contributors xxi

SECTION I World Fisheries for Tarpon and Bonefish

Chapter 1 Indo-Pacific Tarpon *Megalops cyprinoides*: A Review
and Ecological Assessment 3

   Janet A. Ley

Chapter 2 Biology and Ecology of the Recreational Bonefish Fishery at
Palmyra Atoll National Wildlife Refuge with Comparisons to
Other Pacific Islands 27

   Alan M. Friedlander, Jennifer E. Caselle, Jim Beets,
   Christopher G. Lowe, Brian W. Bowen,
   Tom K. Ogawa, Kevin M. Kelley, Todd Calitri,
   Matt Lange, and Bruce S. Anderson

Chapter 3 The Louisiana Recreational Tarpon Fishery 57

   William Dailey, André M. Landry, Jr.,
   and F. Leonard Kenyon II

Chapter 4 Insight into the Historical Status and Trends of Tarpon in
Southwest Florida through Recreational Catch Data Recorded
on Scales 69

   Stephen A. Bortone

Chapter 5 Ecology and Management of Bonefish (*Albula* spp.) in the
Bahamian Archipelago 79

   Andy J. Danylchuk, Sascha E. Danylchuk, Steven J. Cooke,
   Tony L. Goldberg, Jeff Koppelman, and David P. Philipp

Chapter 6 Coastal Ecosystem Management to Support Bonefish and Tarpon
Sportfishing in Peninsula de Zapata National Park, Cuba.................93

Lázaro Viñola Valdez, Lázaro Cotayo Cedeño, and Natalia Zurcher

Chapter 7 Tarpon and Bonefish Fishery on Turneffe Atoll, Belize 99

J. Travis Pritchett

Chapter 8 Aspects of the Biology and Recreational Fishery of Bonefish
(*Albula vulpes*) from Los Roques Archipelago National Park,
Venezuela 103

Juan M. Posada, Denise Debrot, and Constanza Weinberger

Chapter 9 The Nigerian Tarpon: Resource Ecology and Fishery.................... 115

Patricia E. Anyanwu and Kola Kusemiju

SECTION II Biology and Life History Dynamics

Chapter 10 Studies in Conservation Genetics of Tarpon (*Megalops atlanticus*): Microsatellite Variation across the Distribution of the Species... 131

Rocky Ward, Ivonne R. Blandon, Francisco García de León, Sterling J. Robertson, André M. Landry, Augustina O. Anyanwu, Jonathan M. Shenker, Miguel Figuerola, Theresa C. Gesteira, Alfonso Zerbi, Celine D. Acuña Leal, and William Dailey

Chapter 11 Resolving Evolutionary Lineages and Taxonomy of Bonefishes
(*Albula* spp.) 147

Brian W. Bowen, Stephen A. Karl, and Edward Pfeiler

Chapter 12 Ecology of Bonefish during the Transition from Late Larvae to
Early Juveniles 155

Craig Dahlgren, Jonathan M. Shenker, and Raymond Mojica

Chapter 13 Physiological Ecology of Developing Bonefish Larvae 179

Edward Pfeiler

Chapter 14 Reproductive Biology of Atlantic Tarpon *Megalops atlanticus* 195

John D. Baldwin and Derke Snodgrass

Contents

Chapter 15 Rethinking the Status of *Albula* spp. Biology in the Caribbean and Western Atlantic 203

Aaron J. Adams, R. Kirby Wolfe, Michael D. Tringali, Elizabeth M. Wallace, and G. Todd Kellison

SECTION III Population Dynamics and Resource Ecology

Chapter 16 Population Dynamics and Resource Ecology of Atlantic Tarpon and Bonefish 217

Jerald S. Ault, Robert Humston, Michael F. Larkin, Eduardo Perusquia, Nicholas A. Farmer, Jiangang Luo, Natalia Zurcher, Steven G. Smith, Luiz R. Barbieri, and Juan M. Posada

Chapter 17 A Preliminary Otolith Microchemical Examination of the Diadromous Migrations of Atlantic Tarpon *Megalops atlanticus* ... 259

Randy J. Brown and Kenneth P. Severin

Chapter 18 Seasonal Migratory Patterns and Vertical Habitat Utilization of Atlantic Tarpon (*Megalops atlanticus*) from Satellite PAT Tags 275

Jiangang Luo, Jerald S. Ault, Michael F. Larkin, Robert Humston, and Donald B. Olson

Chapter 19 Tagging of Bonefish in South Florida to Study Population Movements and Stock Dynamics 301

Michael F. Larkin, Jerald S. Ault, Robert Humston, Jiangang Luo, and Natalia Zurcher

SECTION IV Lore and Appeal of Fishing for Tarpon and Bonefish

Chapter 20 Bonefish Are without Question My Favorite Fly Rod Quarry 323

Sandy Moret

Chapter 21 Record Tarpon on a Fly Rod 329

Stuart C. Apte

Chapter 22 Learning from History 339

Bob Stearns

Chapter 23 Memories of the Florida Keys: Tarpon and Bonefish Like It Used to Be! 345

Mark Sosin

SECTION V Ecosystem-Based Management and Sustainable Fisheries

Chapter 24 National Parks and the Conservation of Bonefish and Tarpon Fisheries 351

James T. Tilmant

Chapter 25 Improving the Sustainability of Catch-and-Release Bonefish (*Albula* spp.) Fisheries: Insights for Anglers, Guides, and Fisheries Managers... 359

Steven J. Cooke and David P. Philipp

Chapter 26 Florida Keys Bonefish Population Census 383

Jerald S. Ault, Sandy Moret, Jiangang Luo, Michael F. Larkin, Natalia Zurcher, and Steven G. Smith

Chapter 27 Science in Support of Management Decision Making for Bonefish and Tarpon Conservation in Florida 399

Luiz R. Barbieri, Jerald S. Ault, and Roy E. Crabtree

Chapter 28 How to Get the Support of Recreational Anglers 405

Doug Kelly

Chapter 29 Sustaining Tarpon and Bonefish Fisheries: Scientists and Anglers Working Together.. 415

Mark Sosin

Chapter 30 Incorporating User-Group Expertise in Bonefish and Tarpon Fishery Research to Support Science-Based Management Decision Making 419

Robert Humston, Jerald S. Ault, Jason Schratwieser, Michael F. Larkin, and Jiangang Luo

Index 429

Foreword

When Captain Billy Smith caught the first bonefish on fly in the Florida Keys in 1939, he no doubt had a sure sense of the fun and excitement that it provided. However, as one of less than a handful of guides scratching out a living at $2.00 per day chasing bonefish and tarpon at the time, I am sure he had no idea of the proportions the sport of fishing for bonefish, permit, and tarpon would reach today.

Today, shallow water–sports fishing is in high gear and is still growing rapidly. Dedicated anglers tour the globe with a rod and reel, something akin to a buggy whip in their hands in pursuit of these species. They have created a multibillion dollar industry that reaches from the Florida Keys to the Seychelles off eastern Africa and back again via such mid-Pacific fisheries as Palmyra Atoll and Kiritimati (Christmas) Atoll. An industry with seminal roots in the Florida Keys that once had only two or three guides with experimental saltwater Orvis rods is now supported by dozens of rod and reel manufacturers, scores of lodges, and thousands of guides, skiff manufacturers, and tackle shop owners and their staffs.

Regrettable in many of the older and more developed fisheries, as the number of fishermen in pursuit increased, the number of fish caught per angler decreased. This is not only evidenced in just the numbers caught, which might be attributed to angling pressure, but there have also been significant drop-offs in the number of fish seen in some of the more popular fisheries.

There are many possible explanations for this, including:

- Change of habitat
- Juvenile mortality
- Catch and release mortality
- Fishing or boating pressure
- Commercial fishing
- Commercial by-catch
- Increased predation

and the list goes on.

The sad reality is that we do not have sufficient knowledge of species behavior, responses to fishing, or of critical habitat changes to make anything more than intelligent guesses as to causes of declines and possible fixes.

The good news is that through proper science and proper management other species fisheries not only have been stabilized but also have been brought to new heights of excellence. The efforts of those dedicated scientists included in this book, and no doubt many more to come, and the leadership of Bonefish & Tarpon Unlimited (BTU; www.tarbone.org) are dedicated to just that—stabilizing and enhancing bonefish and tarpon fisheries worldwide through scientific knowledge, education, and regulation.

BTU offers itself as a fiscal point of leadership to coordinate these efforts and hopefully can attract sufficient funding to allow the necessary research and management to take place. We applaud the leadership and efforts of Dr. Ault and his associates, who have been the pioneers in these efforts, and we are encouraged by the quantity and quality of other scientists joining the fight.

However, to succeed in this mission, we will need not only the efforts of the best and the brightest scientists, but we will also need a substantial and sustained source of funding and the technical leadership to coordinate these efforts. BTU hopes to provide this oversight. Also, each of us as anglers can be part of the solution by staying informed, being careful anglers, and supporting the mission with our financial support.

We can also all start doing our part today by being more careful anglers. Even though we do not know the exact extent to which catch-and-release mortality is a factor (e.g., Bartholomew and Bohnsack, 2005), we do know that whatever it is, it is preventable or can be mitigated with better and more careful release techniques.

The general rule is the less handling the better, and to avoid removing the fish from the water if practical. Through the use of a device like the Boga-grip one can stabilize the fish in the water and remove a barbless lip-hooked fly without ever touching the fish. Afterward use the Boga-grip to move the fish gently through the water while it recovers and revives, and then release.

If the fish must be touched, use your bare hand but wet your hand before making contact. To be avoided is the double-hand death grip photo shot that we have all done and second worse to that is to dangle the fish vertically from a Boga-grip for a photo (also guilty).

If it is a very special fish or a first fish and a photo is just in the cards, consider getting in the water with the fish or leaning down near the water and holding the fish horizontal with one hand at the tail and a Boga-grip in the mouth. Then hold your breath and remember that you need oxygen, and so does the fish.

Special thanks to Dr. Ault and all those participating in this important work.

I wish you and your grandchildren many years of enjoyment of this wonderful sport and pastime.

Thomas N. Davidson
Chair, Bonefish & Tarpon Unlimited, Inc.
North Key Largo, Florida

REFERENCE

Bartholomew, A. and J.A. Bohnsack, 2005. A review of catch-and-release angling mortality with implications for no-take reserves. Reviews in Fish Biology and Fisheries 15: 129–154.

Preface

What species could be better suited for an integrative systems approach that links all aspects of biology, ecology, resource management, and human-use perspectives to build sustainable fisheries? This book is geared to be a comprehensive reference for the economically and ecologically important tarpon and bonefish species. It is an outgrowth of two international symposia convened in Florida in 2003 and 2006 by Bonefish & Tarpon Unlimited, the International Game Fish Association, Tarpon Tomorrow, the University of Miami Rosenstiel School of Marine and Atmospheric Sciences, and the Florida Fish and Wildlife Conservation Commission. These premier scientific and public awareness organizations recognized the importance of consolidating the science for fishery management to include aspects of coastal marine environment, fishery sectors, population dynamics, stock assessment, and environmental policy. The symposia featured internationally recognized researchers, managers, and sportsfishers united in an effort to conserve and sustain these fisheries worldwide.

When I proposed this book to John Sulzycki, senior editor at Taylor & Francis, he enthusiastically endorsed its generation, and since the project's inception, John has been an amazing and enthusiastic stalwart. I thank him from the bottom of my heart for all his help and support. I am also greatly appreciative of the expert publication production assistance provided by Christine Andreasen, David Fausel, and the Taylor & Francis staff. Thanks also to the staff of Macmillan India for their production work. A long list of peer reviewers helped provide keen insights and focus attention to detail, all of which greatly strengthened the manuscript. These include Aaron Adams, Aaron Bartholomew, Theresa Bert, Jim Bohnsack, Steve Bortone, Steven Cooke, Bill Dailey, Bob Diaz, Jim Franks, Alan Friedlander, Martin Grossell, Kathy Guindon, Scott Holt, Ed Houde, Todd Kellison, Doug Kelly, Richard Kraus, Mike Larkin, Janet Ley, Karin Limbaugh, Ken Lindeman, Jiangang Luo, Behzad Mahmoudi, Sandy Moret, Juan Posada, Dave Philipp, Dave Secor, Steven G. Smith, Ron Taylor, Pat Walsh, and Natalia Zurcher. In addition, I am grateful for key technical support provided by Scott Alford, Stu Apte, Luiz Barbieri, Ivonne Blandon, Curtis Bostick, Roberto Bradley, Tadd Burke, Jim & Pam Callender, Billy Causey, Roy Crabtree, Richard Curry, Bill Curtis, Jack Curlett, Yusso Barquet, Chico Fernandez, Russ Fisher, Tom Gibson, Lisa Gregg, Jeff Harkavy, Joan Holt, George Hommell, Larry Kanitz, Glenda Kelley, Kenny Knudsen, Rob Kramer, Reuben Lee, Bill Legg, Alberto Madaria, Steve Martin, Larry McKinney, Barc Morley, Mike Myatt, George Neugent, Billy Pate, Glenn Patton, Eric Prince, Angel Requejo, Dick Robins, Lance "Coon" Schoest, Joel Shepherd, Mike Smith, Mark Sosin, Roe Stamps, Nancy Swakon, Mike Tringali, Bruce Ungar, Steve Venini, and Jeff "Gator" Wilson. Special thanks go to Capt. Joel Kalman and all participating Florida Keys guides who have helped shape my thinking on tarpon and bonefish biology and fishery dynamics. Robert Humston served as my sounding board over the years, and I sincerely thank him for his keen insights and intellectual acumen, which helped move this project to its successful completion.

Finally, the development of this book would not have been possible without Tom Davidson, Chairman of Bonefish & Tarpon Unlimited—a great friend and mentor since our first meeting on his back porch at the Ocean Reef Club in North Key Largo more than a decade ago. Since then Tom has provided his unique and indispensable blend of strategic guidance and direction that has set into motion an incredibly exciting intellectual endeavor that will, in fact, last a lifetime.

As interest in tarpon and bonefish continues to grow, for perhaps the most important catch-and-release sport fisheries in the world in terms of their ecological and economic value, I trust that this volume will help draw attention to the issues and focus development of coherent and prudent strategies that will build sustainable fisheries for generations to come.

Jerald S. Ault
Key Biscayne, Florida

Introduction

Tarpon and bonefish are two of fishing's supreme challenges. Few species can match the burst speeds of bonefish or the airborne acrobatics and raw power and fighting strength of a tarpon. These fishes, which share an ancient lineage with seemingly disparate fishes like ladyfish and eels, have endured eons of severe environmental changes and eluded the best natural predators. Hooked up with either, you feel the unbridled survival instincts of two of the Earth's oldest creatures that have survived 100 million years of evolution. However, the complicated early life history, biology, and population dynamics of these species make their study a real scientific challenge. Because their fisheries generally lack coherent strategies for study or management—either at regional or global scales—collection of relevant and accurate data to assess and predict stock responses to exploitation and environmental impacts remains an enigma.

FOCUS ON PROTECTION AND CONSERVATION

The seascape for tarpon and bonefish has changed dramatically over the past 50 years. In the Florida Keys, an area widely considered as the birthplace of shallow-water and "flats fishing," some noted fishing guides and experienced anglers have suggested that the bonefish population has declined some 90–95% since the 1940s (e.g., Curtis, 2004). Tarpon populations have experienced obvious and precipitous declines in portions of their historic U.S. range. Port Aransas, Texas, was once the "1950s tarpon capital of the world" and attracted presidents and potentates to catch a "silver king." It is now virtually devoid of the tarpon numbers that made it so famous. There is serious speculation concerning the root fishing and environmental causes for these declines, and whether these could occur elsewhere. The greatest challenge that lies ahead concerns sustainability of these precious fisheries. With a catch-and-release ethic becoming more commonplace, the impacts of recreational fishing for tarpon and bonefish are potentially minimal. However, increased exploitation, shoreline development and habitat degradation, pollution, and other environmental impacts from rapidly growing human populations may threaten critical food supplies and upset a tenuous balance in the ecosystems that support these resources. Despite the economic value of the industry and wide popularity of these precious fishery resources, very little is known about the movements and migrations, population dynamics, life histories, and reproduction that are needed to sustain fisheries for these amazing species. This is clear cause for concern.

Similar patterns of decline have also been noted in Florida and, in fact, throughout the world. Florida accounts for more than two thirds of the standing world records for tarpon and bonefish published by the International Game Fish Association in 2006. South Florida's tarpon and bonefish fisheries alone support a multibillion dollar annual regional economy. However, many of Florida's premier marine fisheries are undergoing extensive changes due to explosive regional growth in human populations, fleet sizes and fishing intensity, habitat losses, and other environmental degradation (e.g., Porter and Porter, 2001; Ault et al., 2005a, 2005b). This is a dire condition for a state that promotes itself as the "fishing capital of the world."

Continued growth of human populations in the coastal margins will compound pressures on the already stressed resources from both directed and nondirected fisheries, incidental mortality from catch-and-release fishing, unreported harvests, loss of key spawning and nursery areas, and pollution. These and other fishery ecosystem effects threaten the viability and longer-term sustainability of these fisheries. For example, as tarpon are still actively harvested in various parts of their range, accurate knowledge of their migration patterns and spawning areas is a keystone piece of information critical to ensure their protection. Unfortunately, much of the critical population-dynamic information needed for sound fishery management decision making is virtually lacking. Even more distressing is the fact that little attention has been paid to the design of scientific and management programs to support conservation of these incredibly important species.

KEY ASPECTS OF THIS BOOK

Because of the ever-increasing relationships between scientific and sport fishing interests, this book is organized to provide discussion and broad communication between scientists, managers, professional guides, anglers, and the public about the past, present, and future of these magnificent sport fishes. The focus of this book is to promote better understanding of the biological and fishery management issues that are paramount to the sustainable future of these valuable fisheries resources. But it is surprising how little is known about the two economically important game fishes. There are less than 100 scientific papers in total that have much of anything to do with research on the life history, population dynamics, and resource ecology of bonefish and tarpon. "We probably know more about the moon than we know about tarpon and bonefish in our waters," said Mark Sosin, a noted outdoors writer and 2004 inductee into the International Game Fish Association Hall of Fame. This is despite the fact that tarpon and bonefish fisheries support a multibillion dollar recreational fishery in south Florida alone. Some important unanswered questions that arise are, for example, "How do they survive and thrive among burgeoning coastal development?" and "How to manage recreational fisheries that are catch-and-release?"

This book summarizes existing scientific literature and presents new perspectives and syntheses on scientific research to guide fishery management and conservation efforts for building sustainable tarpon and bonefish fisheries. The book consists of five major sections:

1. World Fisheries for Tarpon and Bonefish
2. Biology and Life History Dynamics
3. Population Dynamics and Resource Ecology
4. Lore and Appeal of Fishing for Tarpon and Bonefish
5. Ecosystem-Based Management and Sustainable Fisheries

Introduction

Within this framework, this book contains 30 chapters that present an up-to-date summary of what is known about the life history, fishery biology, population dynamics, and management of these important game fish species. Each major section of the book focuses on a series of comparative syntheses, providing historical and current perspectives on the fisheries, population biology, stock assessment, modeling, and management by the foremost experts in their respective fields. Some of these chapters highlight aspects of continuing debates, that rather than providing disagreements, serve to channel a healthy discourse on topics of scientific and management interests. Individual chapters are designed to summarize original research or synthesize the scientific and technical literature, and discuss important issues such as scientific knowledge gaps, resource concerns, and research necessary to support evolving conservation and management strategies for these ecologically and economically important fishery resources. As such, the book centralizes the scientific and institutional knowledge of internationally recognized researchers and foremost authorities, managers, guides, and sport fishermen who share their unique knowledge and concerns for these magnificent game fishes.

PRICELESS INFORMATION

Bonefish and tarpon conservation research programs supported by groups like Bonefish & Tarpon Unlimited (BTU) are now providing unique, baseline datasets that could not be obtained in any other way. These data will be indispensable in determining the extent and sustainability of the unit stocks (on which fishery management is based) for both species, and key population-dynamic data and environmental preferences from which management policy is based. This information is critical because it drives the decision-making process in the regional, national, and international fishery management councils and commissions. Key issues in the success of such programs are the relatively high costs associated with the use of sophisticated technologies, and the willingness of anglers and guides alike to fully participate in reporting results to the scientific research community. History has proven to our BTU founders that many of the world's greatest fisheries have faced near collapse before any proactive intervention took place. Rather than risk a critical or perhaps irreversible decline of these two extraordinary species, BTU members are making a stand today to preserve bonefish and tarpon fishing for many generations to come. These results are yielding huge scientific results, and revolutionizing the way we think about fish, fishing, and the environment. Continued progress in these areas will take all of us working together—scientists, fisheries managers, conservation organizations, and saltwater anglers—to ensure the future of sustainable tarpon and bonefish fisheries. Progress toward that goal is clearly reflected in this comprehensive volume on the biology and management of the world tarpon and bonefish fisheries, but this is a window of opportunity that we cannot afford to miss.

REFERENCES

Ault, J.S., J.A. Bohnsack, and S.G. Smith. 2005a. Towards sustainable multispecies fisheries in the Florida USA coral reef ecosystem. Bulletin of Marine Science 76(2): 595–622.

Ault, J.S., S.G. Smith, and J.A. Bohnsack. 2005b. Evaluation of average length as an indicator of exploitation status for the Florida coral reef fish community. ICES Journal of Marine Science 62: 417–423.

Curtis, B. 2004. Not exactly fishing (alligator fishing). in Bonefish B.S. and Other Good Fish Stories, T.N. Davidson (ed.). Hudson Books, Whitby, Ontario, Canada, pp. 167–171, chapter 11.

Porter, J.W. and K.G. Porter (eds). 2001. The Everglades, Florida Bay, and Coral Reefs of the Florida Keys. CRC Press, Boca Raton, FL. 1000p.

The Editor

Jerald S. Ault, Ph.D., is a professor of marine biology and fisheries at the University of Miami's Rosenstiel School of Marine and Atmospheric Science. An avid lifelong fisherman, he is an internationally recognized leader in fisheries science for his research on theoretical and applied fish population and community dynamics for assessment and management of tropical marine fishery ecosystems, particularly for his development and applications of large-scale spatial ecosystem simulation models to assess the response of multispecies coral reef fisheries to exploitation and environmental changes.

Dr. Ault has published more than 60 peer-reviewed scientific journal papers and an additional 100-plus book chapters, technical memoranda, book reviews, and computer software in other literature. Each year, Dr. Ault leads a multi-institutional (federal, state, and university) research team that conducts a marine-life census in the Florida Keys–Dry Tortugas coral reef ecosystem. His research has been featured on the *CBS Evening News with Dan Rather,* the National Geographic channel, CNN, *Animal Planet, PBS Waterways,* and regional TV news, and in the *Miami Herald, Los Angeles Times, Chicago Tribune,* the *New York Times,* and Reuters International stories.

Dr. Ault earned his B.S. (1979) and M.S. (1982) in fisheries and natural resources, respectively, from Humboldt State University and his Ph.D. (1988) in fishery management science and applied statistics from the University of Miami. He conducted postdoctoral studies in fish population dynamics at the University of Maryland. He is a member of numerous national and international technical advisory committees on fisheries ecosystem management. He was elected to the Sigma Xi National Honor Society in 1986. He was named a Certified Fisheries Professional of AFS in 1998 and a fellow of the American Institute of Fisheries Research Biologists in 1999. He was a recipient of the Best Publication Award from the U.S. Department of Commerce, NMFS Scientific Publications Office, and was named a Hero of Conservation by *Field & Stream* magazine in 2006.

Contributors

Celine D. Acuña Leal
Laboratorio de Biologia Integrativa
Instituto Tecnologico de Cd. Victoria
Ciudad Victoria
Tamaulipas, Mexico

Aaron J. Adams
Mote Marine Laboratory
Center for Fisheries Enhancement
Charlotte Harbor Field Station
Pineland, Florida, U.S.A.
aadams@mote.org

Bruce S. Anderson
The Oceanic Institute
Waimanalo, Hawaii, U.S.A.
banderson@oceanicinstitute.org

Augustina O. Anyanwu
Nigerian Institute for Oceanography
  and Marine Research
Lagos, Nigeria
augustina_anyanwu@yahoo.com.uk

Patricia E. Anyanwu
Nigerian Institute for Oceanography
  and Marine Research
Lagos, Nigeria
akuchinyere@yahoo.com

Stuart C. Apte
Stu Apte Productions
Islamorada, Florida, U.S.A.
stuwho@bellsouth.net

Jerald S. Ault
Division of Marine Biology and Fisheries
Rosenstiel School of Marine and
  Atmospheric Science
University of Miami
Miami, Florida, U.S.A.
jault@rsmas.miami.edu

John D. Baldwin
Department of Biological Sciences
Florida Atlantic University
Davie, Florida, U.S.A.
jbaldwin@fau.edu

Luiz R. Barbieri
Florida Fish and Wildlife
  Conservation Commission
Florida Marine Research Institute
St. Petersburg, Florida, U.S.A.
luiz.barbieri@myfwc.com

Jim Beets
Department of Marine Science
University of Hawaii
Hilo, Hawaii, U.S.A.
jbeets@hawaii.edu

Ivonne R. Blandon
Texas Parks and Wildlife
  Department
Corpus Christi, Texas, U.S.A.
ivonne.blandon@tpwd.state.tx.us

Stephen A. Bortone
Minnesota Sea Grant College
University of Minnesota
Duluth, Minnesota, U.S.A.
sbortone@umn.edu

Brian W. Bowen
Hawaii Institute of Marine Biology
University of Hawaii at Manoa
Kaneohe, Hawaii, U.S.A.
bbowen@hawaii.edu

Randy J. Brown
U.S. Fish and Wildlife Service
Fairbanks, Alaska, U.S.A.
randy_j_brown@fws.gov

Todd Calitri
125 Longmarsh Road
Durham, New Hampshire, U.S.A.
tcalitri@hotmail.com

Jennifer E. Caselle
Marine Science Institute
University of California
Santa Barbara, California, U.S.A.
caselle@msi.ucsb.edu

Lázaro Cotayo Cedeño
Peninsula of Zapata
  National Park
Playa Larga, Cuba
pnacionalcz@enet.cu

Steven J. Cooke
Institute of Environmental Science
Carleton University
Ottawa, Ontario, Canada
scooke@connect.carleton.ca

Roy E. Crabtree
Southeast Regional Office
National Marine Fisheries Service
St. Petersburg, Florida, U.S.A.
roy.crabtree@noaa.gov

Craig Dahlgren
Perry Institute of Marine Science
Jupiter, Florida, U.S.A.
cdahlgren@perryinstitute.org

William Dailey
Department of Wildlife and Fisheries
  Sciences
Texas A&M University
Galveston, Texas, U.S.A.
daileyw@tamug.edu

Andy J. Danylchuk
Flats Ecology and Conservation
  Program
Cape Eleuthera Institute
Fort Lauderdale, Florida, U.S.A.
andydanylchuk@ceibahamas.org

Sascha E. Danylchuk
Flats Ecology and Conservation Program
Cape Eleuthera Institute
Fort Lauderdale, Florida, U.S.A.
saschaclark@ceibahamas.org

Thomas N. Davidson
Bonefish & Tarpon Unlimited, Inc.
North Key Largo, Florida, U.S.A.
tndd@bellsouth.net

Denise Debrot
Departamento de Biología de
  Organismos
Universidad Simón Bolívar
Caracas, Venezuela
debrot@usb.ve

Nicholas A. Farmer
Division of Marine Biology and
  Fisheries
Rosenstiel School of Marine and
  Atmospheric Science
University of Miami
Miami, Florida, U.S.A.
nfarmer@rsmas.miami.edu

Miguel Figuerola
Fisheries Research Laboratory
Puerto Rico Department of Natural
  Resources
San Juan, Puerto Rico
m.figuerola@gmail.com

Alan M. Friedlander
Biogeography Branch
Center for Coastal Monitoring and
  Assessment
National Center for Coastal Ocean
  Science
National Ocean Service
National Oceanic and Atmospheric
  Administration and
The Oceanic Institute
Waimanalo, Hawaii, U.S.A.
afriedlander@oceanicinstitute.org

Contributors

Francisco García de León
Centro de Investigaciones Biologicas de Noroeste, S.C.
La Paz, Mexico
landrya@tamug.edu

Theresa C. Gesteira
Universidad Federal do Ceara
Labomar, Brazil
cvgesteira@secrel.com.br

Tony L. Goldberg
Department of Natural Resources and Environmental Sciences
University of Illinois
Urbana, Illinois, U.S.A.
tlgoldbe@uiuc.edu

Robert Humston
Department of Biology
Virginia Military Institute
Lexington, Virginia, U.S.A.
humstonr@vmi.edu

Stephen A. Karl
Hawaii Institute of Marine Biology
University of Hawaii
Kaneohe, Hawaii, U.S.A.
skarl@hawaii.edu

G. Todd Kellison
National Marine Fisheries Service
Southeast Fisheries Science Center
Miami, Florida, U.S.A.
Todd.kellison@noaa.gov

Doug Kelly
Florida Outdoor Writers Association
Tampa, Florida, U.S.A.
doug7kelly@yahoo.com

Kevin M. Kelly
Department of Biological Sciences
California State University, Long Beach
Long Beach, California, U.S.A.
kmkelly@csulb.edu

F. Leonard Kenyon II
Consultant
Dayton, Ohio, U.S.A.
kenyonl@sbcglobal.net

Jeff Koppelman
Illinois Natural History Survey
Center for Aquatic Ecology and Conservation
Champaign, Illinois, U.S.A.
koppej@msn.com

Kola Kusemiju
Department of Marine Sciences
University of Lagos
Akoka, Lagos, Nigeria
kkush1@yahoo.com

André M. Landry
Department of Wildlife and Fisheries Sciences
Texas A&M University
Galveston, Texas, U.S.A.
landrya@tamug.edu

Matt Lange
The Nature Conservancy–Palmyra Atoll
Sacramento, California, U.S.A.
palmyra@tnc.org

Michael F. Larkin
Division of Marine Biology and Fisheries
Rosenstiel School of Marine and Atmospheric Science
University of Miami
Miami, Florida, U.S.A.
mlarkin@rsmas.miami.edu

Janet A. Ley
Faculty of Fisheries and Marine Environment
Australian Maritime College
Beauty Point, Australia
j.ley@fme.amc.edu.au

Christopher G. Lowe
Department of Biological Sciences
California State University,
  Long Beach
Long Beach, California, U.S.A.
clowe@csulb.edu

Jiangang Luo
Division of Marine Biology and
  Fisheries
Rosenstiel School of Marine and
  Atmospheric Science
University of Miami
Miami, Florida, U.S.A.
jluo@rsmas.miami.edu

Raymond Mojica
Brevard County Environmentally
  Endangered Lands Program
Melbourne, Florida, U.S.A.
rmojica@brevardparks.com

Sandy Moret
Florida Keys Outfitters
Islamorada, Florida, U.S.A.
flkeyout@bellsouth.net

Tom K. Ogawa
The Oceanic Institute
Waimanalo, Hawaii, U.S.A.
togawa@oceanicinstitute.org

Donald B. Olson
Division of Meteorology and Physical
  Oceanography
Rosenstiel School of Marine and
  Atmospheric Science
University of Miami
Miami, Florida, U.S.A.
dolson@rsmas.miami.edu

Eduardo Perusquia
Mexican Sport Fishing
  Federation
Mexico City, Mexico
snookmachine@hotmail.com

Edward Pfeiler
Centro de Investigación en
  Alimentación y Desarrollo A.C.
Guaymas, Mexico
epfeiler@asu.edu

David P. Philipp
Illinois Natural History Survey
Center for Aquatic Ecology
Champaign, Illinois, U.S.A.
philipp@uiuc.edu

Juan M. Posada
Departamento de Biología de
  Organismos
Universidad Simón Bolívar
Caracas, Venezuela
jposada@usb.ve

J. Travis Pritchett
Nicholas School of the Environment
Duke University
Durham, North Carolina, U.S.A.
TPritchett@Harbert.net

Sterling J. Robertson
Biological Sciences Department
Florida Institute of Technology
Melbourne, Florida, U.S.A.
srobertson@interstatecustomcrushing.
  com

Jason Schratwieser
International Game Fish Association
Dania Beach, Florida, U.S.A.
jschratwieser@igfa.org

Kenneth P. Severin
Department of Geology and Geophysics
University of Alaska
Fairbanks, Alaska, U.S.A.
fnkps@uaf.edu

Jonathan M. Shenker
Department of Biological Sciences
Florida Institute of Technology
Melbourne, Florida, U.S.A.
shenker@fit.edu

Contributors

Steven G. Smith
Division of Marine Biology
 and Fisheries
Rosenstiel School of Marine and
 Atmospheric Science
University of Miami
Miami, Florida, U.S.A.
steve.smith@rsmas.miami.edu

Derke Snodgrass
NOAA Fisheries
Southeast Fisheries Science Center
Miami, Florida, U.S.A.
derke.snodgrass@noaa.gov

Mark Sosin
Mark Sosin's Saltwater Journal
Boca Raton, Florida, U.S.A.
sosinmark@aol.com

Bob Stearns
Salt Water Sportsman Magazine
Miami, Florida, U.S.A.
rds9k@bellsouth.net

James T. Tilmant
Water Resources Division
National Park Service
Fort Collins, Colorado, U.S.A.
jim_tilmant@nps.gov

Michael D. Tringali
Florida Fish and Wildlife
 Conservation Commission
Molecular Genetics Laboratory
Fish and Wildlife Research Institute
St. Petersburg, Florida, U.S.A.
mike.tringali@myfwc.com

Lázaro Viñola Valdez
Peninsula of Zapata National Park
Playa Larga, Cuba
pnacionalcz@enet.cu

Elizabeth M. Wallace
Florida Fish and Wildlife
 Conservation Commission
Molecular Genetics Laboratory
Fish and Wildlife Research Institute
St. Petersburg, Florida, U.S.A.
liz.wallace@myfwc.com

Rocky Ward
USGS Biological Resources
 Division
Northern Appalachian
 Research Station
Wellsboro, Pennsylvania, U.S.A.
rward@usgs.gov

Constanza Weinberger
Departamento de Biología de
 Organismos
Universidad Simón Bolívar
Caracas, Venezuela
conyw@hotmail.com

R. Kirby Wolfe
Mote Marine Laboratory
Center for Fisheries Enhancement
Charlotte Harbor Field Station
Pineland, Florida, U.S.A.
kwolfe@mote.org

Alfonso Zerbi
Université de Montpellier II
Montpellier, France
alfzerb@yahoo.com

Natalia Zurcher
Division of Marine Biology
 and Fisheries
Rosenstiel School of Marine
 and Atmospheric Science
University of Miami
Miami, Florida, U.S.A.
nzurcher@rsmas.miami.edu

Section I

World Fisheries for Tarpon and Bonefish

1 Indo-Pacific Tarpon *Megalops cyprinoides*: A Review and Ecological Assessment

Janet A. Ley

CONTENTS

Introduction 3
Literature Synthesis 7
    Larval and Juvenile Stages 7
    Adult Stages 8
    Swim Bladder Function 9
    Ecology 10
Ecological Analysis of Northeastern Queensland Populations 11
    Study Area 11
    Fish Data and Analysis 11
    Characterization of Tarpon Catch: Fishery-Independent Surveys 13
    Abiotic Factors Influencing Distribution Patterns 15
    Biotic Interactions 17
Discussion 22
    Life History Summary 22
    Ecological Summary 23
    Assessment of Vulnerability 23
Acknowledgments 24
References 24

INTRODUCTION

Indo-Pacific tarpon (*Megalops cyprinoides*; Broussonet, 1782) occur between 28° N (Japan) and 35° S latitude (southern Australia and South Africa), and from 25° E longitude (eastern African coast) eastward to 171° W (Samoa) (Figure 1.1). From depths to 50 m in coastal waters, they range inland to hundreds of kilometers upstream in rivers and floodplains (Pusey et al., 2004). A comprehensive list included 255 records of occurrence from 1830 to 2001 (FishBase; Froese and Pauly, 2006). While 20% of these records are from the Philippines, followed by 15% from India,

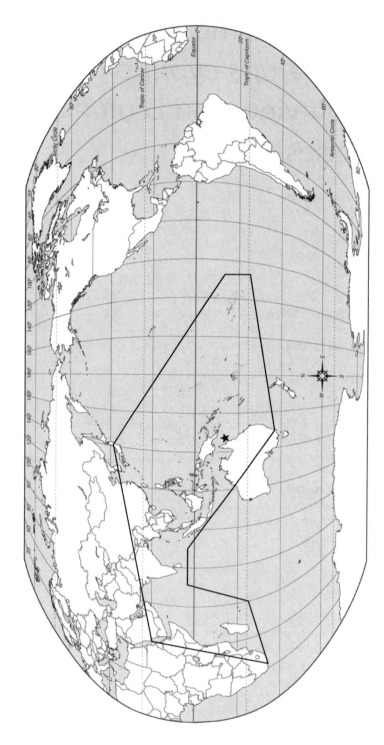

FIGURE 1.1 Generalized map of the worldwide distributional range of *Megalops cyprinoides*. The star indicates the northeastern Queensland study area.

M. cyprinoides also occurs near high oceanic islands (e.g., Ovalau Island, Fiji; Bab-el-Thuap Island, Palau; and Malekula Island, Vanuatu). In Australia, *M. cyprinoides* commonly occurs from the Fitzroy River near Broome, Western Australia, around the northern coast of the nation, and along the eastern shore to Moreton Bay near Brisbane, Queensland. In terms of habitats, *M. cyprinoides* has been collected on coral reefs (Madagascar), and in billabongs (Australia), mangrove swamps (Micronesia), rivers (Mozambique), reservoirs (Papua New Guinea), floodplains (South Africa), coastal bays (South Africa), and man-made canals (Tanzania).

Megalops cyprinoides (also referred to as tarpon, oxeye herring, and other common names) is not known to be an important component of the commercial fisheries of any nation in its range. Artisanal fisheries in Papua New Guinea (Coates, 1987) and other countries have been reported (e.g., India; Rao and Padmaja, 1999), but no catch data are available. Only the Philippines and Malaysia include data on commercial fishery landings of *M. cyprinoides* in their annual reports to the Food and Agricultural Organization of the United Nations (FAO, 2006). Consumed as a food fish by the growing human population in the Philippines, between 1994 and 1995 a threefold increase in catch was observed (Figure 1.2).

In Queensland, Australia, *M. cyprinoides* has been recorded as a bycatch species in estuary set gill net fisheries, which primarily target barramundi (*Lates calcarifer*). However, of the 381 commercial net sets recorded in observer surveys, only three *M. cyprinoides* were netted and these fish were discarded (Halliday et al., 2001). In contrast, commercial surrounding-net fisheries targeting mullet (*Mugil cephalus*) and whiting (*Sillago* sp.) in sheltered embayments netted a relatively greater proportion of *M. cyprinoides*, that is, nine were caught in 110 net sets witnessed by observers. These observer surveys by Halliday et al. (2001) are apparently the only bycatch data available for *M. cyprinoides*.

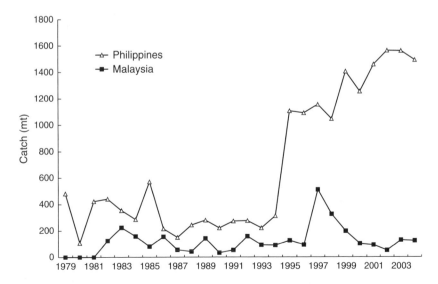

FIGURE 1.2 A time series of available capture fisheries data for *Megalops cyprinoides*. (Adapted from FAO, *Fisheries Global Information System*, http://www.fao.org, 2006.)

Comprehensive data on recreational catches of Indo-Pacific tarpon are equally scarce. Throughout Australia, telephone- and field-based surveys conducted for the National Recreational and Indigenous Fishing Survey (Henry and Lyle, 2003) did not document catch rates of *M. cyprinoides*. However, in northern Australia, *M. cyprinoides* is frequently taken by sport anglers fishing in habitats such as billabongs during the dry season (Wells et al., 2003). Guided recreational fishing enterprises commonly advertise tarpon as an attractive species for customers seeking a satisfying fishing experience. For example, a website for one Cairns, Queensland, operation referred to *M. cyprinoides* as "little balls of silver muscle" (www.fishingcairns.com.au) (see Figure 1.3). Another recreational fishing website indicates that "on a scale awarded in relation to the difficulty of capture (compiled by the Australian National Sportfishing Association (ANSA)), tarpon rate as one of the highest with a fighting factor of 2.0" (www.fishn4.com.au). ANSA maintains a comprehensive database with records of fish tagged and recaptured by trained recreational anglers, including 1383 *M. cyprinoides* caught between 1985 and 2006 in Queensland (www.info-fish.net). Because recreational anglers do not consider tarpon to be a desirable fish for human consumption, most fishermen employ catch-and-release practices when landing *M. cyprinoides*. No studies exist on the survival rate of released individuals.

Just as comprehensive fisheries catch information for *M. cyprinoides* is lacking, little biological information exists in the published primary literature. Four main avenues of research have been summarized in the section on literature synthesis (see below). First, several investigators have studied the larval biology of *M. cyprinoides*, focusing primarily on the leptocephalus stage. A second relatively well-studied topic is the biological function of the swim bladder, which facilitates survival in waters

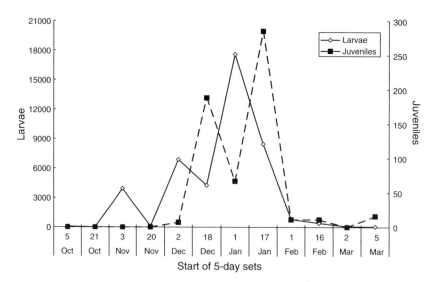

FIGURE 1.3 Numbers of larval and juvenile *Megalops cyprinoides* captured during each set of a tidal trap during wet season, spring tides—Leanyer, Northern Territory, Australia. (Adapted from Davis, T.L.O., *Environ. Biol. Fishes*, 21, 161–172, 1988.)

with low dissolved oxygen concentrations through air gulping. Third, however, basic biological and demographic parameters such as growth rates, length- or age-at-maturity, and average maximum size remain in doubt for *M. cyprinoides*. In fact, to the author's knowledge, such parameter values have not been established through systematic analyses for any population of *M. cyprinoides*. Fourth, while few ecological studies have focused on *M. cyprinoides*, they have been incidentally recorded in numerous surveys throughout their range providing insights into habitat use patterns and functional role.

As a contribution to ecological knowledge on *M. cyprinoides*, this chapter also draws on fishery-independent surveys, which were conducted along the northeastern coast of Queensland, Australia from 1995 to 2000 (Ley et al., 2002; Ley, 2005). The data from these surveys have been used in this chapter to explore distributional patterns of *M. cyprinoides* relative to abiotic factors and associated species. Thus, the objectives of the current study are to

1. Summarize literature published on *M. cyprinoides*, including food habits, biology and ecology;
2. Characterize *M. cyprinoides* data from a series of fishery-independent surveys along northeastern coast of Australia; and
3. Identify management implications and research needs.

LITERATURE SYNTHESIS

LARVAL AND JUVENILE STAGES

Megalops cyprinoides are believed to spawn offshore, but the locations of spawning grounds and subsequent dispersal of larvae remain uncertain. In surveys conducted in 1983–1984, covering a large area of the Great Barrier Reef (GBR) lagoon and Coral Sea, larval *M. cyprinoides* were collected in November and December between Lizard Island and the outer barrier reef (see Leis and Reader, 1991 for further details on methods and sampling sites). Thus, spawning may occur during these early summer months in the GBR offshore lagoon, between 25 and 45 km from the mainland of Australia (J. Leis, personal communication). Assuming that most spawning does occur offshore throughout their range, larvae apparently swim or drift with tidal currents, entering estuaries at 20 to 39 days old (Tzeng et al., 1998). Larval recruitment occurs in strong seasonal peaks. For example, Australian investigators collected enormous numbers of *M. cyprinoides* larvae and juveniles in tidal traps located in ephemeral creeks in Leanyer Swamp (Darwin, Northern Territory) on high spring tides (Davis, 1988). Ranking third in abundance of all species sampled, peak recruitment of *M. cyprinoides* larvae occurred from December to January (rainy season) in the 6-month study (Figure 1.3). Larval *M. cyprinoides* was also among the most abundant species collected on nighttime flood tides in Taiwan estuarine creeks (Tzeng et al., 2002).

Larval biology includes a leptocephalus stage, which has been investigated by several authors. Complete metamorphosis into juveniles occurs in estuaries in approximately 10 days, after drastic shrinkage during the first several days

(Tsukamoto and Okiyama, 1993, 1997; Chen and Tzeng, 2006). Although the size of juvenile fish is almost stable for approximately 1 month, laboratory studies revealed that otolith rings continue to be formed daily. However, due to internal physiological conditions, chemical changes in otolith composition (e.g., strontium/calcium ratios) occur during metamorphosis, regardless of whether the fish is in freshwater or seawater (Shiao and Hwang, 2004; Chen and Tzeng, 2006).

ADULT STAGES

Megalops cyprinoides has been known to live up to 44 years in the wild (Kulkarni, 1983). Maximum sizes up to 1500 mm have been reported, but uncertainty exists with regard to this value (Pusey et al., 2004). The maximum caudal fork length (CFL) recorded in the current study was 525 mm for an individual netted in the Russell River estuary (May 4, 1996). The ANSA database includes an individual measured at 610 mm CFL caught in the Calliope River (near Gladstone, Queensland). The International Game Fish Association (IGFA) reports 2.99 kg (estimated length of 611 mm CFL) as the all-tackle world record for *M. cyprinoides*, caught May 14, 2000, also near Gladstone (International Game Fish Association, 2006). Thus, the likelihood of *M. cyprinoides* attaining a length near 1500 mm is low; rather, a maximum of 610 mm CFL seems more probable.

In FishBase (Froese and Pauly, 2006), an estimate of the von Bertalanffy growth rate parameter K is given as 0.14 per year. However, this value may not be applicable to *M. cyprinoides* since it was derived from studies of *M. atlanticus*. From the tag-recapture data provided by ANSA, out of 1381 *M. cyprinoides* tagged by recreational anglers, one recaptured fish was at large for 625 days and had grown 90 mm (240–330 mm CFL), while two others were recaptured within 27 and 28 days of tagging and had grown 5 mm each. Growth rate was calculated using the forced Gulland–Holt method (King, 1995) with a fixed value of length-at-infinity (L_∞) of 610 mm, an average growth rate ($\bar{Y}$) of 0.17, and an average length ($\bar{X}$) of 287 mm:

$$K = \frac{\bar{Y}}{(L_\infty - \bar{X})} = 0.19 \, \text{year}^{-1}. \tag{1.1}$$

If K is close to this value, *M. cyprinoides* exemplifies those species that grow at an intermediate rate. Tarpon do appear to be sensitive to tagging (e.g., through loss of scales), possibly leading to the very low long-term return rate observed.

Length-at-maturity as reported in FishBase (Froese and Pauly, 2006) is 767 mm, but this estimate was also derived from studies of *M. atlanticus*, which grows to 2222 mm CFL (Froese and Pauly, 2006). Pusey et al. (2004) suggest that *M. cyprinoides* probably achieves sexual maturity in the second year of life when lengths in excess of 300 mm are attained. In another investigation, Coates (1987) found no mature fish in surveys in the Sepik River, Papua New Guinea, and suggested that fish above 400 mm in length return permanently to coastal waters to mature and breed. Until systematic analyses of reproductive biology are conducted for this species, crucial biological parameters such as length-at-maturity, fecundity, and related reproductive behavior will remain unknown. Clearly, comprehensive studies of age, growth, and reproductive biology are important gaps in knowledge for this species throughout its range.

Swim Bladder Function

With a sleek fusiform shape, tarpon are built for sustained swimming and speed (Figure 1.4). The caudal fin of *M. cyprinoides* has a high aspect ratio (2.19) (Froese and Pauly, 2006), and a high proportion of red muscle occurs in the body (Wells et al., 2003). They have a larger gill surface area than many other fishes, and the aerobic capacity of *M. cyprinoides* is also supported by air breathing (Wells et al., 2003). *Megalops cyprinoides* is a facultative bimodal breather that exchanges respiratory gases through gills and a physostomous, vascular swim bladder, which is in contact with the skull (Seymour et al., 2004). Air is taken into the fish's air bladder at the water surface on an often spectacular and rapid roll that presumably minimizes exposure to predators such as raptors (Wells et al., 2003). Recent experiments have revealed that the rapidity of postexercise restoration of routine hematological values in *M. cyprinoides* is linked to the air-breathing trait (Wells et al., 2003). This finding implies that recovery following strenuous angling exercise may be assisted by air breathing. In experiments, oxygen uptake through the swim bladder was found to be a small proportion of the total oxygen uptake in well-oxygenated water (3–7 breaths per hour), where the purpose of rare air-breathing events appears to be for buoyancy control. In contrast, the frequency of breaths increased to 29–37 per hour in hypoxic water.

This facility for air breathing may contribute to behavioral flexibility and survival of *M. cyprinoides* under a variety of environmental conditions. For example, during the wet season, the fish are widely distributed across floodplains, but as the dry season progresses, water levels drop, confining the fish to isolated pools that are progressively warmer, stratified, and hypoxic (Russell and Garrett, 1983; Davis, 1988). Hypoxia can develop further when eutrophication caused by nutrients in runoff at the beginning of the wet season draws down oxygen (Seymour et al., 2004). While Centropomidae and other water-breathing fishes often perish under severe anoxic conditions (e.g., Bishop, 1980), *M. cyprinoides* may have a survival advantage.

FIGURE 1.4 *Megalops cyprinoides* (Indo-Pacific tarpon). (Photo courtesy of J.E. Randall.)

ECOLOGY

Among the few published ecological studies that have included *M. cyprinoides*, the most comprehensive was based on populations in the Sepik River of northern Papua New Guinea (Coates, 1987). In the fully fresh waters of the Sepik, Coates found only immature fish, primarily caught in deeper (>15 m) oxbow portions of the river. Coates also noted that large mats of *Salvinia molesta* occurred in the oxbows, causing reduced oxygen levels. *Megalops cyprinoides* was one of the few fishes besides large Ariid catfishes to be caught beneath such mats. *Megalops cyprinoides* has also been incidentally recorded in other surveys throughout its huge distributional range, providing fragmented insights into its habitat preferences (Blaber, 2000).

In four studies that focused on trophic ecology, gut contents of 441 individual *M. cyprinoides* were examined (Figure 1.5). Food habits appear to vary with habitat and season, but may include fishes, crustaceans, and terrestrial insects (Table 1.1). Composition of the *M. cyprinoides* diet indicates opportunistic feeding and a tendency toward pelagic prey. Based on the percentages given in Figure 1.5, the estimated trophic level would be 3.4, an intermediate carnivore. A similar value is provided in FishBase, that is, 3.3 (Froese and Pauly, 2006). However, in a recent stable isotopic analysis, the basic source of primary production supporting the diets of *M. cyprinoides* collected near mangrove creeks could not be ascertained even though all obvious sources were tested (e.g., marine phytoplankton, mangroves, seagrass, terrestrial plants) (Benstead et al., 2006). Thus, questions remain concerning ecological processes supporting production of *M. cyprinoides*.

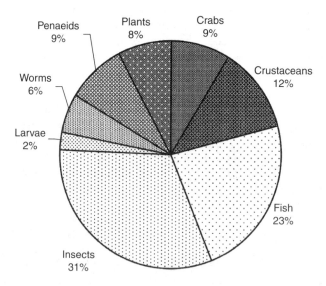

FIGURE 1.5 Chart summarizing the gut contents composition of *Megalops cyprinoides* derived from four studies (Table 1.1) standardized to 100%.

TABLE 1.1
Summary of Food Items Consumed by *Megalops cyprinoides* in Four Studies

Foods	Ley, 2005	Coates, 1987	Pusey et al., 2004	Rao and Padam, 1999	Average Percent
Crabs	17	–	–	–	17
Crustaceans	–	45	10	17	24
Fish	83	24	29	–	45
Insects	83	95	40	31	62
Larvae	–	–	–	5	5
Worms	–	–	–	11	11
Penaeid shrimp	17	–	–	–	17
Plants	–	26	5	–	15

Samples

Habitat	Tropical Estuary	Tropical River	Tropical River	Tropical Bay	
Minimum fish length (mm)	372	103	150	N/A	103
Maximum fish length (mm)	477	440	640	N/A	640
Number of fish	6	94	107	240	441 (total)

Note: Data are percent frequency of occurrence averaged for the studies (see Figure 1.5).

ECOLOGICAL ANALYSIS OF NORTHEASTERN QUEENSLAND POPULATIONS

STUDY AREA

Fishery-independent surveys were conducted in 11 Queensland estuaries from the northeastern tip of Australia (19°50′ S, 147°45′ E), south–eastward 1400 km to Cape Upstart, near Bowen (10°48′ S, 142°33′ E) (Figure 1.6). This tropical coast has a distinct rainy season (November/December–February/March). The GBR lies adjacent to the study area on the east, while the Great Dividing Range (maximum elevation 1700 m) extends the length of the study area, with the main ridge of mountains varying in proximity to the shoreline. Regulated commercial gill-net fishing is permitted in 3 (Nobbies, Barrattas, and Hull) of the 11 estuaries that were sampled, while limited recreational line fishing is permitted in all estuaries. However, the five northern estuaries are very remote and only accessible via off-road vehicles along rugged developmental roads or through a long journey by sea. Details of the study area are described in previous publications (Ley et al., 2002; Ley and Halliday, 2003; Ley 2005).

FISH DATA AND ANALYSIS

Upstream (2–10 km from the mouth) and downstream (within 1 km of the mouth) sites were sampled with groups of monofilament gill nets having stretched mesh sizes of 152, 102, and 51 mm. Multipanel nets, 30 m long by 2 m deep, with stretched

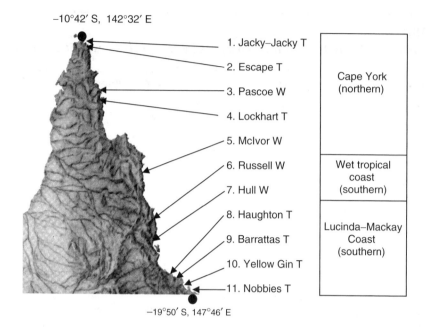

FIGURE 1.6 Study sites and location of the study area in northeastern Queensland, Australia. Estuary names are followed by a letter indicating the type (T = tide; W = wave; see text). The three bioregions and study phase (northern, southern) are listed in boxes to the right of the map.

mesh sizes of 19, 25, and 32 mm were also deployed. Sampling trips were conducted in two phases: (1) five northern rivers in which sampling was carried out in February and June 1996 and (2) six southern rivers in which sampling was carried out bimonthly between March 1998 and March 2000 (Figure 1.6). For the northern rivers, nets were set for up to 11 daylight hours, while for the southern rivers nets were set for up to 7 h day and night (i.e., between 1500 and 2200 hours). During the soak periods, nets were checked hourly and fish captured were measured, recorded, and if in good condition, released. Raw data collected by species for each net set and check were converted to catch per hour (CPUE) to account for variations in duration of periods between checks.

Most statistical tests, including *t*-tests, principal component analysis, and analyses of variance, were facilitated by the use of Statistica 6.0 (Statsoft 2001). Multivariate analyses of 78 nighttime net sets from the southern systems were facilitated by use of PRIMER V6 (Primer-E Ltd., Plymouth, 2004). Owing to the dominance of zero counts, the Bray–Curtis index of similarity was used to derive a matrix of similarity values between pairs of samples (Clarke, 1993). Relationships among groups were ascertained using cluster analysis and nonmetric multidimensional scaling (MDS). For the northern rivers phase of the study only, *in situ* measurements and water samples were collected by a team operating in a separate boat. Samples were stored and analyzed in the Australian Institute of Marine Science laboratories using standard methods (see Furnas, 2003).

CHARACTERIZATION OF TARPON CATCH: FISHERY-INDEPENDENT SURVEYS

A total of 305 *M. cyprinoides* were captured (Table 1.2). In the six southern estuaries, where samples were collected both day and night, 84% of the 208 *M. cyprinoides* were captured at night, indicating strong nocturnal activity. Catch rates were greater during the period just after sunset, curtailing later in the night (Ley and Halliday, in press). For all the months, night catch rates were greater than day except during the wet season when numbers of *M. cyprinoides* were lower overall, possibly because of increased access to freshwater flooded habitat. Thus, in the wet season, *M. cyprinoides* may have spent less time foraging along the mangrove fringe where the research gill nets were deployed; in contrast, during the dry season, catch rates were consistently higher.

A total of 162 *M. cyprinoides* (19.0–52.0 cm CFL) were weighed and the data were fitted to the allometric relationship:

$$W = \alpha L^\beta, \tag{1.2}$$

where W is the weight in g and L the CFL in cm (Figure 1.7). Fitted parameters ($p < 0.0001$) were $\alpha = 0.0152$ (95% confidence interval 0.0062–0.0242) and $\beta = 2.98$ (2.83–3.14). For the Alligator Rivers region of Australia's Northern Territory, Equation 1.2 was estimated based on data for 155 individuals (13.7–41.0 cm CFL) (Bishop et al., 2001) as

$$W = 0.0242 L^{2.83}. \tag{1.3}$$

Thus, the parameter estimates of the Alligator Rivers *M. cyprinoides* were just barely within the 95% confidence intervals of the current study. In a New Caledonia study (Kulbicki et al., 2005), for 35 *M. cyprinoides* ranging from 17.0 to 47.0 cm CFL, Equation 1.2 was estimated as

$$W = 0.0122 L^{3.03}. \tag{1.4}$$

Discrepancies between parameters derived for the three equations indicate that growth rates may vary among stocks. For example, applying the equations, the average 30-cm tarpon from the east coast of Queensland would be expected to weigh 383 g; 363 g (Equation 1.3) from the Alligator Rivers; and 368 g from the

TABLE 1.2
Total *Megalops cyprinoides* Catch by Diel Period, Position in the Estuary, Season, and Stretched Mesh Size of Research Gill Nets

Phase / Estuary	Diel Period		Position		Season		Mesh (mm)				Total
	Day	Night	Upstream	Downstream	Wet	Dry	19/25/32	51	102	151	
Northern											
Jacky Jacky	6		5	1	0	6	0	0	3	3	6
Escape	4		3	1	0	4	0	1	3	0	4
Pascoe	52		43	9	8	44	0	0	51	1	52
Lockhart	2		1	1	1	1	0	0	2	0	2
McIvor	33		33	0	31	2	0	0	33	0	33
Southern											
Russell	5	91	65	31	20	76	0	0	95	1	96
Hull	9	49	21	37	10	48	0	1	50	7	58
Haughton	0	2	1	1	0	2	0	0	2	0	2
Barrattas	5	9	10	4	9	5	0	1	13	0	14
Yellow Gin	8	19	20	7	11	16	0	3	23	1	27
Nobbies	6	5	6	5	8	3	0	0	9	2	11
Total			208	97	98	207	0	6	284	15	305

Note: Data are abundances netted in the 11 northeastern Queensland estuaries (Figure 1.6) during the two study phases (northern, southern) and 16 trips, as listed below:

Phase: Trips:
Northern   Wet season trip = February 1996; Dry season trip = June 1996.
Southern   Wet season trips = March 1998; January 1999; March 1999; January 2000; March 2000.
Southern   Dry season trips = May 1998; July 1998; September 1998; November 1998; May 1999; July 1999; September 1999; November 1999.

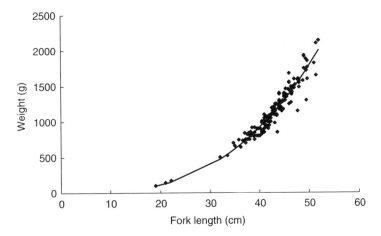

FIGURE 1.7 Length–weight relationship for 162 *Megalops cyprinoides* taken in 11 northeastern Queensland estuaries.

New Caledonian lagoons. Variation in growth rates can be expected owing to factors such as the length range of the specimens analyzed, genetics, foods, and seasonality (King, 1995).

The lengths of all individuals measured ($n = 302$) ranged from 175 to 530 mm CFL in the current study. Although nets with a range of mesh sizes were deployed, 93% of the *M. cyprinoides* were caught in the 102 mm mesh net. Thus, the size range present in the 11 estuaries along the main channels and creeks where the research nets were deployed was apparently well represented by the length–frequency histogram with a median of 430 mm CFL (Figure 1.8).

ABIOTIC FACTORS INFLUENCING DISTRIBUTION PATTERNS

For the six southern estuaries, analysis of variance (ANOVA) revealed that two general factors significantly influenced the catch rates of *M. cyprinoides*: estuary type (i.e., wave dominated vs. tide dominated) and diel activity period (day vs. night) (Table 1.3). In fact, these two factors interacted with each other such that the greatest catch rates occurred in the wave-dominated estuaries during the period after sunset (Figure 1.9). Higher catch rates in passive gear such as gill nets indicate fish activity periods by mobile species such as *M. cyprinoides*, probably associated with feeding (Ley and Halliday, in press). Thus, the main active feeding period for *M. cyprinoides* was apparently at night.

Tide-dominated systems are located in drier catchments and tend to have expansive mangrove areas, broad deltaic mouths, and muddy substrate (Ley, 2005). However, the preferred habitat of *M. cyprinoides*, wave-dominated systems, tend to be located in higher rainfall catchments, have narrow mouths, less mangrove area, and sandy/rocky substrate. Wave-dominated systems had low catch rates of all species combined, and may have become stratified, perhaps leading to oxygen depletion. Thus, the air-breathing ability may be advantageous in wave-dominated systems. While these systems may be relatively low-quality habitats for most estuarine fishes

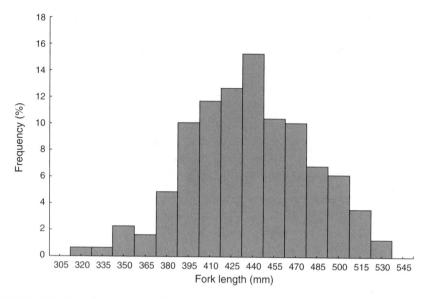

FIGURE 1.8 Length–frequency histogram for 302 *Megalops cyprinoides* greater than 305 mm fork length taken in 11 northeastern Queensland estuaries.

TABLE 1.3
Summary of Results for Analysis of Variance (ANOVA) Based on 331 Gill-Net Samples (both day and night) in the Six Southern Estuaries

Factor	Sum of Squares	df	Mean Squares	F	p
Intercept	22.09	1	22.09	152.70	<0.0001
Estuary type	8.50	1	8.50	58.76	<0.0001
Day–night	8.51	1	8.51	58.80	<0.0001
Estuary type* day–night	5.38	1	5.38	37.21	<0.0001
Error	47.30	327	0.14		

Note: See Figure 1.9: data are CPUE of *Megalops cyprinoides*.

in terms of food, cover, and aquatic conditions, *M. cyprinoides* apparently thrives in them.

Water conditions associated with 21 gill-net samples taken by day in the five northern estuaries were measured during the wet (February) and dry (June) seasons (Table 1.4). A database consisting of 12 physicochemical variables was developed for both sampling trips in each estuary. Owing to low sample size, a subset consisting of five of the measured variables was included in the principal component (PC) analysis (Table 1.4). The first two PCs explained a cumulative total of 65.7% of the variation in the physicochemical data (Figure 1.10). Furthermore, PC 1 ("salinity gradient") was moderately well correlated with catch rates of

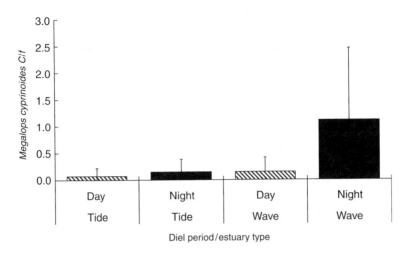

FIGURE 1.9 Mean catch rates for *Megalops cyprinoides* taken in the six southern estuaries illustrating the interaction between diel period (day–night), and estuary type (wave dominated, tide dominated) in the ANOVA (Table 1.3). Error bars indicate 95% confidence limits. c/f = catch per unit effect (CPUE).

M. cyprinoides ($r = 0.49$), while PC 2 ("nutrient gradient") explained very little of the variation in catch rates (Table 1.5). Thus, samples represented by PC 1 had lower salinity (<15 ppt) and temperature (27–29°C), but higher silicate concentrations (200 µM/L average). Silicate concentrations may be greater in naturally flowing rivers with few upstream man-made diversions and dams (Humborg et al., 1997; Ittekkot et al., 2000). These results suggest that greater riverine inflow under natural flow conditions may be an important determinant of *M. cyprinoides* abundance.

BIOTIC INTERACTIONS

The first MDS analysis showed a clear separation between wave-dominated (W) and tide-dominated (T) estuaries for the nighttime samples in the six southern rivers (Figure 1.11A). One-way analysis of similarity (ANOSIM) detected a significant but moderate separation between T and W sample groups (Global-R of 0.219, $p < 0.0001$). In the similarity percentages (SIMPER) analysis, the top families were ranked by discrimination index, that is, the ratio of the average contribution to similarity between groups (W, T) to the standard deviation of similarity between groups. The higher the value of the discrimination index, the more informative the family was for discriminating between the groups of samples (Clarke and Warwick, 2001). Clearly, the family most responsible for discrimination between W and T estuaries was Megalopidae (Table 1.5). Leiognathidae, Clupeidae, and Ambassidae were among several other families consistently netted in greater abundances in wave-dominated estuaries.

A second MDS analysis detected a cluster of families that occurred together in 60% of the nighttime samples (Figure 1.11B). In fact, three highly abundant families

TABLE 1.4
Factors, CPUE, and Abiotic Variables for Five Northern Estuaries by Type

Observation	Season	Estuary	Position	Type	Meg cyp CPUE (×100)	All Fish CPUE (×100)	Temp PC	pH	SpCon
1	Wet	Escape	D	T	0	511	30.3	7.6	49.4
2	Wet	Escape	U	T	0	85	29.1	7.5	30.3
3	Wet	Jacky Jacky	D	T	0	165	30.2	7.4	46.0
4	Wet	Jacky Jacky	U	T	0	244	30.3	7.9	44.6
5	Wet	Lockhart	D	T	0	151	30.1	7.1	39.9
6	Wet	Lockhart	U	T	2	27	28.1	6.9	11.1
7	Wet	McIvor	D	W	0	91	29.8	8.0	44.4
8	Wet	McIvor	U	W	96	344	29.1	6.3	13.7
9	Wet	Pascoe	D	W	0	55	30.6	7.5	45.6
10	Wet	Pascoe	U	W	33	79	28.4	6.6	0.2
11	Dry	Escape	D	T	10	2307	26.8	7.8	50.7
12	Dry	Escape	U	T	13	822	27.2	7.5	39.9
13	Dry	Jacky Jacky	D	T	3	309	26.5	7.5	47.8
15	Dry	Jacky Jacky	FU	T	14	321	26.8	7.1	30.3
14	Dry	Jacky Jacky	U	T	0	490	26.9	7.4	46.9
16	Dry	Lockhart	D	T	3	289	25.9	7.3	50.2
17	Dry	Lockhart	U	T	0	126	25.7	7.0	22.3
18	Dry	McIvor	D	W	0	118	24.6	7.8	40.6
19	Dry	McIvor	U	W	7	179	24.4	6.8	5.3
20	Dry	Pascoe	D	W	113	2663	26.7	7.5	22.4
21	Dry	Pascoe	U	W	113	313	26.2	6.9	2.3
		Mean			19	461	27.4	7.3	32.6

Sal PC	DO%	Turb	TN PC	TP PC	PO$_4$	Si PC	NH$_4$	NO$_2$	NO$_3$	DOC	Tanacd
32.4	100.0	12.7	18.6	0.22	0.02	64.3	0.46	0.19	0.66	–	0.32
18.9	66.4	43.8	19.6	0.14	0.01	41.8	0.49	0.06	0.14	–	1.02
29.9	97.6	–	15.3	0.28	0.02	37.0	0.70	0.10	0.25	–	0.09
28.9	80.0	12.3	24.8	0.36	0.02	53.8	2.80	0.41	1.16	–	–
25.5	74.4	3.5	10.0	0.15	0.01	97.5	0.25	0.16	0.59	–	–
6.7	72.6	40.0	23.9	0.36	0.01	118.5	1.06	0.08	0.84	–	0.97
28.8	85.3	17.0	13.0	0.03	0.01	5.8	0.02	0.06	0.28	–	0.00
8.0	66.2	21.1	22.9	0.56	0.09	120.0	3.82	0.15	0.82	–	0.60
29.7	100.0	17.1	12.3	0.41	0.01	28.5	1.34	0.08	0.25	–	0.00
0.1	88.3	20.0	14.1	0.17	0.01	69.1	1.28	0.06	1.35	–	0.42
33.3	100.0	8.7	7.2	0.01	0.01	194.9	2.31	0.07	0.19	1.51	0.24
25.5	73.5	5.5	14.6	0.37	0.06	7.5	2.77	0.06	0.06	1.16	0.01
31.2	63.9	11.8	29.8	0.42	0.12	43.2	4.56	0.19	0.24	1.99	0.15
18.8	60.0	30.3	12.0	0.05	0.03	165.7	1.35	0.03	0.07	2.01	0.80
30.5	79.9	30.3	10.4	0.18	0.01	26.4	1.87	0.05	0.12	2.19	0.02
33.0	64.8	1.1	13.9	0.00	0.01	88.4	2.89	0.05	0.10	1.52	0.24
13.6	72.6	1.0	14.5	0.48	0.04	276.0	1.63	0.09	0.39	2.29	–
26.0	100.0	0.8	29.3	0.31	0.13	232.6	3.00	0.30	0.50	15.55	2.90
2.9	84.1	3.5	15.3	0.47	0.06	110.2	2.33	0.14	0.25	3.78	0.22
14.1	100.0	–	11.7	0.31	0.01	290.4	1.54	0.05	0.22	1.10	0.22
1.2	93.2	1.8	11.2	0.04	0.04	315.3	1.38	0.03	0.46	1.21	0.14
19.1	82.0	14.7	16.4	0.25	0.03	113.7	1.8	0.11	0.42	3.1	0.96

Note: W = wave; T = tide. Values are given by position in the estuary (D = downstream; U = upstream; FU = far upstream) and season (dry = June; wet = February).

Abbreviations and units: Meg cyp = *Megalops cyprinoides*; Temp = water temperature (°C); SpCon = specific conductivity; Sal = salinity parts per thousand; DO% = dissolved oxygen; Turb = turbidity NTU; TN = total nitrogen; TP = total phosphorus; PO$_4$ = phosphate; Si = silicate; NH$_4$ = ammonium; NO$_2$ = nitrite; NO$_3$ = nitrate; DOC = dissolved organic carbon; Tanacd = tannic acid. Blank cells indicate missing data. PC = variables included in the principal components analysis.

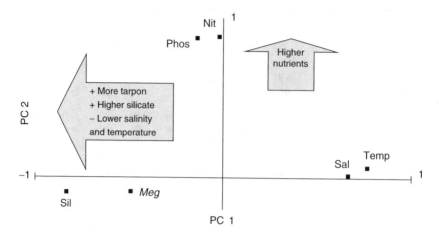

FIGURE 1.10 Results of the principal components (PC) analysis based on five abiotic attributes measured in the five northern estuaries (data in Table 1.4). Abbreviations: Phos = total phosphorus; Nit = total nitrogen; Sil = silicate; Meg = *Megalops cyprinoides* CPUE; Temp = water temperature; Sal = salinity (for full PC results see Table 1.5).

TABLE 1.5
Coefficients for Linear Combinations of Abiotic Variables Making up the Principal Components (PCs) Derived from Data Presented in Table 1.4 and the Values of the Percentage Variation Explained by the First Two PCs

Variable	PC 1	PC 2
Water temperature	**0.77**	0.06
Salinity	**0.66**	0.01
Total nitrogen	−0.02	**0.88**
Total phosphorus	−0.14	**0.87**
Silicate	**−0.83**	−0.09
Percentage variation	34.60	31.10
Cumulative percentage variation	34.60	65.70
Correlation with *Megalops cyprinoides* CPUE	−0.49	−0.09

Note: Values in bold type were correlated with the principal component at greater than 0.40.

in this group—Centropomidae, Mugillidae, and Ariidae—occurred together in >95% of the night samples. Thus, most nighttime samples consistently included the 11 families in this group ("the Centropomidae group"). In contrast, Megalopidae were only observed in 46% of the samples (Table 1.6). Because of the many zero-catch values, Megalopidae were distinguished at the 60% similarity level from the cluster of Centropomidae families. For Megalopidae, most of the samples with catch rates of zero were in the tide-dominated systems.

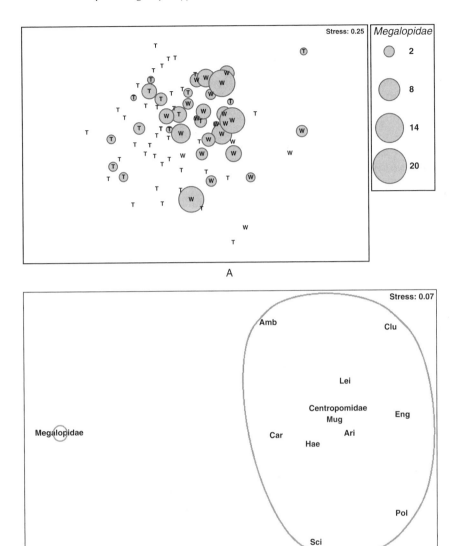

FIGURE 1.11 (A) Multidimensional scaling (MDS) diagram based on ordination of nighttime *samples by family*. Data are CPUE of families contributing at least 10% to one or more of the 78 night samples in the six southern estuaries (see Table 1.6). Labels indicate samples from wave-dominated (W) or tide-dominated (T) estuaries superimposed by circles sized in proportion to CPUE of Megalopidae (all *Megalops cyprinoides*). (B) MDS diagram derived by ordination of *families by sample* based on the same data as (A). Ellipses indicate families that were 60% similar in distribution based on a cluster analysis (group average clustering). Amb = Ambassidae; Ari = Ariidae; Car = Carangidae; Clu = Clupeidae; Eng = Engraulidae; Hae = Haemulidae; Lei = Leiognathidae; Mug = Mugillidae; Pol = Polynemidae; Sci = Sciaenidae.

TABLE 1.6
Comparison of Families Discriminating Two Estuarine Types (T = tide dominated; W = wave dominated) in the MDS Analysis (Figures 1.11A and 1.11B)

Family (Common Name)	% Samples >0	Average CPUE T	Average CPUE W	Discrimination Index
Megalopidae (tarpon)	46	0.01	1.69	1.56
Polynemidae (threadfin)	72	2.68	0.08	1.39
Leiognathidae (ponyfish)	91	2.68	19.08	1.37
Clupeidae (herrings)	94	14.76	45.70	1.32
Engraulidae (anchovies)	88	8.55	3.52	1.31
Ambassidae (glassfish)	71	0.89	2.14	1.28
Carangidae (trevally)	85	1.26	2.36	1.24
Ariidae (sea catfish)	97	9.82	4.18	1.23
Sciaenidae (drums)	68	1.08	0.30	1.22
Haemulidae (grunts)	90	2.22	1.63	1.11
Centropomidae (barramundi)	96	8.55	6.89	1.07
Mugilidae (mullet)	95	9.38	2.52	1.05

Note: Data are average CPUE (backtransformed) for families contributing at least 10% to one or more of the 78 nighttime samples from six southern estuaries. The families are ranked by discrimination index derived from the SIMPER analysis.

DISCUSSION

A comprehensive stock assessment of *M. cyprinoides* is not possible in the absence of data on the status of stocks anywhere in its range. Not surprisingly, then, an ecological assessment would also be unachievable at this time, requiring a thorough understanding of life history stages, as well as knowledge of temporal variation, habitat, functional role, and preferences along physicochemical gradients for each stage. The current state of knowledge as presented in this chapter is summarized in the following discussion.

LIFE HISTORY SUMMARY

Based on larval surveys, broadcast spawning by *M. cyprinoides* is most likely to occur in outer coral reef lagoons. Egg and early larval development apparently begin in these offshore lagoons, with later stage larvae moving inshore to estuaries. Settlement and metamorphosis into juveniles is known to occur in shallow estuarine habitats, following inundation during high spring tides and floods. Given these life history processes, *M. cyprinoides* may be considered a marine transient species (Day et al., 1989) in tropical and subtropical estuaries. However, critical biological factors such as growth rates and age-at-maturity remain unknown for adult phases in particular.

ECOLOGICAL SUMMARY

The favored inshore habitat of *M. cyprinoides* may be wave-dominated estuaries, having lower temperature and salinity but higher silicate levels. Thus, *M. cyprinoides* may be more common in systems under the strong influence of unmodified river flows. Furthermore, characteristics of wave-dominated estuaries include narrow entrances (<500 m) opening directly into open waters of the open ocean, interior basins with reduced area of water and mangrove forests, and sandy or hard-rock substrate (Ley, 2005). Analyses of distributional patterns in other regions are required to determine the generality of these observations throughout the range of *M. cyprinoides*.

Along permanently inundated estuarine shorelines, *M. cyprinoides* are more active early in the night than during the day. Morphologically adapted for high speed, and having a wide protrusible jaw, with an eye structured to capture light at low light levels, *M. cyprinoides* is equipped for foraging on water-column fishes and emergent benthic invertebrates such as penaeid shrimps under low light levels. During the dry season, *M. cyprinoides* is more likely to be present in permanently inundated creeks, channels, and billabongs; in the wet season, it may move into temporarily flooded salt flats.

The trophic role of *M. cyprinoides* is that of an intermediate carnivore, feeding on a highly diverse array of prey, including many insects. This diet and its ability to survive in low oxygen conditions through air breathing give *M. cyprinoides* great flexibility in survival capacity relative to other fishes. However, *M. cyprinoides* is not as ubiquitous in distribution as species such as *L. calcarifer* (Centropomidae), which was abundant in all estuaries closed to commercial fishing, whether wave- or tide-dominated (Ley et al., 2002). Thus, questions remain about the possible competitive interactions between *M. cyprinoides* and other estuarine carnivores and other potential factors influencing distribution.

ASSESSMENT OF VULNERABILITY

Potential vulnerability to detrimental effects of fishing remains unknown since condition of individual *M. cyprinoides* postnetting or postangling has not been studied to date. Scale loss owing to handling can be severe (pers. obs.) and may promote mortality in captured and released fish. A related issue concerns unknown levels of stock resiliency owing to a lack of information on such critical factors such as age and growth, reproductive biology, life history stages and habitat use, and migration patterns. In countries such as the Philippines and India, where exploitation of *M. cyprinoides* for food occurs, effective stock assessments are not possible, given the current state of knowledge. In developed countries, recreational fishing levels may lead to overexploitation as angler numbers increase. Management decisions can be best evaluated based on comprehensive monitoring data for stocks and knowledge of basic biological parameters.

Since larval and juvenile *M. cyprinoides* appear to rely on access to periodically flooded habitats adjacent to mangrove-lined creeks and rivers, loss of these habitats may limit production. Larger *M. cyprinoides* is more abundant in wave-dominated systems where freshwater flow from riverine tributaries is both substantial and unaltered by diversions or dams. Changes to freshwater delivery systems through diversion projects or dams may reduce habitat quality for *M. cyprinoides*. These observations suggest that *M. cyprinoides* may be an indicator species for ecosystem effects of freshwater flow modification.

ACKNOWLEDGMENTS

I am extremely grateful for larval fish data provided by Mark McGrouther and Jeff Leis from the Ichthyology Group of the Australian Museum, and also for recreational fish records provided by Mr. Bill Sawynok of Info-fish. I would also like to acknowledge the contributions of the many individuals involved in the surveys of northeastern Queensland estuaries, particularly P. Dixon, S. Boyle, and R. Partridge (Australian Institute of Marine Science [AIMS]) and I. Halliday, R. Garrett, A. Tobin, and N. Gribble (Queensland Department of Primary Industries). Funding for this study was provided by AIMS and the Fisheries Research and Development Corporation of Australia.

REFERENCES

Benstead, J.P., March, J.G., Fry, B., Ewel, K.C., and Pringle, C.M. 2006. Testing isosource: stable isotope analysis of a tropical fishery with diverse organic matter sources. Ecology 87: 326–333.

Bishop, K.A. 1980. Fish kills in relation to physical and chemical changes in Magela Creek (East Alligator River system, Northern Territory) at the beginning of the tropical wet season. Australian Zoologist 20: 485–500.

Bishop, K.A., Allen, S.A., Pollard, D.A., and Cook, M.G. 2001. Ecological studies on the freshwater fishes of the Alligator Rivers region, Northern Territory: autecology. Office of the Supervising Scientist, Environment Australia, Darwin, Northern Territory, Australia. 29pp.

Blaber, S.J.M. 2000. Tropical estuarine fishes: ecology, exploitation and conservation. Blackwell Sciences Ltd., London. 372pp.

Chen, H.L. and Tzeng, W.N. 2006. Daily growth increment formation in otoliths of Pacific tarpon *Megalops cyprinoides* during metamorphosis. Marine Ecology Progress Series 312: 255–263.

Clarke, K.R. 1993. Non-parametric multivariate analyses of changes in community structure. Australian Journal of Ecology 18: 117–143.

Clarke, K.R. and Warwick, R.M. 2001. Change in marine communities: an approach to statistical analysis and interpretation, 2nd ed., Primer-E Ltd. Plymouth Marine Laboratory, Plymouth, UK.

Coates, D. 1987. Observations on the biology of Tarpon, *Megalops cyprinoides* (Broussonet) (Pisces: Megalopidae) in the Sepik River, Northern Papua New Guinea. Australian Journal of Marine and Freshwater Research 38: 529–535.

Davis, T.L.O. 1988. Temporal changes in the fish fauna entering a tidal swamp system in tropical Australia. Environmental Biology of Fishes 21: 161–172.

Day, J.W., Hall, C.A.S., Kemp, W.M., and Yanez-Arancibia, A. 1989. Estuarine ecology. John Wiley & Sons, New York. 558pp.

FAO. 2006 (Food and Agricultural Organization of the United Nations), Fisheries Global Information System, http://www.fao.org.

Froese, R. and Pauly, D. (eds). 2006. FishBase. World wide web electronic publication. www.fishbase.org.

Furnas, M. 2003. Catchments and corals: terrestrial runoff to the Great Barrier Reef. Australian Institute of Marine Science, Townsville, Queensland. 334pp.

Halliday, I.A., Ley, J.A., Tobin, A.J., Garrett, R.N., Gribble, N.A., and Mayer, D.A. 2001. The effects of net fishing: addressing biodiversity and bycatch issues in Queensland inshore waters. Queensland Department of Primary Industries, Deception Bay, Queensland, Australia. 95pp.

Henry, G.W. and Lyle, J.M. 2003. The national recreational and indigenous fishing survey. Australian Department of Agriculture, Fisheries and Forestry, Canberra ACT, Australia. 190pp.

Humborg, C., Ittekkot, V., Cociasu, A., and Bodungen, B. 1997. Effect of Danube River dam on Black Sea biogeochemistry and ecosystem structure. Nature 386: 385–388.

International Game Fish Association. 2006. Database of IGFA angling records until 2006. IGFA, Fort Lauderdale, FL.

Ittekkot, V., Humborg, C., and Schafer, P. 2000. Hydrological alterations and marine biogeochemistry: a silicate issue? Bioscience 50(9): 776–782.

King, M. 1995. Fisheries biology, assessment and management. Fishing News Books, Oxford, UK. 341pp.

Kulbicki, M., Guillemot, N., and Amand, M. 2005. A general approach to length-weight relationships for New Caledonian Lagoon fishes. Cybium 29(3): 235–252

Kulkarni, C.V. 1983. Longevity of fish *Megalops cyprinoides*. Journal of the Bombay Natural History Society 80: 230–232.

Leis, J.M. and Reader, S.E. 1991. Distributional ecology of milkfish, *Chanos chanos*, larvae in the Great Barrier Reef and Coral Sea near Lizard Island, Australia. Environmental Biology of Fishes 30: 395–405.

Ley, J.A. 2005. Linking fish assemblages and attributes of mangrove estuaries in tropical Australia: criteria for regional marine reserves. Marine Ecology Progress Series 305: 41–57.

Ley, J.A. and Halliday, I.A. 2003. A key role for marine protected areas in sustaining a regional fishery for barramundi *Lates calcarifer* in mangrove-dominated estuaries? Evidence from northern Australia. American Fisheries Society Symposium 42: 225–236.

Ley, J.A. and Halliday, I.A. (In press). Diel variation in mangrove fish abundance and trophic categories in six northeastern Australian estuaries and a conceptual model. Bulletin of Marine Science 80(3).

Ley, J.A., Halliday, I.A., Tobin, A.J., Garrett, R.N., and Gribble, N.A. 2002. Ecosystem effects of fishing closures in mangrove estuaries of tropical Australia. Marine Ecology Progress Series 245: 223–238.

Pusey, B., Kennard, M. and Arthington, A. 2004. Freshwater fishes of north-eastern Australia. CSIRO Publishing, Collingwood, Victoria, Australia. 694pp.

Rao, L.M. and Padmaja, G. 1999. Variations in food and feeding habits of *Megalops cyprinoides* from coastal waters of Visakhapatnam. Indian Journal of Fisheries 46: 407–410.

Russell, D.J. and Garrett, R.N. 1983. Use by juvenile barramundi, *Lates calcarifer* (Bloch), and other fishes of temporary supralittoral habitats in a tropical estuary in northern Australia. Australian Journal of Marine and Freshwater Research 34: 805–811.

Seymour, R.S., Christian, K., Bennett, M.B., Baldwin, J., Wells, R.M.G., and Baudinette, R.V. 2004. Partitioning of respiration between the gills and air-breathing organ in response to aquatic hypoxia and exercise in the Pacific tarpon, *Megalops cyprinoides*. Physiological and Biochemical Zoology 77: 760–768.

Shiao, J. and Hwang, P. 2004. Thyroid hormones are necessary for teleostean otolith growth. Marine Ecology Progress Series 278: 217–278.

Tsukamoto, Y. and Okiyama, M. 1993. Growth during the early life history of the Pacific tarpon, *Megalops cyprinoides*. Japanese Journal of Ichthyology 39: 379–386.

Tsukamoto, Y. and Okiyama, M. 1997. Metamorphosis of the Pacific tarpon, *Megalops cyprinoides* (Elopiformes, Megalopidae) with remarks on development patterns in the Elopomorpha. Bulletin of Marine Science 60: 23–36.

Tzeng, W.N., Wang, Y., and Chang, C. 2002. Spatial and temporal variations of the estuarine larval fish community on the west coast of Taiwan. Marine and Freshwater Research 53: 419–430.

Tzeng, W.N., Wu, C.E., and Wang, Y.T. 1998. Age of Pacific tarpon, *Megalops cyprinoides*, at estuarine arrival and growth during metamorphosis. Zoological Studies 37: 177–183.

Wells, R.M.G., Baldwin, R.S., Baudinette, R.V., Christian, K., and Bennett, M.B. 2003. Oxygen transport capacity in the air-breathing fish, *Megalops cyprinoides*: compensations for strenuous exercise. Comparative Biochemistry and Physiology Part A 134: 45–53.

2 Biology and Ecology of the Recreational Bonefish Fishery at Palmyra Atoll National Wildlife Refuge with Comparisons to Other Pacific Islands

Alan M. Friedlander, Jennifer E. Caselle, Jim Beets, Christopher G. Lowe, Brian W. Bowen, Tom K. Ogawa, Kevin M. Kelley, Todd Calitri, Matt Lange, and Bruce S. Anderson

CONTENTS

Introduction 28
Palmyra Bonefish Fishery 30
Biology and Ecology of Bonefish at Palmyra Atoll 30
    Fish Stock Definition 30
    Allometric Growth 31
    Sex Ratios and Reproductive Condition 32
    Growth and Mortality 33
Stomach Contents 34
Larval Biology 35
    Genetic Analysis 36
        Population Isolation 37
        Recruitment 37
        Genetic Diversity 38
        Effective Population Size 39
        Age of Population 40

Fisheries Information from Palmyra Atoll 40
  Spatial and Temporal Trends in Bonefish Catch 40
  Tagging Program 41
  Remote Monitoring of Bonefish Movement 42
  Bonefish Physiological Responses to Catch-and-Release Stress 43
Bonefish Information from Other Pacific Islands... 44
  Hawaii... 45
    Stock Identification 46
    Oahu Catch-and-Release Fishery... 46
    Juvenile Recruitment 47
    Feeding 48
  Kiritimati (Christmas) Atoll 49
Tarawa Atoll, Kiribati 50
Discussion 52
Acknowledgments 54
References 54

INTRODUCTION

Bonefish support important recreational fisheries in subtropical and tropical regions worldwide (Crabtree et al., 1996; Kaufmann, 2000). At most locations in the Pacific Ocean, the recreational fishery resource is shared with subsistence and small-scale commercial fisheries that harvest with spears, nets, traps, and lines. As a result of intensive fishing effort, bonefish and many of their predators have been overharvested in many locations (Ault et al., 1998, 2002, 2005; Beets, 2000), making it difficult to obtain baseline data on their unexploited ecology and population biology.

Palmyra Atoll is situated in the Line Islands at 5°53′ N, 162°5′ W in the central Pacific, approximately 1600 km south of Hawaii (Figures 2.1 and 2.2). Palmyra Atoll is a relatively pristine environment located in the Intertropical Convergence Zone that receives approximately 450–500 cm of rainfall per year (Vitousek et al., 1980), a wet climate somewhat atypical of most Pacific atolls. This weather results in lush vegetation and beach forests, at one time supporting one of the largest remaining stands of *Pisonia* in the Pacific (Maragos, 2000). The atoll also has healthy nesting populations of at least 10 seabird species, including some of the largest-known colonies of red-footed boobies and black noddies (Fish and Wildlife Service, 2001). Palmyra contains approximately 4 km^2 of emergent land with three large lagoons and extensive sandy reef flats (Figure 2.2). The western lagoon is nearly 2 km^2 and as deep as 55 m. The central and eastern lagoons are each 1 km^2 and up to 30 m deep, approximately. The lands of Palmyra were acquired in late 2000 by The Nature Conservancy (TNC), and in 2001, the atoll's submerged lands were designated a National Wildlife Refuge (NWR) under the jurisdiction of the U.S. Fish and Wildlife Service. Palmyra Atoll possesses a lightly exploited bonefish population owing to its unique status as a NWR. The nearly pristine environment of Palmyra Atoll, coupled with the protection from heavy subsistence or commercial exploitation, renders it one of the few places on Earth left to examine the status and dynamics of a "natural" bonefish population.

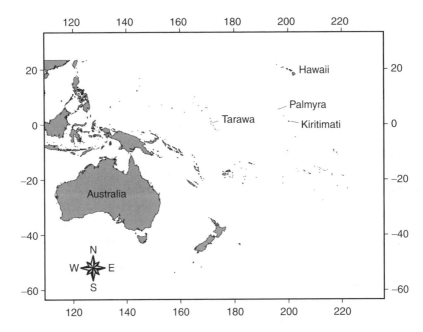

FIGURE 2.1 Map of the central and western Pacific showing bonefish locations examined in this study.

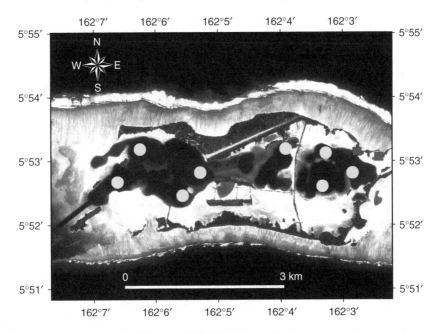

FIGURE 2.2 Palmyra Atoll National Wildlife Refuge. Circles represent locations of acoustic receivers in the lagoons used to examine bonefish movement. Squares show locations where channel nets were deployed to collect bonefish larvae. On the IKONOS satellite image, land areas appear dark and the sand flats appear white.

PALMYRA BONEFISH FISHERY

Palmyra, like Kiritimati (Christmas) Atoll to the south, is world renowned for its bonefishing—the lagoon sand flats provide excellent habitat and support a large population of bonefish (*Albula glossodonta*). Since May 2000, TNC has been operating small-scale experimental ecotourism activities at Palmyra including fishing, diving, kayaking, and wildlife photography. Lessons learned during this experimental period include (1) sportfishing opportunities are varied and the fishing is excellent; (2) fly-fishing for bonefish on the flats is an attraction that is rarely available elsewhere in the U.S. Pacific; and (3) sportfishing is a potential activity that, if well managed, would be compatible with, and could perhaps help financially support, conservation goals for Palmyra. These findings are reflected in the "interim compatibility determination" for bonefish fishing by the Fish and Wildlife Service (FWS; Fish and Wildlife Service, 2001). Bonefishing was found to be compatible with the conservation purposes of the refuge, if fishing were subjected to the following regulations: (1) limited to eight rods (persons) in the lagoon at any given time; (2) catch-and-release fishing with barbless hooks only; (3) limited to certain seasons; (4) limited to certain areas; and (5) if fishing success (e.g., catch rates) and stock status were rigorously monitored and assessed through logbook records and tagging studies.

Following these guidelines, we initiated a conservation research program focused on bonefish designed to provide critical biological, ecological, and fishery information for resource management at Palmyra Atoll. In addition, the research has provided some new and generally applicable information about Pacific bonefish, particularly *A. glossodonta*, a regionally important species that has received little scientific attention.

BIOLOGY AND ECOLOGY OF BONEFISH AT PALMYRA ATOLL

The low exploitation of the bonefish population at Palmyra provided a baseline of biological and ecological measures that may be useful to assess bonefish populations at Pacific locations subjected to varying exploitation levels and environmental stress.

FISH STOCK DEFINITION

Meristic measurements of 250 bonefish collected for biological sampling and 65 fish that were tagged and released indicated that all the individuals examined belong to *A. glossodonta*. Results from all biological sampling revealed the presence of only *A. glossodonta* at Palmyra Atoll. Shaklee and Tamaru (1981) used morphological and electrophoretic data to demonstrate the presence of two cyptic bonefish species in the central Pacific: *A. glossodonta* and *A. forsteri*. The two species are difficult to distinguish based on external morphology, but are highly distinct in mtDNA cytochrome *b* sequences ($d = 0.24$ sequence divergence). Based on a limited survey of Palmyra specimens ($N = 65$), we detected only the *A. glossodonta* mtDNA lineage identified by Colborn et al. (2001). These data demonstrate that the dominant species in the Palmyra fishery, and perhaps the only bonefish in Palmyra, is *A. glossodonta*. However, given our limited sampling, it is possible that *A. forsteri* occurs at low frequency.

Results from nearby Kiritimati (Christmas) Atoll yielded the same outcome; all bonefish specimens from the lagoon ($N = 52$) had mtDNA sequences characteristic of *A. glossodonta*. At Tarawa Atoll, investigation of bonefish biology also found only *A. glossodonta* to be present (Beets, 2000). However, due to the low sample sizes of these studies, it is possible that *A. forsteri*, a species that is present in Hawaii and elsewhere in the Pacific occurs in small numbers or occupies unsurveyed habitats.

ALLOMETRIC GROWTH

Bonefish collected from biological sampling ($n = 249$) and from tagging ($n = 890$) ranged in size from 15.5 to 67.0 cm fork length (FL), with a mean length of 41.5 cm FL (± 7.6 sd) (Table 2.1). The mean weight of specimens was 723.5 g (± 327.7 sd), the smallest individual weighed 9 g and the largest 1920 g. From these data, length–weight and length–length relationships were developed (Table 2.2). The largest tagged fish were

TABLE 2.1
Sex, Sample Size, Fork Length (FL) and Weight of Bonefish Collected at Palmyra Atoll between 2002 and 2003

Sex	Number	Length (cm FL)	Weight (kg)
Male	121	39.07 (4.16)	0.81 (0.23)
Female	93	38.41 (5.53)	0.79 (0.33)
Unknown	9	31.21 (7.13)	0.43 (0.26)
Juvenile	26	23.76 (3.81)	0.19 (0.86)
Total	249	36.94 (6.67)	0.72 (0.33)

Note: Standard deviation of the mean is in parentheses.

TABLE 2.2
Linear Length–Length and Log-Linear Length–Weight Relationships from Bonefish Collected at Palmyra Atoll Parameters for the Generalized Linear Regression Model $Y_i = b_0 + b_1 X_i + \varepsilon_i$

X	Y	b_0	b_1	r^2
FL	SL	−4.825 (2.959)	0.925 (0.008)	0.99
TL	SL	−12.975 (3.529)	0.794 (0.008)	0.99
SL	FL	7.346 (3.134)	1.074 (0.009)	0.99
TL	FL	−7.977 (0.433)	0.858 (0.008)	0.99
FL	TL	12.114 (3.903)	1.161 (0.011)	0.99
SL	TL	19.718 (4.251)	1.250 (0.012)	0.99
$\mathrm{Log}_{10}\mathrm{FL}$	$\mathrm{Log}_{10}\mathrm{WT}$	−4.914 (0.073)	2.968 (0.028)	0.99

Note: FL = fork length (cm); SL = standard length (cm); TL = total length (cm); WT = weight (g). Standard error of the mean is in parentheses.

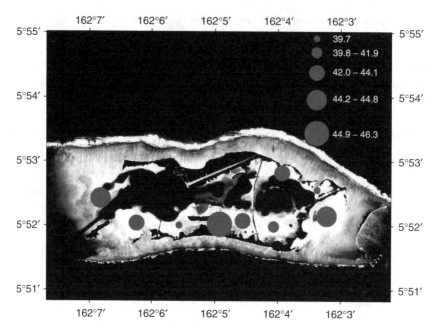

FIGURE 2.3 Average size (cm FL) of bonefish at Palmyra Atoll from tagging data. Proportional bubbles indicate differences in sizes among all flats containing 20 or more fish sampled.

generally associated with seaward reef flats, while smaller fish were more often found on protected, inshore lagoon flats (Figure 2.3).

SEX RATIOS AND REPRODUCTIVE CONDITION

The sex ratio and reproductive condition of bonefish were assessed during all moon phases over a period of 2 years. Males comprised 51% of the 241 fish for which sex could be determined, females accounted for 39%. The remaining 11% were classified as juveniles as the gonads appeared undifferentiated.

On average, males were slightly larger than females, but the difference was not statistically significant (t-test $= 1.02$, $P = 0.31$) (Table 2.1). The largest female was 58.0 cm FL, while the largest male was 48.7 cm FL. The smallest distinguishable mature male in all samples was 24.5 cm FL, and the smallest distinguishable mature female was 28.7 cm FL.

The gonadosomatic index (GSI) was calculated as the ratio of gonad weight (g) to somatic body weight (total body weight − gonad weight). Fatty tissue was sometimes associated with gonadal tissue, typically when the gonads were not reproductively active. For all samples combined, the mean GSI of males was 0.81 (±0.77 sd) and was significantly greater (t-test $= 2.90$, $P = 0.004$) than females ($\bar{X} = 0.51 \pm 0.71$ sd). GSI for females was significantly higher ($P < 0.05$) during the full-moon period, but did not differ significantly ($P > 0.05$) between new- and quarter-moon phases ($F_{2,90} = 4.13$, $P = 0.02$; Figure 2.4). GSI for males was apparently lowest during

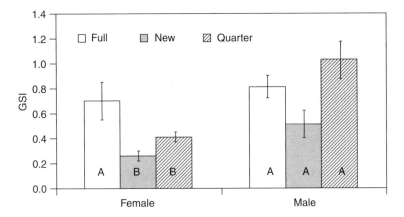

FIGURE 2.4 Mean gonadosomatic index (GSI) by lunar phase for female (one-way ANOVA $F_{2,90} = 4.13$, $P = 0.019$) and male (one-way ANOVA $F_{2,118} = 2.90$, $P = 0.06$) bonefish at Palmyra Atoll. Moon phases with the same letters are not significantly different (Tukey–Kramer HSD $P > 0.05$). Error bars are standard error of the mean.

the new moon ($F_{2,118} = 2.90$, $P = 0.06$). These results suggest that females are most reproductively active around full moons, but males are apparently ripe throughout the lunar cycle.

Growth and Mortality

We combined all length samples ($N = 1140$) to approximate a steady-state population. Ultimate maximum length L_∞ of 67.3 cm FL was calculated using a Powell–Wetherall plot (Table 2.3) (Wetherall et al., 1987). Age-at-length data was obtained from otolith annuli counts from a sample of 33 bonefish collected at Palmyra in March 2003. These fish ranged in size from 15.5 to 52.5 cm FL ($\bar{X} = 34.1 \pm 8.4$ sd), and in age from 3 to 11 years ($\bar{X} = 5.9 \pm 1.8$ sd). From these data, assuming von Bertalanffy growth, a Brody growth coefficient K of 0.3 was obtained. Instantaneous total mortality rate Z (0.2686/year) was derived using a length-based mortality function:

$$\left(\frac{L_\infty - L_\lambda}{L_\infty - L_c}\right)^{Z/K} = \frac{Z(L_c - \bar{L}) + K(L_\infty - \bar{L})}{Z(L_\infty - \bar{L}) + K(L_\infty - \bar{L})}, \tag{2.1}$$

where L_∞ is the ultimate length; L_λ is the length at maximum age; $\bar{L}$ is the mean length in the sample; and L_c is the length at first capture (Ault and Ehrhardt 1991; Ehrhardt and Ault 1992). This estimate of Z (0.2686) is very close to the estimate of natural mortality ($M = 0.27$) provided by Alagaraja (1984):

$$M = \frac{-\ln(S)}{t_\lambda}, \tag{2.2}$$

TABLE 2.3
Population Dynamics Parameter Estimates for Growth and Mortality of Bonefish at Palmyra Atoll

Ultimate Length ($L\infty$)	Brody Growth Coefficient (K)	Mortality/Growth (Z/K)	Total Mortality Rate Z	Natural Mortality Rate M
67.28	0.3	0.8953	0.2686	0.27

where S = survival (0.05) and t_λ = longevity. Since fishing mortality is likely negligible, instantaneous total mortality is equal to natural morality ($Z = M$). The value for $Z/K = 0.2686/0.3 = 0.8953$.

Estimates of growth of bonefish at Palmyra are slightly higher than those reported in the Florida Keys (K = 0.24–0.28) by Crabtree et al. (1996) and off Mexico's Pacific coast (K = 0.275) by Morales-Nin (1994). This may reflect warmer year-round water temperatures and the lack of seasonality in growth of bonefish at Palmyra. Estimates of natural mortality rate for bonefish from Palmyra were at the high end compared to Crabtree et al. (1996) data from the Florida Keys (M = 0.2–0.3).

STOMACH CONTENTS

Stomach contents were identified to resolve the prey composition and feeding habitats of Palmyra bonefish. Of the 160 stomachs examined in this study, 66 (41%) were empty. Crabs, primarily ghost crabs (*Macrophthalmus* spp.), made up 33% of the total weight and 41% of the total volume of prey consumed by bonefish (Table 2.4). Acorn worms (Sipunculids) accounted for 29% of both weight and volume of prey consumed by bonefish at Palmyra. The remainder of the prey items consisted of various crustaceans (e.g., shrimp, isopods) and polychaete worms, with a few small fishes and one terrestrial beetle. Coral rubble comprised 19% of the weight and 9% of the volume of stomach contents. Numerically, peanut worms were the most abundant taxa, accounting for 16% of the total number of prey items. The two species of ghost crabs comprised an additional 14% of the prey items by number, followed by crawling crustaceans with 12% the total prey items encountered in the stomachs of bonefish.

Weight of stomach contents for bonefish caught in gill nets on the falling tide was significantly greater than stomach content weight from fish collected on incoming tides (H = 44.02, $P < 0.001$; Figure 2.5). Bonefish caught on the falling tide were full of small crabs (>80% by weight), and these fish were likely feeding on the flats prior to capture. Stomachs of fish captured on the incoming tide contained (by weight) mainly peanut worms (32%), coral rubble (22.0%), and sand mixed with organic material (14.9%). These fish were likely feeding in the lagoon prior to capture.

TABLE 2.4
Prey Items Identified from Stomachs of 160 Bonefish at Palmyra

	Taxon	N	Wet Wt (g)	Percent Wet Wt (g)	Volume (mL)	Percent Volume (mL)
Crabs						
Ghost crab	*Macrophthalmus convexus*	20	8.58	11.58	9.00	13.87
Ghost crab	*Macrophthalmus telescopicus*	11	12.66	17.09	12.40	19.11
Swimming crab	Portunidae	3	0.20	0.27	0.30	0.46
Crawling crustaceans	Reptania	26	2.06	2.78	3.40	5.24
Swimming crabs	*Thalamita* spp.	2	0.61	0.82	0.60	0.92
Mud crab	Xanthidae	2	0.61	0.82	0.60	0.92
Worms						
Fire worms	Amphinomidae	1	0.01	0.01	0.10	0.15
Lugworms	Polychaeta	2	0.32	0.43	0.40	0.62
Peanut worms	Sipuncula	35	21.49	29.01	18.70	28.81
Shrimp						
Snapping shrimp	Alpheidae	17	1.70	2.29	2.80	4.31
Swimming shrimp	Natantia	20	0.76	1.03	2.20	3.39
Mysid shrimp	Mysidacea	5	0.05	0.07	0.50	0.77
Mantis shrimp	Stomatopoda	12	0.69	0.93	1.50	2.31
Fishes						
Goby	Gobiidae	1	0.11	0.15	0.10	0.15
Lizardfishes	Synodontidae	1	0.16	0.22	0.20	0.31
Bony fishes	Teleostei	2	0.27	0.36	0.40	0.62
Terrestrial beetle		1	0.06	0.08	0.10	0.15
Isopods	Isopoda	6	0.06	0.08	0.60	0.92
Other crustaceans		7	1.06	1.43	1.10	1.69
Debris		5	0.28	0.38	0.50	0.77
Unidentified		23	8.16	11.01	3.40	5.24
Rubble		13	14.19	19.15	6.00	9.24

LARVAL BIOLOGY

Bonefish are primitive teleost fishes that have the unusual leptocephalus larva found only in bonefish, tarpon, ladyfish, and eels. The bonefish leptocephalus is long (ca. 50–70 mm TL) and ribbon-shaped. After the planktonic larval stage, larvae move into shallow habitats where they metamorphose into juveniles. Recruitment of larvae to inshore areas has been found to have seasonal peaks, as well as strong lunar-month, tidal, and diurnal signals (Mojica et al., 1995).

We employed fixed channel nets (1-m × 1-m-square opening, 1000-µm mesh size, 3-m length) in two major channels that drain into the easternmost lagoon from eastern (windward) reefs (Figure 2.2). The channels are several hundred meters from the fringing reef; the area in between consists of coral reef flats and can be

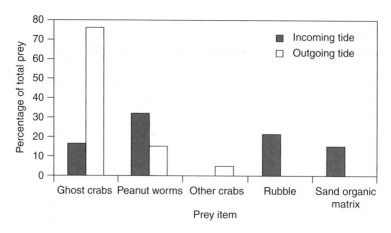

FIGURE 2.5 Percentage of prey items consumed by bonefish on incoming and outgoing tides at Palmyra.

exposed on low tides. Nets were fished on nighttime flood (incoming) tides, and the majority of our sampling took place during the week surrounding the new moon, an interval correlated with high catches of bonefish larvae in the Bahamas (Mojica et al., 1995). There appeared to be a seasonal pattern to recruitment. In all, 15 bonefish larvae (and many other leptocephali) were captured at Palmyra during March and August. No bonefish larvae (and very few other leptocephali) were captured during three sampling trips in November and February.

Sagittal otoliths were removed from all bonefish larvae and examined to determine planktonic larval duration (PLD). Larval bonefish otoliths were generally clear and did not require any grinding or other preparation for aging. Daily growth bands of otoliths were counted using a compound microscope (40–100×), and measurements were made using image analysis software (Image-Pro, MediaCybernetics Co.).

PLDs of bonefish larvae settling at Palmyra ranged from 48 to 72 days with an average of 57.2 days (±6.1 sd). Average length of larvae collected was 51.4 mm TL (±4.3 sd), and ranged from 43.3 to 58.7 mm TL. The average maximum otolith radius from core to edge was 118.2 µm (range [89.5, 188.2]). Despite the geographic isolation of Palmyra, the PLDs found in this study were comparable to those of bonefish larvae from the Bahamas (range [41, 71]; mean = 56 days; Mojica et al., 1995).

GENETIC ANALYSIS

To maximize the conservation dividends from our bonefish study, molecular genetic tools were employed to answer questions about genetic diversity, population isolation, recruitment, effective population size, and age of the Palmyra population. Sampling for these studies was accomplished primarily with nondestructive collections of fin clips (about 1 cm^2 per fish). The molecular marker of choice was mtDNA cytochrome *b*, which has been used to estimate bonefish population structure and cryptic evolutionary units on a global scale (Colborn et al., 2001).

Population Isolation

Three locations in the central Pacific [Palmyra; Kiritimati (Christmas) Atoll; and Oahu, Hawaiian Islands] and the Seychelles in the Indian Ocean were surveyed to address the geographic scale of population structure (Table 2.5). The Seychelles, at the presumed far western end of the species distribution, serves as an "outgroup" population to assess species-wide genetic diversity. We observed significant population structure overall (ϕ_{st} = 0.133, $P < 0.001$), and all pairwise comparisons were significant, except for Palmyra vs. Kiritimati (Table 2.6). The two Line Islands locations (Palmyra and Kiritimati) are nearly identical in haplotype diversity, nucleotide diversity, and other population parameters (Table 2.6). We concluded that the primary fishery locations in the Line Islands are part of a single large genetic population, probably connected by the relatively long PLD of 42–78 days.

It is notable that, even with a small sample size ($N = 10$), the Hawaiian location is highly differentiated from the Line Islands. Perhaps population divisions and corresponding management units occur on a scale of the Line Islands vs. the Hawaiian Islands, separated by approximately 2000 km. Additional samples across these two archipelagos, and elsewhere in the Pacific, are desirable to test this hypothesis.

Recruitment

Larvae of the two bonefish species, *A. glossodonta* and *A. forsteri*, are morphologically indistinguishable, and it is possible that some bonefish recruits to Palmyra are from spawning populations located elsewhere in the central Pacific. Here we analyzed 14 larvae and compared them to 51 juveniles and adults, all collected from Palmyra, to determine whether the recruits to Palmyra are from the same population (and species) as the adults in the fishery (*A. glossodonta*). Our results show no significant genetic differences between larvae and juveniles–adults ($\phi_{st} < 0.001$,

TABLE 2.5
Pairwise Estimates of Population Differentiation Based on ϕ_{st}, a Molecular Genetic Analog to F_{st}

Location	Palmyra	Kiritimati	Oahu	Seychelles
Palmyra	–	0.000	0.204	0.212
Kiritimati (Christmas)	0.828	–	0.215	0.252
Oahu, Hawaii	<0.001	0.001	–	0.786
Seychelles (Indian Ocean)	<0.001	<0.001	<0.001	–

Note: The ϕ_{st} values are above the diagonal; *P* values indicating level of significance (based on 100,000 Markov steps) are below the diagonal.

TABLE 2.6
Estimates of mtDNA Diversity and Demographic Parameters for the Bonefish (*Albula glossodonta*) in the Central Pacific

Location	N	h	π	θ_0	N_{f0}	θ_1	N_{f1}	τ (age)
Palmyra	65	0.753	0.0028	0.000	0	83.75	2,080,000	1.99 (248,000)
Kiritimati (Christmas)	52	0.759	0.0027	0.000	0	66.25	1,639,000	2.02 (251,000)
Oahu, Hawaii	10	0.511	0.0019	0.001	25	1.82	45,200	2.66 (331,000)
Seychelles (Indian Ocean)	18	0.000	0.000	–	–	–	–	–

Note: N = sample size; h = haplotype diversity; π = nucleotide diversity; θ_0 and θ_1 = expected pairwise differences before and after the most recent population expansion; N_{f0} and N_{f1} = female effective population size before and after the most recent population expansion, estimated from theta (θ) values; τ(age) = mutational timescale (translated into a time estimate in years before present). Because of the lack of diversity at Seychelles, population demographic parameters could not be estimated. All population parameters were calculated with ARLEQUIN version 2.0 (Schneider, S., Roessli, D., and Excoffier, L., *ARLEQUIN, Version 2.0: A Software for Population Genetics Data Analysis*, Genetics and Biometry Lab, University of Geneva, Geneva, Switzerland, 2000).

$P = 0.576$). We concluded that larvae may be drawn from the Line Islands' population of *A. glossodonta* that encompasses Palmyra and Kiritimati.

Genetic Diversity

The bonefish population showed genetic diversity in previous allozyme and mtDNA surveys (Shaklee and Tamaru, 1981, Colborn et al., 2001). However, isolated island populations may have reduced diversity, an indication of vulnerability to environmental stress. The mtDNA data alone are not sufficient to address this issue, but can indicate whether further investigations of genetic diversity are warranted. In the three central Pacific locations we surveyed, 12 haplotypes were observed, indicating a normal level of haplotype diversity (h values; Table 2.6). However, these haplotypes were closely related, as indicated by low levels of nucleotide diversity (π values; Table 2.6). A parsimony network illustrates this feature, wherein no individual is more than four nucleotide differences away from any other individual (Figure 2.6). Considering that population sampling spans the central Pacific and western Indian Ocean, this is a notably low level of intraspecific divergence (see Grant and Bowen, 1998).

Among the 14 haplotypes identified in this survey of *A. glossodonta*, haplotype ALB101 is the hub of the parsimony network in Figure 2.6, and designated by the program TCS version 1.13 (Clement et al., 2000) as the ancestral state. It is also the

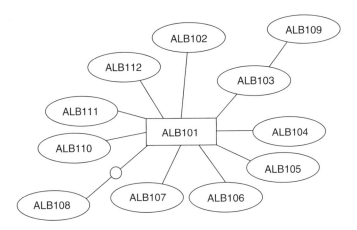

FIGURE 2.6 Relationships among haplotypes observed at Palmyra and other locations for *A. glossodonta*. The box surrounding haplotype ALB101 indicates a designation of ancestral condition by the program TCS version 1.13. Each line between haplotypes indicates a single mutation, with the interrupted line between ALB101 and ALB108 indicating two mutations. (Redrawn from Clement, M., Posada, D., and Crandall, K.A., *Molecular Ecology*, 9, 1657–1659, 2000.)

most common haplotype (70 out of 145 individuals), and observed at every location, albeit at significantly different frequencies. This is the only haplotype observed at the Seychelles ($N = 18$), possibly indicating a population bottleneck or recent founder event at the western fringe of the range for *A. glossodonta*.

Effective Population Size

Genetic effective population size (N_e) is the number of successful spawning adults averaged across the last few thousand generations. It is usually much smaller than the current population abundance, indicating historical bottlenecks or high variance in reproductive success, that is, "sweepstake recruitment" following Hedgecock (1994).

Based on the pattern of mutational differences among individuals, we estimated effective population size at the beginning of the most recent population expansion (either from colonization event or crash and recovery) and the current effective population size. These estimates are subject to numerous caveats (see Lecomte et al., 2004), especially that the maternally inherited mtDNA measures the female effective populations size (N_f). Assuming a 1:1 sex ratio, the values of N_f in Table 2.6 would be doubled to account for the entire effective population size. These values also depend on mutation rate for cytochrome *b* in bonefishes, estimated at 1.5% per million years between lineages, or 0.75% per million years within each lineage (Colborn et al., 2001). All of these qualifications and assumptions indicate that estimates of genetic effective population size are qualitative values, not specific quantitative values. With these caveats in mind, the effective population size in the Line Islands (including Palmyra and Kiritimati) is on the order of 2 million females, perhaps 4 million individuals in total. The difference in estimates from Palmyra ($N_f = 2$ million) and Kiritimati ($N_f = 1.6$ million) is probably not significant, given the high variance and uncertainty in parameter estimates (Table 2.6). It is sufficient to conclude that the Line Islands have a bonefish population that numbers in millions. In contrast, the estimate from Hawaii is two orders of magnitude lower, possibly indicating a lower effective population size that is perhaps due to reduction from exploitation. However, larger sample sizes and more thorough surveys are necessary to confirm these preliminary results, which are based on only 10 individuals.

Age of Population

Using a mismatch analysis (Rogers and Harpending, 1992), we estimated the age of bonefish populations. As with the estimates of effective population size, age estimates are subject to several caveats, including the mutation rate (see above) and approximation of the generation time of bonefish, which we provisionally placed at 5 years. For these reasons, we regard the age estimates as first-order approximations. With these limitations in mind, the populations at Palmyra and Kiritimati coalesce to a common ancestor on the order of 250,000 years ago (Table 2.6). The Hawaii population coalesces on a somewhat longer timescale (331,000 years ago), but given the small sample sizes and uncertainties about population parameters, it is unlikely that these differences were significant. It is sufficient to conclude that the central Pacific populations at the Line and Hawaiian Islands coalesce to a common ancestor in the late pleistocene, during the intervals characterized by upheavals in sea level associated with glacial maxima.

A previous phylogenetic survey indicates that *A. glossodonta* is several million years apart from other bonefish species (Colborn et al., 2001), yet the mtDNA diversity in contemporary populations coalesces to a common ancestor on the order of a quarter million years. The reasons for this may include a selective sweep for a superior mtDNA type, but also must consider connectivity across a wide range from the western Indian Ocean to the central Pacific in the late pleistocene. All surveyed populations share the putative ancestral haplotype ALB101, and this haplotype has been detected in west Pacific populations as well (unpublished data).

Management units for *A. glossodonta* seem to emerge on the archipelago scale, rather than for specific islands. Hawaii and the Line Islands appear to be significantly differentiated, but the Line Island locations have uniform population parameters (Table 2.6). The genetic effective population size for Palmyra and the Line Islands is on the order of millions of individuals, in what appears to be a group of shallow habitats connected by perhaps larval dispersal. This is consistent with the finding of no discernable population structure among Caribbean locations in a related species, *A. vulpes* (Colborn et al., 2001). Additional sampling in the Line and Hawaiian Islands will be necessary to test the hypothesis of management units on a scale of island archipelagos.

FISHERIES INFORMATION FROM PALMYRA ATOLL

Spatial and Temporal Trends in Bonefish Catch

A catch-and-effort logbook program was established to determine trends in fishing success and bonefish population abundance. Owing to the size of Palmyra and the nature of the fishery, the logbook program provided nearly 100% coverage of fishing activities. Project information fliers were developed and distributed to anglers, and guides on Palmyra were trained in data-collection techniques. Information was collected on (1) time and location of caught-and-hooked fish; (2) angler experience level (novice, intermediate, advanced); (3) fork lengths of fish caught and numbers of hooked fish; (4) tide, moon, and weather conditions; and (5) observations on school size, movement patterns, etc.

Angler skill level had a significant effect on fishing success, with mean catch per unit effort (CPUE) of 2.10 for advanced anglers, 1.89 for intermediate anglers, and 1.23 for novices ($F_{2,276} = 6.33$, $P = 0.002$, advanced = intermediate > novice). From April 2002 to November 2003, overall mean CPUE, excluding novice anglers, was 2.03 (±1.49 sd) fish per rod hour. Overall daily catch rates (novices excluded) varied greatly among dates, but no significant CPUE trend was observed from April 2002 to November 2003 (Figure 2.7; $P > 0.05$, $N = 86$ days). Catch rates were highest within ±3 days of the full moon, and lowest around the new moon. No significant differences in catch rates were found among lunar phases (Figure 2.8, $F_{2,90} = 2.20$, $P = 0.12$).

TAGGING PROGRAM

Guides and anglers were trained in tagging and data-collection methods, and a flier about the tagging program was created to provide additional information. Three visual tagging methods were used on bonefish at Palmyra for different applications. The majority of tagging was conducted using t-bar anchor tags and tagging guns (Floy Tag and MFG. Co., Inc.). Tags were 6–8 cm long and clear white in color to reduce possible predator detection. Tag shedding was detected, with some fish

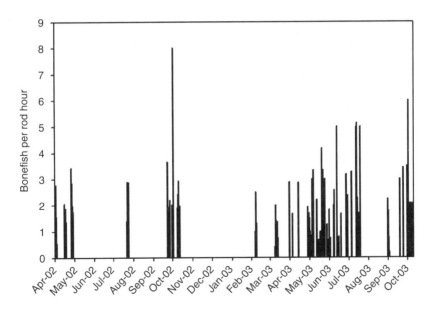

FIGURE 2.7 Mean daily bonefish catch rates at Palmyra Atoll from April 2002 to November 2003. Novice anglers excluded from the analysis.

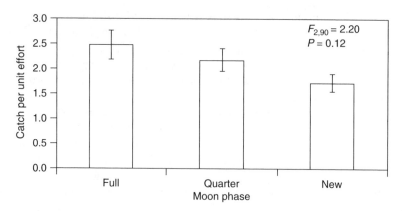

FIGURE 2.8 Mean catch per unit effort (number of bonefish per line hour) by lunar phase for bonefish at Palmyra Atoll, novice anglers excluded. Moon phases are not significantly different ($F_{2,90} = 2.20$, $P = 0.12$); quarter—first- and third-quarter moon phases. Full and new moon phases include dates 3 days on either side of moon. Error bars are standard error of the mean.

showing signs of previous tagging. To reduce tag shedding and predator detection, small (4–6 cm) dart tags (Hallprint Pty Ltd.) were also used. These tags have been reported to have a better retention rate than t-bar tags but are more difficult to apply, and most anglers and guides preferred using tagging guns with t-bar tags. Visible implant alphanumeric tags (VI Alpha) were applied beneath the transparent tissue (usually in the adipose eyelid) of bonefish after acoustic tag surgery in an attempt to reduce stress and predator detection.

Between May 2001 and March 2004 861 bonefish were tagged, with only 2 recaptures and 10 resightings during this time period. In Hawaii, 10 (1.3%) of 791 bonefish were recaptured in a similar study (see section on Hawaii below). The first recaptured bonefish was recorded on November 8, 2002. The 48-cm FL fish was tagged at the eastern end of the atoll on 6 November at 15:25 and recaptured at the western end at 17:00 on 8 November. The fish moved 3.7 km (2.3 mi) in slightly less than 50 h. A second tagged 48-cm FL fish was recaptured on October 17, 2003, but the tag number was not readable at the time of recapture. Low recapture and resighting rates may likely result from at least five nonexclusive factors: (1) large population size and low fishing effort; (2) tag loss; (3) mortality from tagging; (4) high natural mortality (i.e., predation); and (5) emigration.

REMOTE MONITORING OF BONEFISH MOVEMENT

Movement of fishes is important to both anglers and scientists. We collected data on short- and long-term movements of bonefish using acoustic telemetry methods to better understand fish site fidelity and behavior. For short-term movement studies, VEMCO V8SC-1L pingers (3.0 × 0.8 cm^2) were epoxied to 1.2-cm laminated disk tags and implanted into the dorsal musculature of bonefish using two 7.5-cm nickel pins. Several fish were tracked continually for short durations (1–4 h) using a manual receiver (model VR60, VEMCO, Ltd.) mounted on a kayak. Tracking showed initial movement into the deep lagoon following tagging, which was likely a stress response. Once fish moved back onto the flats they were difficult to follow and were quickly lost.

We deployed single-channel automated acoustic receivers (model VR2, VEMCO, Ltd.) at eight locations within the three lagoons adjacent to major flats utilized by bonefish to examine longer-term movements of bonefish. These omnidirectional receivers recorded the identification number and time stamp from the coded acoustic transmitters as tagged bonefish travel within receiver range, which was determined to be between 400 and 500 m. Based on preliminary range detection studies, our receivers could detect acoustic tags in the majority of lagoonal habitats.

For long-term tracking of bonefish, we surgically implanted VEMCO V8SC-1L pingers into their stomach cavities (see Lowe et al., 2003 and Humston et al., 2005 for details). Bonefish were caught using hook and line at various flats and quickly placed in a tub of seawater where they were rolled over with their ventral surface facing upward. This induced tonic immobility and eliminated the need for anesthesia. A 1-cm incision was made 1 cm off-center from the ventral midline between the pelvic fins and the anus, and a small acoustic transmitter (V8SC-1L) was placed within the visceral cavity. Battery life for these transmitters ranged from 1 to 1.5 years, on average. Acoustic transmitters were coated in a combination of beeswax and paraffin (1:2.33) to reduce immunorejection. The incision was closed with two to three surgical sutures (Ethicon Chromic Gut 2-0) and the fish were observed to ensure adequate recovery. The time from initial capture to the time of release ranged from 6 to 10 min. During recovery, each fish was measured to SL and then tagged externally with a dart tag.

A total of 40 fish were tagged between November 6, 2002 and August 27, 2003. Days at large ranged from 1 to 24 (mean 5.3 days). Fish moved freely among all lagoons. Some fish were observed to move between lagoons and then back to the original location within several days. The lagoons at Palmyra possess large numbers of blacktip reef sharks (*Carcharhinus melanopterus*) and predation on bonefish released with acoustic transmitters was observed on several occasions by these predators. Considering the common use and success of this method in areas with lower predator abundance (specimens retain tags for a year or more), we suspect that the small number of days at large is likely a result of high mortality associated with predators.

BONEFISH PHYSIOLOGICAL RESPONSES TO CATCH-AND-RELEASE STRESS

A number of tag-and-recapture studies have examined post-release mortality in bonefish (Crabtree et al., 1996; Cooke and Philipp, 2004), but very little is known regarding the subsequent physiological effects impact of angling that can significantly impact post-release performance and long-term survivorship (Bartholomew and Bohnsack, 2005). Therefore, there is a significant need to define and understand the impact of catch stress on the physiology of bonefish and other marine fishes, such that best fishing practices can be identified to reduce postrelease stress and mortality.

Stress responses in fishes, as in all vertebrates, are characterized by rapid increases in circulating levels of catecholamines (epinephrine) and cortisol (e.g., Sumpter, 1997). In contrast to the changes in catecholamines, which are typically transitory, elevation in cortisol has been observed to last up to several days, and possibly longer, even after exposure to single stressors (Carragher and Pankhurst, 1991). Chronically elevated plasma cortisol leads to fuel mobilization (e.g., increased glucose production) and inhibition of energy-expensive physiological processes such as growth, reproduction, or immune function (see Kelley et al., 2001, 2006; Wendelaar Bonga et al., 1997). Thus, the secondary physiological impacts of stress (cortisol) can have important and long-lasting deleterious impacts (e.g., on growth) in post-catch-and-release fish.

Our preliminary work on bonefish was initially directed at defining baseline controls ("pre-stress") with respect to different endocrine and biochemical markers in blood plasma. Such controls are obtained by rapid catching and blood sampling, preferably within 3 min, such that the neuroendocrine stress response has not yet had the time to express the expected surges in plasma levels of cortisol and metabolites like glucose. Bonefish were captured by hook-and-line and blood sampled via syringe and needle at the cardiac sinus within 2 min. In these fish, plasma cortisol concentrations were found to be around 1.5 ng/mL (Figure 2.9A), while glucose and lactate were 4 and 1.5 mmol/L, respectively (Figure 2.9B), all within the typical ranges of other "unstressed" fish and vertebrates (see Schreck et al., 1997; Kelley et al., 2001, 2006; Wendelaar Bonga et al., 1997). In contrast, caught bonefish showed a 12-fold increase in plasma cortisol within 24 h of being placed into 4-m-diameter pens, and >25-fold increases after 36 and 72 h. Plasma glucose concentrations increased in association with the increasing cortisol, exhibiting >twofold higher levels by 24 and 36 h, results consistent with cortisol's well-known hyperglycemic actions (see above citations). Plasma concentrations of lactate (Figure 2.9B), a marker for increased muscular activity, were elevated sevenfold by 24 h after catching, but showed levels nearing that of controls by 36 and 72 h, indicating recovery from oxygen debt. Measurements of the insulinlike growth factor (IGF) system (see Kelley et al., 2006) are now under way to determine possible impacts on the growth endocrine system of these fish.

Therefore, our findings to date indicate that bonefish exhibit substantial stress-induced hormonal and metabolic responses to catching-related activities. Caging effects may have contributed to these elevated stress levels and work is currently under way to conduct similar studies with bonefish in shore-based holding tanks to eliminate the potential effects of caging. Future work must now be directed toward understanding the specific impacts of variables such as the degree of physical exertion during capture, hook type and placement/removal effects, handling and confinement effects, and behavioral responses to captive conditions. By understanding such variables, fishing practices may be devised that reduce the deleterious impacts of the resulting stress.

BONEFISH INFORMATION FROM OTHER PACIFIC ISLANDS

Bonefish are targeted by commercial, recreational, and subsistence fishers throughout the Pacific. Subsistence fishing for bonefish in locations such as Tarawa consists of one of the most important protein sources for the island's human population. In other locations like Kiritimati Atoll and Hawaii, recreational anglers compete with commercial and subsistence fishers.

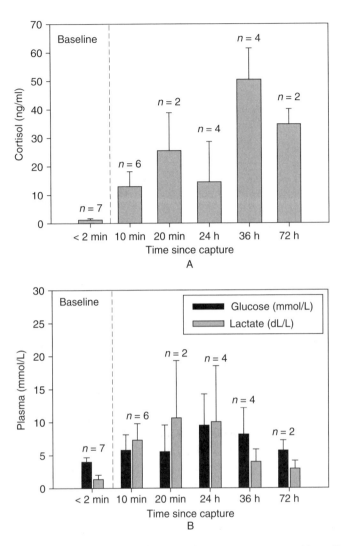

FIGURE 2.9 (A) Plasma cortisol concentrations in bonefish, as measured by radioimmunoassay. (B) Plasma glucose and lactate concentrations in bonefish, as measured by colorimetric assays. Fish were caught by hook and line, rapidly retrieved, and blood sampled within 2 min from initial hooking ("baseline controls," $n = 7$), or they were caught and placed into 4 m diameter pens for periods of 10–20 min ($n = 8$), 24 h ($n = 4$), 36 h ($n = 4$), or 72 h ($n = 2$). Bars indicate mean ± standard error.

HAWAII

Bonefish were an important food resource for early Hawaiians and are targeted today by a mix of commercial, recreational, and subsistence fishers. Commercial landings of bonefish in Hawaii have declined from over 136.4 mt in 1900 to only 1.2 mt in 2001 (Figure 2.10). Most recreational anglers in Hawaii used cut bait to catch bonefish, although a small fly-fish fishery exists on Oahu. State regulations have

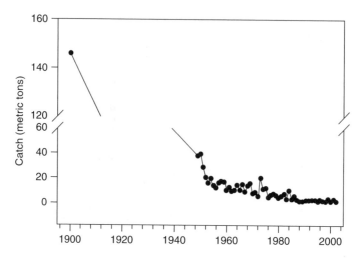

FIGURE 2.10 Commercial catch data for bonefish 1900–2004 in Hawaii. Data from 1900 were based on market surveys. Data since 1946 were from commercial logbook catch data reported to the Hawaii Department of Land and Natural Resources, Division of Aquatic Resources (DAR, unpublished data; Cobb, 1905.)

recently raised minimum size from 23 cm (9 in.) TL to 36 cm (14 in.) FL, although there is neither a closed season nor a bag limit.

Stock Identification

Shaklee and Tamaru (1981) used morphological and electrophoretic data to demonstrate the presence of two cyptic bonefish species in Hawaii that range elsewhere in the Indo-Pacific region as *A. glossodonta* and *A. neoguinaica*. Randall and Bauchot (1999) have recently regarded *A. forsteri* as the senior synonym of *A. neoguinaica* and described a method for rapidly distinguishing between these two species externally. The distance from the tip of the snout to the end of the maxilla (upper jaw) of *A. glossodonta* is shorter relative to the length of the head (measured from the tip of the snout to the end of the opercular membrane) than in *A. forsteri*. The ratio of head length to this snout–upper jaw measurement for *A. glossodonta* is 3.03–3.31 compared to 2.67–2.87 for *A. forsteri*. The broadly rounded lower jaw on *A. glossodonta* distinguished it in the field from *A. forsteri*, where the lower jaw tended to be more angular with a more or less pointed symphysis.

Oahu Catch-and-Release Fishery

In 2003, a bonefish-tagging program was initiated to characterize the resource for the purpose of supporting appropriate resource management and conservation programs, as well as helping to encourage a catch-and-release ethic among fishermen. Volunteers reported that they were able to clearly distinguish the two species of bonefish in Hawaii based on the descriptions and photographs provided in the tagging instructions.

Of the 538 bonefish tagged between September 2003 and June 2004, 186 (35%) were identified only as bonefish. Of the remaining fish, 72% were *A. glossodonta*

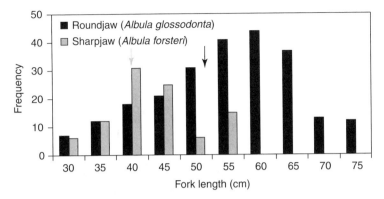

FIGURE 2.11 Size frequency distribution of roundjaw (*Albula glossodonta*) and sharpjaw (*A. forsteri*) bonefish tagged around the island of Oahu, Hawaii. Arrows denote means.

(roundjaw) and 28% were *A. forsteri* (sharpjaw). As of March 2005, a total of 791 bonefish were tagged with 10 recaptures recorded. Nearly all recaptures were within 1 km of their original capture site where time at large ranged from 7 days to nearly 21 months.

Based on the results from tagging data, mean FL for *A. glossodonta* (mean = 51.93 cm ± 10.99 sd, $n = 236$) was significantly greater (t-test = 8.97, $p < 0.001$) than that observed for *A. forsteri* (mean = 41.02 cm ± 6.99 sd, $n = 95$) (Figure 2.11). Participants in the study reported that *A. glossodonta* typically travels across sand and coral flats in loose schools or pairs, with larger individuals (>60 cm FL) often seen traveling alone. *Albula forsteri* is most often caught in deeper water (10–15 m), and therefore less is known regarding their movement and behavior.

Differences in morphology of the lower jaw suggest differing food preferences. Catch data show very little mixing of the two species, supporting the presumption of subtle habitat segregation. Beach seine sampling of juveniles (100–400 mm SL) in Kailua Bay, Oahu resulted in catches of only one of the two species in any given haul, although a series of hauls will result in collection of both species (Shaklee and Tamaru, 1981). Hence, the two species may be schooling separately even when cooccurring in the same area.

Juvenile Recruitment

Monthly beach seining (24 × 1.8 m² with a 1.3-cm mesh) was conducted along windward Oahu from 1994 to 2004. A total of 793 beach seine trips with an average of 9.7 (±3.0 sd) hauls per trip yielded 874 bonefish (mean = 13. 87 cm FL [±40.4 sd]; range [2.5, 33.0]). The small size of the individuals precluded the separation of species and all bonefish were classified as *Albula* spp. Mean monthly CPUE (number per seine haul) of juvenile bonefish (<30 cm) was highest from mid-summer through the fall (July–December), while the mean size of juvenile bonefish was larger during the winter and spring months (Figure 2.12A). Kahana Bay on windward Oahu had the longest time-series to examine annual trends (Figure 2.12B). Capture of recruits in Kahana was highest in 1999 and has declined by 79% since that time. Small bonefish (mean = 10.7 cm FL) were also found to utilize the surf zone in

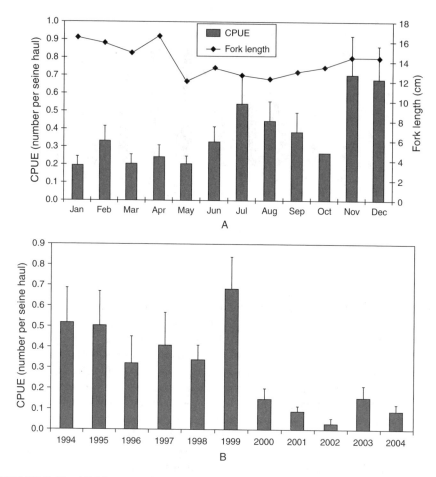

FIGURE 2.12 (A) Mean monthly CPUE (number per seine haul) and fork length (cm) of juvenile bonefish (<30 cm) captured in beach seines along windward Oahu from 1994 to 2004. (B) Mean annual CPUE (number per beach seine haul) of juvenile bonefish (<30 cm) captured in Kahana Bay, Oahu. Error bars are one standard error of the mean.

Hanalei Bay, Kauai, as juvenile nursery habitat during both daytime and nighttime periods (Friedlander et al., 1997).

Feeding

A total of 25 adult *Albula* spp. (26.6–41.3 cm FL) were collected in Hanalei Bay, Kauai, by line fishing at night (1900–0600) for stomach content analysis (Friedlander et al., 1997). Three fish had empty stomachs. Of the remaining 22 fish, small shrimp, mainly in the family Ogyrididae, were numerically (45.3%) the most important component of the diet (Table 2.7). Crabs (mainly *Portunus* spp.) were the most important taxa by volume (37.8%). Both groups were eaten by a large fraction of all *Albula* specimens. Small polychaetes, especially Opheliidae, were also numerically

TABLE 2.7
Diet of *Albula* spp. Captured in Hanalei Bay, Kauai ($n = 22$)

Prey Category	Numerical Percent	Volume Percent	Frequency Percent
Shrimp	45.3	23.4	86.4
Ogyrididae	38.2	18.8	81.8
Pasiphaeidae	5.9	2.6	18.2
Other shrimp	1.0	2.0	27.3
Crabs	7.4	37.8	68.2
Portunidae	5.3	35.1	59.0
Other crabs	0.8	1.9	22.7
Crab megalops	1.3	0.8	4.5
Polychaetes	33.1	4.9	27.3
Bivalves	2.6	6.0	31.8
Echinoderms	5.9	2.1	22.7
Amphipods	3.3	0.1	50.0
Fish	0.8	5.7	22.7
Stomatopods	0.8	5.6	18.2
Gastropods	0.7	0.1	22.7
Cephalochordates	0.3	0.1	9.1
Unidentified material	–	13.3	36.4

Source: Friedlander, A.M., et al., *Habitat Resources and Recreational Fish Populations at Hanalei Bay, Kauai*, Project report by the Hawaii Cooperative Fishery Research Unit to Hawaii Department of Land and Natural Resources, 1997, 296. With permission.

important (33.1%) in the diet. Bivalves and the small irregular urchin, Clypeasteridae, were of about equal importance as prey items.

KIRITIMATI (CHRISTMAS) ATOLL

Kiritimati (Christmas) Atoll is world renowned for its bonefishing and is the closest location to Palmyra that has active recreational and commercial/subsistence bonefish fisheries as well as local knowledge concerning the life history of the species. The Fisheries Division of Kiribati conducted a household survey of artisanal fishing activities in 1995 on Kiritimati. Milkfish (*Chanos chanos*) was the dominant species caught (76%), while bonefish was the next most common species caught, accounting for 7% of the total catch (Kamatie et al., 1995). Gill nets caught 82% of the bonefish with handlines providing the remainder (18%) of the catch.

Most of the recreationally caught bonefish at Kiritimati Atoll are released after capture, and the government has established "no-kill" areas on a number of popular sand flats. Some of the lagoonal ponds have been designated conservation areas where all fishing except catch-and-release recreational fishing is prohibited. Despite these conservation measures, a survey conducted by the fisheries division found a highly skewed sex ratio (15 males:1 female) in one closed area, and no females were observed from bonefish taken adjacent to Tabakea village. In areas with fewer predators and lower natural mortality, females are larger and therefore selectively removed first by the fishery. These results indicate potential overfishing, and efforts should be made to develop additional conservation areas and management strategies.

Interviews with guides, fishermen, and others with local knowledge indicate that fish spawn monthly during the full moon, with Paris Flats being one of the major prespawning staging locations (Figure 2.13A). Fishermen on Kiritimati maintained that bonefish do not spawn in the lagoon, but they were unsure of the actual spawning location. Recently, aggregations of >100 larger bonefish (50–70 cm FL) have been discovered and harvested just north of a newly constructed pier on the seawardside of the island, near the village of Tabakea (Figure 2.13A, K. Andersen, personal communication). These aggregations occur monthly just after the full moon in about 10–15 m of water at the reef/sand interface. The recent discovery and exploitation of this aggregation has potential implications for the reproductive success of bonefish at Kiritimati. It is reported that bonefish may have previously formed prespawning aggregations near London, but increased human population and the creation of seaweed farms may have disrupted this prespawning staging site. It was also reported that fishing tends to be poorer during El Niño years.

Tarawa Atoll, Kiribati

Bonefish are the most important fish harvested in Tarawa Lagoon, Kiribati, but recent studies have demonstrated significant declines in abundance and average size of bonefish in the catch between 1977 and the late 1990s (Beets, 2000). Of great concern is the shift in sex ratio that may be indicative of stressed populations. Since male bonefish mature at smaller size than females, the intensive fishing effort, especially for larger fish, may have depleted females in the population, similar to the trend at Kiritimati. Loss of spawning stock biomass and egg production by large females could result in spawning failure and population collapse.

The fishermen of Tarawa Atoll have a historical perspective that is highly relevant to the management of these fish (Johannes and Yeeting, 2000). All but one of its known spawning runs have been eliminated according to fishermen, and this last remaining run is showing signs of severe depletion (Figure 2.13B). Fishermen in Tarawa believe that the method known as "splash fishing," where bonefish are chased into gill nets by splashing 2-m-long crowbars in the water, had disturbed bonefish to the point that spawning runs were disrupted and reproductive migrations may have shifted to deeper water (Johannes and Yeeting, 2000). Spawning runs in Tarawa may also be impacted by the construction of causeways, as fishermen reported that a causeway (although fitted with a culvert) had effectively destroyed a spawning run. Regulations were imposed in 1994, in north Tarawa, to prohibit bonefish fishing during the 3 days on either side of the full moon and to restrict certain fishing methods. In 1999, fishers reported that the catch-per-unit effort and the average size of bonefish were both increasing, and a bonefish-spawning run was reported outside the reef of South Tarawa. The annual take of bonefish from Tarawa Lagoon is between 1,000,000 and 5,000,000 fish per year (Yeeting, unpublished data), but no stock assessment has ever been conducted. Although heavily exploited, bonefish at Tarawa appear to be somewhat resilient to intense fishing pressure, and the life history of the species (early age at sexual maturity and a protracted spawning season) may result in rapid recovery from overfishing if proper management strategies are initiated.

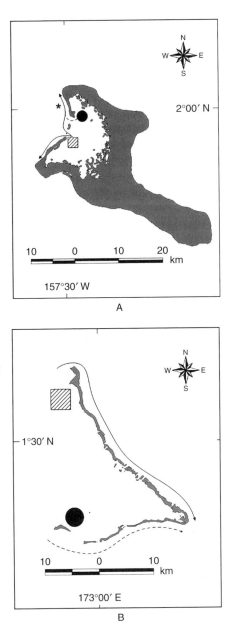

FIGURE 2.13 Bonefish prespawning aggregation locations and suggested migratory routes at (A) Kiritimati (Christmas) Atoll and (B) Tarawa Atoll. Hatched squares show location of existing prespawning aggregation locations. Hypothesized existing (solid lines) and previous (dashed lines) show spawning routes. Solid circle and arrows show location of former prespawning locations. Asterisk on Kiritimati denotes location of forereef aggregation location (see text for details).

DISCUSSION

Palmyra hosts one the largest lightly fished populations of bonefish in the Pacific, representing a unique opportunity to gather critical baseline information. The overall goals of the Palmyra bonefish conservation research program were to provide (1) basic information on the biology and ecology of a relatively undisturbed population of bonefish and associated ecosystem and (2) scientific foundations for establishing a sustainable catch-and-release bonefish fishery. Fly-fishing for bonefish on the flats is an attraction that is not readily available elsewhere in the U.S. Pacific, and sportfishing is an activity that, if well managed, may be compatible with, and financially supportive of, conservation goals for Palmyra Atoll NWR. The lightly exploited bonefish population of Palmyra is also of profound scientific importance, especially as a control in comparative studies to other heavily exploited populations of *A. glossodonta* elsewhere in the Pacific.

Because of the elusive nature of bonefish, much of the information necessary to manage this species can only be obtained through the efforts of anglers and guides. Analyses of catch data from Palmyra showed large variations in catch rates by location, angler expertise, and time of year, but no apparent trends in catch rates were detected during the period of study. This is not surprising, considering the relatively large bonefish population size and brief history of the recreational fishery. Collaboration with anglers and guides provided valuable data as well as helped formulate recommendations for bonefish management at Palmyra.

Biological data provided critical information for the management of bonefish at Palmyra and elsewhere. Reproductive condition data indicate that spawning occurs around the full moon and feeding and movement patterns are strongly influenced by tidal fluctuations. Larvae are most frequently encountered around the new moon period and larval duration was similar to bonefish examined in other locations.

Sex ratios of bonefish at Palmyra were assumed to represent those for a lightly exploited population compared to the highly skewed sex ratios observed at both Tarawa and Kiritimati Atolls where exploitation pressures are relatively intense (Table 2.7). Heavy fishing pressure and degradation of habitats at Tarawa and Kiritimati Atolls have resulted in the loss of pre-spawning staging sites and spawning migration routes, which may be responsible for the observed declines in bonefish catches, average size, and sex ratios at these locations.

The average size of fishes among these atolls was similar but smaller than those observed on Oahu, a high volcanic island in Hawaii (Table 2.8). Genetic isolation and varying environmental factors likely explain these differences. Female bonefish have been observed to be larger than males in Tarawa Atoll (Beets, 2000) and Kiritimati Atoll (Kamatie, 1995). It is therefore interesting that there does not seem to be any difference in size of bonefish between sexes at Palmyra or Florida (Ault et al., Chapter 16, this volume). The large number of apex predators at Palmyra (sharks and jacks) likely results in high natural mortality, and this in turn may help explain the differences in sizes of bonefish at Palmyra compared with other locations in the Pacific. Intense predation pressure of blacktip reef sharks may restrict bonefish foraging, thereby reducing their growth rates and maximum size. In addition, it is possible that juvenile blacktip reef sharks also feed on the same prey and

TABLE 2.8
Comparisons of Mean Size $\bar{L}$, Maximum Observed Size L_λ (cm FL), and Sex Ratio (male:female) for *Albula glossodonta* among Several Locations in the Pacific

Location	$\bar{L}$	L_λ	Sex Ratio	Source
Palmyra	41.5 (7.6)	67.0	1:1.25	This study
Tarawa	40.7 (5.7)	55.7	1:6.75	Beets, 2000
Kiritimati	43.6 (2.1)	60.0	1:15 (one area—no females)	Kamatie et al., 1995
Hawaii				
Windward and southshore Oahu	51.9 (11.0)	73.7		Oahu tagging study

Note: Standard error of the mean is in parentheses.

therefore compete with bonefish. The high density of bonefish at Palmyra also may result in competition for resources and therefore lower growth rates compared to other bonefish populations in the Pacific, but estimates of growth are lacking for *A. glossodonta* from other locations in the Pacific. The lack of life-history information on bonefish in the Pacific represents an important knowledge gap for science-based management and conservation of these stocks.

Bonefish exhibit similar nonstressed (baseline) physiological levels to those of other species such as California sheephead (*Semicossyphus pulcher*) studied to date (Lowe, unpublished data). They exhibit a profound and significant increase in blood cortisol, glucose, and lactate that are indicative of fight exertion and in some cases exhaustion. Owing to the high density of blacktip reef sharks at Palmyra and the physiological state of bonefish after angling with light tackle, released fish may be at greater risk of predation. Strategies to improve survivorship may include translocating fish to areas where there are fewer sharks (e.g., deeper parts of the lagoons), where fish may have time to recover without the presence of sharks. Future research should examine not only time to recovery of bonefish caught and released, but also the physiological impacts of repeated recapture.

Genetic data indicate that Palmyra bonefish are part of a large, genetically diverse archipelago-wide population that may encompass the entire Line Islands. In contrast, Hawaii is less diverse genetically, and that may be due in part to overexploitation.

Remote locations such as Palmyra, with limited fishing pressures, are among the few remaining examples of marine ecosystems without major anthropogenic influence (Friedlander and DeMartini, 2002). Our research at Palmyra has offered an important opportunity to understand how unaltered ecosystems are structured, how they function, and how they can most effectively be managed. Scientists must work with anglers and resource managers to develop a viable catch-and-release fishery for bonefish that is sustainable and compatible with the objectives of Palmyra Atoll National Wildlife Refuge.

ACKNOWLEDGMENTS

We thank The Nature Conservancy and The U.S. Fish and Wildlife Service for supporting this project. Financial support was provided by the Gordon and Betty Moore and David and Lucile Packard Foundations, as well as contributions from anonymous donors. Dan Lindstrom, Mark Monaco, Chris Gotschalk, Michael Sheehy, and Robert Warner assisted with fieldwork. The genetic data were generated by Marc Crepeau and financed by the National Science Foundation (NSF Grant OCE-0453167) and the International Game Fish Association. Aaron Adams and Teresa DeBruler of Mote Marine Laboratory assisted with ageing of bonefish. Lisa Wedding of the University of Hawaii at Manoa provided assistance with maps. This project could not have been conducted without the help of Elizabeth Lange and The Nature Conservancy staff at Palmyra.

REFERENCES

Alagaraja, K., Simple methods for estimation of parameters for assessing exploited fish stocks, Indian Journal of Fisheries, 31, 177–208, 1984.

Ault, J.S. and Ehrhardt, N.M., Correction to the Beverton and Holt Z-estimator for truncated catch length-frequency distributions, ICLARM Fishbyte, 9(1), 37–39, 1991.

Ault, J.S., Bohnsack, J.A., and Meester, G.A., A retrospective (1979–1996) multispecies assessment of coral reef fish stocks in the Florida Keys, Fishery Bulletin, 96(3), 395–414, 1998.

Ault, J.S., Humston, R, Larkin, M.F., and Luo, J., Development of a bonefish conservation program in South Florida, Final report to National Fish and Wildlife Foundation on grant No. 20010078000-SC, 2002.

Ault, J.S., Smith, S.G., and Bohnsack, J.A., Evaluation of average length as an estimator of exploitation status for the Florida coral reef fish community, ICES Journal of Marine Science, 62, 417–423, 2005.

Bartholomew, A. and Bohnsack, J.A., A review of catch-and-release angling mortality with implications for no-take reserves, Reviews in Fish Biology and Fisheries, 15, 129–154, 2005.

Beets, J., Declines in finfish resources in Tarawa Lagoon, Kiribati, emphasize the need for increased conservation effort, Atoll Research Bulletin, 490, 1–14, 2000.

Carragher, J.F. and Pankhurst, N.W., Stress and reproduction in a commercially important marine fish, *Pagrus auratus* (Sparidae), in *Proceedings of the Fourth International Symposium in Reproductive Physiology of Fish*, Scott, A.P., Sumpter, J.P, Kime, D.E. and Rolfe, M.S., eds., Fishery Symposium, 91, 253–263, 1991.

Clement, M., Posada, D., and Crandall, K.A., TCS: a computer program to estimate gene genealogies, Molecular Ecology, 9, 1657–1659, 2000.

Colborn, J., et al., The evolutionary enigma of bonefishes (*Albula* spp.): cryptic species and ancient separations in a globally distributed shorefish, Evolution, 55(4), 807–820, 2001.

Cooke, S.J. and Philipp, D.P., Behavior and mortality of caught-and-released bonefish (*Albula* spp.) in Bahamian waters with implications for a sustainable recreational fishery, Biological Conservation, 118, 599–607, 2004.

Crabtree, R.E., et al., Age, growth, and mortality of bonefish, *Albula vulpes*, from the waters of the Florida Keys, Fishery Bulletin, 94, 442–451, 1996.

Ehrhardt, N.M. and Ault, J.S., Analysis of two length-based mortality models applied to bounded catch length frequencies, Transactions of the American Fisheries Society, 121(1), 115–122, 1992.

Friedlander, A.M. and DeMartini, E.E., Contrasts in density, size, and biomass of reef fishes between the northwestern and the main Hawaiian islands: the effects of fishing down apex predators, Marine Ecology Progress Series, 230, 253–264, 2002.

Friedlander, A.M., et al., *Habitat Resources and Recreational Fish Populations at Hanalei Bay, Kauai*, Project report by the Hawaii Cooperative Fishery Research Unit to Hawaii Department of Land and Natural Resources, 1997, 296.

Fish and Wildlife Service (FWS), *Conceptual Management Plan for the Proposed Palmyra Atoll National Wildlife Refuge, Line Islands, Central Pacific Ocean*, U.S. Fish and Wildlife Service, Honolulu, 2001.

Grant, W.S. and Bowen, B.W., Shallow population histories in deep evolutionary lineages of marine fishes: insights from the sardines and anchovies and lessons for conservation, Journal of Heredity, 89, 415–426, 1998.

Hedgecock, D., Temporal and spatial genetic structure of marine animal populations in the California Current, Reports of the California Cooperative Oceanic Fisheries Investigations (CalCOFI), 35, 73–81, 1994.

Humston, R., et al., Movements and site fidelity of bonefish (*Albula vulpes*) in the northern Florida Keys determined by acoustic telemetry, Marine Ecology Progress Series, 291, 237–248, 2005.

Johannes, R.E. and Yeeting, B., I-Kiribati knowledge and management of Tarawa's lagoon resources, Atoll Research Bulletin, 498, 1–24, 2000.

Kamatie, M., Tekinaiti, T., and Uan, J., *Fisheries Research Surveys of Kiritimati Island*, Fisheries Division, Ministry of Natural Resources Development, Republic of Kiribati, 1995, 17.

Kaufmann, R., *Bonefishing*, Western Fisherman's Press, Moose, Wyoming, 2000, 390.

Kelley, K.M., et al., Serum insulin-like growth factor-binding proteins (IGFBPs) as markers for anabolic/catabolic condition in fishes, Comparative Biochemistry and Physiology Part B, 129, 229–236, 2001.

Kelley, K.M., et al., Insulin-like growth factor-binding proteins (IGFBPs) in fish: beacons for (disrupted) growth endocrine physiology, in *Fish Endocrinology*, Reinecke, M., ed., Science Publishers, Plymouth, UK, 2006.

Lecomte, F.L., et al., Living with uncertainty: genetic imprints of climate shifts in East Pacific anchovy (*Engraulis mordax*) and sardine (*Sardinops sagax*), Molecular Ecology, 13, 2169–2182, 2004.

Lowe, C.G., Topping, D.T., Cartamil, D.P., and Papastamatiou, Y.P., Movement, home range and habitat utilization of adult kelp bass (*Paralabrax clathratus*) in a temperate no-take reserve, Marine Ecology Progress Series, 256, 205–216, 2003.

Maragos, J.E., Palmyra Atoll: jewel of America's Pacific Coral Reef, in *Status of Coral Reefs of the World: 2000*, Wilkinson, C., ed., Australian Institute of Marine Science, Cape Ferguson, Queensland, and Dampier, Western Australia, 2000, 1.

Mojica, R., Jr., et al., Recruitment of bonefish, *Albula vulpes*, around Lee Stocking Island, Bahamas, Fishery Bulletin, 93, 666–674, 1995.

Morales-Nin, B., Growth of demersal fish species of the Mexican Pacific Ocean, Marine Biology, 121, 211–217, 1994.

Randall, J.E. and Bauchot, M., Clarification of the two Indo-Pacific species of bonefishes, *Albula glossodonta* and *A. forsteri*, Cybium, 23(1), 79–83, 1999.

Rogers, A.R. and Harpending, H., Population growth makes waves in the distribution of pairwise genetic differences, Molecular Biology and Evolution, 9, 552–569, 1992.

Schneider, S., Roessli, D., and Excoffier, L., *Arlequin, Version 2.000: A Software for Population Genetics Data Analysis*, Genetics and Biometry Lab, University of Geneva, Geneva, Switzerland, 2000.

Schreck, C.B., Olla, B.L., and Davis, M.W., Behavioral responses to stress, in *Fish Stress and Health in Aquaculture*, Iwama, G.K., Pickering, A.D., Sumpter, J.P., and Schreck, C.B., eds., Cambridge University Press, London, 1997, 145–170.

Shaklee, J.B. and Tamaru, C.S., Biochemical and morphological evolution of Hawaiian bonefishes (*Albula*), Systematic Zoology, 30(2), 125–146, 1981.

Sumpter, J.P., The endocrinology of stress, in *Fish Stress and Health in Aquaculture*, Iwama, G.K., Pickering, A.D., Sumpter, J.P., and Schreck, C.B., eds., Cambridge University Press, London, 1997, 95–118.

Vitousek, M.J., Kilonsky, B., and Leslie, W.G., Meteorological observations in the Line Islands, 1972–1980, Hawaii Institute of Geophysics, HIG-80-7, Honolulu, Hawaii, 1980.

Wendelaar Bonga, S.E., The stress response in fish, Physiological Reviews, 77, 591–625, 1997.

Wetherall, J.A., Polovina, J.J., and Ralston, S., Estimating growth and mortality in steady-state fish stocks from length-frequency data, ICLARM Conference Proceedings, 13, 53–74, 1987.

3 The Louisiana Recreational Tarpon Fishery

*William Dailey, André M. Landry, Jr.,
and F. Leonard Kenyon II*

CONTENTS

Introduction 57
Louisiana's Dynamic Recreational Tarpon Fishery 59
Louisiana's 200 lb Tarpon Club and Record Tarpon 62
The International Grand Isle Tarpon Rodeo 63
Management and Conservation 66
Acknowledgments 67
References 67

INTRODUCTION

The Atlantic tarpon (*Megalops atlanticus*) is a highly prized sport fish, widely distributed in warm temperate, subtropical, and tropical waters ranging from Nova Scotia to Argentina in the western Atlantic Ocean, from Senegal to Angola in the eastern Atlantic Ocean, and more recently proximate to the Panama Canal terminus in the eastern Pacific Ocean (Hildebrand, 1939; Murdy et al., 1997; Nelson, 1994; Wade, 1962; Zale and Merrifield, 1989). This magnificent fish is esteemed for its incredible leaps, aerial acrobatics, tremendous strength, and powerful runs. Often referred to reverently as the "silver king," tarpon currently support valuable recreational fisheries in Florida, Louisiana, and Mexico.

Seasonal migrations account for recreational fisheries in Texas, Alabama, Georgia, South and North Carolina, and Virginia, while resident and migratory stocks contribute to healthy fisheries in Mexico, Belize, Nicaragua, Costa Rica, and a host of Caribbean nations. In the southern hemisphere, Trinidad, Venezuela, French Guiana, and Brazil have significant seasonal fisheries during the austral summer. Gabon, Guinea-Bissau, Sierra Leone, Angola, and Liberia are increasingly popular west African destinations for anglers seeking world record tarpon. The world record (130 kg [286 lb 9 oz]) was landed by Frenchman Max Domecq at Rubane, Guinea-Bissau on March 19, 2003 (International Game Fish Association, 2006). Prior to the new record, two 128.4-kg (283-lb) tarpon landed from Lake Maracaibo, Venezuela, and

Sherbro Island, Sierra Leone, in 1956 and 1991, respectively, accounted for the record (Crawford, 2003).

Management practices for tarpon across state and international boundaries vary from limited to none, despite the undeniable value of its recreational fishery. State fishing regulations are commonplace and are usually in the form of minimum sizes (i.e., total lengths), bag limits, or requiring possession of trophy tags. In general, state regulations attempt to manage the fishery for quality tarpon, but arguably these restrictions occur where and how the fishing public perceives the management needs rather than based on tarpon biology or life history. Management of highly migratory tarpon by state and federal agencies has been complicated by lack of data on movements and migrations, and spatial distribution related to landings and fishing effort. Regarded as nonpalatable in the United States, tarpon and their highly esteemed roe are usually available in the fish markets of developing and third-world countries with artisanal or subsistence fisheries. Negligible commercial value is a primary reason why tarpon have historically not been included in U.S. federal management plans. The National Oceanic and Atmospheric Administration (NOAA) Fisheries' Marine Recreational Fisheries Statistics Survey (MRFSS) treats the recreational U.S. tarpon fishery as a specialized event or catch-and-release fishery for which specific landings data are not collected. Failure of federal and state fishery surveys to document tarpon landings and catch-per-unit effort (CPUE) renders management of this species problematic. Historical and current documentation of tarpon landings throughout its range is largely restricted to tournament weighmaster logbooks, local newspaper sports pages, and regional natural history books.

Tarpon in excess of 91 kg (200 lb) are regarded as the "holy grail" among many saltwater fly-fishing anglers. The world saltwater fly rod record was set in 2002 at 91.9 kg (202 lb) by Jim Holland of Vancouver, Washington, while fishing in Homosassa Springs, Florida. Numerous fly fishermen have pursued tarpon records by state and line/tippet class including fishing legends Stu Apte (see Chapter 6, this volume), Billy Pate, and Ted Williams. Fly records seem to be set each season, including Robert Cunningham's landing of a 58.9-kg (130-lb) fish in Louisiana from Capt. Lance Schouest, Sr.'s *Mr. Todd*, in 2003 and Guide Scott Graham's capture and release of a silver king in nearshore waters at Port O'Connor, Texas, taping to 220.9 cm total length (TL) (87 in.) and 109.2 cm girth (43 in.) with an estimated weight of 91.1 kg (201 lb) (Lance Schouest, Sr., Houma, Louisiana, and Ted Baker, Angler's Edge, Houston, Texas, personal communication). Silver king anglers have included Presidents Theodore Roosevelt, Herbert Hoover, and Franklin Roosevelt; Boston Red Sox Hall of Famer Ted Williams; and Dallas Cowboy Hall of Famer Bob Lilly.

Tarpon have been primary target species of tournaments in Louisiana, Texas, Florida, and Mexico for more than a half century. Anecdotal evidence indicates populations in the northwest Gulf of Mexico, particularly Texas, have declined substantially since the 1960s (Sutton, 1937; Roberts, 1970; Kuehne, 1973; Farley, 2002). Prior to 1960, tarpon arrived in Texas estuaries and nearshore waters from Boca Chica at the mouth of the Rio Grande to Sabine Pass at the Texas–Louisiana stateline as early as May, and were plentiful as late as November. Internationally renowned "tarpon rodeos" at Port Aransas and South Padre Island, Texas, were renamed in the early 1970s to reflect this collapse of the Texas tarpon fishery in the early 1960s, and subsequent elimination of the fish as a category or division. In contrast, tarpon tournaments in Boca Grande, Florida, Grand Isle, Louisiana, and Veracruz, Mexico, continue to generate economic impacts in the millions of dollars. Florida is the primary destination of tarpon anglers today and hosts more than a dozen tarpon tournaments statewide, including Suncoast Tarpon Roundup, Boca Grande World's Richest Tarpon Tournament, and Gold Cup Invitational Tarpon Fly Tournament. Approximately three decades ago, there were as many as a dozen tarpon rodeos in Louisiana from the Sabine River east to the Mississippi River. Four prominent rodeos continue today: Golden Meadow Tarpon Rodeo, International Grand Isle Tarpon Rodeo, Terrebonne Sportsman's League Annual Rodeo, and Empire-South Pass Tarpon Rodeo.

In Louisiana coastal waters, tarpon primarily feed on Gulf menhaden (*Brevoortia patronus*), small clupeid fishes that form large, dense, near-surface schools in coastal waters of the northern Gulf of Mexico from spring through fall (Lassuy, 1983; Smith et al., 2002; Whitehead, 1985). Louisiana and the mouth of the Mississippi River are a center of abundance for the Gulf menhaden resource, and this high-energy food source coupled with the overall productivity of the region undoubtedly contributes strongly to the tarpon's seasonal migrations to the Mississippi delta.

LOUISIANA'S DYNAMIC RECREATIONAL TARPON FISHERY

Occasionally referred to by its Cajun French common name *grande ecaille*, the tarpon has a significant social and cultural history in fishing communities of Louisiana's Mississippi River delta and coast. Despite this tremendous influence, tarpon anglers, their numbers, demographics, motivations, and preferences are no better known or understood by fishery managers and social scientists today than they were more than a half century ago. Estimates of the numbers of anglers continue to rely on tournament participation records. In contrast to several Florida tournaments (e.g., World's Richest Tarpon Tournament), there is little to no prize money associated with angling tarpon at Louisiana rodeos. A primary motivation for Louisiana tarpon anglers is socialization. Leaderboards at Louisiana rodeos reflect both genealogy and fishing expertise with listings of winners including fathers, sons, and grandsons (Figure 3.1). The golden era of tarpon fishing in Louisiana was arguably the late 1960s, with both spectacular angler participation and a tremendous number of silver king landings. Participants in the Golden Meadow Tarpon Rodeo numbered approximately 3000 in 1967, with some unidentified proportion of these being tarpon anglers who entered 45 silver kings. The 1966 Abbeville Tarpon Rodeo drew in excess of 3100 participants and 110 tarpon were landed at its Intracoastal City weigh station. In a similar fashion, the 1966 Grand Isle rodeo entrants, totaling approximately 2000, established its landings record of 48 silver kings (Falkner, 1967). The specific number of tarpon anglers in these rodeos is not documented; however, tarpon anglers at the Grand Isle rodeo have waned from as many as 500 in the 1960s and early 1970s to fewer than 100 in 2003 (Grady Lloyd and Marty Bourgeois, Grand Isle Rodeo weighmasters, personal communication).

The historical Louisiana recreational tarpon fishery grounds spanned from Lake Pontchartrain and the Pearl River delta around the birdfoot delta of the Mississippi River and west to Atchafalaya and Vermilion Bays, a linear distance of more than

FIGURE 3.1 Three generations of the Schouest family at the 1st Annual Coon-pop Tarpon Classic (September 15, 2002). Capt. Lance "Coon" Schouest, Sr. (kneeling with granddaughter) has tagged more than 700 tarpon; seven have been recaptured. (Courtesy of William Dailey.)

300 km. Areas west of Grand Isle, particularly Marsh Island, Little Pass Timbalier, and Timbalier Bay, were popular hotspots prior to the 1970s, and the fishing destinations of anglers in the Abbeville Tarpon and New Iberia Rod and Gun Rodeos. The decline in tarpon fishing success in the Vermilion–Atchafalaya Bay complex coincided with reduced freshwater inflows into this brackish estuary associated with the 1963 construction of the Old River Control Structure (ORCS). This system of floodgates and diversion canal, built proximate to the confluence of the Mississippi and Red Rivers, was designed to reverse a century-long trend of increasing Mississippi discharge into the Atchafalaya River.

The historical tarpon season began in April and ran through early November. Tarpon were primarily taken on large spoons (e.g., Pet 21) trolled at six to seven knots proximate to pass fishing grounds and rolling tarpon pods or small schools. "Rolling pods" refer to tarpon's tendency to form small aggregations and slowly roll at the surface to gulp air in a fashion similar to dolphins and other marine mammals. Traditional catch rates languished at 3 fish landed for every 10 tarpon "jumped" or hook-ups.

The current fishery is relatively localized and operates primarily within and just west of the Mississippi River delta. These fishing grounds consist of approximately

100 km of nearshore and offshore sites stretching from the delta's Northeast Pass to the barrier island of Grand Isle and Barataria Bay. Recent restoration efforts in Lake Pontchartrain have produced a dramatic recovery in water quality and resurgence in its fishery including a 94.3-kg (208-lb) silver king landed in August 2004. Anglers target waters from 6 to >40 m deep where tarpon aggregate to feed on an abundant and diverse array of prey ranging from anchovies to menhaden to mullet. The Coon-Pop® was introduced by tarpon guide Lance Schouest, Sr., of Venice, Louisiana, in 1987. It consisted of a brightly colored BB-filled lead head and soft plastic lure wired to large circle hook (Figure 3.2). Catch rates doubled from 3 to 6 or more tarpon for every 10 jumped, and the landing frequency of tarpon that weighed 90.7 kg (200 lb) or more increased substantially with the introduction of the CoonPop and integration of the circle hook. Advent of the CoonPop precipitated the fishery's emphasis on reducing vessel-related disruption of rolling schools, intercepting these rolling schools, and using bait-casting and drifting techniques in lieu of trolling. This fishing strategy resulted in increased dependence on calm seas to facilitate sighting of rolling tarpon. Capt. Lance Schouest, Sr. reported that successful tarpon seasons were often coincidental with Bermuda High weather systems, which settle Louisiana delta waters and create nearly windless and flattened seas.

Similar to many marine fish, distribution of tarpon throughout their range is strongly influenced by seasonal temperature regimes. Monthly median water temperatures at the NOAA Grand Isle Station (GDIL1) ranged from 31.6°C in July to 27°C in October during the seasonal fishery in 2004 (National Data Buoy Center).

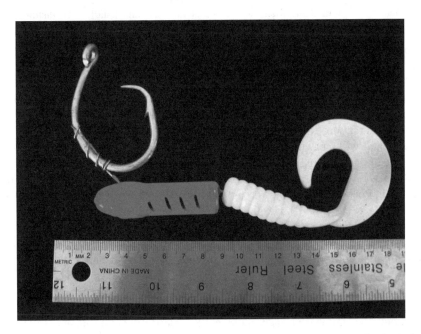

FIGURE 3.2 The CoonPop lure, consisting of a BB-filled leadhead and soft plastic lure wired to circle hook, was introduced by tarpon guide Capt. Lance "Coon" Schouest, Sr., Venice, Louisiana, in 1987. (Courtesy of Heidi Amin.)

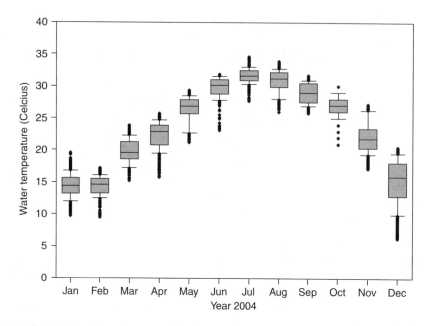

FIGURE 3.3 Hourly water temperatures by month at the NOAA Grand Isle Station (GDIL1) in 2004. Lower and upper boundaries of the box indicate the 25th and 75th percentiles; line within the box represents the median water temperature; and error bars represent the 10th and 90th percentiles. (Data compiled from National Data Buoy Center, http://www.ndbc.noaa.gov/data/download_data.php?filename=gdil1h2004.txt.gz\&dir=data/historical/stdmet/.)

Applying the historical season to 2004 water temperature data, median temperatures for April arrivals and November departures would have been 22.9 and 21.8°C, respectively (Figure 3.3). Preliminary results from pop-up archival tags on tarpon indicate migration patterns between tropical and subtemperate waters reflect thermal preferences (Luo et al., Chapter 18, this volume; Michael Domeier, unpublished data). The extended duration of the historical fishery and a portion of the current season are coincidental with spawning period in the Gulf of Mexico and Caribbean. Tarpon reproductive activity in Florida occurs in offshore waters from May through August, while spawning occurs year-round in Costa Rica (Crabtree et al., 1992, 1997). Reproduction in Puerto Rico extends throughout the year with seasonal peaks in April and August (Zerbi et al., 2001). Despite the rare capture of young-of-the-year tarpon, its waters, Louisiana's delta, and coastal wetlands are thought to serve only as foraging grounds during the annual migrations of adults and subadults in the Gulf of Mexico, with minimal contribution as spawning grounds or nursery habitat.

LOUISIANA'S 200 lb TARPON CLUB AND RECORD TARPON

The Louisiana Tarpon Club and Capt. Lance Schouest, Sr. have documented 40 tarpon >90.7 kg (200 lb) landed in state waters since 1973. Twenty-five of these "giant" silver kings were landed since 1990. "Trophy" tarpon are arguably 63.5 kg (140 lb) and more. Nine of eleven of the Louisiana's largest tarpon were landed since the

TABLE 3.1
Top 10 Tarpon Landed and Weighed in Louisiana Including Angler, Weight of Tarpon (lb/kg), Vessel, Date (Month and Year), and Location

Rank	Angler	Weight (lb/kg)	Vessel	Date	Location
1	Tom Gibson	230/104.3	Anticipation	August 1993	Grand Isle
2	Pat Parra	222.8/101	Bandit	June 1979	West Delta 58
3	Jessica Barkhurst	221.5/100.5	Argonaut	August 1993	West Delta 58
4	Joshua Tanner	220.5/100	Lil Moon	August 1997	Southwest Pass
5	Lance Schouest, Sr.	219.5/99.6	Mr. Todd	October 1989	Grand Bayou
6	James Eichorn	218/98.9	Mr. Todd	October 1984	West Delta 58
7	Joe Roberts	216.8/98.3	Bambo Bernie	August 1990	West Delta 58
8	Chris Schouest	215.5/97.7	Crawdaddy	September 1995	Southwest Pass
9	John Deblieux	215.2/97.6	Rock-n-Roll	July 2004	Grand Bayou
t10	Debbie Ballay	214.5/97.3	Aw Heck	August 1990	West Delta 58
t10	Buddy Hebert	214.5/97.3	Fru Fru Maru	September 1997	Southwest Pass

Note: t = tie.
Source: Lance Schouest, Sr. and Jeff Deblieux, Louisiana Tarpon Club, unpublished data.

CoonPop's introduction into the fishery. Retired National Aeronautics and Space Administration (NASA) engineer and avid international tarpon angler Tom Gibson landed the state record near Grand Isle, in August 1993, during slow trolling a CoonPop (Table 3.1). The overwhelming majority of tarpon greater than 200 lb were captured in late summer or early fall. Four silver kings were landed prior to July 30, and 30 were landed in August (19) and September (11) (Figure 3.4). Whether these landings data are more representative of fishing pressure or peak migratory abundance is pure conjecture without a thorough survey of tarpon anglers and their catches. The state's record tarpon (104.3 kg [230 lb]) ranks second behind Florida (110.2 kg [243 lb]) among the nine southeastern U.S. states with recreational tarpon fisheries. Tarpon records for each of the current Louisiana rodeos are in excess of 90.7 kg (200 lb), and if the state has failed to rival Florida in sheer tarpon abundance, Louisiana has a strong argument for its trophy fishery.

THE INTERNATIONAL GRAND ISLE TARPON RODEO

Grand Isle is a barrier island west of the Mississippi River delta at the southern terminus of Louisiana State Highway 1 in Jefferson Parish, Louisiana. The International Grand Isle Tarpon Rodeo (IGITR) is conducted annually, Thursday through Saturday during the third or final weekend in July. The tournament, the oldest contest of its kind, hosted its 84th annual event in July 2005. The first IGITR was held in 1928 with 25 anglers from New Orleans. Seven of these fishermen landed a tarpon, and five tarpon weighed in excess of 45.4 kg (100 lb). The rodeo was suspended in 1930 because of the Great Depression and from 1942 to 1945 during World War II (Crawford, 2001). Angler

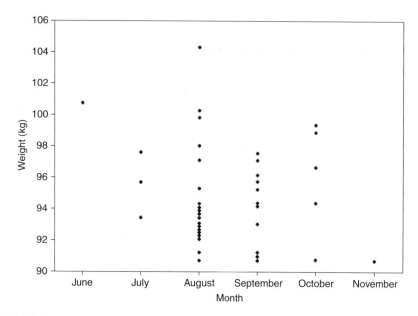

FIGURE 3.4 Temporal distribution of 90.7-kg (200-lb) tarpon landings by month. Nearly one half of 40 trophy silver kings were landed in the month of August. (Data compiled from Jeff Deblieux and Lance Schouest, Sr., personal communication.)

preferences for target species other than tarpon were recognized early by the tournament committee, and by 1948, categories were expanded to include tripletail (*Lobotes surinamensis*), cobia (*Rachycentron canadum*), Atlantic bonito (*Sarda sarda*), crevalle jack (*Caranx hippos*), king mackerel (*Scomberomorus maculatus*), dolphinfish (*Coryphaena hippurus*), sheepshead (*Archosargus probatocephalus*), spotted seatrout (*Cynoscion nebulosus*), and red drum (*Sciaenops ocellatus*). Many other finfish species have been added to the tournament through the years. Currently, the four divisions of the IGITR are tarpon, big game, shoreline and fly-fishing, with categories for more than 25 species. The tarpon division consists of "harvest" and tag-and-release subdivisions.

Several conservation measures related to tarpon were initiated for the 1993 rodeo including (1) a minimum entry weight of 50 lb (22.7 kg) for the tarpon harvest or "kill" division; (2) winning entries, that is, award recognition, reduced from 10 to 5 largest tarpon (by weight); and (3) introduction of a tag-and-release division. This division was established to discourage harvest of smaller noncompetitive tarpon, provide conservation-minded anglers with a nonharvest option regardless of weight, and assist NOAA Fisheries in its cooperative tagging program related to migratory behavior of recreational finfish species. With the 2001 tournament, award recognition was further reduced from five to three largest tarpon, and minimum weight was increased to 100 lb (55.4 kg) in 2001. There is no minimum weight restriction in the tag-and-release division, and arguably, all tarpon captured in the rodeo are subsequently accounted for in either the harvest or tag-and-release divisions. The number of tarpon caught and released prior to the introduction of the tag-and-release division in 1993 is unknown.

Tournament weighmaster logs and programs, local newspapers, and library archives were reviewed for historical and current documentation of tarpon landings at the IGITR. For the purpose of this review, a "landing" is defined as a tarpon entered in either the harvest or tag-and-release divisions. Rodeo results were collected and analyzed for 47 tournaments dating from 1957 through 2003. During this period, 691 silver kings were landed in the Grand Isle Rodeo. Four tarpon exceeding 90 kg were landed and included the event winners in 1973 and 2001, and the win and place silver kings in 2002. John Guidry of Galliano, Louisiana, set the rodeo record with a 93.4-kg (206-lb) silver king in 1973, and Lee Schouest of Houma, Louisiana, won "most outstanding fish in the rodeo" in 2001 with his 90.7-kg specimen. Six of the ten largest entries since 1957 were landed subsequent to the introduction of the CoonPop in 1987. Rodeo entries exceeded 70 kg (154.3 lb) twice prior to 1975. Since the mid-1970s, tarpon have exceeded 70 kg every tournament except four: 1975, 1976, 1992, and 2000. Gradual increases in winning-entry weight and mean weight since 1957, and especially post-1993, can be attributed in part to the introduction of conservation measures introduced by the rodeo committee and the growing conservation ethic among anglers (Figure 3.5). The increase in winning-entry weight is likely attributable to introduction of the CoonPop lure into the tarpon fishery, growing integration of circle hooks into this fishery, and emergence of expert anglers. Four of the top five rodeos in number of tarpon landings occurred prior to 1975. Tournament landings peaked at 48 in 1966, and exceeded 30 in 1962, 1965, 1966, and 1988. In contrast, five or fewer tarpon were entered in 1964, 1967, 1970, 1974, and 1995. Silver king landings have not exceeded 15 fish since 1988. There is a general trend in reduced landings for the period 1957–2003 (Figure 3.6).

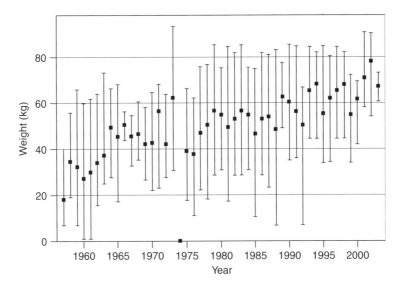

FIGURE 3.5 Maximum, minimum, and mean weight and 95% CI of tarpon landed annually at the International Grand Isle Tarpon Rodeo, Grand Isle, Louisiana, for the period 1957–2003.

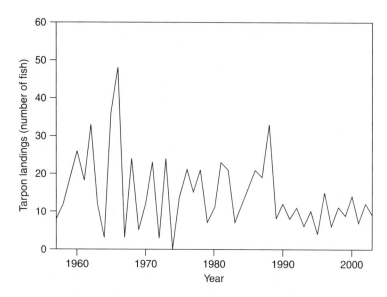

FIGURE 3.6 Number of tarpon landed or leadered annually at the International Grand Isle Tarpon Rodeo, Grand Isle, Louisiana for the period 1957–2003.

Since the early to mid-1970s, the rodeo has been characterized by a reduction in tarpon angler numbers from nearly 500 in the 1960s, to approximately 200 in the 1970s and 1980s, and approximately to 100 from 1990 to present. The downward trend in rodeo landings is more likely the result of reduced fishing pressure rather than declining tarpon abundance.

MANAGEMENT AND CONSERVATION

Angling opportunities for tarpon in U.S. waters are largely restricted to nearshore waters (i.e., state territorial waters); consequently, state management impacts the fishery and the fish itself. The saltwater fishing license for residents in Florida, Louisiana, and Texas costs $13.50, $15, and $33, respectively. Florida and Texas require possession or trophy tags to take a tarpon at a cost of $51.50 and $120, respectively. The Florida possession tag has no size restriction, while its Texas counterpart has a minimum length of 80 in. (203.2 cm TL). At the conclusion of the 2003 Texas season, their trophy tag fee increased by 20% from $100 to $120. Tag sales in Florida are approximately 400 annually, while sales in Texas have languished at less than 20 tags annually since its inception in 1995. Louisiana currently has no size restriction, tags, or stamps associated with its recreational tarpon fishery.

Research is needed to better understand aspects of tarpon biology and life history, as well as tarpon angler demographic characteristics such as level of fishing participation, fishing experience and socialization, motivations and attitudes, and expenditures of tarpon and tournament anglers in Louisiana, and furthermore, throughout the northwest Gulf of Mexico. Tarpon and the tarpon sport fishery are likely to benefit as NOAA Fisheries and Louisiana Department of Wildlife and Fisheries make a transition from single-species management to an ecosystem approach.

ACKNOWLEDGMENTS

This chapter would not have been possible without the expertise and experience of Grady Lloyd, chief weighmaster of the Grand Isle and Empire-South Pass Tarpon Rodeos; Lance "Coon" Schouest, Sr., tarpon guide and captain of the *Mr. Todd*; and Steve Hein and Marty Bourgeois, Louisiana Department of Wildlife and Fisheries biologists and weighmasters of the Golden Meadow, Grand Isle, and Terrebonne Sportsman's League Tarpon Rodeos. In addition, significant contributions were made by numerous Louisiana tarpon anglers including, but not limited to, Jeff Deblieux and John Deblieux of Houma; David and Debbie Ballay, former owners and operators of the Venice Marina; Chris Schouest, tarpon guide and captain of the *Crawdaddy*; and all members of the Louisiana Tarpon Club. Alberto Madaria and Luis Turribiates Jimenez, Tampico, Tamaulipas, Mexico provided invaluable support at tarpon rodeos in Tamaulipas and Veracruz, Mexico. We appreciate the generosity of Captains Jim Leavell, Corpus Christi, Texas, and James Trimble, Tiki Island, Texas, for matters related to the Texas recreational tarpon fishery. John Sumich, Darryl Bubrig, and Dominick Scandurro, owners–operators of the Empire Inn, Buras, Louisiana, and Venice Inn, Venice, Louisiana; Jimmy LeBlanc, Baton Rouge, Louisiana; and Ronald Bartels, Grand Isle, Louisiana contributed complimentary accommodations for fieldwork at the Louisiana tarpon tournaments. Former Texas A&M–Galveston students Molly Dillender, Brandi Gates, Ryan Carrico, Tom Sullivan, and Vu Nguyen were relentless and invaluable in their field efforts at the Grand Isle and Empire-South Pass rodeos. Finally, we would like to thank all anglers, attendees, and organizers of the Golden Meadow, Grand Isle, Terrebonne Sportman's League Annual Rodeo, and Empire-South Pass Tarpon Rodeo.

REFERENCES

Crabtree, R.E., E.C. Cyr, R.E. Bishop, L.M. Falkenstein, and J.M. Dean. 1992. Age and growth of tarpon, *Megalops atlanticus*, larvae in the eastern Gulf of Mexico, with notes on relative abundance and probable spawning areas. *Environ. Biol. Fish.*, 35: 361–370.

Crabtree, R.E., E.C. Cyr, D.C. Chaverri, W.O. McLarney, and J.M. Dean. 1997. Reproduction of tarpon, *Megalops atlanticus*, from Florida and Costa Rican waters and notes on their age and growth. *Bull. Mar. Sci.*, 61: 271–285.

Crawford, A. 2001. *Tarpon rodeo stands test of time.* 81st Annual International Grand Isle Tarpon Rodeo Program. 79pp.

Crawford, R. 2003. *World record game fishes.* International Game Fish Association, Dania Beach, FL. 352pp.

Falkner, J. 1967. Abbeville rodeo largest in the state. *Baton Rouge Sunday Advocate*, 30 July, Sect A: 12.

Farley, B. 2002. *Fishing yesterday's Gulf Coast.* Texas A&M University Press, College Station, TX. 149pp.

Hildebrand, S.F. 1939. The Panama Canal as a passageway for fishes, with lists and remarks on the fishes and invertebrates observed. *Zoologica*, 24: 15–45.

International Game Fish Association. 2006. *World record game fishes.* International Game Fish Association, Dania Beach, FL. 384pp.

Kuehne, C.M. 1973. *Hurricane junction; a history of Port Aransas.* St. Mary's University, San Antonio, TX. 216pp.

Lassuy, D.R. 1983. *Species profiles: life histories and environmental requirements (Gulf of Mexico). Gulf menhaden.* United States Fish and Wildlife Service Biological Report 82 (11.2). United States Army Corps of Engineers, TR EL-82-4. 13pp.

Murdy, E.O., R.S. Birdsong, and J.A. Musick. 1997. *Fishes of Chesapeake Bay.* Smithsonian Institution Press, Washington, DC. 324pp.

National Data Buoy Center, http://www.ndbc.noaa.gov/data/download_data.php?filename=gdil1h2004.txt.gz\ &dir=data/historical/stdmet/.

Nelson, J.S. 1994. *Fishes of the world.* 3rd ed., John Wiley & Sons, New York. 600pp.

Roberts, E.M. 1970. *The stubborn fisherman.* Creighton Publishing, Port Aransas, TX. 234pp.

Smith, J.W., E.A. Hall, N.A. McNeill, and W.B. O'Bier. 2002. The distribution of purse-seine sets and catches in the Gulf menhaden fishery in the Northern Gulf of Mexico. *Gulf Mex. Sci.*, 20 (1): 12–24.

Sutton, R.L. 1937. *The silver kings of Port Aransas and other stories.* Brown-White Company, Kansas City, MO. 338pp.

Wade, R.A. 1962. The biology of the tarpon, *Megalops atlanticus*, and the ox-eye, *Megalops cyprinoides*, with emphasis on larval development. *Bull. Mar. Sci.*, 12: 545–622.

Whitehead, P.J.P. 1985. FAO species catalogue. Vol. 7. Clupeoid fishes of the world. An annotated and illustrated catalogue of the herrings, sardines, pilchards, sprats, anchovies and wolf-herrings. Part 1—Chirocentridae, Clupeidae and Pristigasteridae. FAO Fisheries Synopses. Rome, Italy. 303pp.

Zale, A.V. and S.G. Merrifield. 1989. *Species profiles: life histories and environmental requirements of coastal fishes and invertebrates (South Florida). Ladyfish and tarpon.* United States Fish and Wildlife Service Biological Report 82 (11.104). United States Army Corps of Engineers, TR EL-82-4. 17pp.

Zerbi, A., C. Aliaume, and J.C. Joyeux. 2001. Growth of juvenile tarpon in Puerto Rican estuaries. *ICES J. Mar. Sci.*, 58: 87–95.

4 Insight into the Historical Status and Trends of Tarpon in Southwest Florida through Recreational Catch Data Recorded on Scales

Stephen A. Bortone

CONTENTS

Introduction 69
    Background 69
    Justification 70
    Methods and Materials 70
    Results 71
Discussion 72
Acknowledgments 76
References 76

INTRODUCTION

BACKGROUND

Tarpon, *Megalops atlanticus* Valenciennes 1847, is broadly distributed in coastal areas in the western North Atlantic from Virginia (with occasional records from as far north as Nova Scotia) southward along the Gulf of Mexico, throughout the Caribbean Sea, and extending as far south as Rio de Janeiro, Brazil (Zale and Merrifield, 1989; Crabtree et al., 1995). Tarpon are also found along the west coast of Africa (Migdalski and Fichter, 1976), where the world record of 130 kg was captured (IGFA, 2005). Beginning around 1900, a recreational fishery for this species developed, especially in Florida where early sportfishing clubs were directed almost exclusively toward the recreational capture of tarpon during the late winter to spring months.

Jordan and Evermann (1923, p. 85) indicated that excellent tarpon fishing was known in Florida from Punta Gorda and Fort Myers where "on the west coast of Florida has, perhaps, been the most popular resort."

A characteristic of these early recreational fishers was to remove a scale from a landed tarpon and record the catch data directly on the scale. These data most often included weight (in pounds), length (presumed to be total length in feet and inches), and girth (in inches). Other data recorded were date of capture, angler's name, hometown of the angler, name of the fishing guide, and place of capture. Thus, the data inscribed by participants in the historical recreational fishery provide a virtual "treasure chest" of basic biological features of tarpon caught during the early twentieth century.

Justification

Crabtree (2002) indicated that although the recreational tarpon fishery off Florida was well developed, there was a paucity of data on its historical population structure and abundance, thus disallowing an examination into long-term trends of the species in areas where it has been highly exploited for over a century. A similar situation prompted Holt et al. (2005) to examine data recorded on tarpon scales as part of the historical recreational fishery off Texas. Their presentation and analyses of the data recorded on scales allowed some insight into changes noted in the Texas tarpon fishery.

The utility of historical data on the tarpon fishery prompted this author to investigate the possible presence of tarpon-scale collections from Florida's southwest coast, an area acknowledged as the center for the early recreational tarpon fishery in the United States (Oppel and Meisel, 1987). Three locations were identified as having a substantial number of tarpon scales from the early fishery. These included two establishments near Boca Grande Pass: the Gasparilla Inn on Gasparilla Island, and the Collier Inn and Tarpon Bar on Useppa Island. A third location, the Olde Marco Inn, was identified on Marco Island (about 100 km southwest of Boca Grande Pass). The information inscribed on the scales served as a basis for this investigation into the population structure of the historical tarpon fishery off Florida.

Methods and Materials

Visits to the aforementioned establishments led to obtaining data from 1027 tarpon scales: 166 from Gasparilla Island; 464 from Useppa Island; and 397 from Marco Island. Additional scales were observed at each location but were not included in the analyses, as the data were unreadable (i.e., faded ink or illegible handwriting). Many of the scales (especially on Useppa Island) were embedded in resin on wall plaques or laminated into tabletops. All data were recorded and entered into a spreadsheet, which served as the database for analyses. It should be noted that scales rarely had complete capture data. Weights and lengths were converted into metric units (kg and cm, respectively). Lengths were presumed to be recorded as total length (TL). To allow for comparisons with other studies (e.g., Crabtree et al., 1995; Holt et al., 2005), all lengths were converted to fork length (FL) using the equation of $FL = -10.8096 + 0.8967\ TL$ from Crabtree et al. (1995). Many scales included only weight and not length.

To allow a more expanded database for some analyses, weights were converted to FL based on the equation presented below in the length/weight analysis.

A condition index (Fulton's K = weight/length^3) was calculated following Holt et al. (2005). Values of $K > 1$ indicate that fish were in "good" condition while values of $K < 1$ were indicative of fish in "poor" condition.

The data were assumed to be nonparametric, that is, not randomly chosen from a normally distributed population. It was also assumed that the decision to include a tarpon scale in the available collections was equally selective over time, that is, scales from larger fish were generally retained for inclusion in the series. Data from all locations were pooled. An analysis of the length/weight data from the two areas (and three locations) indicated that there was no significant difference ($p < 0.05$) in this basic aspect of their population structure. Consequently, it was assumed that tarpon from either of these areas were from the same unit stock.

RESULTS

Data were available from scales retained from tarpon caught from 1902 through 1998, but the majority of scales were retained from 1910 to 1930 (Figure 4.1). Of the 1027 scales examined, 926 had sufficient information to determine length, either measured directly or calculated from the length/weight relationship equation offered below. The smallest and largest tarpon analyzed here were 69 and 210 cm FL, respectively. Evidence indicates that a decline in tarpon captures began after the mid-1930s with a notable decrease during World War II (1941–1945). Tarpon landings by month display a notable, consistent, seasonal pattern. Fish were generally caught beginning in March through May (Figure 4.2). Modern local fishers indicate

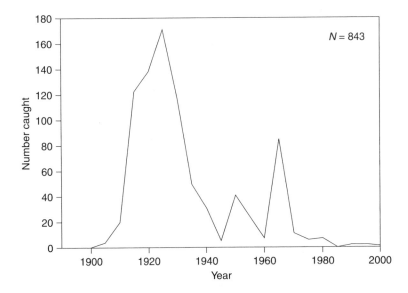

FIGURE 4.1 Year class (5-year intervals) vs. number of tarpon scales examined from 1902 to 1998.

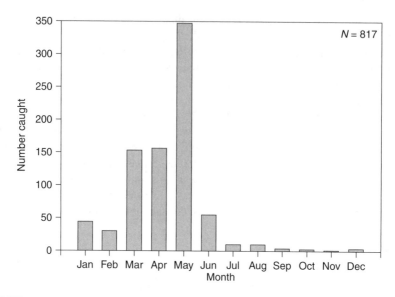

FIGURE 4.2 Number of tarpon caught (as evidence from scales) according to month of capture.

that this pattern has only altered slightly, but the peak fishery season is delayed by 1 month, beginning in April and lasting until the end of June. Based on tarpon lengths, a higher number of tarpon caught were between 140 and 170 cm FL (measured; Figure 4.3, top) or between 110 and 190 cm FL (combined measured and calculated; Figure 4.3, bottom). Modal sizes were 140 and 160 cm FL, respectively. An examination of tarpon size over time indicates a slight but significant ($p < 0.05$) decline (Figure 4.4).

The relationship between actually recorded (not calculated) length and weight is presented in Figure 4.5. The regression coefficient of +0.89 is high and significant ($p < 0.05$) for the log FL/log weight relationship. Using the length–weight data, the condition factor (Fulton's K) was calculated for each fish for which data were available. The relationship between condition factors and years is depicted in Figure 4.6. Most tarpon had a condition factor >1, indicating they were in "good" condition. Although there was a positive relationship between year and condition factor ($Y = -3.5682 + 2.5629X$), this relationship was not significant ($p = 0.305$), indicating that there was no long-term change in the condition factor among these fish.

DISCUSSION

Regarding the erratic pattern in the number of tarpon scales available after 1945, it should be noted that this may reflect changes in the social/recreational aspect of anglers recording data on a tarpon scale. Alternatively, the pattern could depict a decline in tarpon landed by this fishery. There are no data currently available that allow testing of either hypothesis. It should be noted that historically it was "fashionable" to make an extended annual vacation to these resorts to fish for tarpon.

Insight into the Historical Status and Trends of Tarpon 73

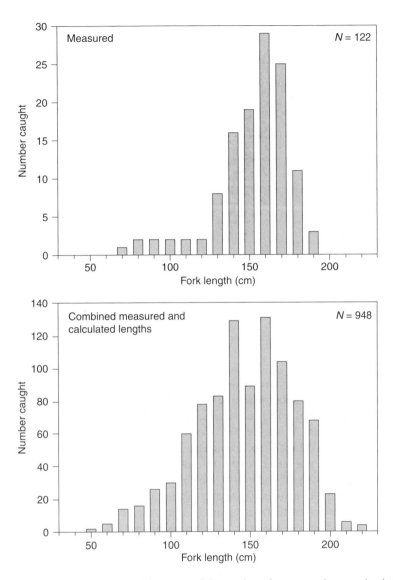

FIGURE 4.3 Length–frequency histogram of the number of tarpon scales examined. (Top) Includes only those fish for which fish lengths were recorded on the scale. (Bottom) Includes lengths from all fish for which measured lengths were recorded and for which lengths were calculated from reported weights.

Hence, anglers were housed in the very lodges or inns where the scales were retained, inscribed, and mounted. More recently, anglers are likely to fish the same areas via a private boat launched some distance from the location of the present locations of the inns and lodges. Thus, few tarpon captured as part of the present-day recreational tarpon fishery are brought to the sites where historical tarpon scales are currently on display.

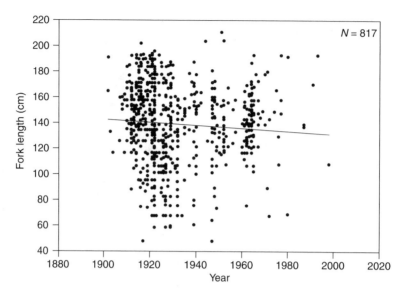

FIGURE 4.4 Scatter plot of length vs. year for tarpon based on scales examined. The regression line is represented by FL = 366.0479 − 0.1173 * year. The correlation coefficient is −0.069 and is significant at $p < 0.05$.

Interestingly, smaller tarpon were often recorded during the earlier years for which there are data. This may have been related to the types of gear used for fishing. Possibly, anglers during more recent times either do not catch smaller-sized tarpon or do not report them. Assuming there was no long-term trend in human behavior to retain scales from smaller fish (or a trend among anglers to overestimate size earlier in the twentieth century or underestimate size later in the same century!), the long-term, downward trend in tarpon size may be biologically meaningful. Overfishing usually results in a decline of a species' average size in the catch over time. Haedrich and Barnes (1997) indicated that a reduction in size structure and catch-per-unit-effort over time are indicative of a stock under exploitation. Holt et al. (2005) presented evidence that there was no obvious decline in length modes with time among tarpon caught off Texas. They also indicated that larger tarpon tended to be caught in more recent years, but conceded that this could also be evidence of size selectivity by the fishing public to retain larger fish for display or acknowledgment. Holt et al. (2005) concluded with the hypotheses that there may be a lack of recruitment of tarpon into the Texas fishery, especially from Mexico, after 1960, perhaps indicative of a decline in nursery habitat. Here it should be noted that, while the conclusions reached by Holt et al. (2005) could also be operating among tarpon populations along Florida's southwest coast, evidence of overexploitation off Florida is more apparent given the decline in average length in the catches and a potential increase in condition factor over time. It would be reasonable to assume, however, that both features may be operating on the recreational tarpon fishery along Florida's southwest coast—overexploitation of adults and recruitment limitations on juveniles could be co-occurring. The mutual occurrence of these features may be a more reasonable explanation.

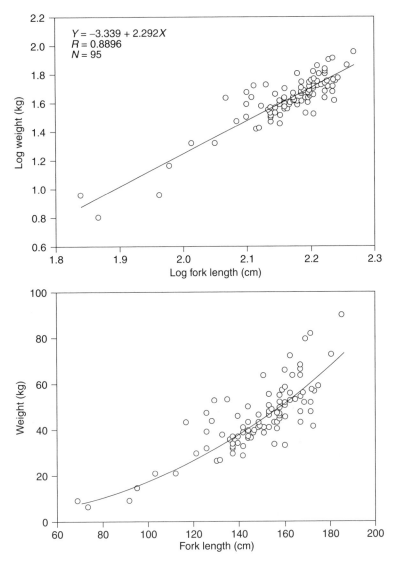

FIGURE 4.5 Length/weight plots and calculated regression lines for those fish for which both length and weight were available. (Top) Log weight (kg) vs. log FL (cm). (Bottom) Weight (kg) vs. FL (cm); line derived from equation (top).

First, while most adult tarpon are released live, some degree of mortality still occurs upon release. Shark attacks on newly released tarpon are particularly frequent off Boca Grande Pass. Second, inshore waters in which juvenile tarpon occur (Shenker et al., 2002) are becoming more subject to degradation due to increased development along many coastal areas (Bortone, 2005), especially in Florida.

Evidence presented here indicates that the condition factor (based on an assumption of isometric growth) has been stable among tarpon caught as part of the

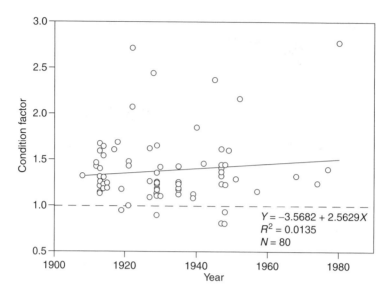

FIGURE 4.6 Condition factor (Fulton's K) vs. year for tarpon examined. The dashed line at 1.0 is a boundary line indicating that fish with $K > 1$ were in "good" condition and fish with $K < 1$ were in "poor" condition.

nearshore recreational fishery over the past 100 years. While the data certainly have an element of error included in them because of the lack of control in the information recording process (i.e., no certified/calibrated weighing scales, no verification of lengths, nonrandom sampling design, etc.), this study (along with studies such as recently completed by Holt et al., 2005) is potentially useful in assessing the causes of observed changes in tarpon population structure (abundance and size). It is hoped that this effort will serve to bring other sources of historic data on tarpon fisheries to the attention of the scientific community for further evaluation.

ACKNOWLEDGMENTS

Staff at the Sanibel-Captiva Conservation Foundation Marine Laboratory were diligent in their efforts recording the information on the "sometimes difficult to read" tarpon scales. Staff individuals included J. Greenawalt, J. Guinn, M. Hannan, B. Klement, E. Lindland, A.J. Martignette, E. Milbrandt, and J. Spinelli. I thank Mary Krutchen of the Olde Marco Inn, Andy Nagle of the Gasparilla Inn, and the management of the Tarpon Bar and Collier Inn on Useppa Island for allowing access to the tarpon scales in their care. Lastly, I thank Dr. G. Joan Holt, who initially introduced me to the concept of using tarpon scales from the historic recreational fishery as a source of information.

REFERENCES

Bortone, S.A. The quest for the "perfect" estuarine indicator: an introduction, in *Estuarine Indicators*, Bortone, S.A., Eds., CRC Press, Boca Raton, FL, chap. 1, 2005.

Crabtree, R.E. A review of recent tarpon research: what is known and what is not, *Contributions in Marine Science*, 35, 101–102, 2002.

Crabtree, R.E., Cyr, E.C., and Dean, J.M. Age and growth of tarpon, *Megalops atlanticus*, from South Florida waters, *Fishery Bulletin*, 93, 619–628, 1995.

Haedrich, R.L. and Barnes, S.M. Changes over time of the size structure in an exploited shelf fish community, *Fisheries Research*, 31, 229–239, 1997.

Holt, G.J., Holt, S.A., and Frank, K.T. What can historic tarpon scales tell us about the tarpon fishery collapse in Texas? *Contributions in Marine Science*, 37, 65–76, 2005.

IGFA. *2005 World Record Game Fishes*, International Game Fish Association, Dania, FL, 2005.

Jordan, D.S. and Evermann, B.W. *American Food and Game Fishes*. Doubleday, Page and Company, 1923. Read as an unabridged republication produced in Dover Publications, Inc., New York, 1969.

Migdalski, E.C. and Fichter, G.S. *The Fresh & Salt Water Fishes of the World*, Alfred A. Knopf, New York, 1976.

Oppel, F. and Meisel, T. *Tales of Old Florida*, Castle Press, Seacaucus, NJ, p. 477, 1987.

Shenker, J.M., Cowie-Mojica, E., Crabtree, R.E., Patterson, H.M., Stevens, C., and Yakubic, K. Recruitment of tarpon (*Megalops atlanticus*) leptocephali into the Indian River Lagoon, Florida, *Contributions in Marine Science*, 35, 55–69, 2002.

Zale, A.V. and Merrifield, S.G. Species profiles: life histories and environmental requirements of coastal fishes and invertebrates (South Florida)—ladyfish and tarpon. *U.S. Fish Wildl. Serv. Biol. Rep.*, 82(11.104), U.S. Army Corps of Engineers, TR EL-82-4, 17pp., 1989.

5 Ecology and Management of Bonefish (*Albula* spp.) in the Bahamian Archipelago

*Andy J. Danylchuk, Sascha E. Danylchuk,
Steven J. Cooke, Tony L. Goldberg,
Jeff Koppelman, and David P. Philipp*

CONTENTS

Introduction 79
History of the Bonefish Fishery 80
    Subsistence Fishery 80
    Recreational Fishery 80
Ecology of Bonefish in The Bahamas 81
    General Applicability 81
    Distribution and Abundance 83
    Habitat Use and Movements 83
    Feeding Ecology 85
    Population Dynamics 85
Bonefish Conservation and Management Strategies 87
Research and Conservation Needs 88
Acknowledgments 89
References 90

INTRODUCTION

Bonefish (*Albula* spp.) are an important group of fishes inhabiting shallow, nearshore marine environments worldwide. Historically, bonefish have played a strong role in supporting local and regional economies of the Bahamian Archipelago (i.e., The Bahamas and the Turks and Caicos Islands) (Alexander, 1961; BEST, 2002, 2005), an extensive expanse of shallow bank environments that comprise nearly 90% of the 300,000 km^2 archipelago (Sealey, 1994, Buchan, 2000). The ample nearshore habitats of the Bahamian Archipelago make bonefish readily accessible to local residents and visitors of this unique island chain (Kaufmann, 2000).

Despite their regional, economic, and ecological importance, relatively little scientific information exists to assist assessment or conservation management of bonefish in the Bahamian Archipelago. The purpose of this chapter is to review the history of the Bahamian bonefish fishery, and to highlight ecological and fishery research that has been conducted on bonefish in the Bahamian Archipelago, either as a target species or incidentally as part of other studies. This synthesis and analysis will help identify information gaps in the Bahamian Archipelago that need to be filled before bonefish stocks can effectively be managed and conserved.

HISTORY OF THE BONEFISH FISHERY

SUBSISTENCE FISHERY

For generations, bonefish have been the focus of subsistence and artisanal fisheries in The Bahamas and the Turks and Caicos Islands (Olsen, 1986; BEST, 2002; BEST, 2005). Catches of bonefish tend to be sold to individuals or small restaurants in rural communities, where bonefish were a favored species of finfish for consumption (Olsen, 1986). Subsistence and small-scale commercial harvesting was traditionally conducted in relatively shallow waters using handlines or by "hauling" seine nets (Olsen, 1986). Recently, monofilament gill nets have been employed for the harvest of bonefish in some areas. Unfortunately, these gears are nonselective, resulting not only in excessive harvests of bonefish, but also in substantial bycatch of other important species (e.g., turtles, barracudas, dolphins, sharks) (Clark and Danylchuk, 2003).

Use of bonefish as a subsistence food item has declined in recent decades (Rudd, 2003). Attrition of old-time "haulers" and the increased availability of commercially produced food items to local communities have contributed to the decreased reliance on bonefish as a staple food. In addition, the social stigma of bonefish as a "poor man's" food to some extent has reduced its popularity among islanders (Rudd, 2003).

RECREATIONAL FISHERY

As subsistence and small-scale commercial fisheries for bonefish in the Bahamian Archipelago have subsided, bonefish have gained importance as a target species for specialized recreational anglers. Angling for bonefish has become extremely popular because their wary nature and powerful swimming abilities when hooked make them a challenge to catch using lightweight fly-fishing and conventional hook-and-line gears (Kaufmann, 2000; Davidson, 2004; Fernandez, 2004). In addition, the remoteness and tranquil beauty of subtropical and tropical locales and serene qualities of the "flats" environment has turned bonefishing into a highly sought-after "holistic" angling experience. The clear, unpolluted waters of the Bahamian Archipelago with abundant bonefish and proximity to the United States are all draws for well-healed recreational anglers (BEST, 2002).

Interest in sportfishing has influenced the development of tourism-based industries specifically focused on recreational angling for bonefish (Figure 5.1).

Ecology and Management of Bonefish in The Bahamas

FIGURE 5.1 A beautiful Bahamas bonefish, a real focus of the region's high-value tourism industry. (Photo courtesy of Bob Stearns.)

From fishing tackle and guiding fees to travel and accommodations, the amount of direct and indirect revenues from the bonefishing industry can be high (Humston, 2001). For example, in the Florida Keys, regional economic contributions of the recreational industry centered on bonefishing generate a billion dollars in revenue per annum (Humston, 2001; Ault et al., Chapter 26, this volume). In developing countries, such as The Bahamas and Turks and Caicos Islands, local communities can be solely reliant on revenues generated by recreational bonefishing, especially when there is a paucity of alternative sources of revenue.

In The Bahamas, tourism represents more than 50% of the annual gross domestic product, making tourism the largest single contributor to the country's economy (Buchan, 2000; BEST, 2002). Recreational angling is a popular activity for tourists visiting The Bahamas and the Turks and Caicos Islands, many of whom dedicate their entire trip to fishing for bonefish. Of the 1.5 million tourists in 2004 who filled out immigration departure forms in The Bahamas, 5000 (0.3%) of these individuals stated that the purpose of their trip was for "bone/fly-fishing" (Government of The Bahamas, unpublished data). Most of these tourists who visited primarily for angling responded that their "bone/fly-fishing" trip targeted the "family islands" such as Abaco, Andros, and Eleuthera (Figure 5.2). Almost all respondents (92%) were from the United States, and most of these individuals were from the southern (41.1%) or northeastern areas of the country (28.8%) (Government of The Bahamas, unpublished data).

ECOLOGY OF BONEFISH IN THE BAHAMAS

GENERAL APPLICABILITY

Research conducted on the ecology of bonefish in the waters of the Bahamian Archipelago has been relatively limited. Fewer than 10 peer-reviewed scientific publications

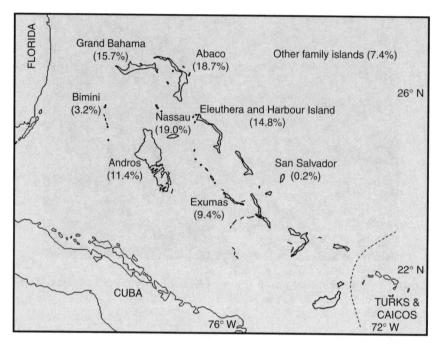

FIGURE 5.2 Proportion of immigration departure cards collected throughout The Bahamas in 2004, whose respondents indicated that the purpose of their visit was "bone/fly-fishing" ($n = 5000$). (From Government of The Bahamas, unpublished data.)

have been produced that specifically focus on bonefish (e.g., Colton and Alevizon 1983a, 1983b; Clark and Danylchuk, 2003; Cooke and Philipp, 2004; A.J. Danylchuk et al., 2007; S.E. Danylchuk et al., 2007) or that sampled bonefish as part of broader research questions (e.g., Layman and Silliman, 2002; Layman et al., 2004; Nero and Sullivan-Sealey, 2005). Although research conducted on bonefish in other parts of the world can provide some insights into the ecology and management of bonefish inhabiting the Bahamian Archipelago, studies by Pfeiler et al. (2000) and Colborn et al. (2001) have revealed the potential existence of multiple species of bonefish across several spatial scales. This brings into question the legitimacy of extrapolating results across geographic regions because different species may have vastly different life histories and behavioral patterns. Most accounts in the scientific literature refer to bonefish inhabiting the Bahamian Archipelago as *Albula vulpes*; however, rarely has the species identity of these populations been confirmed through genetic and morphometric analyses. An exception is a study by Bowen et al. (2003), which identified both *A. vulpes* and a second species (nova or species b) in a sample of bonefish collected from Bimini, although their overall sample size was relatively small. Regardless, comparative studies of genetics and morphometrics for bonefish may help to clarify whether distinct stocks occur across the Bahamian Archipelago and help lay the framework for ecological studies and management plans.

Understanding the ecology of bonefish in the Bahamian Archipelago could be complicated by the sheer size and unique oceanographic features characteristic of the region (Buchan, 2000). The Bahamian Archipelago is made up of a series of banks distributed on a southeast to northwest axis to the north of Cuba and Hispaniola that are separated by deep expansive oceanic toughs (Sealey, 1994; Buchan, 2000). The strong northward-flowing oceanic currents of the Gulf Stream to the west and the Antilles current to the east interact with these banks and troughs to generate complex patterns of water circulation that can influence the recruitment, distribution, abundance, and genetic differentiation of marine organisms on a regional scale (Gunn and Watt, 1982; Colin, 1995; Almada et al., 2001; Floeter et al., 2001). Similarly, variation in bathymetry and tidal currents generated by the close proximity of landmasses and the influence of wind on water movement across the shallow banks (Smith, 2004a, 2004b) could influence the ecology of bonefish populations. As such, extrapolating the results of other studies to bonefish in the Bahamian Archipelago (or vice versa) should be done with caution until the extent of the variation in systematics and ecology of bonefish populations in the region are more thoroughly examined.

DISTRIBUTION AND ABUNDANCE

Anecdotal observations by subsistence fishers, recreational anglers, and guides indicate that bonefish are widely distributed throughout the Bahamian Archipelago. Popular media articles and books on bonefish often provide extensive detail as to the regional and local distribution and relative abundance of bonefish the Bahamian Archipelago (e.g., Kaufmann, 2000). For instance, Kaufmann (2000) highlights the Abacos, Andros Island, Berry Islands, Bimini, Crooked and Acklins Islands, Eleuthera including Spanish Wells and Harbour Island, the Exumas, Grand Bahama Island, Great Inagua, Long Island, and the Turks and Caicos as prime destinations for recreational angling for bonefish. Undoubtedly, bonefish reside in the waters adjacent to other islands in the Bahamian Archipelago; however, their presence, distribution, and relative abundance are not generally known.

Although it is recognized that bonefish are distributed throughout the Bahamian Archipelago, no formal studies have been conducted to determine their distribution and relative abundance across a range of spatial and temporal scales. Recreational anglers and guides often comment on relative differences in the abundance and size structure of bonefish inhabiting different islands in the Bahamian Archipelago and during different seasons (Kauffmann, 2000); however, there has been no formal study or population census quantitatively assessing the abundance of bonefish in the region or whether spatial and temporal patterns in abundance do indeed exist. Spatial and temporal variation in the abundance of bonefish both within and among distinct regions of the Bahamian Archipelago could be related to intrinsic (e.g., reproductive ecology, species distribution) or extrinsic factors (e.g., oceanography, predation); and identifying the relative influence of such factors on bonefish abundance is crucial to developing reliable conservation management plans.

HABITAT USE AND MOVEMENTS

Bonefish in the Bahamian Archipelago generally inhabit shallow, nearshore waters (Kaufmann, 2000). Studies on the localized movements of bonefish in The Bahamas have suggested that bonefish utilize a range of nearshore habitats, including seagrass beds, mangrove creeks, and even coral reefs (Colton and Alevizon, 1983a, 1983b; Cooke and Philipp, 2004; A.J. Danylchuk et al., 2007; S.E. Danylchuk et al., 2007). Articles in popular angling publications and ancillary information garnished from recreational anglers and guides indicate that bonefish in the Bahamian Archipelago are also often observed and caught from other habitat types within the nearshore flats environment (Kaufmann, 2000), including sandy flats devoid of benthic vegetation (Layman and Silliman, 2002; Layman et al., 2004; Nero and Sullivan-Sealey, 2005).

Nearshore movements of bonefish within the Bahamian Archipelago have received some attention. Colton and Alevizon (1983a) used ultrasonic telemetry to examine the activity and daily movements of bonefish in waters near Deep Water Cay off Grand Bahama Island. Of the 13 fish surgically implanted with transmitters, only 3 were relocated more than 24 h post-release. The inability to detect 10 of the transmitter-implanted bonefish could have been attributed to predation following release or their movement out of reception range. The three remaining bonefish were tracked for between 5 and 100 days post-release, and their movements tended to be synchronous with the ebbing and flooding tides (moving into deeper water with ebbing tides and moving into shallow flats on flooding tides). On Andros Island, Nero and Sullivan-Sealey (2005) attributed variability in fish abundance among sites, including bonefish, to tides as well as to season; however, their data were not sufficient to determine if specific coastal or benthic factors were driving observed differences. Bohlke and Chaplin (1993) also reported that bonefish move into deeper water at slack low tides, with large schools being observed at depths of over 15 m below the edge of the drop-off in the Tongue of the ocean near Green Cay. Such movements are similar to the reoccurring localized pattern observed by Humston et al. (2005) for bonefish studied with acoustic telemetry in the Florida Keys, and by Colton and Alevison (1983a) for bonefish at Deep Water Cay in The Bahamas. Both studies inferred that the bonefish movement into deeper channels was attributed to avoidance of high water temperatures associated with shallow flats. In the case of Deep Water Cay, Colton and Alevison (1983a) noted that the proportion of large fish (>555 mm fork length, FL) was inversely correlated with inshore water temperatures, and that these observations were supported by anecdotal information provided by guides, anglers, and lodge owners.

Bonefish movement and migration patterns in the Bahamian Archipelago may also reflect the distribution and abundance of predators (Cooke and Philipp, 2004; Humston et al., 2005; A.J. Danylchuk et al., 2007). Although Humston et al. (2005) suggested that Florida Keys bonefish may avoid deep channels frequented by sharks; several recent studies in The Bahamas demonstrated that even bonefish in shallow waters (i.e., <0.5 m depth) are susceptible to predation, particularly following catch-and-release angling (Cooke and Philipp, 2004; A.J. Danylchuk et al., 2007; S.E. Danylchuk et al., 2007). Predation may have also affected the observations made by Colton and Alevizon (1983a) about the long-term movement patterns of bonefish at Deep Water Cay, since their lack of detection of transmitter-implanted fish or the recapture of externally tagged fish may have been caused by bonefish migrating out of the study area, or by predation by sharks or barracudas following release.

Movements of bonefish in the Bahamian Archipelago may be related to body size, reproductive maturity, spawning migrations, or ontogenetic shifts in feeding habits (Colton and Alevizon, 1983a; Bohlke and Chaplin, 1993). According to anecdotal accounts by Bahamian fishermen, large bonefish appear to return to tidal creeks in the fall where they aggregate in large numbers prior to spawning (Colton and Alevizon, 1983a). It is also commonly observed that schools of bonefish are generally composed of small- to medium-sized fish, while larger individuals tend to be more solitary, at least outside the spawning season (Bohlke and Chaplin, 1993). In the Turks and Caicos Islands, Clark and Danylchuk (2003) collected a total of 120 bonefish ranging in size from 28 to 72 cm total length (TL) as part of a tag-and-release study to determine movements on the Caicos Bank. During the course of the study, only one tagged bonefish was recaptured, with the fish being caught by a local hauler using a seine net (Clark and Danylchuk, 2003). They noted that the mean size of bonefish increased from west to east across Caicos Bank, potentially indicating ontogenetic shifts in habitat use. Local fishermen from South Caicos have also reported schools of large bonefish over offshore patch and coral reefs close to the wall of the Columbus Passage during winter months, and they believe that these aggregations might be related to spawning activity.

FEEDING ECOLOGY

Several diet studies, which examined stomach contents, have been conducted on bonefish in the western Atlantic (e.g., Warmke and Erdman, 1963; Crabtree et al., 1998b), with two of these in The Bahamas (Colton and Alevison, 1983b; Layman and Silliman, 2002). In all studies, bonefish were found to feed predominately on benthic invertebrates, but occasionally on small fishes. In Deep Water Cay, Colton and Alevizon (1983b) examined the stomach contents of 393 bonefish that ranged from 25 to 69 cm FL. Only 7% of stomachs were empty. Over 88% of the diet was comprised of invertebrates, with bivalves and crabs making up the majority of the biomass (dry weight) consumed (Colton and Alevizon, 1983b). Other prey items included small benthic fishes, such as gobies. Colton and Alevizon (1983b) also indicated that the dietary composition of bonefish differed among sand and seagrass habitats, likely related to the availability of prey items. Layman and Silliman (2002) examined the diet of considerably smaller bonefish (mean size of 13.8 ± 0.4 cm) in creek systems on Andros Island and found that 90% had eaten crustaceans, with 40% being decapod crabs. The majority of the diet by volume was composed of crustaceans (48%), mollusks (17%), and insects (18%) (Layman and Silliman, 2002). Although their sample size was relatively small ($n = 10$), Layman and Silliman (2002) did find that these small bonefish were most abundant over sand flats.

POPULATION DYNAMICS

No formal studies on population dynamics of bonefish (e.g., age and growth, reproduction, survivorship) have been conducted in the Bahamian Archipelago. Only incidental accounts of body size for bonefish in the Bahamian Archipelago have been reported in the scientific literature (Table 5.1). Those collected by scientific studies, in general, tend to be smaller than those caught by anglers (Kaufmann, 2000). For instance, bonefish exceeding 5 kg have been reported by guides and anglers

TABLE 5.1
Body Size of Bonefish Reported in Studies Conducted for Populations across the Bahamian Archipelago

Location	Length	N	Capture Method	Purpose of Study	Source
Andros Island	13.8 ± 0.4 mm SD	10	Cast net	Diet	Layman and Silliman, 2002
Deep Water Cay	50.5 – 61.0 cm FL	13	Angling (3), gill net (10)	Movement	Colton and Alevizon, 1983a
Deep Water Cay	25 – 69 cm FL	393	Not stated	Diet	Colton and Alevizon, 1983b
Deep Water Cay	50.2 ± 1.4 cm TL SE	18	Angling	Post-release mortality	Cooke and Philipp, 2004
San Salvador	51.2 ± 1.4 cm TL SE	17	Angling	Post-release mortality	Cooke and Philipp, 2004
Eleuthera	48.2 ± 5.0 cm TL SE	87	Angling	Post-release mortality	S.E. Danylchuk et al., 2007
Eleuthera	50.0 ± 8.4 cm TL SE	14	Seine	Post-release mortality	S.E. Danylchuk et al., 2007
Eleuthera	47.1 ± 1.2 cm TL SE	12		Post-release mortality	A.J. Danylchuk et al., 2007
Turks and Caicos Island	28 – 72 cm TL	120	Angling, seine	Movement	Clark and Danylchuk, 2003

across the Bahamian Archipelago, but not in primary scientific research. Nevertheless, if age and growth patterns can be generalized across regions in the western Atlantic, bonefish in the 10–12 lb range inhabiting the Bahamian Archipelago could easily be over 12 years old (Bruger, 1974; Crabtree et al., 1996).

All information on the seasonal timing of bonefish reproduction in the Bahamian Archipelago is based on anecdotal observations made by local fishers, recreational anglers, and fishing guides. Anglers often comment on the release of milt or eggs when fish are handled, especially between January and May. Anecdotal observations made in the Bahamian Archipelago suggest that bonefish aggregate and spawn in the fall, winter, and early spring (November–April). Mojica et al. (1995) studied larval recruitment patterns of *Albula* spp. near Lee Stocking Island and found leptocephali during fall and early winter, in agreement with anecdotal observations and with maturation patterns for bonefish in the Florida Keys (Crabtree et al., 1997). However, Mojica et al. (1995) also noted a large pulse of recruitment during a single 72-day sampling period in the summer months, indicating that spawning may occur year-round in The Bahamas. Otolith analysis of larval duration for specimens collected near Lee Stocking Island ranged from 41 to 71 days. Almost all leptocephali were collected at night in the upper 1 m of the water column, and inshore movement was strongly associated with flooding tides and the new moon (Mojica et al., 1995).

BONEFISH CONSERVATION AND MANAGEMENT STRATEGIES

Despite their ecological and economic importance, fishery regulations for bonefish across the Bahamian Archipelago are limited. In The Bahamas, the capture of bonefish using nets and the commercial trade of bonefish are prohibited (Bahamas Department of Fisheries, 1986). In the Turks and Caicos Islands, there are no specific regulations for bonefish (Turks and Caicos Islands Government, 1998a). At the same time, fishing guides in the Turks and Caicos Islands state that monofilament gill nets are being deployed across tidal creeks, resulting in the mortality of large numbers of juvenile and adult bonefish, as well as the bycatch of other important species such as marine turtles (Clark and Danylchuk, 2003).

In an effort to conserve fish stocks and their habitats, both countries are using marine protected areas in conjunction with existing fisheries regulations to build sustainable fisheries and protect marine biodiversity (Turks and Caicos Islands Government, 1998b). Although a marine reserve was established in the Turks and Caicos Islands in 1992 with bonefish conservation specifically in mind, no formal scientific information was used in its design and implementation. Only recently has there been any effort to assess the efficacy of this particular marine reserve, or whether marine protected areas in general are useful for conserving bonefish stocks (Clark and Danylchuk, 2003; Cooke et al., 2006).

One potential way in which bonefish in the Bahamian Archipelago are partially protected is through voluntary catch-and-release efforts (Cooke et al., 2006). Catch-and-release is commonly practiced by recreational anglers with a strong conservation ethic who travel to The Bahamas and the Turks and Caicos Islands. Catch-and-release angling can be an effective way to help maintain bonefish stocks only if the postrelease mortality is minimized (Cooke and Suski, 2005). When a fish is hooked by an angler, many factors affect the outcome of the event for the fish (Cooke et al., 2002; Cooke and Philipp, Chapter 25, this volume). At best, the fish will survive the event. At worst, the fish will not survive. Although anglers strive for the former outcome, an intermediate outcome in which the fish suffers transient physiological and behavioral impacts is probably more likely (Cooke and Philipp, 2004; Cooke and Suski, 2005; Bartholomew and Bohnsack, 2005), can increase the susceptibility of released fish to predation (Cooke and Philipp, 2004), and may ultimately lead to population-level effects.

Recently, Bartholomew and Bohnsack (2005) highlighted a number of factors related to recreational angling that influenced the mortality of released fish. They concluded that catch-and-release angling was not compatible with the conservation objectives of no-take marine protected areas. In a response, Cooke et al. (2006) indicated that the effects of the factors identified by Bartholomew and Bohnsack (2005), such as hooking in vital organs and angling duration and handling, could be reduced to the point where the fishing mortality rate approached zero, increasing the likelihood of integrating catch-and-release angling with no-take reserves. Determining whether catch-and-release is a useful tool for bonefish conservation requires more attention, especially as there is an increase in the demands of recreational anglers seeking bonefish along with the associated tourist operations supporting this activity (Crabtree et al., 1998a; Cooke and Philipp, 2004; Bartholomew and Bohnsack, 2005; Cooke et al., 2006).

Some studies have examined the short-term (24–48 h) mortality of bonefish following catch-and-release angling. In The Bahamas, these studies have found that predation of bonefish by lemon sharks (*Negaprion brevirostris*) and barracuda (*Sphyraena barracuda*) can range from 0 to 39%, with predation rates being correlated with the relative abundance of predators (Cooke and Philipp, 2004; A.J. Danylchuk et al., 2007) and the handling practices of anglers (S.E. Danylchuk et al., 2007). Post-release predation rates on bonefish could be regulated by the actions of anglers, potentially reducing the impacts of catch-and-release angling and making this activity more compatible with the conservation goals of no-take reserves (Cooke et al., 2006).

RESEARCH AND CONSERVATION NEEDS

A systematic, integrative, and cooperative approach is clearly needed to better understand and manage bonefish populations in the Bahamian Archipelago. Developing effective ecosystem management plans depends greatly on a comprehensive understanding of the systematics, biology, ecology, and population dynamics of bonefish throughout the region. Identifying if unique bonefish stocks occur (by compatible genetic and morphometric methods) in the Bahamian Archipelago is of primary importance, since stock mixing could significantly complicate management of the species. To determine whether traits in bonefish populations vary significantly across the large spatial scale of the Bahamian Archipelago, basic information on the genetic identity, age, growth, and reproductive potential (e.g., size and age at maturity, fecundity) needs to be collected at multiple locations across the region as part of a coordinated Bahamian Archipelago–wide sampling (monitoring) and assessment effort. Such an archipelago-wide program would help encompass potential variation in bonefish populations associated with different properties of individual shallow water banks (e.g., degree of physical isolation, interactions with major oceanographic currents, and latitude). Such sampling should occur at regular intervals throughout the year to determine whether the population structure of bonefish varies temporally and is potentially related to spawning migrations, recruitment, or climatic patterns. Sampling the age, growth, and reproduction of bonefish populations at multiple locations throughout the year will allow for the examination of age- and size-specific trends in the allocation of energy to gonad development that, in turn, would help quantify the spatial and temporal patterns in the phenology of reproduction for bonefish across the Bahamian Archipelago. At selected focal research sites, the input of bonefish leptocephali could be monitored using channel nets or light traps as a way to cross-validate the seasonal timing of reproduction inferred through the direct examination of gonad development. In addition, movement studies of bonefish using remote acoustic telemetry could be conducted in concert with the examination of gonad development and larval input to help determine where spawning activity actually occurs.

Given that the nearshore environment of the Bahamian Archipelago is relatively diverse at both the local and regional scales and that the region is prone to environmental extremes (e.g., high summer water temperatures, freshwater input, hurricanes), understanding how natural variation and natural disturbance regimes shape bonefish populations will allow for a more thorough evaluation of how anthropogenic disturbances may affect bonefish stocks (Cooke and Philipp, 2004; Sealey, 2004).

Such comparisons could be facilitated through before-after-control-impact studies (Underwood, 1994), empirical studies on bonefish populations subjected to a range of natural and anthropogenic disturbances, and experimental or manipulative studies that target particular disturbances. For instance, the tourist industry is steadily increasing throughout the Bahamian Archipelago, often resulting in anthropogenic disturbances such as dredging and coastal eutrophication (Rudd, 2003; Sealey, 2004). The potential effects of such disturbances on bonefish populations could be examined by monitoring bonefish populations before and after dredging or shoreline development has occurred in a particular area, specifically to test if modifying or eliminating foraging habitat has cascading impacts on bonefish distribution, life history traits, and ultimately abundance (Syms and Jones, 2000; Gust et al., 2001; Hixon et al., 2001; Sadovy, 2005). Similarly, comparative and manipulative studies may help differentiate the effects of recreational activities or if angling-related activities such as wading have detrimental effects on the integrity of nearshore habitats (Cooke and Suski, 2005).

The interdependence of coastal environments of the small islands and the dependence of local communities on bonefish for income in the Bahamian Archipelago calls for a holistic and comprehensive management strategy to conserve and protect bonefish stocks. Although marine protected areas are often advocated and used throughout the Bahamian Archipelago as a low-cost tool for protecting habitats and species (BEST, 2005; Dahlgren, 2002; Danylchuk, 2003; Lubechenco et al., 2003), they will only be effective if they balance the needs of society with the needs of the local marine resources (Murray et al., 1999; Hanna, 2001; Roberts et al., 2001; Sealey, 2003). With this in mind, determining whether or not catch-and-release angling is compatible with the conservation goals of marine-protected areas is important (Bartholomew and Bohnsack, 2005; Cooke et al., 2006). If recreational angling for bonefish is deemed compatible with marine-protected areas, then the development of locally based tourism focused on this activity could be promoted as part of a larger integrative management plan without disrupting the overall level of protection offered to the ecosystem (Cooke et al., 2006).

An effective archipelago-wide sampling and management program for bonefish will depend greatly on collaborative partnerships between scientific institutions, pertinent local and regional governments, conservation organizations, and stakeholders. Integrating cooperative research with education and outreach programs throughout the Bahamian Archipelago will also instill the importance for marine conservation, including the protection of bonefish stocks. Only through such partnerships and education programs will realistic conservation management plans be developed that adequately encompass the needs of bonefish stocks, as well as the sustainable development of local communities in the Bahamian Archipelago.

ACKNOWLEDGMENTS

We gratefully acknowledge Chris Maxey and the Cape Eleuthera Foundation for financial and logistical support during the preparation of this chapter. We thank Earlston McPhee and Garry Young from The Bahamas Ministry of Tourism for providing statistics on bonefishing-based tourism in The Bahamas. Thanks also to J. Ault and two anonymous reviewers for their comments and suggestions on an earlier version of this chapter. We would also like to thank the many anglers, guides, and local fisherman who have provided invaluable anecdotes on bonefish, including S. Gardiner (Silver Creek Adventures), B. Jayne and G. Lockhart (Beyond the Blue Charters), B. Gardiner (Bonefish Unlimited), A. Dean (Silver Deep), F. Lockhart, T. Morris, S. Jennings, C. Leathen, R. Reckley, H. Rolle, A. McKinney, and D. Rankin. A special thanks also goes out to T. Davidson and R. Fisher (Bonefish & Tarpon Unlimited) for their guidance and support.

REFERENCES

Alexander, E.C., A contribution to the life history, biology and geographical distribution of bonefish, *Albula vulpes* (Linnaeus), Dana-Report, *Carlsberg Found.*, 53, 1, 1961.

Almada, V.C., Oliveria, R.F., Gonçalves, E.J., Almeida, A.J., Santos, R.S. and Wirtz P., Patterns of diversity of the north-eastern Altantic blenniid fish fauna (Pisces: Blenniidae), *Global Ecol. Biogeo.*, 10, 411, 2001.

Bahamas Department of Fisheries, *Bahamian Fisheries Regulations and Reports*, 1986.

Bahamas Environment, Science and Technology Commission (BEST), *Bahamas Environment Handbook*, Government of The Bahamas, 2002.

Bahamas Environment, Science and Technology Commission (BEST), *State of the Environment*, Government of The Bahamas, 2005.

Bartholomew, A. and Bohnsack, J.A., A review of catch-and-release angling mortality with implications for no-take reserves. *Rev. Fish Biol. Fish.*, 15, 129, 2005.

Bohlke, J.E. and Chaplin, C.C.G., *Fishes of The Bahamas and Adjacent Tropical Waters*, 2nd edition, University of Texas Press, Austin, TX, 1993.

Bowen, B.W., Colborn, J., Karl, S.A. and Curtis, C., Systematics and ecology of bonefish (*Albula* spp.) in Florida waters, in *Investigations into Nearshore and Estuarine Gamefish Behavior, Ecology, and Life History in Florida*, Five year Performance Report to the US Fish and Wildlife Service, Sport Fish Restoration Project F-59, Florida Marine Research Institute, St. Petersburg, FL, 14, 2003.

Bruger, G.E., Age, growth, food habits and reproduction of bonefish (*Albula vulpes*) in South Florida waters. *Mar. Res. Pub.*, 3, Florida Department of Natural Resources, St. Petersburg, FL, 1974.

Buchan, K.C., The Bahamas. *Mar. Pollut. Bull.*, 41, 94, 2000.

Clark, S.A. and Danylchuk, A.J., Introduction to the Turks and Caicos Islands bonefish research project tagging program, *Proc. Gulf Carib. Fish. Inst.*, 54, 396, 2003.

Colborn, J., Crabtree, R.E., Shaklee, J.B., Pfeiler, E. and Bowen, B.W., The evolutionary enigma of bonefishes (*Albula* spp.): cryptic species and ancient separations in a globally distributed shorefish, *Evolution*, 55, 807, 2001.

Colin, P.L., Surface currents in Exuma Sound, Bahamas, and adjacent areas with reference to potential larval transport, *Bull. Mar. Sci.*, 56, 48, 1995.

Colton, D.E. and Alevizon, W.S., Movement patterns of the bonefish (*Albula vulpes*) in Bahamian waters, *Fish. Bull.*, 81, 148, 1983a.

Colton, D.E. and Alevizon, W.S., Feeding ecology of bonefish in Bahamian waters, *Trans. Am. Fish. Soc.*, 112, 178, 1983b.

Cooke, S.J. and Philipp, D.P., Behavior and mortality of caught-and-released bonefish (*Albula* spp.) in Bahamian waters with implications for a sustainable recreational fishery, *Biol. Conserv.*, 118, 599, 2004.

Cooke, S.J. and Suski, C.D., Do we need species-specific guidelines for catch-and-release recreational angling to conserve diverse fishery resources? *Biodivers. Conserv.*, 14, 1195, 2005.

Cooke, S.J., Schreer, J.F., Dunmall, K.M. and Philipp, D.P., Strategies for quantifying sublethal effects of marine catch-and-release angling—insights from novel freshwater applications, *Am. Fish. Soc. Symp.*, 30, 121, 2002.

Cooke, S.J., Danylchuk, A.D., Danylchuk, S.A., Suski, C.D. and Goldberg, T.L., Is catch-and-release recreational fishing compatible with no-take marine protected areas? *Ocean Coastal Manage.*, 49, 342, 2006.

Crabtree, R.E., Harnden, C.W., Snodgrass, D. and Stevens, C., Age, growth, and mortality of bonefish, *Albula vulpes*, from the waters of the Florida Keys, *Fish. Bull.*, 94, 442, 1996.

Crabtree, R.E., Snodgrass, D. and Harnden, C.W., Maturation and reproductive seasonality in bonefishes, *Albula vulpes*, from the waters of the Florida Keys, *Fish. Bull.*, 95, 456, 1997.

Crabtree, R.E., Snodgrass, D. and Harnden, C., Survival rates of bonefish, *Albula vulpes*, caught on hook-and-line gear and released based on capture and release of captive bonefish in a pond in the Florida Keys, in *Investigation into Nearshore and Estuarine Gamefish Abundance, Ecology, and Life History in Florida*, Five year Technical Report to the US Fish and Wildlife Service, Sport Fish Restoration Project F-59, Florida Marine Research Institute, St. Petersburg, FL, 252, 1998a.

Crabtree, R.E., Stevens, C., Snodgrass, D. and Stengard, F.J., Feeding habits of bonefish, *Albula vulpes*, from the waters of the Florida Keys, *Fish. Bull.*, 96, 754, 1998b.

Dahlgren, C., Marine protected areas in The Bahamas, *Bahamas J. Sci.*, 9, 41, 2002.

Danylchuk, A.J., Fisheries management in South Eleuthera: can a marine reserve help save the "holy trinity," *Proc. Gulf Carib. Fish. Inst.*, 56, 169, 2003.

Danylchuk, A.J., Danylchuk, S.E., Cooke, S.J., Goldberg, T.L., Koppelman, J. and Philipp, D.P., Post-release mortality of bonefish (*Albula* spp.) exposed to different handling practices in South Eleuthera, Bahamas, *Fish. Manage. Ecol.*, 14, 149–154, 2007.

Danylchuk, S.E., Danylchuk, A.J., Cooke, S.J., Goldberg, T.L., Koppelman, J. and Philipp, D.P., Effects of recreational angling on the post-release behavior and predation of bonefish (*Albula vulpes*): the role of equilibrium status at the time of release. *J. Exper. Mar. Biol. Ecol.*, 346, 127–133, 2007.

Davidson, T., *Bonefish B. S. and Other Good Fish Stories*, Hudson Books, Toronto, 2004.

Fernandez, C., *Fly-Fishing for Bonefish*, Stackpole Books, Mechanicsburg, PA, 2004.

Floeter, S.R., Guimaraes, R.Z.P., Rocha, L.A., Ferreira, C.E.L., Rangel, C.A. and Gasparini, J.L., Geographic variation in reef fish assemblages along the Brazilian coast, *Global Ecol. Biogeogr.*, 10, 423, 2001.

Gunn, J.T. and Watt, D.R., On the currents and water masses north of the Antilles/Bahamas Arc, *J. Mar. Res.*, 40, 1, 1982.

Gust, N., Choat, J.H. and McCormick, M.I., Spatial variability in reef fish distribution, abundance, size and biomass: a multi-scale analysis, *Mar. Ecol. Prog. Ser.*, 214, 237, 2001.

Hanna, S., Managing the human-ecological interface: marine resources as example and laboratory, *Ecosystems*, 4, 736, 2001.

Hixon, M.A., Boersma, P.D., Hunter, M.L. Jr., Icheli, F., Norse, E.A., Possingham, H.P. and Snelgrove, P.V.R., Oceans at risk: research priorities in marine conservation biology, in *Conservation Biology, Research Priorities for the Next Decade*, Soulé M.E. and Orians G.H., Eds., Island Press, Washington, DC, 125, 2001.

Humston, R., *Development of movement models to assess the spatial dynamics of fish populations*, Ph.D. dissertation, Rosenstiel School of Marine and Atmospheric Science, University of Miami, FL, 2001.

Humston, R., Ault, J.S., Larkin, M.F. and Luo, J., Movements and site fidelity of the bonefish *Albula vulpes* in the northern Florida Keys determined by acoustic telemetry, *Mar. Ecol. Prog. Ser.*, 291, 237, 2005.

Kaufmann, R., *Bonefishing*, Western Fisherman's Press, Moose, WY, 2000.

Layman, C.A. and Silliman, B.R., Preliminary survey and diet analysis of juvenile fishes of an estuarine creek on Andros Island, Bahamas, *Bull. Mar. Sci.*, 70, 199, 2002.

Layman, C.A., Arrington, D.A., Langerhans, R.B. and Silliman, B.R., Degree of fragmentation affects fish assemblage structure in Andros Island (Bahamas) estuaries. *Carib. J. Sci.*, 40, 232, 2004.

Lubechenco, J., Palumbi, S.R., Gaines, S.D. and Andleman, S., Plugging a hole in the ocean: the emerging science of marine reserves, *Ecol. Appl.*, 13, S3, 2003.

Mojica, R., Shenker, J.M., Harnden, C.W. and Wanger, D.E., Recruitment of bonefish, *Albula vulpes*, around Lee Stocking Island, Bahamas, *Fish. Bull.*, 93, 666, 1995.

Murray, S.N., Ambrose, R.F., Bohnsack, J.A., Botsford, L.W., Carr, M.H., Davis, G.E., Dayton, P.K, Gotshall, D., Gunderson, D.R., Hixon, M.A., Lubchenco, J., Mangel, M., MacCall, A., McArdle, D.A., Ogden, J.C., Roughgarden, J., Starr, R.M., Tegner, M.J. and Yoklavich, M.M., No-take reserve networks: sustaining fishery populations and marine ecosystems, *Fisheries*, 24, 11, 1999.

Nero, V.L. and Sullivan-Sealey, K., Characterization of tropical near-shore fish communities by coastal habitat status on spatially complex island systems, *Environ. Biol. Fishes*, 73, 437, 2005.

Olsen, D.A., *Fisheries assessment for the Turks and Caicos Islands*, Food and Agriculture Organization of the United Nations, Rome, 1986.

Pfeiler, E., Padron, D. and Crabtree, R.E., Growth rate, age and size of bonefish from the Gulf of California, *J. Fish Biol.*, 56, 448, 2000.

Roberts, C.M., Bohnsack, J.A., Gell, F., Hawkins, J.P. and Goodridge, R., Effects of marine reserves on adjacent fisheries, *Science*, 294, 1920, 2001.

Rudd, M.A., Fisheries landings and trade of the Turks and Caicos Islands, *Univ. Br. Columb. Fish. Cent. Res. Rep.*, 11, 149, 2003.

Sadovy, Y., Trouble on the reef: the imperative for managing vulnerable and valuable fisheries, *Fish Fish.*, 6, 167, 2005.

Sealey, K.S., Balancing development and the environment in the Bahamian Archipelago, *Bahamas J. Sci.*, 5, 2, 2003.

Sealey, K.S., Large-scale ecological impacts of development on tropical island systems: comparison of developed and undeveloped islands in the Central Bahamas, *Bull. Mar. Sci.*, 75, 295, 2004.

Sealey, N.E., *Bahamian Landscapes: An Introduction to the Geology of The Bahamas*, Media Enterprises Ltd., Nassau, Bahamas, 1994.

Smith, N.P., Transport over a narrow shelf: Exuma Cays, Bahamas, *Ocean Dyn.*, 54, 435, 2004a.

Smith, N.P., Transport processes linking shelf and back reef ecosystems in the Exuma Cays, Bahamas, *Bull. Mar. Sci.*, 75, 269, 2004b.

Syms, C. and Jones, G.P., Disturbance, habitat structure and the dynamics of a coral-reef fish community, *Ecology*, 81, 2714, 2000.

Turks and Caicos Islands Government, *Fisheries Protection Ordinance*, CAP 104, 1998.

Turks and Caicos Islands Government, *National Park Ordinance*, CAP 80, 1998.

Underwood, A.J., On beyond BACI: sampling designs that might reliably detect environmental disturbances, *Ecol. Appl.*, 4, 3, 1994.

Warmke, G.L. and Erdman, D.S., Records of marine mollusks eaten by bonefish in Puerto Rican waters, *Nautilus*, 76, 115, 1963.

6 Coastal Ecosystem Management to Support Bonefish and Tarpon Sportfishing in Peninsula de Zapata National Park, Cuba

Lázaro Viñola Valdez, Lázaro Cotayo Cedeño, and Natalia Zurcher

CONTENTS

Introduction 93
Las Salinas 94
      Seasonal Fishing Weather and Catch Rates 95
The Hatiguanico River 96
      Seasonal Fishing Weather and Catch Rates 97
Discussion 97
Acknowledgment 98

INTRODUCTION

Located in the southern Matanzas province, the National Park of the Peninsula of Zapata is part of the protected area of Peninsula of Zapata wetland in Cuba (Figure 6.1). The wetland has been a protected area since 1995 (Cuban legislation, Executive Committee of the Council of Ministers, January 1995), and it was declared a biosphere reserve by the United Nations Educational, Scientific and Cultural Organization (UNESCO) in January 2000. It is both the largest and most ecologically important wetland in the Caribbean. Owing to its vast area and the importance of the ecosystem, the Peninsula of Zapata is one of the most remarkable geographic units of the Cuban territory. The natural resources of this large insular wetland are of vital importance for the livelihood of the locals, mainly the extraction of wood and production of charcoal. The forests are also used for tourism and as a source of food for local communities. A small fishing port in the area supplies the needs of southern Matanzas province.

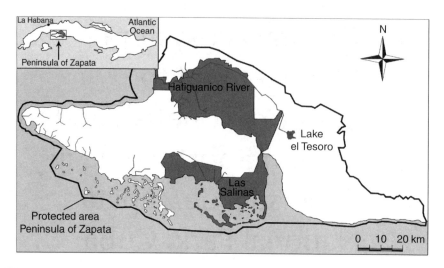

FIGURE 6.1 Map showing the location and boundary of the protected area of the Peninsula of Zapata. The dark gray area illustrates the National Park regions of the Hatiguanico River and Las Salinas.

The Peninsula of Zapata has unique vegetative ecosystems and a system of surface drainage characterized by several rivers, lagoons, swamps, channels, and artificial canals of medium to small flow with an important hydrological function. Intricate hydrological systems support a large diversity of habitats and variable climatic conditions that maintain a great diversity of species and provide an ideal habitat for bonefish and tarpon.

The National Park of the Peninsula of Zapata includes the salt marshes of Las Salinas and the basin of the Hatiguanico River (Figure 6.1). Among the most important activities in the park is sportfishing for bonefish (*Albula vulpes*) and tarpon (*Megalops atlanticus*). Since 2002, in an effort to maintain greater control and ecological sustainability of the ecosystem for the future, sportfishing for bonefish in the area of Las Salinas and for tarpon in the Hatiguanico River have been monitored by the National Park authorities. All fishing activities in the park are led by professional guides working for the National Park, whereas local and nationwide travel agencies handle outreach programs to attract tourists to the area.

LAS SALINAS

Las Salinas, with an area of 35,000 ha, is located in the southern center of the peninsula. The different ecosystems in this area vary as a result of the presence or absence of fresh, brackish, or salt surface waters, which directly affects the vegetation and the terrestrial and aquatic fauna. Salt marshes are the most important ecosystem in Las Salinas. This saltwater ecosystem has many shallow lagoons linked by very small channels that run from the coastline toward the inland for an approximate distance of 8–10 km. The channels are connected through an underground hydrologic system, and tidal variations are extremely low, making it ideal for sight casting most of the day. The primary type of vegetation surrounding these lagoons is mangroves. In winter, parts of the lagoons dry out, creating concentrations of natural salt that give the area its name.

A trail of 21 km provides terrestrial access to approximately 90 km² of fishing area. It is impossible and forbidden to use motorized boats at Las Salinas. The lagoons average 0.3 m in depth. The main species caught include bonefish (*A. vulpes*), permit (*Trachinotus falcatus*), snook (*Centropomus undecimalis*), barracuda (*Sphyraena barracuda*), and horse-eye jack (*Caranx latus*). The area has an outstanding and abundant bonefish population, with an average size ranging from 1.1 to 1.8 kg.

Las Salinas has been declared by the chamber of commerce as an ideal area for bonefish fishing because of its size, suitable depths, great preservation of its natural pristine environment, and seclusion from other human activities. The variety of bottom habitat types includes extensive sand bars; open mangrove areas; rocky-bottom lagoons; and sandy, muddy, and mixed bottoms, which create a challenging environment for both guides and anglers. In contrast, shallow depths and clear water allow fish to be easily seen while feeding and tailing, creating exceptional conditions for sportfishing of the species. Only boats that are flat-bottomed, non-motorized, have no keel, have a freeboard height of 20 cm, and a capacity for only two persons (i.e., one guide and one angler) are allowed in the area. Only catch-and-release fly-fishing is allowed in Las Salinas. The park allows entrance only to licensed fishing guides within a designated fishing zone per day. The carrying capacity has been defined as six guides per 4 days per week or the equivalent of 24 fishing sessions per week. Bonefish guides are assigned different zones, changing periodically to ensure maintenance of the appropriate conditions of the fisheries. This zonation was implemented after monitoring fish behavior to guarantee optimal catch rates of fish of a large average size and favorable environmental conditions. The regular rotation changes only in cases of extreme weather conditions. Each guide is in charge of collecting a series of data and information on the progress of his/her daily fishing activities. This information is crucial for both scientific understanding and management decision making required for the sustainability of the bonefish fisheries of Las Salinas.

SEASONAL FISHING WEATHER AND CATCH RATES

There is not a significant annual difference in the daily catch and effort in Las Salinas; however, there is a clear seasonal correlation between weather conditions of a given area and catch rates.

- *December to February*: These months are notorious for being the coldest of the year, with temperatures ranging between 15 and 25°C, lower water levels, and the passage of periodic cold fronts with winds in excess of 30 km/h. Therefore, during these months, a fishing day generally starts early at 0900, coinciding with the time when bonefish are actively feeding. Fish are found in the shallow lagoons outside the mangroves and the limiting factors are the cold fronts that create cloudy skies, which in turn create difficult conditions for sight casting.

- *March to May*: During this period temperatures are higher, between 20 and 30°C, and fishing starts at 0800 and continues throughout the entire day. Water reaches the lowest levels and weather conditions are more stable, favoring bonefishing. This is the best time of the year for the sportfishing of bonefish in the Salinas.
- *June to August*: These are the warm and rainy months, with temperatures ranging between 25 and 35°C. Higher water levels cause the mangroves to be flooded, creating extensive feeding grounds for bonefish. Fish usually move into these areas in the early hours of the morning, thus fishing days must begin as early as possible and practically end early in the afternoon. Owing to the presence of fresh water in the lagoons, the water coloration is dark and stained with tannic acid. Winds have little or no influence on fishing; however, the low winds contribute to a greater presence of mosquitoes.
- *September to November*: Known to be the active hurricane season, temperatures vary between 23 and 33°C. Water levels are usually high with abundant fresh water that mixes in the lagoons and contributes to darker water colorations. Fishing in the mangroves takes place during the early hours of the morning. If hurricanes are not present, general weather conditions are favorable for fishing.

THE HATIGUANICO RIVER

The basin of the Hatiguanico River has been described as the Amazon of Cuba. It is located on the west side of the National Park of the Peninsula of Zapata. The river is 30 km long and crosses the marsh and drains off the surface water to the Broa Cove. Depths vary between 4 and 6 m. The river width is about 20 m inland and approximately 300 m at the mouth. A well-protected mangrove forest and distinctive swamp grasslands grow on the edges of the river. More than 80 species of birds have been reported in this area, several of which are common to the entire country and three that are endemic to the area. In addition, it provides habitat to the Cuban crocodile, manatees, and the jutía (hutia). The main species of fish that can be found here include tarpon (*M. atlanticus*), snook (*C. undecimalis*), cubera snapper (*Lutjanus cyanopterus*), and horse-eye jack (*C. latus*). The fishing area includes the main river and its tributaries, Rios Negros, Guareira, and Gonzalo.

Magnificent natural conditions and scenic beauty make the Hatiguanico River a perfect place for tarpon fishing. There is an abundant population of small-size tarpon, ranging from 1.8 to 5.5 kg; larger tarpon of about 45 kg or more are occasionally caught in the river. The park allows only licensed fishing guides to fish in the river. The carrying capacity has been defined as four guides per 3 days per week, or the equivalent of 12 fishing sessions per week. The distribution of guides is random, adjusted only to the daily movements of the fish. The boats used in the Hatiguanico River have outboard engines with a maximum capacity of one guide and one angler.

Seasonal Fishing Weather and Catch Rates

There are two well-defined seasons for tarpon fishing in the Hatiguanico River.

- *December to May*: This is the cold and dry season. It is characterized by clear waters with cold temperatures, which contribute to low catch rates and the lowest numbers of sightings of tarpon.
- *June to November*: This period is known to be the warm and rainy season. The river water is dark and favorable for the sighting of big schools of fish. This is also the period of highest catches for fly-fishing in the area.

DISCUSSION

The National Park of the Peninsula of Zapata is responsible for any management decision making related to human activities in the park, including any decisions about carrying capacity of the region, zonation, environmental conditions, and conservation of the area during the fishing season. Although the National Park has implemented fishing programs to manage and monitor the bonefish and tarpon fisheries of the area, much is unknown. To date, limited work has been done studying population dynamics in the park. All management decision making is based on carrying capacity, catch rates, and environmental conditions. Preliminary results of the management program have led to new information about the behavior of the species, previously not available or gathered from reports from other places, and on the effects of water temperatures and lunar phase on catch rates and average sizes of fish caught. Area closures have only been implemented if after permanent monitoring of the resources there is an indication of overload and excessive fishing pressure.

Carrying capacity has been defined for each area by taking the natural and biological factors of the species under consideration. Knowledge has been acquired by previous observations and through consultation with experts on the subject and the area. In the future, carrying capacity will be evaluated by permanently monitoring and recording catch rates and condition of the fisheries, which will indicate if any changes are necessary.

Knowledge of the existing natural conditions and their evolution is a key element to the efficient management of the area. Understanding how fishing impacts the normal functioning and natural conditions of the area has led to enforcement of certain regulations in the park. These regulations are intended to control and sustain the resources and performance of the guides and anglers during the fishing, optimizing their experience when they visit the park. To maintain the existing natural conditions and to ensure a positive fishing experience, park personnel continuously take care of cleaning the narrow channels, removing any obstacles that will hinder fishing, and monitoring any invading species.

To maintain and expand their management program, the National Park of the Peninsula of Zapata hopes to collaborate with other national and international institutions to establish a research program that will allow them to thoroughly study fishing impacts on the biodiversity and the functioning of the ecosystem, and to guide implementation of a management plan that will reduce or mitigate these effects.

The main management objectives are (1) to study fish behavior, feeding, and reproduction in the area and changes during the different seasons and weather conditions; (2) to study movements in and out of the area; (3) to manage the catch rate per session; (4) to monitor ecological and weather conditions specific to the area; (5) to monitor water quality and feeding grounds in the fishing areas; (6) to monitor the tarpon population in the river and its movements to other zones; (7) to periodically conduct inventories of sportfishing species in the area; and (8) to implement systematic regulations to manage the fisheries during the entire season. This information will also provide the park with more complete and accurate population dynamics parameters and other critical biological information that will facilitate management decision making to build sustainable fisheries.

Outreach to support the sportfishing activities and management decision making required for sustainable bonefish and tarpon fisheries in the park is a central and essential component of the program. The Ministry of Tourism of Cuba is in charge of marketing programs and promoting and selling packages for sportfishing for bonefish and tarpon, an indispensable feature for attracting tourists to the park. Public presentations and workshops will be organized to provide opportunities to establish dialogue with other institutions, fishing guides, and the general public.

ACKNOWLEDGMENT

This document was developed from an original Spanish language report provided in 2006 by the National Park of the Peninsula of Zapata: L. V. Valdez and L. C. Cedeño, *Proyección y experiencias preliminares del manejo de ecosistemas marino costeros, para la pesca deportiva del macabí y el sábalo por el Parque Nacional Península de Zapata, Cuba*, National Park of the Peninsula of Zapata, Playa Larga, Matanzas, Cuba.

7 Tarpon and Bonefish Fishery on Turneffe Atoll, Belize

J. Travis Pritchett

CONTENTS

The Environment 99
The Fishery 100
The Opportunity 101
Acknowledgments 102
References 102

THE ENVIRONMENT

Turneffe Atoll is located 30 km east of Belize City, Belize in the western Caribbean Sea (Figure 7.1). This atoll stretches 48 km long and about 16 km across at the widest point. Almost 80% of the total land area is submerged. The remaining 20% is just 0.5–1.0 m above sea level. A wall of living coral reef surrounds Turneffe Atoll, a series of mangrove stands and cays formed from sand, mud, and coral rubble. Inside the reef, over 200 cays make up the landmass that surrounds two central lagoons, the Northern Lagoon and the Southern Lagoon. The lagoons average ≤4 m in depth with the areas between the reef system and the cays averaging considerably less. These systems provide nursery and feeding grounds for myriad fish species and critical habitat for many species of wildlife (CZMAI, 2001). This atoll provides a rare and unique opportunity to catch bonefish, permit, and tarpon in the same ecosystem.

Turneffe Atoll has also long been recognized as a premier destination for saltwater fly-fishing, scuba diving, and marine ecotourism. The atoll has three tourism-based lodges—Turneffe Flats, Blackbird Cay, and Turneffe Island Lodge—that support tourists with diving, snorkeling, ecotours, and catch-and-release fishing. The atoll is also home to a few private residences, a research outpost, and a number of commercial fishing camps. During peak times, as many as 300 people can be on the atoll. However, the area remains relatively pristine, thanks to its geographic isolation from the mainland.

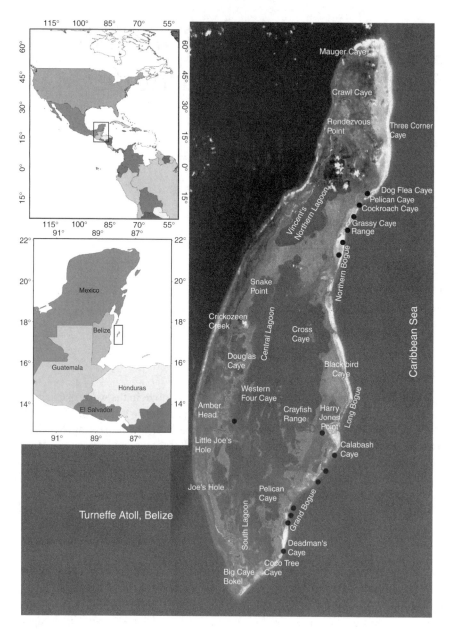

FIGURE 7.1 Location map showing Turneffe Atoll, Belize.

THE FISHERY

Historically, the atoll was used by commercial fishermen who would set up camps on stilts. They would spend days at a time fishing for lobster, conch, and finfish, but they were limited to only a few days due to the difficulties of getting their product back to the markets in Belize City. Today the fishermen are able to get back and forth between the mainlands much more easily, but they only travel back to the markets after a big catch. They typically spend the entire season on the atoll, sending fish back to the market in Belize City as needed. Most of the commercial fish camps are on the central lagoon in deeper water.

Early recreational fishermen came to the atoll and camped in pursuit of bonefish, permit, and tarpon. Later, several mothership operations formed out of Belize City that allowed anglers to fish Turneffe and the surrounding cays while staying on a larger ship. Today there are three fishing lodges that also have dive operations on the east side of the atoll, as well as a few mothership operations from surrounding areas.

The majority of anglers are fly fishermen who enjoy the challenge of multiple species in one location. Turneffe is one of the few places in the world that offers the opportunity to catch bonefish, tarpon, and permit in the same place. The eastern flats located between the cays and the reef system provide excellent habitat for wading for bonefish. The flats are generally composed of hard bottom coral rubble, sand, or seagrass. Bonefish average around 1.35 kg and appear to be a resident population. The bonefish are abundant (>3000 km^2), but finicky.

The flats, channels, and deeper lagoons hold permit year-round. While little research has been done on either bonefish or permit at Turneffe, permit are believed to be a resident population as well. Tarpon move into the atoll in the early summer and usually stay for around 3 months. Summers are an angler's best chance to catch all three species. There are a few tarpon that will stay year-round, but the majority of fish migrate in during the summer months.

THE OPPORTUNITY

Belize is one of the world's most biologically diverse nations with the integrity of its natural resources still very much intact. It boasts the largest barrier coral reef and three of the four coral atolls in the western hemisphere. Belize has been a leader in environmental protection with nearly one-third of the country under national park authority or some other protected status. Belize is home to 14 marine-protected areas (MPAs), of which 8 are marine reserves, 2 are natural monuments, 1 is a national park, and 2 are wildlife sanctuaries (CZMAI, 2001). Turneffe Atoll, the largest and most diverse atoll, has one small reserve put in place to protect the American alligator nesting sites of Cockroach Cay.

Turneffe is one of the few remaining environments that experiences very few user conflicts. The government of Belize has a tremendous opportunity to take advantage of a progressive group of commercial fishermen working with recreational users to protect this very diverse, pristine environment. Turneffe needs greater resource protection for its endangered resident manatee population, the alligator nesting sites, untouched mangroves and the economies this ecosystem supports, commercial, and recreational (Jacobs, 1998). Belize depends on tourism and commercial fishing for roughly 20% of its economy (World Tourism and Travel Council, 2005). Turneffe Atoll supports both of these endeavors in the form of commercial fishing for conch and lobster, as well as recreational pursuits such as fishing and diving.

The institutions in place to provide this protected status include necessary legislation, active NGO and volunteer network, and committed stakeholders. Belize already has the framework and precedent for setting up MPAs. The country has several local case studies on the effectiveness of providing protection while still providing access.

There are many NGOs actively working on coastal issues in Belize. They range from world-wide organizations like World Wildlife Fund to local, grassroots organizations. Nearly all of the activities are funneled through a committee of private and public sector stakeholders known as the Turneffe Island Coastal Advisory Committee.

The Turneffe Island's Coastal Advisory Committee was formed to link the tourism operators on Turneffe with the commercial fishing cooperatives, the Government of Belize, and the University of Belize. The goal of the committee is to institute a conservation plan resulting in sustainable tourism and commercial fishing at Turneffe. The current members represent all of the stakeholder groups on the island.

The Turneffe Atoll ecosystem and its associated fisheries appear to be stable at the present time. However, as Belize grows alongside its burgeoning tourism industry, access to the atoll will become easier. Plans must now be put in place to protect this marine gem, as well as preserve the vital economies that are supported by the natural resources of the atoll. To effectively manage this fishery, baseline data are needed on the current bonefish, tarpon, and permit populations. In addition, research should also be focused on assessment of the potential impacts of future development, to establish the angler carrying capacity, to determine life history and movement patterns, and to identify spawning grounds of these economically and ecologically important fish populations. A plan for the future will ensure that this unique and world-class fishery remains viable for generations to come.

ACKNOWLEDGMENTS

I would like to thank Craig and Karen Hayes of Turneffe Flats for making this research possible.

REFERENCES

CZMAI (Coastal Zone Management Authority and Institute). 2001. *State of the Coast Report, Belize*. Belize City, Belize.

Jacobs, N.D. 1998. Assessment and analysis of the fisheries sector and marine coastal areas. Belize National Biodiversity Strategy and Action Plan (UNDP/GEF Project No. BZE/97/G31).

World Tourism and Travel Council. 2005. *The 2005 Travel and Tourism Economic Research, Belize*. London.

8 Aspects of the Biology and Recreational Fishery of Bonefish (*Albula vulpes*) from Los Roques Archipelago National Park, Venezuela

Juan M. Posada, Denise Debrot, and Constanza Weinberger

CONTENTS

Introduction 103
Reproductive Biology 105
Feeding Habits 108
Age and Growth 111
The Fishery 111
    Pre-Hispanic and Artisanal Fishery 111
    Recreational Fishery 112
Conclusions 113
Acknowledgments 113
References 113

INTRODUCTION

Bonefish, *Albula vulpes*, is a common species in coastal marine environments along the north coast of Venezuela (southern Caribbean Sea), particularly in Los Roques Archipelago National Park (Cervigón, 1991), where large schools inhabit the clear waters of this marine-protected area.

Los Roques Archipelago (LRA) is an insular reef platform located 157 km north off the central coast of Venezuela (Figure 8.1), encompassing an area of 1250 km^2, with a maximum depth of 50 m. The archipelago is composed of 42 islands and 200 sand banks, distributed in an irregular oval shape around an inner lagoon with an average depth of 5 m. In 1972, the LRA was declared a national park to

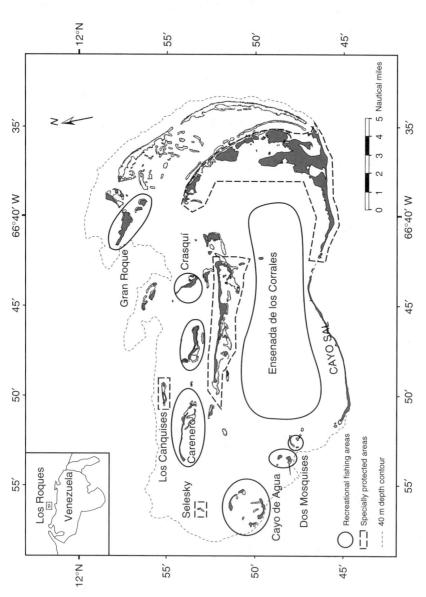

FIGURE 8.1 Map of Los Roques Archipelago National Park, showing the study area (Dos Mosquises Island), and the location of the main flats for bonefish recreational fishing and the areas under special protection ("no-take" zones). (From Debrot, D. and Posada, J.M., *Cont. Mar. Sci.*, 37, 60–64, 2005. With permission.)

protect a range of highly diverse marine habitats dominated by coral reefs, seagrass meadows, sand beaches, salt and brackish lagoons, and mangrove forests (Schweizer et al., 2005). Since 1991, LRA has implemented a resource management plan in which seven management zones were outlined, including a "no-take" marine-protected area closed to fishing and human visitation (Figure 8.1).

Los Roques has a permanent human population of more than 1200 residents, mostly settled on Gran Roque, the archipelago's main island. Tourism is the most important economic activity, and more than 50,000 tourists visit LRA every year, providing direct employment to 40% of the residents. An artisanal lobster fishery is second most important economic activity in LRA, accounting for more than US$300,000 annually (Méndez, 2002).

Over the past decade, LRA has become one of the most popular recreational fishing destinations in the Caribbean for bonefish, receiving an annual average of 400 anglers. This fishery provides an important income to the local economy, and a lucrative alternative to traditional artisanal fisheries, such as lobster and other commercially valuable fish (Debrot and Posada, 2005). Bonefish are not valued for their meat in LRA, and most of the fish caught by recreational anglers are released. However, bonefish are occasionally caught with pocket nets in shallow waters by local artisanal fishermen and used as bait to fish for commercially valuable species (Debrot and Posada, 2005). Currently, there are no regulations that guarantee a sustainable bonefish fishery in this protected ecosystem; in fact, there are no limits on the number or minimum size of bonefish that are captured and not released.

Despite the economic importance of recreational fisheries for bonefish in the wider Caribbean Sea, studies on its biology and population dynamics throughout the region have been limited. South Florida, the Bahamas, and Los Roques are the three areas that have received most of the scientific effort. Significant gaps in knowledge on aspects such as early life history, recruitment, population, and fishery dynamics exist. In this chapter, we attempt to summarize the information on bonefish reproductive biology, feeding habits, and recreational fishery, as described in our previous studies carried out in LRA, as well as preliminary results on its longevity and growth. The information provided in this chapter will contribute to the scientific understanding of bonefish biology and population dynamics in the Caribbean region, which is critical for the development of a sustainable recreational fishery for bonefish in LRA.

REPRODUCTIVE BIOLOGY

To examine length at sexual maturity and reproductive seasonality, Debrot and Posada (2003) collected adult bonefish from waters surrounding the island of Dos Mosquises, in the southwestern portion of LRA (Figure 8.1). Based on macroscopic and microscopic examinations of 440 gonads from adult bonefish ranging in size from 286 to 717 mm fork length (FL), they found that the smallest sexually mature female and male in this sample were 351 and 424 mm FL, respectively. The length at 50% sexual maturity (M_{50}), estimated as the inflection point of a fitted logistic curve, was 456 mm FL for females (95% confidence interval (CI) was 446–466 mm) and 467 mm FL for males (95% CI was 454–479 mm) (Figure 8.2).

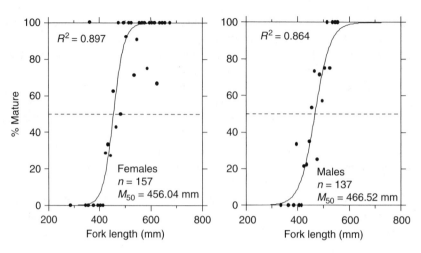

FIGURE 8.2 Length–maturity relationship for bonefish (*A. vulpes*) in 10-mm size classes sampled in Los Roques Archipelago National Park. The line is the fitted logistic equation: % mature = $100/(1 + \exp(-a\,(FL - b)))$. Parameter b in the equation is the inflection point and represents the length at which 50% of the individuals have mature gonads (M_{50}). All specimens collected in the postreproductive months (February–April) were excluded from this analysis. (From Debrot, D. and Posada, J.M., *Proc. Gulf. Carib. Fish. Inst.*, 54, 506–512, 2003. With permission.)

Bonefish from LRA reached sexual maturity at a larger size than reported by Bruger (1974) for the Florida Keys. Bruger reported mature females as young as 1 year old, ranging from 221 to 352 mm FL. His sample size was small ($n = 148$), and most of his small sexually mature females were caught in deep water (9.1–12.2 m). Previous observations (Bruger, 1974; Crabtree et al., 1996) and recent genetic studies (Colborn et al., 2001) have suggested the possible existence of a cryptic bonefish species that inhabits deeper waters of the western Atlantic (Florida and Brazil) as a potential explanation for the presence of exceptionally small and sexually mature bonefish. Crabtree et al. (1997), however, reported minimum lengths at sexual maturity for the Florida Keys similar to those found in LRA. Crabtree et al.'s (1997) study had a relatively large sample size ($n = 437$) and found that the smallest mature female was 358 mm FL and 425 mm FL for males. In contrast, the estimates of M_{50} in LRA differ considerably from those in south Florida. In LRA, females reach M_{50} at a smaller length than males (Debrot and Posada, 2003), while the opposite was reported for the Florida Keys, where females reach M_{50} at 488 mm FL and males at 418 mm FL (Crabtree et al., 1997). These differences could have resulted from the apparent differences between the length distributions of adult females and males from LRA and south Florida. In LRA length–frequency distributions of males and females were significantly different (Kolmogorov–Smirnov two-sample test; $D = 0.0410$; $p < 0.001$), and females were significantly larger than males (Mann Whitney U-test; $p < 0.001$). Females ranged in size from 286 to 717 mm FL (mean = 492.2 mm; SD = 65.36; $n = 255$) and males from 334 to 600 mm FL

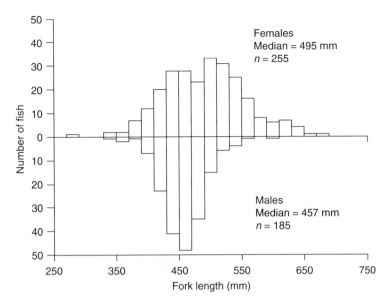

FIGURE 8.3 Length–frequency distributions for bonefish (*A. vulpes*) from Los Roques Archipelago National Park. Lengths are plotted in 20-mm size classes. (From Debrot, D. and Posada, J.M., *Proc. Gulf. Carib. Fish. Inst.*, 54, 506–512, 2003. With permission.)

(mean = 459.36; SD = 35.15; n = 185) (Debrot and Posada, 2003; Figure 8.3). Females in south Florida ranged from 228 to 702 mm FL and males from 322 to 687 mm FL (Crabtree et al., 1997).

Gonadal activity, assessed by monthly mean gonadosamatic index (GSI) of bonefish in LRA, showed clear seasonality, with development occurring from June to January for females, and from May to January for males (Debrot and Posada, 2003) (Figure 8.4). The reproductive seasonality observed in Los Roques is consistent with results reported for bonefish in the Bahamas by Mojica et al. (1995) and for south Florida by Crabtree et al. (1997). Based on back-calculated ages and spawning dates of field-collected larvae in tidal channels around Lee Stoking Island (Bahamas), Mojica et al. (1995) suggested that bonefish in the Bahamas might spawn continuously from mid-October through January, probably extending through May. Crabtree et al. (1997) found that gonadal development of bonefish from south Florida also occurs over a period of 8 months, from November to June, and bonefish are reproductively inactive during a few months in the summer.

The location of bonefish spawning grounds in LRA remains unknown. The absence of females with evidence of eminent spawning, such as fully hydrated oocytes and postovulatory follicles in bonefish caught in shallow waters, suggests that bonefish may migrate to deeper waters in the archipelago to spawn (Debrot and Posada, 2003), as suggested by Crabtree et al. (1997) for bonefish in the Florida Keys. This behavior has been observed in *A. glossodonta* in the Pacific Ocean, where large periodic spawning migrations have been documented for this species (Johannes and Yeeting, 2001).

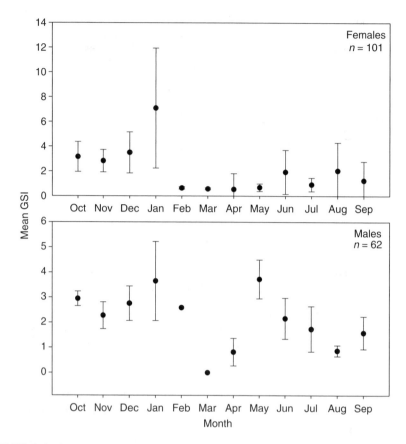

FIGURE 8.4 Seasonal variation in the gonadosomatic index (GSI) for female and male bonefish (*A. vulpes*) larger than M_{50}, collected from October 1999 to September 2000 in Los Roques Archipelago National Park. Values represent mean ± standard deviation.

FEEDING HABITS

Typical for many other areas of the species range, bonefish in Los Roques forage in shallow waters over seagrass or sandy bottoms, mostly on small benthic and epibenthic prey. Weinberger and Posada (2005) examined the stomach contents of 187 bonefish that ranged in size from 336 to 644 mm FL from Dos Mosquises Island in the southwestern portion of LRA (Figure 8.1). They determined that those bonefish feed mainly on crustaceans, teleosts, bivalves, polychaetes, and gastropods, with decapods and cupleiformes forming the most significant portion of the diet (Table 8.1). Among decapods crabs of the subfamily Mithracinae (Majidae) were the most important prey and among the cupleiformes (Table 8.1) *Anchoa* sp. and *Harengula humeralis*.

Similar results were observed for south Florida bonefish (Crabtree et al., 1998), where decapods dominated the diet ($N = 42.1\%$, $F = 88.6\%$, and $W = 67.8\%$). In contrast to LRA, they reported a higher percentage of relative frequency of abundance for mollusks ($N = 51\%$) than for the teleosts ($N = 45\%$). Also, the gulf toadfish, *Opsanus*

TABLE 8.1
Prey Items Found in Stomachs of 187 Bonefish (*A. vulpes*) Caught in Los Roques Archipelago National Park

Prey Items	N	W	F
Plant material	—	7.1	40.4
Unidentified plant material	—	6.0	19.1
Syringodium sp.	—	0.1	11.8
Halodule sp.	—	0	1.5
Thalassia testudinum	—	0.9	8.1
Miscellaneous material	2.9	9.6	17.6
Phylum Chordata			
Class Osteichthyes	19.0	22.2	32.4
Unidentified Osteichtyes	4.6	4.3	6.6
Order Clupeiforme	34.4	31.0	90.4
Engraulidae			
Anchoa sp.	9.2	10.3	15.4
Clupeidae			
Harengula humeralis	4.4	6.9	9.6
Order Anguilliforme	0.7	0.7	0.7
Ophichthidae			
Unidentified Ophichthinae	0.7	0.7	0.7
Subphylum Crustacea			
Class Malacostraca	34.4	30.1	90.4
Order Decapoda	34.4	30.1	90.4
Unidentified Decapoda	0.9	0.3	4.4
Unidentified Brachyura	9.5	9.4	19.1
Xanthidae			
Panopeus sp.	0.2	0.2	0.7
Eurypanopeus abbreviatus	0	0	0.7
Portunidae			
Unidentified Portuninae	2.7	2.1	12.5
Majidae			
Unidentified Mithracinae	12.7	13.7	22.1
Mithrax forceps	0.9	0.3	3.7
Unidentified Anomura	3.5	2.0	12.5
Galatheidae			
Munida sp.	0.1	0.1	1.5
Diogenidae			
Pagurites sp.	1.1	0.6	2.2
Unidentified Penalidae	0.1	0	0.7
Unidentified Penaeidae	2.8	2.2	10.3
Phylum Mollusca			
Unidentified Mollusca	—	0.9	7.4
Class Gastropoda	6.3	2.8	35.3
Unidentified Gastropoda	1.7	1.3	11.0
Order Caenogastropoda	4.3	1.4	22.1
Marginellidae			
Persicula interruptolineata	0.6	0.1	5.9
Cerithidae			
Cerithium litteratum	1.8	1.1	2.9

(Continued)

TABLE 8.1 (continued)
Prey Items Found in Stomachs of 187 Bonefish (*A. vulpes*) Caught in Los Roques Archipelago National Park

Prey Items	N	W	F
Columbellidae			
Unidentified Columbellidae	0.4	0.1	2.2
Cosnioconcha nitens	0.7	0.1	3.7
Columbella mercatoria	0	0	0.7
Olividae			
Oliva australis	0.8	0.1	6.6
Order Vetigastropoda	0.3	0	2.2
Phasianellidae			
Tricolia tessellata	0.1	0	1.5
Naticidae			
Sigatica sp.	0.1	0	0.7
Class Bivalvia	11.3	14.7	44.1
Unidentified Bivalvia	1.6	4.7	16.2
Order Pteroida	0.1	0.1	1.5
Unidentified Pectinidae	0.1	0.1	1.5
Order Veneroida	9.0	9.6	24.3
Lucinidae			
Lucina sp.	0.3	0	0.7
Codakia orbiculatus	7.8	9.0	15.4
Veneridae			
Unidentified Veneridae	0.6	0.1	2.9
Chione sp.	0.2	0.4	1.5
Chione cancellata	0.2	0.1	3.7
Order Arcoida	0.5	0	1.5
Arcidae			
Unidentified Arcidae	0.5	0	1.5
Order Mytiloida	0.1	0.2	0.7
Mytilidae			
Brachidontes sp.	0.1	0.2	0.7
Phylum Annelida			
Class Polychaeta	13.4	10.7	22.8
Unidentified Polychaeta	3.2	0.1	6.6
Unidentified Maldanidae	0.1	0.1	0.7
Unidentified Pectinaridae	1.6	2.2	2.9
Unidentified Oenonidae	0.6	0	2.2
Capitellidae			
Notomastus sp.	7.9	8.0	10.3
Phylum Sipuncula			
Class Sipunculidea	1.0	1.1	2.2
Order Aspidosiphoniforme	1.0	1.1	2.2
Unidentified Aspidosiphonidae	1.0	1.1	2.2

Note: N = percent of numerical abundance; W = percent weight; F = percent frequency of occurrence.

Source: From Weinberger, C. and Posada, J., *Cont. Mar. Sci.*, 37, 30–44, 2005. With permission.

beta (Batrachoidiformes), was the most important teleost in the diet of south Florida bonefish, while it was missing from the stomach contents of bonefish in LRA. However, teleosts of the order Cupleiformes were not represented as an important prey item in south Florida bonefish.

Similarity analysis (ANOSIM), based on the percentage of relative frequency of wet weight (W) of prey items, showed no significant differences in feeding habits between males and females ($R = 0.001; p = 0.406$), or for any of the four 80 mm FL size intervals tested ($R = 0.016; p = 0.135$) (Weinberger and Posada, 2005). In contrast, Crabtree et al. (1988) showed a positive correlation between prey size and fish length for south Florida bonefish and noted that large individuals consumed more decapods and teleosts than bonefish < 440 mm FL.

However, differences in prey items consumption were significant on a seasonal basis in LRA: dry season (November–February vs. rainy season (March–July) ($R = 0.029; p = 0.029$). Bonefish preyed mostly on decapods and teleosts during the dry season when the water temperatures are lower, and more on gastropods during the rainy season when the water is warmer (Weinberger and Posada, 2005). A similar pattern of seasonality in the diet of bonefish was also observed in south Florida by Crabtree et al. (1998). In both cases, these differences may reflect temporal availability of prey.

AGE AND GROWTH

Preliminary analysis of sectioned otoliths (sagittae) from bonefish sampled in shallow waters of LRA showed regular distributions of presumed annuli (opaque bands visible under dissecting microscope and reflected light), only 9% of the otoliths did not displayed legible growth bands. For the 91 bonefish used in the analysis, lengths ranged from 138 mm to 650 mm of FL and ages from 1 to 17 years. An average growth curve was generated from size to age data, and lifetime growth trajectories were estimated by fitting the von Bertalanffy growth function (VBGF). In this sample the VBGF for bonefish (both sexes combined) was

$$L(t) = 517.461(1 - e^{-0.344(t+1.076)})$$

(Debrot et al., unpublished data) (Figure 8.5). These preliminary results indicate that LRA bonefish growth seems rapid until about 5 years, and that asymptotic length is reached at approximately 8 years of age (Figure 8.5). Bonefish from LRA reach a similar maximum age to that of south Florida bonefish ($\geq$20 years, Crabtree et al., 1996). However, south Florida bonefish appear to grow faster and attain larger sizes (Ault et al., Chapter 16, this volume) than in LRA.

THE FISHERY

PRE-HISPANIC AND ARTISANAL FISHERY

Fish remains recovered during systematic archaeological excavations in LRA suggest that bonefish were exploited by Ameridians in the late pre-Hispanic times (AD 1300–1500) (Antczak, 1999). Although the queen conch (*Strombus giigas*) was by far

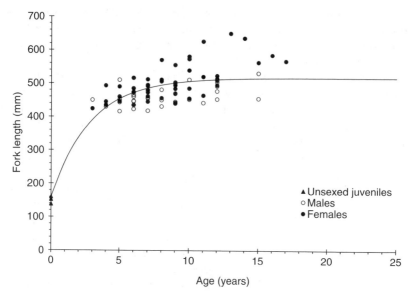

FIGURE 8.5 Size at age plot with fitted von Bertalanffy growth function for male and female bonefish (*A. vulpes*) from Los Roques Archipelago National Park ($n = 91$). (From Debrot et al., unpublished data.)

the most exploited resource by the Ameridians who occupied LRA, the abundance of fish remains found on these archeological sites, strongly suggests that bonefish were also an important fishery resource at that time. In fact, bonefish's otoliths were the second most abundant fish remain found in LRA (Antczak, 1999).

By the 1950s, when artisanal fishers arrived to LRA from other regions of Venezuela, bonefish were also caught in shallow waters with seine nets and used as bait in longlines to fish for sharks and rays. The use of seine nets was banned in the 1980s and since then, bonefish are only caught occasionally by artisanal fishers using small pocket nets.

Recreational Fishery

While recognized as a world-class destination, the LRA recreational fishery is still in development compared to other bonefish fisheries in the Caribbean and Florida Keys. The guided recreational fishing industry of LRA consists of only a few companies that operate from Gran Roque (Debrot and Posada, 2005). By 2005, there were 7 fishing companies and 12 local fishing guides. These companies commonly offer guided fishing tours on 20–30 ft fiberglass boats and accommodations in small lodges in Gran Roque.

Fishing for bonefish in LRA is predominantly catch-and-release and it is a year-round activity; nevertheless, the number of anglers peak from January to June (Debrot and Posada, 2005). Bonefish are frequently caught, using either the spinning or fly-fishing technique, on many of the archipelago's flats, either over sandy or seagrass bottoms. The main flats are located at the center portion of the archipelago, in the Ensanada de Los Corrales, and in the islands located in the southwest. The beaches surrounding Gran Roque also are very popular for fishing solitary bonefish (Figure 8.1). According to the fishing guides, catch rates of 10 fish per day are not unusual, and most of the bonefish caught by the recreational fishery weigh between 1 and 2 kg (Debrot and Posada, 2005). The recreational fishing pressure in LRA is relatively low compared to other bonefish fisheries. However, local fishing guides in LRA are concerned by the increasing fishing pressure, and believe that rotating the fishing areas could prevent the schools from abandoning their usual feeding grounds to move to more remote flats in the no-take area, where no fishing is allowed.

CONCLUSIONS

The protected marine ecosystems of Los Roques Archipelago National Park and its limited recreational fishery offers a unique opportunity to study the biology, ecology, and fishery dynamics of bonefish populations in a relatively unspoiled condition. Further scientific research in LRA should be focused on filling the gaps on critical aspects such as recruitment, movement and spawning behavior, as well as population size and mortality. Monitoring the catches and fishing effort of the bonefish recreational fishery in LRA should figure as a priority for the park's managers, who along with fishing guides and scientists should work together in the development of a management program that guarantees a sustainable catch-and-release bonefish fishery in this protected environment.

ACKNOWLEDGMENTS

We thank Fundación Científica Los Roques for providing logistic facilities at its biological station in Los Roques and Instituto Nacional de Parques (INPARQUES) for issuing the research permits. We also thank Dr. Jerald Ault of the University of Miami RSMAS, Dr. Luiz Barbieri, and the Age and Growth Laboratory at the Florida Fish and Wildlife Research Institute for assistance with the preparation and reading of otoliths. Field and laboratory assistance were provided by M.F. Larkin, P.S. Mata, J.M. Martinez, and C. Pombo. Funding to D. Debrot was provided by Decanato de Estudios de Postgrado de la Universidad Simón Bolívar and the University of Miami RSMAS through Bonefish & Tarpon Unlimited. Two anonymous reviewers made helpful comments that improved the manuscript.

REFERENCES

Antczak, A. 1999. Late prehistoric economy and society of the islands off the coast of Venezula: a contextual interpretation of the non-ceramic evidence. Ph.D. Thesis, Institute of Archaeology, University College London.

Bruger, G.E. 1974. Age, growth, food habits and reproduction of bonefish, *Albula vulpes*, in South Florida waters. Florida Marine Research Publications 3, St. Petersburg.

Cervigón, F. 1991. Los peces marinos de Venezuela. Vol. I, 2da ed. Fundación Científica Los Roques, Caracas, 24.

Colborn, J, R.E. Crabtree, J.B. Shaklee, E. Pfeiler and B.W. Bowen. 2001. The evolutionary enigma of bonefish (*Albula* spp.): Cryptic species and ancient separations in a globally distributed shorefish. *Evolution* 55 (4): 807–820.

Crabtree, R.E., D. Snodgrass, C.H. Harnden and C. Stevens. 1996. Age, growth and mortality of bonefish, *Albula vulpes*, from the waters of Florida Keys. *Fish. Bull.* 94: 442–451.

Crabtree, R.E., D. Snodgrass and C.H. Harnden. 1997. Maturation and reproductive seasonality in bonefish, *Albula vulpes*, from the waters of the Florida Keys. *Fish. Bull.* 95: 456–465.

Crabtree, R.E., D. Snodgrass, C. Stevens and F. J. Stengard. 1998. Feedings habits of bonefish, *Albula vulpes*, from the Florida Keys. *Fish. Bull.* 96: 754–766.

Debrot, D. and J.M. Posada. 2003. Reproductive biology of the bonefish, *Albula vulpes*, at Los Roques Archipelago National Park, Venezuela. *Proc. Gulf. Carib. Fish. Inst.* 54: 506–512.

Debrot, D. and J.M. Posada. 2005. A brief description of bonefish recreational fishery in Los Roques Archipelago National Park, Venezuela, in *Tarpon and other fishes of the western Gulf of Mexico. Proceedings from the Third International Tarpon Forum Cont. Mar. Sci.* 37: 60–64.

Johannes, R.E and B. Yeeting. 2001. Kiribati knowledge and management of Tarawa's lagoon resources. *Atoll Res. Bull.* 489: 1–25.

Méndez, M.G. 2002. La aventura de poblar Los Roques: El Siglo XX, in *Guía del Parque Nacional Archipélago Los Roques,* Zamarro, J., ed. Agencia Española de Cooperación. International y Ministerio del Ambiente y de los Recursos Naturales. Caracas, Venezuela, pp. 39–54.

Mojica, R., Jr., J.M. Shenker, C.W. Harnden and D.E. Wagner. 1995. Recruitment of bonefish, *Albula vulpes*, around Lee Stocking Island, Bahamas. *Fish. Bull.* 93: 666–674.

Schweizer, D., R.A. Armstrong and J. Posada. 2005. Remote sensing characterization of benthic habitats and submerged vegetation biomass in Los Roques Archipelago National Park, Venezuela. *Int. J. Remote Sensing* 26: 2657–2667.

Weinberger, C. and J.M. Posada. 2005. Analysis on the diet of bonefish, *Albula vulpes*, in Los Roques Archipelago National Park, Venezuela, in *Tarpon and other fishes of the western Gulf of Mexico. Proceedings from the Third International Tarpon Forum Cont. Mar. Sci.* 37: 30–44.

9 The Nigerian Tarpon: Resource Ecology and Fishery

Patricia E. Anyanwu and Kola Kusemiju

CONTENTS

Introduction 115
Historical Background on the African Tarpon Fishery 116
Methods 117
Results 118
    Resource Ecology 118
        Study Area 118
        Physicochemical Parameters of Study Area 118
        Tarpon Distribution 118
        Morphology and Meristic Characters of *M. atlanticus* 121
    Fishery for *M. atlanticus* in Nigeria 122
        Capture Fishery 122
        Culture Fishery 125
        Ecotourism Attractions 126
Discussion 126
Acknowledgments 128
References 128

INTRODUCTION

Atlantic tarpon *Megalops atlanticus* in the eastern Atlantic coast of Africa ranges from Mauritania to Angola, concentrating around the Gulf of Guinea (Figure 9.1). In the western Atlantic, the species occurs from Nova Scotia to Brazil with centers of high abundance in the warm coastal waters of Florida, Gulf of Mexico, and the West Indies (Irvine, 1947; Fischer et al., 1981; Whitehead et al., 1984). *Megalops atlanticus* is of great commercial importance with a special fishery in the central and southwestern Atlantic Ocean where it is also a very important game fish (Stamatopoulus, 1993; Crabtree et al., 1997; and Zerbi, 1999). In the eastern Atlantic, especially in the coastal waters of western Nigeria, *M. atlanticus* is an important aquatic resource and a delicacy food served on special occasions like marriage ceremonies or festivals. Fishing villages in the coastal areas of southwest

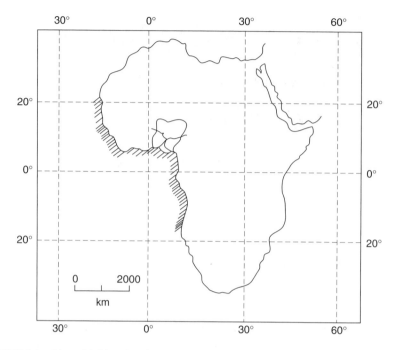

FIGURE 9.1 Map of Africa showing the range of tarpon.

Nigeria have been identified to be involved in a booming tarpon-fingerling trade, and served as supply centers for tarpon juveniles to many parts of the country (Ezenwa et al., 1985). Owing to the high demand and exploitation of tarpon fry, there appears to be a relatively high fishing pressure on tarpon population in the coastal waters of western Nigeria.

Unfortunately, coastal wetlands and swamps in Nigeria are frequently reclaimed and modified for the establishment of housing estates, industries, oil pipelines, or for other purposes. These habitat modifications, according to Zerbi (1999), affect survival and recruitment of many estuarine-dependent species, including *M. atlanticus*. The destruction of these critical nursery habitats may deleteriously affect recruitment into the mature phase of the population, which could lead to disappearance of the species.

HISTORICAL BACKGROUND ON THE AFRICAN TARPON FISHERY

The "silver king" known as the tarpon *M. atlanticus* is called the greatest of the game fishes (IGFA, 1987). However, there is a dearth of historical data on the tarpon fishery in Africa. Early records, as reported by Irvine (1947), indicated that Governor Hosdon regularly fished for tarpon during the months of June and July at Ada, the junction of the mouth of River Volta and the sea. Some of the reported specimens landed measured as large as 188, 122, and 107 cm fork length (FL). In the estuary of Kouilou in Ponte Noire, Irvine (1947) reported that the administrator of the port caught a tarpon weighing 101 kg and measuring 230 cm FL using a fly on nylon 80/100 (mesh size).

IGFA (1987) reported that a tarpon weighing 102 kg caught at Port Michel, Gabon on December 22, 1985 took first place in the 11th Annual IGFA Fishing Contest. Other notable tarpon caught that year at Port Michel weighed 112.4, 112.6, 102, 100.3, 99.1, and 87.5 kg.

In Nigeria, there is also a dearth of historical data on tarpon fishery. There is a tarpon club in the Ikoyi area of Lagos State, but it was established mainly for sailing and occasionally for sportfishing. Club records indicate that a 21-kg tarpon was caught in March 1988 along the West Mole of Lagos Harbor. In the coastal areas of Ondo State, the "Ilaje" people, who are renowned fishermen, reported that their grandfathers caught tarpon as far back as 1908 and that specimens measuring over 180 cm FL and weighing more than 75 kg have been reported landed. Unfortunately, the "Ilajes" do not hold organized tarpon fishing tournaments, but rather celebrate fishing festivals during the period of Christmas. During fishing festivals, all the fish species caught, including tarpon, were cooked and consumed for entertainment by the natives and visitors. The fishermen reported that large numbers of tarpon ranging from 15 to 60.8 kg were caught during the festivals, which coincided with peak abundance of adult tarpon during spawning season.

Presently, fishing festivals are not organized on a large scale and fewer tarpon are caught. This change in fishery dynamics has been attributed to the high cost of fishing; hence, most tarpon fishermen have switched over to crayfish (lobster) fishing and additional means of livelihood other than fishing. Provision of subsidies for fishing inputs (i.e., nets, boats, and outboard engines) will provide gainful employment for these fishermen.

METHODS

Field sampling trips to the coastal areas of Lagos and Ondo States were undertaken monthly from January 1996 to December 1997. Fish landing sites, coastal fishing villages, and shore frontiers along the Atlantic Ocean were surveyed, particularly for the nature of the shoreline, vegetation, and estuarine systems. Water samples were collected monthly from the sampling stations. Physicochemical parameters including air and surface water temperatures, pH, dissolved oxygen, salinity, and water transparency were determined by methods reported by Boyd (1979) and Ugwumba (1984).

Information on the distribution of *M. atlanticus*, based on their abundance, size, and maturity stages, was obtained from the relative occurrence of the species caught and landed throughout the year. Fishermen specializing in the capture and culture of *M. atlanticus* in the study areas were interviewed. Fishing methods for *M. atlanticus* were studied and catches by fishermen were recorded. Traditional culture methods of the species were identified. Principal morphometric measurements taken included total length, standard length, head length, body depth, eye diameter, and head depth (Fischer et al., 1981). Other measurements used were fork length, preorbital length, snout-to-dorsal fin origin, and snout-to-anal fin origin. The meristic characters were obtained by counting the number of dorsal, pelvic pectoral, and anal fin rays, as well as scales along lateral line, transverse rows of scale, branchiostegal rays on the left opercular bone, and gill rakers (Fischer et al., 1981).

RESULTS

RESOURCE ECOLOGY

Study Area

The study area was located between 3°10' and 4°52' E longitude and 6°02' and 6°28' N latitude (Figure 9.2). This part of the Nigerian coastline in Lagos State is characterized by erosive sandy beaches (Ibe et al., 1985; Awosika et al., 1994). The Ondo State coastal area is characterized by low-lying muddy flat beaches with gentle slopes and absence of sand (Figure 9.3). There was great tidal influence as the low gradient of the intertidal zone exposed wide expanses of land of over 1 km to tidal influence. This coastal ocean tide constituted an environmental force responsible for the movement of various stages of *M. atlanticus* from the open Atlantic Ocean to the shoreline and into tidal creeks. The presence of Avon and Mahin canyons offshore near the muddy beaches of the Ondo State coastal area may also be responsible for channeling sand brought into the area away from the coast into the deep waters (Ibe et al., 1985; Awosika et al., 1994). The Lagos area, however, has sandy beaches with steep slopes and is dominated by strong wave action (Ibe et al., 1985; Awosika et al., 1994).

The regional climate is tropical with two main seasons: (1) a rainy season lasting from April to October and (2) a dry season from November to March. Mean monthly rainfall ranged from 0.0 to 599.5 mm. The heaviest monthly rainfall of 599.5 mm was recorded in June 1996. High rainfall was coincidental with the migration pattern of adult *M. atlanticus*, as well as the abundance and seasonality of juvenile tarpon. The vegetation in this region is composed mainly of mangrove trees (*Rhizopora* sp.), coconut palms (*Cocos nucifera*), and oil palms (*Elaeis guineensis*), as well as sedges, herbaceous plants, and climbers. Mangrove trees dominated the coastal swamps in Ondo State and coconut palms in Lagos State.

Physicochemical Parameters of Study Area

Air temperatures ranged between 26.3 and 31.6°C, while surface water temperature ranged from 26.0 to 33.5°C. Hydrogen ion concentration (pH) ranged between 6.1 and 8.9, while dissolved oxygen content ranged from 2.4 to 6.9 mg/L. Salinity ranged from 0.2 to 32.1 ppt. Higher salinities ranging from 25.3 to 32.1 ppt were recorded for the beaches throughout the study period. In contrast, salinity of the creeks and lagoons fluctuated between 0.2 and 28.0 ppt, indicating a range of both fresh and brackish water conditions. Rainfall affected salinity levels in the creeks and lagoons with lower salinities in the rainy season. Water transparency varied from 20.0 to 65.6 cm. Higher water turbidity (20–30 cm maximum transparency) occurred mainly during the rainy season from July to October.

Tarpon Distribution

Megalops atlanticus were regularly encountered in the coastal waters of the study area from Ilepete in Ondo State to Badagry in Lagos State. The different life stages obtained were floating fertilized eggs, fry, fingerlings, subadults, and adults.

The Nigerian Tarpon: Resource Ecology and Fishery

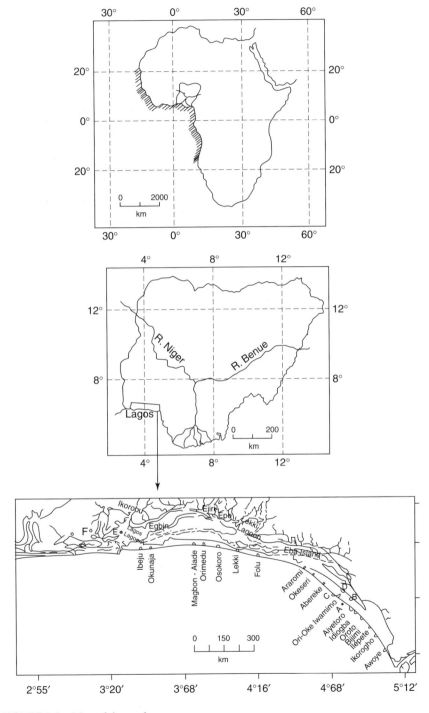

FIGURE 9.2 Map of the study area.

FIGURE 9.3 Flat muddy beach of Aiyetoro coast in Ondo State.

FIGURE 9.4 Mass of floating fertilized eggs of *M. atlanticus*.

Fertilized eggs (Figure 9.4) hatched into tarpon fry when stocked in nursery ponds. Leptocephalus larval stages were not encountered, probably due to the selectivity of the sampling equipment. The mesh size of the lace net may have been too wide for capture of tarpon larvae.

Adult *M. atlanticus* ranging from 20 to 35 kg body weight were landed mainly by fishermen in Ondo State coastal areas rather than fishermen in Lagos State. The adults were found mainly at sea off the Atlantic coast at depths ranging from 5 to 39 m, and were obtained throughout the year, although peak abundance occurred from June to July and November to December. The species are pelagic and hence their absence in the catches of Nigerian fishing companies, who primarily use bottom trawls. The fry, fingerlings, and subadults, which moved in shoals, were obtained mainly in the brackish water environment, especially in creeks. Unlike the adults, subadults were not obtained throughout the year. Fertilized egg masses were found attached to floating vegetation and could be obtained at sea, as well as along the shorelines of the study area. The seasonal abundance and habitat of the different stages of *M. atlanticus* are presented in Table 9.1.

Morphology and Meristic Characters of *M. atlanticus*

The body of *M. atlanticus* is long, herringlike, compressed, and covered with thick large cycloid scales (Figure 9.5). The flank is silvery, while the back is gray-black. The mouth is superior and the lower jaw protrudes due to the presence of a bony gular

TABLE 9.1
Seasonal Abundance and Habitat of the Different Life Stages of Atlantic Tarpon (*M. atlanticus*) in Nigeria

Life Stage	Location	Availability	Peak Season
Eggs	Open sea and shorelines	October–March	December–January
Fry	Tidal pools, creeks, and canals	November–May	February–April
Fingerlings	Creeks, canals, lagoons	February–August	April–July
Subadults	Creeks, canals, and lagoons	June–September	August–September
Mature adults	Open sea (inshore waters)	January–December	June–July and November–December

FIGURE 9.5 External features of *M. atlanticus*.

plate, which projects between the arms of the lower jaws. The snout is conical and two supramaxillaries are present. There are bands of villiform teeth on the jaws.

A total of 518 tarpon specimens were used for meristic analysis. The standard length (SL) was approximately five times the head length and three and half times the body depth. The head increased with increase in SL. Correspondingly, the ratios of the snout to origin of the dorsal fin were 1.86 for Lagos and 2.0 for the Ondo region. This indicated that the dorsal fin was located about the middle of the SL. In addition, the ratio of the snout to anal fin in the SL was approximately 1.5. Similarly, the ratio of the head length to snout length was approximately 4.7, while that of the caudal peduncle was 1.20. The fins of *M. atlanticus* were paired and lacked spines. The dorsal fin was located about the middle of the body, the last ray being projected into a long filament. The pelvic fin was abdominally placed. The pectoral and pelvic fins had accessory scales. The caudal fin was widely forked and the lateral line almost straight. The dorsal fin rays of *M. atlanticus* from the Lagos and Ondo regions ranged from 11 to 14. The pelvic fin rays ranged from 8 to 12 in Lagos, and 8 to 10 in Ondo, with a mean of 9.6 and 8.9, respectively. The anal fin ray count for the Lagos area ranged from 18 to 25 with a mean of 20.1, while those for Ondo ranged from 18 to 22 with a mean of 19.9. The left gill rakers from Lagos Lagoon ranged from 51 to 66, while those of Aiyetoro Creek ranged from 50 to 62. The lateral line scales ranged from 41 to 49 while the branchiostegal rays ranged from 20 to 28.

Fishery for *M. atlanticus* in Nigeria

Capture Fishery

Tarpon *M. atlanticus* were mainly exploited by artisanal fishermen in the study areas. In the Lagos region, there was no targeted fishery of the species, but in the Ondo State coastal area, a special fishery of the species was observed. The most commonly used gears by small-scale artisanal fishermen in the estuaries included traps, beach seines, cast nets (Figure 9.6), gill nets, and hook-and-line. For the capture of the adults at sea, wounding gear like spears (Figure 9.7) were sometimes employed in combination with related fishing gears. An estimated catch of 80,000–150,000 tarpon juveniles (5–20 cm total length (TL)) was landed annually by the local fishermen in the Ondo State coastal area during the study period. The juveniles were targeted mainly for aquaculture, and they are the primary species cultured traditionally in the region. High fishing pressure on the juveniles of the species was observed. There were no conservation or regulatory measures for the fishery. Tarpon catch rates in the study area were irregular. Only a few fishermen actually targeted adult tarpon due to the high costs of fishing.

Total cost of investment (craft and gear) for capture of adult *M. atlanticus* at sea was N640,215 (US$4750). The required break-even period as reported by fishermen ranged from 1 to 2 years. The major problem encountered by fishermen was the high cost of fishing, especially outboard engines and fishing nets. Major fish species associated with the tarpon fishery and caught by small-scale artisanal fishermen in the study area included bonga (*Ethmalosa fimbriata*), giant African threadfin (*Polydactylus quadrifilis*), barracudas (*Sphyraena* spp.), sharks (*Sphyrna* sp.), tilapia (*Sarotherodon melanotheron*), mullets (*Liza* sp.), and crayfish (*Nematopaleamon hastatus*). A typical fish-landing site in the coastal areas of Ondo State is shown in Figure 9.8.

FIGURE 9.6 Cast net used for capture of juveniles *M. atlanticus*.

FIGURE 9.7 Wounding gear (spear) used for the capture of *M. atlanticus*.

FIGURE 9.8 Tarpon, bonga, and crayfish fisheries at a coastal beach in Ondo State, Nigeria.

Culture Fishery

The traditional culture method of *M. atlanticus* in Nigerian coastal fishing villages was carried out in earthen ponds and fish pens installed in the creeks. The ponds ranged from 0.002 to 0.25 ha. Large ponds were observed only at Orioke and were built specifically for natural spawning of *M. atlanticus*. Pond depths varied from 0.7 to 1.8 m. Fish pens were constructed of fine mosquito netting and installed in creeks and lagoons. Polyculture of *M. atlanticus* and tilapia species was carried out with juvenile tilapia serving as food for tarpon. Collection of *M. atlanticus* fry (3–4 cm TL) commenced in March or April. Mosquito nets and lace materials were used to collect the fry (Figure 9.9). Fry were reared in nursery ponds to fingerlings

FIGURE 9.9 Collection of *M. atlanticus* fry using lace net material.

measuring 12.0–15.0 cm TL. The coastal villages of Ondo, particularly Aiyetoro, Araromi, Ilepete, Orioke, and Okesiri, served as supply centers for fry and fingerlings of *M. atlanticus* for other parts of Nigeria.

Stocking densities ranged from 1 to 7 fish m^{-2} in the ponds and 1–3 fish m^{-2} in the fish pens. In earthen ponds, *M. atlanticus* were stocked at a tarpon: tilapia ratio ranging from 1:87 to 1:23, primarily as a predator to control overreproduction of tilapia. Tilapia fry served as prey for tarpon *M. atlanticus* were fed a variety of food items. Crayfish, miscellaneous fish (fresh or smoked), crabs, and bread were commonly administered as fish feeds. Sometimes poultry feed was fed to the early fry. Commercial fish pellet feeds were not used. The water in most of the ponds was turbid with Secchi disk visibility ranging from 9.0 to 27.9 cm due to phytoplankton blooms. Dissolved oxygen content was low, ranging from 0.8 to 2.34 mg/L. However, survival rates of tarpon in ponds were relatively high and ranged from 73 to 86.7%. For example, out of 617 tarpon stocked in a 0.1-ha pond, 535 specimens were recovered at harvest.

Grow out periods extended from 1.5 to 3 years to enable the fish to attain a size of over 100 cm FL and 5 kg. Most consumers preferred large specimens of *M. atlanticus* for entertaining guests during marriage ceremonies and other festivals. Large specimens were sometimes restocked in large ponds for natural spawning. Tarpon spawned naturally in the brackish water ponds in the study region.

Ecotourism Attractions

Some local fishermen and fish farmers in Lagos State stocked tarpon in their ponds purely for recreational purposes. These tarpon farms were integrated with snack bars and served as tourist centers. Visitors to the fish farm on excursions were charged a fee of N50–N100 per person as an entry permit into the farm. Stamatopoulus (1993) and Zerbi (1999) pointed out that the fishery of *M. atlanticus* in western Atlantic was mainly recreational and fetched millions of U.S. dollars annually.

DISCUSSION

Atlantic tarpon *M. atlanticus* is a single species of the family Megalopidae occurring in the warm temperate, tropical, and subtropical Atlantic Ocean, from Mauritania to Angola in the eastern Atlantic and Nova Scotia to Brazil in the western Atlantic. Water temperature has been implicated in the abundance and distribution of *M. atlanticus* (Twomey and Byrne, 1984; Zale and Merrifield, 1989). These authors observed that tarpon were distinctly thermophilic; for example, annual abundance at Port Aransas, Texas, historically correlated with water temperature. They also observed that the lower lethal temperature for *M. atlanticus* was about 10°C, while the early stage I leptocephalus larvae occurred in oceanic waters of 22.2–30.0°C. In the Nigerian study area, water temperature ranged from 26.0 to 35°C throughout the study. Mean monthly rainfall ranged from 4.8 to 599.5 mm in 1996, to 0.0–454.6 mm in 1997. Rainfall affected salinity levels and thus migratory patterns of *M. atlanticus*. Zale and Merifield (1989) observed that throughout most of its life stages, tarpon tolerated a wide range of salinity, although the early stage I larvae were collected only at oceanic salinities of 28.5–39.0 ppt, indicating that such salinity may be required by eggs, yolk-sac, and stage I larva for proper development. The species is euryhaline and salinity levels obtained in this study (0.2–32.1 ppt) were within the tolerance limits.

Adult *M. atlanticus* were caught mainly at sea, while the juveniles and subadults were found in the creeks and lagoons. Crabtree et al. (1997) reported that adult specimens occurred beyond the continental shelf and have been found as far as 300 km offshore. In Costa Rica and Nicaragua, *M. atlanticus* were frequently caught in freshwater lakes and rivers, far from the coast but linked to the sea (Crabtree et al., 1993). Floating masses of fertilized eggs of *M. atlanticus* were also collected off the Atlantic shorelines of Aiyetoro and Orioke fishing villages in Ondo State between October and March, with peak periods in December to January; however, according to Crabtree et al. (1997), eggs of *M. atlanticus* have not been observed off the American coast, whereas they were common in the Ondo State study area. Meristic counts of *M. atlanticus* from Lagos and Ondo regions did not show wide variations. Zale and Merrifield (1989) observed that the mean dorsal fin ray of *M. atlanticus* from south Florida in the United States was 12 while that of the anal fin was 20. The lateral line scales ranged from 41 to 48. These meristic counts were similar to those obtained from the coastal areas of Lagos and Ondo States.

M. atlanticus species were exploited mainly by artisanal fishermen in the study areas. The high cost of investment for the capture of adult *M. atlanticus* at sea may be the cause of low catches of the species in the areas. The gears used for capture of *M. atlanticus* were also used for catching other large fishes like giant African threadfin (*Polydactylus quadrifilis*), shark (*Sphyrna* spp.), barracudas (*Sphyraena* spp.), etc. Udolisa et al. (1994) made similar observations. The break-even period reported by fishermen ranged from 1 to 2 years.

Annual catches for the period 1970–1990 for *M. atlanticus*, as reported by Stamatopoulus (1993), decreased from 600 to 289 mt for the western central Atlantic (Caribbean), and from 3200 to 1400 mt for the southwest Atlantic Ocean (Brazil area). In the Ondo State coastal area, 2–3.5 mt of adult tarpon were landed annually during the period of this study. Garcia and Solano (1995) reported that *M. atlanticus* were fast disappearing from the Caribbean coast of Colombia, without anyone apparently noticing it. Kusemiju (1973) in his study on the catfishes of Lekki Lagoon with particular reference to *Chrysichthys walkeri* observed that there was overfishing of the catfish and thus the need for conservation. Similarly, there is great need for conservation measures for tarpon in Nigeria.

The culture of *M. atlanticus* in the study areas was not full time and hence not commercialized. This was attributed to the extended culture period. Better pond management and adequate feeding will enhance the aquaculture potentials of the species. A lot of fishing pressure on tarpon juvenile population was observed. Juveniles were targeted mainly for aquaculture purposes. There were no conservation or regulation measures for the fishery. This may lead to stock depletion as recruitment into the adult population could be jeopardized. There is therefore need for regulation of the fishery through adoption of some conservation methods. Spawning of *M. atlanticus* and development of techniques for mass production of the fingerlings may be the best strategy for now to relieve the exploitation pressures from capture of wild tarpon fingerlings.

ACKNOWLEDGMENTS

The authors are grateful to the Nigerian Institute for Oceanography and Marine Research, Lagos, for sponsorship of this study.

REFERENCES

Awosika, L.F., Dublin-Green, C.O., Adegbie, A.T., Ibe, C.E., Folorunsho, R. and Isebor, C.A. (1994). Ondo State coastline, geomorphology and coastal dynamics. NIOMR Special Technical Report. 49p.

Boyd, C.E. (1979). *Water quality in warm water fish ponds*. Agricultural Experimental Station. Auburn University, Auburn, AL. 359p.

Crabtree, R.E., Bert, T.M. and Taylor, R.G. (1993). Research on abundance, distribution and life history of tarpon and bonefish in Florida. *FWS Tech. Rep.* **59**: 1–42.

Crabtree, R.E., Cyr, E.C., Chaverri, D.C., Mclarney, W.O. and Dean, J.M. (1997). Reproduction of tarpon, *Megalops atlanticus* from Florida and Costa Rican waters and notes on their ages and growth. *Bull. Mar. Sci.* **6** (2): 271–268.

Ezenwa, B., Ebietomiye, O. and Anyanwu, P. (1985). Stocking densities of *Megalops atlanticus*. NIOMR Annual Report. 56–57.

Fischer, W., Bianchi, G. and Scott, W.B. (1981). FAO species identification sheets for fishery purposes: east central Atlantic fishing area 34, 47 (in part). Vols. 1–7 (pag.var.). FAO, Rome.

Garcia, C.B. and Solano, O.D. (1995). *Tarpon atlanticus* in Colombia: a big fish in trouble. *Naga, ICLARM Q.* **8** (3): 47–49.

Ibe, A.C., Awosika, L.F., Ihenyen, A.E., Ibe, C.E. and Tiamiyu, A.I. (1985). Coastal erosion at Awoye and Molume, Ondo State Nigeria. A Report for Gulf Oil Company Nig. Ltd. 123p.

IGFA. (1987). *World record of game fishes*. International Game Fish Association, FL. 327p.

Irvine, F.R. (1947). *The fishes and fisheries of Gold Coast*. Crown Agents, London. 243p.

Kusemiju, K. (1973). A study of the catfishes of Lekki Lagoon with particular reference to *Chrysichthys walkeri* (Bagridae). Ph.D. Thesis, University of Lagos, Nigeria. 395p.

Stamatopoulus, C. (1993). Trends in catches and landings of Atlantic fisheries. 1970–1991. *FAO Fish. Cir.* No. 855/1. 223p.

Twomey, E. and Byrne, H. (1984). A new record for tarpon, *Megalops atlanticus* (Val.) (Osteichthyes–Elopiformes–Elopidae) in the eastern north Atlantic. *J. Fish Biol.* **4**: 359–362.

Udolisa, R.E.K., Solarin, B.B., Lebo, P.E., and Ambrose, E.E. (1994). A catalogue of small scale fishing gear in Nigeria. RAFR Publication. RAFR/014/FI/94/02. 142p.

Ugwumba, A.O. (1984). The biology of the ten pounder, *Elops larceta*. Ph.D. Thesis, University of Lagos, Nigeria. 397p.

Whitehead, P.J.P., Bauchot, M-L., Hureau, J-C., Nielsen, J. and Tortonese, E. (1984). Fishes of the north-eastern Atlantic. UNESCO Report, Vol. 1. 510p.

Zale, A.V. and Merrifield, S.G. (1989). Specific profiles: life histories and environmental requirements of coastal fishes and invertebrates (South Florida). Ladyfish and Tarpon. *U.S.F.W.S. Biol. Rep.* **82**, 17p.

Zerbi, A. (1999). Ecology and biology of juveniles of two groups of fishes exploited in sport fisheries in Puerto Rico: the Snook, (*Centropomus*) and the Tarpon (*Megalops atlanticus*). Ph.D. Thesis, Université d'Aix-Marseille II, Marseille, France. 164p.

Section II

Biology and Life History Dynamics

10 Studies in Conservation Genetics of Tarpon (*Megalops atlanticus*): Microsatellite Variation across the Distribution of the Species

Rocky Ward, Ivonne R. Blandon,
Francisco García de León, Sterling J. Robertson,
André M. Landry, Augustina O. Anyanwu,
Jonathan M. Shenker, Miguel Figuerola,
Theresa C. Gesteira, Alfonso Zerbi,
Celine D. Acuña Leal, and William Dailey

CONTENTS

Introduction 131
Methods 133
    Collections 133
    Isolation and Characterization of Microsatellites 134
    Statistical Analyses 134
Results 135
    Within-Population Diversity 135
    Among-Population Divergence 136
Discussion 141
Acknowledgments 143
References 143

INTRODUCTION

Atlantic tarpon (*Megalops atlanticus*) represent a valuable component of the recreational fishery across a distribution that encompasses coastal and nearshore waters of the Caribbean Sea, Gulf of Mexico, and the western Atlantic from Canada to

Brazil. Peripheral populations occur in the eastern Atlantic off Africa and the Pacific of Panama. Tolerant of wide ranges in salinity and oxygen concentrations, tarpon distribution is limited by sensitivity to low temperature at the northern extreme of the range (Zale and Merrifield, 1989). The reproductive cycle of tarpon is complex. Adults spawn offshore (Crabtree et al., 1992), eggs and larvae have an extended planktonic stage followed by recruitment into fresh and brackish water nursery areas, and juveniles spend 4–5 years in rivers, bays, and estuaries before joining offshore aggregations of adults (Crabtree et al., 1995). Adults in some portions of the range are highly migratory. They move from spawning sites to forage areas on a seasonal basis (see Luo et al., Chapter 18, this volume).

Highly migratory marine organisms such as tarpon, which are characterized by offshore spawning and extended planktonic residence of larvae, are not expected to exhibit extensive population subdivision (Gyllensten, 1985), although within-population genetic diversity may be high (DeWoody and Avisc, 2000). Numerous examples support this generalization (e.g., Graves et al., 1992; Gold et al., 1997). However, exceptions are common (e.g., Johnson et al., 1993; Hutchinson et al., 2001; Gold et al., 2002a). Genetic population structure in marine organisms requires biotic or abiotic mechanisms that result in isolation among population subdivisions. Gold et al. (1999) suggested that structuring in estuarine-dependent sciaenids might be explained by factors such as natal bay philopatry or limited migration between neighboring bays. Among open-water spawners, such as tarpon, population structure could be explained by spawning site fidelity (Ward et al., 2002) or by long-term persistence of schools of closely related individuals that are maintained by behavioral or oceanographic factors (Taylor and Hellberg, 2003).

Examinations of genetic variation in tarpon have, in general, suggested limited population structuring. Blandon et al. (2002) examined variation in mtDNA across the distribution of the species. African and Pacific of Panama tarpon exhibited reduced haplotype diversity. Genetic divergence among sites in the western Atlantic and Caribbean was minimal. Within the Gulf of Mexico, among-population genetic divergence was higher, though lacking consistent geographic patterns. García de León et al. (2002) used allozymes and restriction fragment length polymorphisms of the 12S rRNA mtDNA gene to examine variation in the western Gulf of Mexico. A distinct break was found in allele and haplotype frequency distributions between the upper and lower coasts of Texas. Ward et al. (2005) used direct sequencing to examine variation in a 12S rRNA mtDNA fragment among tarpon from the Gulf of Mexico and Chetumal, Mexico in the western Caribbean. Samples from the Gulf of Mexico were found to be distinct from all other samples. McMillan-Jackson et al. (2005) examined allozymes and mitochondrial DNA (mtDNA) variation in tarpon from the Atlantic and Caribbean. African tarpon were found to be genetically depauperate and showed little evidence of genetic exchange with other Atlantic populations. Subtle population structuring was evident in the western Atlantic and Caribbean, where Costa Rica and Florida tarpon showed allele frequency differentiation.

The present study describes the genetic variation across the distribution of tarpon by examining a set of highly polymorphic microsatellite DNA markers (Blandon et al., 2003). Microsatellites facilitate detection of subtle population structuring in marine organisms (e.g., Nielsen et al., 2003; Shaw et al., 1999a) and may help resolve differences noted in earlier tarpon studies of population genetics.

This study is one of a series initiated in response to proposals by the Enhancement Branch of Texas Parks and Wildlife Department (TPWD) to utilize artificial culturing techniques to produce juvenile tarpon for stocking in Texas' marine waters. Marine stockings have been controversial for ecological and genetic reasons (Richards and Edwards, 1986; Grimes, 1998). To be justified, enhancement efforts must be designed to protect the ecological integrity of stocked ecosystems and the genetic integrity of enhanced populations. The data provided by this and earlier studies are intended to provide a scientific basis for broodfish procurement, broodfish management, and stocking site choices.

METHODS

COLLECTIONS

All samples were obtained legally according to the laws and regulations of the various countries. Collection localities are shown in Figure 10.1. Tarpon were sampled during routine resource monitoring by TPWD, Puerto Rico Department of Natural Resources, and Nigerian Institute for Oceanography and Marine Research crews. Other samples were collected by fishermen and guides, including participants in tournaments at Veracruz, Tecolutla, and Tampico in Mexico, in Louisiana, and on Florida's west coast. Tarpon were also sampled at fish markets in Mexico, Columbia, and Brazil. Additional tarpon were collected from angler catches on the Caribbean coast of Costa Rica and from the Pacific Ocean off Panama. A majority of samples were from subadult and adult individuals, though some juveniles were included. Most samples consisted of single scales, allowing release of individuals taken in catch-and-release recreational fisheries. Scales were placed in 70% ethanol, stored at 4°C for a minimum of 24 h, and then transported to the TPWD research station near Palacios, Texas.

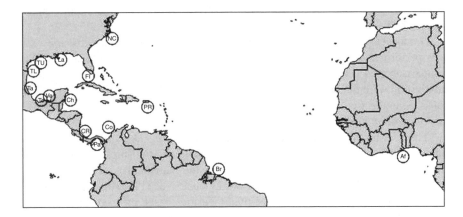

FIGURE 10.1 Collection sites for tarpon included in study of microsatellite variation. Collection site labels are defined in Table 10.1.

Isolation and Characterization of Microsatellites

Genomic DNA was extracted using the Puregene® kit and protocols (Gentra Systems, Minneapolis, MN). Scales of a single individual from the south coast of Texas were used to develop a genomic DNA library prepared following methods described by Estoup and Turgeon (1996). Construction and screening of the library is described by Blandon et al. (2003).

Preliminary assessment of variability and reliability involved screening a minimum of 25 individuals, each from Texas and North Carolina, by PCR amplification of each microsatellite locus (primer sequences provided in Blandon et al., 2003) using Ready-To-Go™ Beads (Amersham Pharmacia Biotech, Inc., Piscataway, NJ), to which were added approximately 100 ng of template in a 25-μL reaction volume. Amplification was carried out using a GeneAmp® PCR System 2400 (Applied Biosystems, Foster City, CA). The amplification protocol used was: 93°C (1 min), 54°C (2 min), 71°C (2 min) repeated for 40 cycles. An extension period of 7 min at 71°C followed the final cycle. PCR products were separated on 10% polyacrylamide gels and visualized by ethidium bromide (0.5 μg/mL in 1X TBE buffer for 20 min). Alleles were scored through comparisons with internal standards.

Statistical Analyses

Summary statistics for mean number of alleles per locus, average heterozygosily, and unbiased genetic diversity (Nei, 1987) were generated using ARLEQUIN version 2.0 software (Schneider et al., 1999). Genotypic frequencies at each locus were tested for deviation from Hardy–Weinberg expectations using exact tests performed with Markov-chain randomization (Guo and Thompson, 1992). Probability values (P) for Hardy–Weinberg tests at each locus within each collection site were estimated by permutation with 100,000 resamplings (Manly, 1991). Sequential Bonferroni corrections (Rice, 1989) were used to adjust significance levels for simultaneous inferential tests in this and other comparisons. Genotypic equilibrium between pairs of loci was used to assess linkage. Exact tests implemented using the statistical program GENEPOP (Raymond and Rousset, 1995) were used to determine significance of probability values.

Homogeneity of allele distributions at each locus was examined using exact tests performed in GENEPOP. Permutation with 1000 resamplings per individual comparison was used to test for significance of exact tests. Levels of population subdivision were quantified by estimation of Weir and Cockerham's (1984) θ, computed using the statistical package ARLEQUIN (Excoffier et al., 1992), with 1000 random permutations.

The Cavalli-Sforza and Edwards' chord distance (D_C; Cavalli-Sforza and Edwards, 1967) was used to reconstruct phylogenetic relationships among sampling sites. Estimations of D_C were obtained using the statistical package GENETIX version 4.04 (Belkhir et al., 2003). Takezaki and Nei (1996) found D_C to be a better estimate of genetic divergence than measures based on the step-wise mutation model. A phenogram was generated from the chord-distance matrix with the neighbor-joining (N-J) algorithm. The N-J phenogram, with bootstrap estimates (as percentage of 10,000 replications) obtained by resampling loci within samples, was generated with the statistical program NJBPOP (Cornuet et al., 1999). Bootstrapped confidence values of branches were generated by resampling loci within samples and are reported as percentages of 10,000 replications. Multidimensional scaling (MDS) was used to search for biologically meaningful patterns in the distance matrix (Pritchard et al., 2000). Analysis of the chord distance matrix was performed using the PROC MDS program contained in SAS (SAS Institute Inc., 1999). The relationship between genetic distances, as estimated by D_C (Cavalli-Sforza and Edwards, 1967), and geographic distance (measured from the approximate center of the collection region) was explored using the Mantel test routine contained in the program Tools for Population Genetic Analyses, version 1.3 (Miller, 1997). This program performs 999 permutations of rows and columns to obtain an estimate of how often the Z-score from the original data is matched or exceeded by the permuted matrices.

An assignment test that utilizes the likelihood-based method (see discussion in Hansen et al., 2001) in the program GENECLASS (Cornuet et al., 1999) was used to examine the utility of individual genotype as a predictor of population affinity and to screen each collection locality for possible dispersers from other sample sites. Log-likelihood estimates were obtained using ARLEQUIN version 2.0 (Schneider et al., 1999). As a follow-up analysis, collections suggested by the MDS analysis to be components of a central panmictic population were combined and compared with collections interpreted as peripheral (i.e., those from the Pacific of Panama, Africa, and Costa Rica).

RESULTS

A total of 328 tarpon from 15 sampling localities were included in the study (Figure 10.1). Sample size ranged from $N = 9$ from Brazil to $N = 39$ on the upper Texas coast.

Fifteen polymorphic microsatellite loci were identified following construction and screening of the genomic library; of these, 6 were chosen for inclusion in the current study on the basis of consistent amplification, informative allele frequencies, and ease of interpretation. Two GT repeats (Mat04 and Mat08) and 2 CA repeats (Mat03 and Mat16) were included in the study along with Mat22, a TCTA tetranucleotide repeat, and Mat11, a compound microsatellite combining a GACA tetranucleotide repeat with a dinucleotide GT repeat. Genebank accession numbers and other details may be found in Blandon et al. (2003).

WITHIN-POPULATION DIVERSITY

Measures of genetic diversity are presented in Table 10.1. The mean number of alleles per locus ranged from 3.3 among Chetumal, Costa Rica, and Brazil tarpon to 5.7 in tarpon from the Texas upper coast. Observed heterozygosity values ranged from $H_O = 0.36$ off Puerto Rico to $H_O = 0.56$ off Brazil. Within-population genetic diversity values ranged from $H_S = 0.31$ for Costa Rica to $H_S = 0.50$ for the Pacific of Panama. Other peripheral populations from North Carolina ($H_S = 0.47$) and Africa ($H_S = 0.37$) exhibited moderate to relatively high genetic diversity.

TABLE 10.1
Within-Population Variation at 6 Microsatellite Loci for 15 Locations Where Tarpon Were Collected

Location	N	$\overline{N}_{\text{allels}}$	H_O	H_S
Veracruz, Mx (Ve)	31	4.50	0.44 (0.11)	0.45 (0.27)
Tecolutla, Mx (Te)	15	3.83	0.43 (0.09)	0.45 (0.28)
Tampico, Mx (Ta)	20	4.67	0.46 (0.13)	0.37 (0.23)
Texas low (TL)	22	5.50	0.38 (0.10)	0.40 (0.25)
Texas up (TU)	39	5.67	0.48 (0.10)	0.49 (0.29)
Louisiana (LA)	24	3.67	0.41 (0.11)	0.37 (0.23)
Florida, Gulf (FL)	20	3.50	0.48 (0.11)	0.49 (0.29)
North Carolina (NC)	25	4.17	0.51 (0.10)	0.47 (0.28)
Puerto Rico (PR)	32	4.00	0.36 (0.12)	0.35 (0.22)
Chetumal, Mx (Ch)	19	3.33	0.55 (0.11)	0.47 (0.28)
Costa Rica (CR)	17	3.33	0.40 (0.13)	0.31 (0.21)
Columbia (Co)	16	3.50	0.40 (0.08)	0.44 (0.27)
Brazil (Br)	9	3.33	0.56 (0.13)	0.42 (0.27)
Africa (Af)	16	4.17	0.43 (0.14)	0.37 (0.23)
Panama Pacific (Pa)	23	4.17	0.49 (0.07)	0.50 (0.30)
Total	328			

Note: N = the number of tarpon included per sample. $\overline{N}_{\text{allels}}$ = the mean number of alleles per locus. H_O = mean observed heterozygosity. H_S = Nei's unbiased gene diversity across all loci. Standard errors are in parentheses. Abbreviations for sampling sites are in parentheses under population.

Of the 96 site/locus comparisons, 2 (locus Mat04 on the Texas lower coast and locus Mat08 off Florida) were not found to be in Hardy–Weinberg equilibrium following Bonferroni correction. In both instances these samples exhibited heterozygote deficiencies. In the absence of the Bonferroni correction, 17 site/locus comparisons approached or reached significance at $\alpha = 0.05$. Of these, 16 exhibited heterozygote deficiencies. Tests for genotypic equilibrium within collection localities were nonsignificant following Bonferroni adjustment except loci Mat03 and Mat04 for the Veracruz sample.

AMONG-POPULATION DIVERGENCE

In the exact tests of allele frequency distributions, five microsatellite loci (all except Mat03) exhibited significant heterogeneity among sampling localities (Table 10.2). Values of θ ranged from 0.003 for Mat03 to 0.098 for Mat11. All loci except Mat03 had θs significantly different from 0 when compared with adjusted levels of alpha ($\alpha = 0.008$). Across all loci, the combined $\theta = 0.038$. The N-J tree depicting relationships based on the Cavalli-Sforza and Edwards chord distance matrix (Table 10.3) provided some suggestion of geographically congruous population structure (Figure 10.2). The two southern samples (Brazil and Africa) were found

TABLE 10.2
Results of Tests for Homogeneity in Allele Distributions and Estimates of Population Structure among Geographic Samples of Tarpon (*Megalops atlanticus*)

Locus	P_{EXACT}	θ	P
Mat03	0.723	0.003	0.451
Mat04	<0.0001	0.043	<0.001[a]
Mat08	<0.0001	0.052	<0.001[a]
Mat11	<0.0001	0.098	<0.001[a]
Mat16	<0.0001	0.037	<0.001[a]
Mat22	<0.0001	0.018	<0.001[a]

Note: P_{EXACT} = probability of allele frequency homogeneity across collection localities based on exact test.
θ = estimated population subdivision (Weir and Cockerham, 1984).
P = probability $\theta = 0$.
[a] Significant at adjusted α-level ($\propto = 0.009$).

to be genetically divergent, as were samples from Panama, Costa Rica, and Florida. Other samples clustered relatively tightly, especially in the Caribbean and the Gulf of Mexico, with little indication of the east-west differentiation noted by García de León et al. (2002) or by Ward et al. (2005). Bootjack support was less than 50% at all nodes.

Multidimensional scaling of the Cavalli-Sforza and Edwards chord distance matrix (Figure 10.3) revealed a tight grouping of most localities in dimension 1. Only Florida and Costa Rica were differentiated on this dimension. The second dimension distinguished the Pacific of Panama sample in one direction and the African sample in the opposite. The African tarpon sample was most closely related to those from Brazil. All other localities formed a relatively tight cluster centrally along both dimensions.

The Mantel test estimated the correlation between the genetic distance matrix (D_C) and a matrix of geographic distances to be $r = 0.28$. The genetic distance matrix and the geographic distance matrix are significantly correlated (the permutation test found a $p = 0.135$ of obtaining a Z-value greater than or equal to the original data).

Correct assignment of individuals to source populations ranged from 6.5% among Veracruz fish to 62.5% among African tarpon (Table 10.4) with an a priori probability of correct assignment of 6.67%. In every population except Veracruz and the lower Texas coast, the largest single group was correctly assigned to their source population. Misassigned individuals were usually not placed with geographically adjacent samples. Collection sites that are not believed to have substantial permanent tarpon populations (e.g., Texas, Louisiana, and North Carolina) have relatively low correct assignment values, supporting the hypothesis that tarpon in these samples may be migrants from other regions.

TABLE 10.3
Pairwise Cavalli-Sforza and Edwards Chord Distances (above Diagonal) and *P*-Values (below Diagonal)

	PW	PE	CH	TA	TU	AF	TL	CO	TE	VE	PA	LA	NC	FL	CR	BR
PW		0.024	0.040	0.030	0.033	0.064	0.028	0.042	0.034	0.021	0.045	0.015	0.033	0.050	0.049	0.074
PE	0.48		0.037	0.032	0.036	0.061	0.035	0.042	0.047	0.030	0.064	0.034	0.025	0.057	0.056	0.066
CH	<0.01	<0.01		0.035	0.040	0.067	0.034	0.043	0.034	0.037	0.062	0.030	0.033	0.046	0.056	0.063
TA	0.40	0.27	0.03		0.037	0.063	0.033	0.053	0.043	0.030	0.058	0.036	0.035	0.059	0.040	0.065
TU	0.06	0.04	<0.01	0.02		0.050	0.022	0.041	0.043	0.024	0.044	0.034	0.025	0.045	0.066	0.074
AF	<0.01	<0.01	<0.01	<0.01	<0.01		0.042	0.040	0.049	0.040	0.099	0.048	0.053	0.063	0.052	0.065
TL	0.75	0.40	0.05	0.51	0.79	0.06		0.027	0.032	0.024	0.056	0.026	0.026	0.045	0.046	0.059
CO	0.11	0.05	0.01	0.01	<0.01	0.07	0.80		0.034	0.025	0.057	0.038	0.023	0.049	0.048	0.041
TE	0.15	0.01	0.02	0.03	<0.01	<0.01	0.55	0.13		0.028	0.070	0.022	0.027	0.039	0.041	0.046
VE	0.38	0.04	0.01	<0.01	<0.01	0.01	0.50	0.22	0.11		0.033	0.024	0.014	0.042	0.037	0.051
PA	0.01	<0.01	<0.01	<0.01	<0.01	<0.01	<0.01	<0.01	<0.01	<0.01		0.060	0.045	0.080	0.082	0.090
LA	0.06	0.02	<0.01	0.01	<0.01	<0.01	0.45	0.01	0.36	<0.01	<0.01		0.034	0.034	0.042	0.070
NC	0.07	0.13	<0.01	0.03	0.04	<0.01	0.31	0.18	0.08	0.40	<0.01	<0.01		0.027	0.057	0.048
FL	<0.01	<0.01	<0.01	<0.01	<0.01	<0.01	0.01	<0.01	<0.01	<0.01	<0.01	<0.01	0.05		0.086	0.080
CR	0.03	<0.01	<0.01	0.03	<0.01	0.02	0.06	0.01	0.01	0.01	<0.01	<0.01	<0.01	<0.01		0.055
BR	0.02	0.03	<0.01	0.02	<0.01	0.04	0.22	0.40	0.16	0.03	<0.01	<0.01	0.01	<0.01	0.05	

Note: Population designations are listed in Table 10.1.

Studies in Conservation Genetics of Tarpon

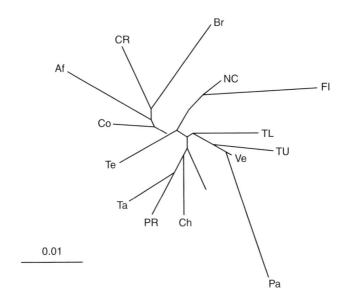

FIGURE 10.2 Unrooted neighbor-joining tree depicting structure found for 16 collections of tarpon using pairwise Cavalli-Sforza and Edwards chord distance matrix for 16 collections of tarpon. Bootstrap support for all nodes were below 50%. Collection site labels are defined in Table 10.1. Puerto Rican samples are labeled "PW" for Puerto Rico west and "PE" for Puerto Rico east.

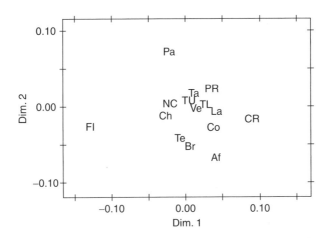

FIGURE 10.3 Plot of the first two dimensions of the multidimensional scaling analysis of genetic affinity among 16 tarpon collecting localities.

Combining sites identified by the MDS analysis as components of a central panmictic cluster produced markedly improved percentage correct assignment (Table 10.5). Individuals from the central sites were correctly assigned 52.9% of the time and correct assignment for the peripheral collection sites ranged from 75.0%

TABLE 10.4
Assignment Test

To	\multicolumn{15}{c}{Assigned from}

To	Ve	Te	Ta	TL	TU	La	Fl	NC	PR	Ch	CR	Co	Br	Af	Pa
Ve	2			1	1	1	1	1			1	1		2	
Te	4	8	1	2	1	2	2	2	3	3	2			1	1
Ta	4		11	1	3	1		3	1	2	1				1
TL	1			2				2							
TU	1			1	9			1		1					
La	1	3		2	1	9			4	1		2			2
Fl			2		2		11							1	1
NC					4		2	9	1	1					4
PR	4	2	2	3	9	3	1	2	17	9		2	1		
Ch	2		1		2	1	1		1		1	2	2		
CR	3		2	2	2	2			2		10	2		1	1
Co	3	2		4	2	2		3	1		1	5			
Br				1	1		1	1				1	4		
Af	2		1	1	2	1	1			1		1	1	10	
Pa	4			1	2		1		2						12
%C	6.5	53.3	55.0	9.1	23.1	37.5	55.0	36.0	53.1	47.4	58.8	31.3	44.4	62.5	52.2

Note: Collection site labels are defined in Table 10.1.

TABLE 10.5
Assignment Test—Allele Frequencies for 12 Central Collection Sites Are Combined

to	Assigned from				
	Central	Pa	Fl	Af	CR
Central	128	1	2	1	1
Pa	38	20	1	0	1
Fl	33	1	17	1	0
Af	22	0	0	12	0
CR	21	1	0	2	13
%C	52.9	87.0	85	75.0	86.7

Note: %C = percentage correctly assigned.

(for Africa) to 87.0% (Panama Pacific) with a 20% a priori probability of correct assignment.

DISCUSSION

Over much of its distribution, the tarpon population is weakly subdivided. Only on the periphery of the distribution do genetically distinguishable subpopulations occur. The plot of the first two dimensions of the MDS analysis of the distance (D_C) matrix demonstrates a clustering of most samples near the midpoints of both dimensions. Exceptions are tarpon from the Pacific of Panama, the Gulf of Guinea (Africa), Costa Rica, and Florida. Pacific Ocean tarpon represent a population established since the opening of the Panama Canal in 1914. Measures of diversity for this population are all above the median for the samples, and H_S among Pacific tarpon is the highest observed for any collection site, suggesting that the number of tarpon transiting the canal has been extensive. Despite the relatively high within-population genetic diversity for Pacific tarpon, the divergence between these fish and tarpon in the Caribbean is considerable as indicated by MDS and assignment test analyses. Gene flow, though apparently extensive, did not prevent divergence. African tarpon separate from all other collection sites on the MDS analysis. The nearest sample on the MDS plot to Africa is Brazil, suggesting current or past gene flow between African and South American populations. Transport of both larval and adult tarpon from Brazil to the Gulf of Guinea would be facilitated by the north equatorial counter current (The Cooperative Institute for Marine and Atmospheric Studies, University of Miami, http://oceancurrents.rsmas.miami.edu/atlantic/north-equatorial-cc.html). African tarpon were found to be genetically diverse, in contrast to studies utilizing allozymes and mtDNA by McMillan-Jackson et al. (2005) and mtDNA by Blandon et al. (2002), which found African samples to be nearly monomorphic. This may be due to the limited sample size of the previous studies ($N = 5$ and $N = 2$, respectively), or the microsatellite markers employed in the present study may resolve variability not detected by other genetic markers. The diversity value for the Costa Rican sample ($H_S = 0.31$) was the lowest observed among all samples included in this study and possibly reflects limited recruitment into the coastal habitat available at this locale. Significant genetic divergence has been noted in similar situations (Planes et al., 1998). The genetically distinct Florida sample is anomalous because it is from a nonperipheral locality. This is in agreement with the findings of McMillan-Jackson et al. (2005); however, Ward et al.'s (2005) analysis of 12S rRNA mtDNA sequences found Florida tarpon clustered with those collected from Louisiana and Texas, which is consistent with satellite PAT tagging data (Luo et al., Chapter 18, this volume).

With some exceptions, tarpon occurring in the Gulf of Mexico, Caribbean Sea, and western Atlantic Ocean are genetically similar, but some substructure is evident. Some level of population structure is necessary to explain the ability of the assignment test to correctly classify individuals to collection localities with greater than chance probabilities. Weak geographic differentiation based on resident subadult tarpon may be explained by factors such as spawning-site fidelity or it may simply represent ephemeral differences based on recruitment of genetically related juveniles to nursery areas. Further studies focusing on juvenile populations sampled among year-classes are required to make this distinction.

The current analysis failed to discern the population structure detected by previous studies in the western Gulf of Mexico (García de León et al., 2002; Blandon et al., 2002; Ward et al., 2005). Different molecular markers may be more or less efficient at detecting population structure (Neigel, 1994); however, microsatellites have often shown superior ability to resolve population-level genetic structure (e.g., Shaw et al., 1999b; Gold et al., 2002b).

Several samples were not in Hardy–Weinberg equilibrium or approached nonequilibrium. In each case but one deviation from Hardy–Weinberg expectations was due to lower than expected levels of heterozygosity. Failure to meet Hardy–Weinberg expectations due to reduced heterozygosity may be caused by a number of factors including misscoring of heterozygous genotypes as homozygous, undetected null alleles, inbreeding, negative heterosis, or samples composed of individuals drawn from more than one population (Wahlund's effect). The heterogeneous composition of adult tarpon samples would explain the observed lower than expected heterozygosity values.

In summary, microsatellite markers demonstrated within- and among-population genetic variability in tarpon. Over most of its range, genetic differentiation was subtle. Geographically isolated and peripheral populations were much more distinct. Populations from Florida, the Pacific of Panama, the Gulf of Guinea, and Costa Rica were genetically differentiated.

Proposals to culture and stock tarpon by various agencies must take two genetic factors into account. First, population structure should be protected by stocking fish that are genetically representative of local populations. Exotic alleles should not be introduced by stocking efforts and local allele frequencies should not be disrupted. Stocking should not be an anthropogenic source of gene flow. Second, an adequate number of broodfish should be utilized to ensure that genetic diversity of the stocking cohort approaches that of the natural population. Analysis of microsatellites, in contrast to allozymes and mtDNA (García de León et al., 2002; Blandon et al., 2002), suggests that broodfish for enhancement programs in the Gulf of Mexico or northern Caribbean may be safely obtained from any nonisolated tarpon population in this region. Of greater concern are the relatively high levels of genetic diversity found in the present study, in agreement with previous studies, suggesting that a large number of broodfish must contribute to stocking cohorts to protect the genetic integrity of enhanced populations.

ACKNOWLEDGMENTS

Funding for this project was provided by TPWD and by the Texas Parks and Wildlife Foundation and valuable support by the Instituto Tecnológico de Cd. Victoria. Tarpon samples were collected by TPWD resource and enhancement staff; by students and staff of Instituto Tecnológico de Cd. Victoria; and by tarpon fishermen, guides, and tournament directors throughout the distribution of this fish. Sr. Mario Cruz-Ayala of the Mexican Sportsfishing Association provided valuable feedback and assistance in contacting Mexican fishermen. Laboratory assistance was provided by William J. Karel and Brandon Mobley of TPWD and Raymond Ary of Oklahoma State University.

REFERENCES

Belkhir K., P. Borsa, L. Chikhi, N. Raufaste, and F. Bonhomme. 2003. GENETIX 4.04, logiciel sous Windows TM pour la génétique des populations. Laboratoire Génome, Populations, Interactions, CNRS UMR 5000, Université de Montpellier II, Montpellier, France.

Blandon, I. R., F. J. García de León, R. Ward, R. A. Van Den Bussche, and D. S. Needleman. 2003. Studies in conservation of tarpon (*Megalops atlanticus*)—V. Isolation and characterization of microsatellite loci. Mol. Ecol. Notes 3: 632–634.

Blandon, I. R., R. Ward, F. J. García De León, A. M. Landry, A. Zerbi, M. Figuerola, T. C. Gesteira, W. Dailey, and C. D. Acuña Leal. 2002. Studies in conservation genetics of tarpon (*Megalops atlanticus*)—I. Variation in restriction length polymorphisms of mitochondrial DNA across the distribution of the species. Contr. Mar. Sci. 35: 1–17.

Cavalli-Sforza, L. L. and A. W. F. Edwards. 1967. Phylogenetic analysis models and estimation procedures. Am. J. Hum. Gen. 19: 233–257.

Cornuet J. M., S. Piry, G. Luikart, A. Estoup, and M. Solignac. 1999. New methods employing multilocus genotypes to select or exclude populations as origins of individuals. Genetics 153: 1989–2000.

Crabtree, R. E., E. C. Cyr, and J. M. Dean. 1995. Age and growth of tarpon, *Megalops atlanticus*, from south Florida waters. Fish. Bull. 93: 619–628.

Crabtree, R. E., E. C. Cyr, R. E. Bishop, L. M. Falkenstein, and J. M. Dean. 1992. Age and growth of tarpon, *Megalops atlanticus*, larvae in the eastern Gulf of Mexico, with notes on relative abundance and probable spawning areas. Envir. Biol. Fishes 35: 361–370.

DeWoody, J. A. and J. C. Avise. 2000. Microsatellite variation in marine, freshwater and anadromous fishes compared with other animals. J. Fish Biol. 56: 461–473.

Estoup, A. and J. Turgeon. 1996. Microsatellite markers: isolation with non-radioactive probes and amplification. Laboratoire de Génetique des Poissons INRA. http://www.inapg.inra.fr/dsa/microsat/microsat/htm.

Excoffier, L., P. E. Smouse, and J. M. Quattro. (1992) Analysis of molecular variance inferred from metric distances among DNA haplotypes: application to human mitochondrial DNA restriction data. Genetics 131: 479–491.

García de León, F. J., C. D. Acuña Leal, I. R. Blandon, and R. Ward. 2002. Studies in conservation genetics of tarpon (*Megalops atlanticus*)—II. Population structure of tarpon of the western Gulf of Mexico. Contr. Mar. Sci. 35: 18–33.

Gold, J. R., F. Sun, and L. R. Richardson. 1997. Population structure of red snapper from the Gulf of Mexico as inferred from analysis of mitochondrial DNA. Trans. Am. Fish. Soc. 126: 386–396.

Gold, J. R., L. R. Richardson, and T. F. Turner. 1999. Temporal stability and spatial divergence of mitochondrial DNA haplotype frequencies in red drum (*Sciaenops ocellatus*) from coastal regions of the western Atlantic Ocean and Gulf of Mexico. Mar. Biol. 133: 593–602.

Gold, J. R., E. Pak, and D. A. DeVries. 2002a. Population structure of king mackerel (*Scomberomorus cavalla*) around peninsular Florida, as revealed by microsatellite DNA. Fish. Bull. 100: 491–509.

Gold, J. R., L. B. Stewart, and R. Ward. 2002b. Population structure of spotted seatrout (*Cynoscion nebulosus*) along the Texas Gulf Coast, as revealed by genetic analysis, pp. 17–29. In: Biology of the spotted seatrout. S. A. Bortone (Ed.). CRC Press, Boca Raton, FL.

Graves, J. E., J. R. McDowell, A. M. Beardsley, and D. R. Scoles. 1992. Stock structure of the bluefish *Pomatomus saltatrix* along the mid-Atlantic coast. Fish. Bull. 90: 703–710.

Grimes, C. B. 1998. Marine stock enhancement: Sound management or techno-arrogance? Fisheries 23(9): 18–23.

Guo, S. and E. Thompson. 1992. Performing the exact test of Hardy-Weinberg proportion from multiple alleles. Biometrics 48: 361–372.

Gyllensten, U. 1985. The genetic structure of fish: differences in the intraspecific distribution of biochemical genetic variation between marine, anadromous, and freshwater species. J. Fish Biol. 26:691–699.

Hansen, M.M., E. Kenchington, and E. E. Nielsen. 2001. Assigning individual fish to populations using microsatellite DNA markers: methods and applications. Fish Fish. 2: 93–112.

Hutchinson, W. F., G. R. Carvalho, and S. I. Rogers. 2001. Marked genetic structuring in localized spawning populations of cod *Gadus morhua* in the North Sea and adjoining waters, as revealed by microsatellites. Mar. Ecol. Progr. Ser. 223: 251–260.

Johnson, A. G., W. A. Fable, Jr., C. B. Grimes, L. Trent, and J. V. Perez. 1993. Evidence for distinct stocks of king mackerel, *Scomberomorus cavalla*, in the Gulf of Mexico. Fish. Bull. 92: 91–101.

Manly, B. J. F. 1991. Randomization and Monte Carlo methods in biology. Chapman & Hall, New York.

McMillan-Jackson, A. B., T. M. Bert, H. Cruz-Lopez, S. Seyoum, T. Orsoy, and R. E. Crabtree. 2005. Molecular genetic variation in tarpon (*Megalops atlanticus* Valenciennes) in the northern Atlantic Ocean. Mar. Biol. 146: 253–261.

Miller, M. P. 1997. Tools for population genetic analysis of allozyme and molecular genetic data version 1.3. Computer software distributed by author.

Nei, M. 1987. Molecular evolutionary genetics. Columbia University Press, New York.

Neigel, J. E. 1994. Analysis of rapidly evolving molecules and DNA sequence variants: alternative approaches for detecting genetic structure in marine populations. CalCOFI Rep. 35: 82–89.

Nielsen, E. E., M. M. Hansen, D. E. Ruzzante, D. Meldrup, and P. Grønkjær. 2003. Evidence of a hybrid-zone in Atlantic cod (*Gadus morhua*) in the Baltic and the Danish Belt Sea revealed by individual admixture analysis. Mol. Ecol. 12: 1497–1508.

Planes, S., P. Romans, and R. Lecomte-Finiger. 1998. Genetic evidence of closed life-cycles for some coral reef fishes within Taiaro Lagoon (Tuamotu Archipelago, French Polynesia). Coral Reefs 17: 9–14.

Pritchard, J. K., M. Stephens, and P. Donnelly. 2000. Inferences of population structure using multilocus genotype data. Genetics 155: 945–959.
Raymond, M. and F. Rousset. 1995. GENEPOP (ver. 1.2): population genetics software for exact tests and ecumenicism. J. Hered. 86: 248–249.
Rice, W. R. 1989. Analyzing tables of statistical tests. Evolution 43: 223–225.
Richards, W. J. and R. E. Edwards. 1986. Stocking to enhance marine fisheries, pp. 75–80. *In*: Fish culture in fisheries management. R. H. Stroud (Ed.). Proceedings of a symposium on the role of fish culture in fisheries management, Lake Ozark, MO. American Fisheries Society, Bethesda, MD.
SAS Institute Inc. 1999. SAS/STAT User's Guide, Version 8. SAS Institute Inc., Gary, NC.
Schneider, S., D. Roessli, and L. Excoffier. 1999. ARLEQUIN version 2.0: a software for population genetic data analysis. Genetics and Biometry Laboratory, University of Geneva, Switzerland.
Shaw, P. W., G. J. Pierce, and P. R. Boyle. 1999a. Subtle population structuring within a highly vagile marine invertebrate, the veined squid *Loligo forbesi*, demonstrated with microsatellite DNA markers. Mol. Ecol. 8: 407–417.
Shaw, P. W., C. Turan, J. M. Wright, M. O'Connell, and G. R. Carvalho. 1999b. Microsatellite DNA analysis of population structure in Atlantic herring (*Clupea harengus*), with direct comparison to allozyme and mtDNA RFLP analyses. Heredity 83: 490–499.
Takezaki, N. and M. Nei. 1996. Genetic distances and reconstruction of phylogenetic trees from microsatellite DNA. Genetics 144: 389–399.
Taylor, M. S. and M. E. Hellberg. 2003. Genetic evidence for local retention of pelagic larvae in a Caribbean reef fish. Science 299(3): 107–109.
Ward, R., M. Figuerola, B. E. Luckhurst, I. R. Blandon, and W. J. Karel. 2002. Genetic characterization of red hind, *Epinephelus guttatus*, collected from three breeding aggregations off western Puerto Rico—variation in allozymes, pp. 460–471. *In*: Proceedings of the fifty-third annual Gulf and Caribbean Fishery Institute. R. L. Creswell (Ed.). November 2000, Biloxi, MS.
Ward, R., I. R. Blandon, A. M. Landry, F. J. García de León, W. Dailey, and C. D. Acuña Leal. 2005. Studies in conservation genetics of tarpon (*Megalops atlanticus*)—III. Variation across the Gulf of Mexico in the nucleotide sequence of a 12S mitochondrial rRNA gene fragment. Contrib. Mar. Sci. 37: 45–59.
Weir, B. S. and C. C. Cockerham. 1984. Estimating F-statistics for the analysis of population structure. Evolution 38: 1358–1370.
Zale, A. V. and S. G. Merrifield. 1989. Species profiles: life histories and environmental requirements of coastal fishes and invertebrates (South Florida)—ladyfish and tarpon. U.S. Fish Wildl. Serv. Biol. Rep. 82(11.104). U.S. Army Corps of Eng., TR EL-82-4.

11 Resolving Evolutionary Lineages and Taxonomy of Bonefishes (*Albula* spp.)

Brian W. Bowen, Stephen A. Karl, and Edward Pfeiler

CONTENTS

Introduction 147
Phylogenetics 149
Taxonomy and Distribution 150
    A. argentea (Forster, 1801 in Bloch and Schneider, 1801) 150
    A. glossodonta (Valenciennes, 1847 in Cuvier and Valenciennes, 1847) 150
    A. oligolepis (Hidaka et al., submitted) 150
    A. nemoptera (Fowler, 1911) 151
    A. virgata (Jordan and Jordan, 1922) 151
    A. vulpes (Linnaeus, 1758) 151
    Albula sp. A (Undescribed) 151
    Albula sp. B (Undescribed) 151
    Albula sp. C (Undescribed) 152
    Albula sp. D (Undescribed) 152
    Albula sp. E (Undescribed) 152
Summary 152
Acknowledgments 153
References 153

INTRODUCTION

In the realm of modern sportsmen, there is no animal so highly prized, yet so poorly known, as the bonefish (Fernandez and Adams, 2004). Since the original description by Linnaeus (1758), there have been many names applied to bonefishes by research groups working in relative isolation in different parts of the globe. Ultimately there were 23 named species of bonefish, but as communication improved over the last century it became apparent that many of these names applied to the same species.

By 1940, most bonefishes were synonymized into a single species, *Albula vulpes* (Linnaeus 1758, reviewed in Whitehead, 1986). The exception, the rare and enigmatic threadfin bonefish *A. nemoptera*, is quite morphologically distinct from the others and has retained species status. This simplicity was shattered by Shaklee and Tamaru (1981), who discovered two genetically distinct bonefishes in Hawaii, both clearly different from the West Atlantic *A. vulpes*. Pfeiler (1996) subsequently demonstrated a deep genetic partition between Caribbean and East Pacific bonefishes. The genetic separation between the two Hawaiian forms indicates a 20 MY separation, while the Caribbean–East Pacific divergence may be about 12 MY old. In addition, *A. nemoptera* includes both West Atlantic and East Pacific populations that likely have been separated by the Isthmus of Panama for at least 3.5 MY (Coates and Obando, 1996).

The case of the two Hawaiian bonefishes illustrates one of the rules of taxonomic nomenclature, wherein the earliest species name linked to a museum specimen is the valid one. Since the discoveries of Shaklee and Tamaru (1981) and Pfeiler (1996), taxonomists have been sorting through species names in old literature to discover the proper nomenclature for Pacific bonefishes. One of these is *A. glossodonta* (Forsskål, 1775), which seems to have a stable nomenclature. The other Hawaiian form, labeled *A. neoguinaica* (Valenciennes, 1847 in Cuvier and Valenciennes, 1847) for the past 25 years, was subsequently revised to *A. forsteri* (Bloch and Schneider, 1801) based on the nomenclatural rule of precedence (Randall and Bauchot, 1999), and recently revised to *A. argentea* (Forster in Bloch and Schneider, 1801; see Randall, 1995; Hidaka et al., submitted).

By the dawn of the 20th century there were four recognized species of bonefish: the widespread *A. vulpes*, the morphologically distinct *A. nemoptera*, and *A. argentea* and *A. glossodonta* in Hawaii.

In the first mitochondrial DNA (mtDNA) survey of bonefishes, Colborn et al. (2001) reported eight deep phylogenetic lineages based on cytochrome *b* gene sequences. Hence, there may be as many as 10 bonefish species if the Atlantic and Pacific *A. nemoptera* prove to be distinct species. The nomenclature of bonefishes is currently in a state of revision, as some of the old names must be resurrected to accommodate the recently confirmed species. In some cases it may prove that no taxonomic names (and corresponding museum specimens) are available, and new names and type specimens must be provided.

Bonefishes represent an evolutionary riddle because they are deeply divergent in mtDNA and allozyme surveys, yet they are very similar (often indistinguishable) in terms of morphology. Further, divergent species pairs overlap and occupy similar or identical habitats. Evolutionary theory maintains that when similar species overlap, either they will diverge in ecological traits, or one will out-compete the other to the point of exclusion or even extinction. In contrast to this rule, bonefish species that are millions of years apart can be caught in the same locations in Hawaii, Florida, Brazil, and many other areas. It is possible that in sympatry (overlapping distributions), these bonefish species occupy cryptic but distinct niches, as this aspect of bonefish biology is understudied (Crabtree et al., 1998).

Linking the eight lineages, in Colborn et al. (2001), to taxonomic identities is a work in progress. Three lineages correspond to the three recognized species

(*A. vulpes* in the Atlantic, *A. glossodonta* and *A. argentea* in the Indo-Pacific). Five additional lineages were provisionally labeled *Albula* sp. A, B, C, D, and E. *Albula* sp. E was a special problem because fin clips were provided in a sample that was predominately *Albula* sp. B from Brazil. Species E was highly divergent but researchers lacked the morphological identity to address this further. Subsequent mtDNA sequence analyses proved it to be a sister group to Pacific *A. nemoptera* (Pfeiler et al., 2006; S.A.K., unpublished data).

PHYLOGENETICS

The phylogeny of *Albula* (Figure 11.1) is based on maximum likelihood divergence estimates of new and previously published sequences of the mtDNA cytochrome *b* gene. New samples were sequenced following Colborn et al. (2001). Haplotypes representative of each species were obtained from GenBank. Most notably, we have

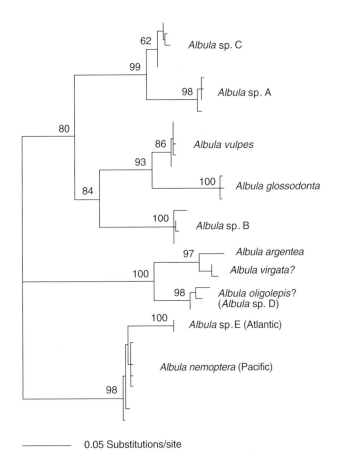

FIGURE 11.1 Phylogenetic relationships of bonefishes based on maximum likelihood analysis of mtDNA cytochrome *b* gene sequences. Numbers above nodes indicate bootstrap support.

sequenced several individuals of *A. nemoptera* unavailable to Colborn et al. (2001), including Pacific samples from El Salvador ($N = 8$) and Costa Rica ($N = 14$). For these analyses, various models of evolution were tested using Modeltest (Posada and Crandall, 1998), and Akaike Information Criterion (AIC; Akaike, 1974) indicated that a TIM+G model (see Posada and Crandall, 1998) is most appropriate. A variety of tree estimating methods (parsimony and distance), however, support the species level relationships in Figure 11.1. The overall strong support for major nodes in the phylogeny is indicated by consistently large bootstrap values.

Using the data from Colborn et al. (2001), and a molecular clock of about 1%/MY (based on mtDNA sequence divergence $d = 0.04$ between Atlantic and Pacific *A. nemoptera*), divergence times among bonefish species are approximately 3–30 MY. The deepest divergence is between the sympatric *A. glossodonta* and *A. argentea*, with a mtDNA cytochrome *b* sequence divergence of $d = 0.26$–0.30, well above the optimal range of resolution. This tree is in substantial agreement with Figures 1 and 2 in Pfeiler et al. (2006). All mtDNA trees for this group indicate sister relationships between *Albula* sp. A and C in the eastern Pacific, *Albula* sp. D and *A. argentea* in the Indo-Pacific, and *A. vulpes* and *A. glossodonta* in the Atlantic and Indo-Pacific, respectively. The difficult lineage in previous analyses is *Albula* sp. E (now identified as Atlantic *A. nemoptera*; Pfeiler et al., 2006), comprising a deep and poorly resolved branch that has been alternately affiliated with *Albula* sp. B, *Albula* sp. D and *A. argentea*, or sister to a cluster of five species (Colborn et al., 2001). The affiliation with *Albula* sp. D/*argentea* is strongly supported by Pfeiler et al. (2006), but is unresolved in the current analysis likely due to the lack of a suitable outgroup.

TAXONOMY AND DISTRIBUTION

A. ARGENTEA (FORSTER, 1801 IN BLOCH AND SCHNEIDER, 1801)

The longjaw bonefish occurs from the Indo-Malayan region to the Marquesas, with an uncertain distribution in the Indian Ocean. Type locality is Tahiti. The Hawaiian longjaw bonefish is recently described as a distinct species, *A. virgata* in an "argentea" complex that encompasses three species (see below).

A. GLOSSODONTA (VALENCIENNES, 1847 IN CUVIER AND VALENCIENNES, 1847)

The shortjaw bonefish occurs from Hawaii and French Polynesia to the Seychelles in the western Indian Ocean. Type locality is the Red Sea. There is shallow but significant population structure between the Pacific and Indian Oceans (Colborn et al., 2001), and between Hawaii and the Line Islands (Friedlander et al., Chapter 2, this volume).

A. OLIGOLEPIS (HIDAKA ET AL., SUBMITTED)

The smallscale bonefish is distributed from the Coral Sea to South Africa, with a type locality in the latter location. As noted by Hidaka et al. (submitted), this member of the "argentea" complex is almost certainly the undescribed *Albula* sp. D in Colborn et al. (2001).

A. NEMOPTERA (FOWLER, 1911)

The threadfin bonefish was formerly in the genus *Dixonina* (Fowler, 1911), subsequently placed in synonymy with *Albula* by Rivas and Warlen (1967). Recent mtDNA data support this synonymy, locating *A. nemoptera* within the phylogenetic tree for *Albula* (Pfeiler et al., 2006). This species occurs in the tropical western Atlantic and eastern Pacific and in deeper water than the typical habitat of *Albula* species (Smith and Crabtree, 2002). Anecdotal evidence and the capture location of some museum specimens indicate an association with river outflows. *A. nemoptera* may eventually be split into Atlantic and Pacific species, pending the outcome of ongoing investigations. The biogeography and genetics support this taxonomic distinction, with a minimum separation of about 3.5 million years, and $d = 0.04$ sequence divergence in cytochrome *b* gene sequences. In this case, the Atlantic form (i.e., *Albula* sp. E) would retain the name *A. nemoptera* (type locality Dominican Republic), and the name *A. pacifica* (shafted bonefish) is available for the East Pacific form (Beebe, 1942, Pfeiler et al., 2006).

A. VIRGATA (JORDAN AND JORDAN, 1922)

The endemic Hawaiian longjaw bonefish is known locally as the $\overline{O}'i\overline{o}$. It is morphologically distinct from *A. argentea* (Hidaka et al., submitted) and has dark longitudinal lines not observed in other species (Jordan and Jordan, 1922). There is provisional molecular support for a distinct Hawaiian clade. With the exception of a single specimen, all Hawaiian individuals comprise a monophyletic mtDNA lineage distinct from West Pacific specimens by $d = 0.03$–0.04 (Colborn et al., 2001).

A. VULPES (LINNAEUS, 1758)

The bonefish occurs in the tropical and subtropical northwest Atlantic. Genetic surveys have detected it only in the Caribbean (Pfeiler, 1996; Colborn et al., 2001). Thus, the species once believed to be global in scope (Briggs, 1960; Alexander, 1961), now has one of the most restricted distributions among bonefishes. No type locality exists, but Eschmeyer (1998) suggests that the type location may have been the Bahamas.

ALBULA SP. A (UNDESCRIBED)

The Cortez bonefish (Nelson et al., 2004) occurs in the Gulf of California and southern California, but the limits to this distribution are unknown. Notably, it is sister taxon to the eastern Pacific bonefish (*Albula* sp. C; Figure 11.1) that occurs to the south along the same coastline (sequence divergence $d = 0.06$–0.07). To our knowledge there is no scientific name available for this species.

ALBULA SP. B (UNDESCRIBED)

The big-eye bonefish was first reported by sportsmen around the Florida peninsula. It occurs as an adult in deeper water than the common *A. vulpes*, and has a slightly larger eye and distinct dentition. Apart from these features, it is difficult to distinguish from the sympatric *A. vulpes*. However, they are deeply divergent in mtDNA assays ($d = 0.12–0.14$). The species range in the western Atlantic includes subtropical United States, the Caribbean Sea, and subtropical Brazil. It is uncertain whether this range is continuous. A divergent population ($d = 0.01–0.02$) occurs in the Gulf of Guinea, eastern tropical Atlantic. The big-eye bonefish has been dubbed *A. garcia*, in honor of Jerry Garcia, the deceased guitar player for the musical group the "Grateful Dead," but no formal description exists. The name *A. goreensis* (Cuvier and Valenciennes, 1847; type locality Senegal) may be available for this species (J. Shaklee, personal communication in Randall, 1995).

ALBULA SP. C (UNDESCRIBED)

The eastern Pacific bonefish occurs from Pacific Panama to southern Mexico (E.P., unpublished), but the geographic limits of its range are unknown (Pfeiler et al., 2002). The name *A. esuncula* (Garman, 1899; type locality, Acapulco) is available for this species.

ALBULA SP. D (UNDESCRIBED)

This species is probably *A. oligolepis* described in Hidaka et al. (submitted). The corresponding mtDNA lineage was detected from the Coral Sea to South Africa, with strong population structure across this range (Colborn et al., 2001). This is a sister lineage to *A. argentea/A. virgata* in the mtDNA phylogeny ($d = 0.08–0.13$).

ALBULA SP. E (UNDESCRIBED)

This species likely is *A. nemoptera*, as noted above.

SUMMARY

While sportsmen and sportswomen may have been amused by the taxonomic circus of the last century, we believe that the parade of nomenclatural extremes has ended. A second round of proliferation (sundering a single species into several) is unlikely. Likewise, synonymizing these species is unlikely given the deep mtDNA divergences among them, indicating ancient evolutionary separations. Additional species may await discovery in the underexplored regions of the planet, and there will be additional population-level subdivisions, especially for those species with broad geographic distributions. We feel that it is important, however, to guard against future application of a geopolitical species concept (Karl and Bowen, 1999) made worse by invalid or outdated nomenclature. At this junction we need formal descriptions of each evolutionary lineage, with type localities and neotypes where necessary. Subsequently the species identifications for bonefishes should be validated with mtDNA cytochrome *b* gene sequences, in view of the many junior synonyms and misidentifications in the scientific literature. This will prevent confusion and anchor future efforts to a robust phylogenetic framework.

ACKNOWLEDGMENTS

For invaluable advice, assistance, and encouragement, we thank J.B. Shaklee, J.E. Randall, A. Adams, J. Ault, J. Beets, S. Brown, J. Carter, D.E. Colton, R.E. Crabtree, M. Crepeau, N. Dixon, R. van der Elst, S. Fennesy, L.T. Findley, J.E. Fitch, A. Friedlander, E. Grant, P. Greenham, R.E. Gillette, C. Harnden, R. Kusack, A. Lewis, R.J. McKay, J. Mortimer, J.R. Paxton, C.R. Rocha, L.A. Rocha, D.R. Robertson, G. Sedberry, D. Snodgrass, C.S. Tamaru, B. Tibbats, R. Toonen, and G. White. B.W.B. is supported by the National Science Foundation (OCE-0454873), the National Marine Fisheries Service, and the International Gamefish Association. S.A.K. is supported by the National Science Foundation (DEB 03-2192) and the International Gamefish Association. This is publication number 1247 from the Hawaii Institute of Marine Biology, and publication number 6977 from the University of Hawaii School of Oceanography and Earth Science and Technology.

REFERENCES

Akaike, H. 1974. A new look at the statistical model identification. IEEE Trans. Autom. Contr. 19: 716–723.

Alexander, E.C. 1961. A contribution to the life history, biology, and geographical distribution of the bonefish, *Albula vulpes* (Linnaeus). Dana-Rep. 53: 1–51.

Beebe, W. 1942. Eastern Pacific expeditions of the New York Zoological Society, XXX. Atlantic and Pacific fishes of the genus *Dixonina*. Zoologica 27: 43–48.

Bloch, M. and J.G. Schneider. 1801. Systema ichthyologiae; iconibus CX illustratum. Sumtibus Austoris Impressum et Bibliopolio Sanderiano Commissum, Berlin, 554pp.

Briggs, J.C. 1960. Fishes of worldwide (circumtropical) distribution. Copeia 1960: 171–180.

Coates, A.G. and J.A. Obando. 1996. The geologic evolution of the Central American Isthmus. In: Evolution and Environment in Tropical America (Eds: J.B.C. Jackson, A.G. Coates, and A. Budd). University of Chicago Press, Chicago, IL, pp. 21–56.

Colborn, J., R.E. Crabtree, J.B. Shaklee, E. Pfeiler, and B.W. Bowen. 2001. The evolutionary enigma of bonefishes (*Albula* spp.): cryptic species and ancient separations in a globally-distributed shorefish. Evolution 55: 807–820.

Crabtree, R.E., C. Stevens, D. Snodgrass, and F.J. Stengard. 1998. Feeding habits of bonefish, *Albula vulpes*, from the waters of the Florida Keys. Fish. Bull. 96: 754–766.

Cuvier, G. and A. Valenciennes. 1847. Histoire naturelle des poissons. Tome dix-neuvième. Suite du livre dix-neuvième. Brochets ou Lucioïdes. Livre vingtième. De quelques familles de Malacoptérygiens, intermédiaires entre les Brochets et les Clupes. Hist. Nat. Poiss. Vol. 1–19. 544pp.

Eschmeyer, W.N. 1998. Catalog of Fishes. California Academy of Sciences, San Francisco, CA.

Fernandez, C. and A.J. Adams. 2004. Fly Fishing for Bonefish. Stackpole Books, Mechanicsburgh, PA.

Forsskål, P. 1775. Descriptiones animalium avium, amphibiorum, piscium, insectorum, vermium; quae in itinere orientali observavit. Hauniae, Moller. 164pp.

Fowler, H.W. 1911. A new albuloid fish from Santo Domingo. Proc. Acad. Nat. Sci. Phila. 62: 651–654.

Garman, S. 1899. The fishes. In: Reports on an exploration off the west coasts of Mexico, Central and South America, and off the Galapagos Islands by the U.S. Fish Commission steamer "Albatross," during 1891. Mem. Mus. Comp. Zool. 24: 1–431.

Hidaka, K., Y. Iwatsuki, and J.E. Randall. Submitted. Redescriptions of the Indo-Pacific bonefishes *Albula argentea* (Forster) and *A. virgata* Jordan and Jordan, with a description of a related new species, *A. oligolepis*. Ichthyol. Res.

Jordan, D.S. and E.K. Jordan. 1922. A list of the fishes of Hawaii, with notes and descriptions of new species. Mem. Carneg. Mus. 10: 6–7.

Karl, S.A. and B.W. Bowen. 1999. Evolutionary significant units versus geopolitical taxonomy: molecular systematics of an endangered sea turtle (genus *Chelonia*). Cons. Biol. 13: 990–999.

Linnaeus, C. 1758. Systema Naturae, Ed. X. (Systema naturae per regna tria naturae, secundum classes, ordines, genera, species, cum characteribus, differentiis, synonymis, locis. Tomus I. Editio decima, reformata.) Holmiae. Systema Nat. Ed. 10: 1–824.

Nelson, J.S., E.J. Crossman, H. Espinosa-Pérez, L.T. Findley, C.R. Gilbert, R.N. Lea, and J.D. Williams. 2004. Common and Scientific Names of Fishes from the United States, Canada, and Mexico, 6th ed. American Fisheries Society, Special Publication, 29, Bethesda, MD, 386pp.

Pfeiler, E. 1996. Allozyme differences in Caribbean and Gulf of California populations of bonefishes (*Albula*). Copeia 1996: 181–183.

Pfeiler, E., J. Colborn, M.R. Douglas, and M.E. Douglas. 2002. Systematic status of the bonefishes (*Albula* spp.) from the eastern Pacific Ocean inferred from analyses of allozymes and mitochondrial DNA. Environ. Biol. Fish. 63: 151–159.

Pfeiler, E., B.G. Bitler, and R. Ulloa. 2006. Phylogenetic relationships of the shafted bonefish *Albula nemoptera* (Albuliformes: Albulidae) from the eastern Pacific based on cytochrome *b* sequence analyses. Copeia 2006: 778–784.

Posada, D. and K.A. Crandall. 1998. MODELTEST: testing the model of DNA substitution. Bioinformatics 14: 817–818.

Randall, J.E. 1995. Coastal Fishes of Oman. University of Hawaii Press, Honolulu, HI.

Randall, J.E. and M.-L. Bauchot. 1999. Clarification of the two Indo-Pacific species of bonefishes, *Albula glossadonta* and *A. forsteri*. Cybium 23: 79–83.

Rivas, L.R. and S.M. Warlen. 1967. Systematics and biology of the bonefish, *Albula nemoptera* (Fowler). Fish. Bull. 66: 251–258.

Shaklee, J.B. and C.S. Tamaru. 1981. Biochemical and morphological evolution of the Hawaiian bonefishes (*Albula*). Syst. Zool. 30: 125–146.

Smith, D.G. and R.E. Crabtree. 2002. Order Albuliformes. In: The Living Marine Resources of the Western Central Atlantic, Vol. 2. (Ed: K. Carpenter). FAO Species Identification Guide for Fishery Purposes. F.A.O., Rome, pp. 683–684.

Whitehead, P.J.P. 1986. The synonymy of *Albula vulpes* (Linnaeus, 1758) (Teleostei, Albulidae). Cybium 10: 211–230.

12 Ecology of Bonefish during the Transition from Late Larvae to Early Juveniles

Craig Dahlgren, Jonathan M. Shenker, and Raymond Mojica

CONTENTS

Introduction 155
The Planktonic Larval Duration of Bonefish 156
Timing of Onshore Larval Migrations and Settlement 157
Behavioral Adaptations of Bonefish Larvae 160
Physical Transport Processes and Bonefish Larval Influx 161
Interannual Variability in Larval Influx 168
Spatial Variability in Recruitment 169
Settlement and Juvenile Habitats 173
Research Priorities and Application for Management 174
Acknowledgments 175
References 175

INTRODUCTION

Despite the importance of bonefish (*Albula* spp.) fisheries, relatively little is known about the biology and ecology of most bonefish species. This is particularly true of early life history stages of bonefish, including both larval and early juvenile stages. Early life stages are often of critical importance to fish populations. The larval stage is when dispersal occurs over the greatest distances, connecting populations that may otherwise be separated by barriers to juvenile or adult movements. Larval and early juvenile stages are also the ones in which mortality rates are greatest and can vary considerably in response to a variety of environmental, biological, and anthropogenic influences (Chambers and Trippel, 1997). Physical transport of larvae by ocean currents, larval behavior, varying environmental conditions that affect larval growth, trophic interactions that affect larval survival, and both quantitative and qualitative differences in settlement and nursery habitats can cause larval influx to vary (reviewed by Cowen, 2002; Leis and McCormick, 2002).

While humans may be able to influence some of these factors affecting larval survival and the availability of high-quality settlement habitat, many of these processes are beyond our ability to control. Nevertheless, effective management of fish stocks must account for spatial and temporal variability in the supply of larvae and the recruitment of individual juveniles into the population since variability in the level of recruitment can influence fishery productivity and sustainability in subsequent years. Variability in the number of larvae migrating onshore and recruiting to the population can have a significant influence on the size, structure, and distribution of fish populations. A particularly large number of recruits into an area in a given year may result in year classes that dominate the population for several years (e.g., Rothschild, 1986; Doherty and Fowler, 1994; Russ et al., 1996). Under these circumstances, a single recruitment year or even a single recruitment event may be responsible for supporting fisheries and population reproductive outputs in subsequent years. Similarly, differences in the number of recruits between locations may have a significant impact on catches of fish from those locations in subsequent years (e.g., Doherty and Fowler, 1994).

Few studies exist on the early life history of bonefishes (*Albula* spp.). Like eels, tarpon, and ladyfish, bonefish have a leptocephalus larval stage. The research presented in Chapter 13 shows the progress that has been made in understanding bonefish leptocephalus biology and physiological ecology. Nevertheless, we still have much to learn about larval behavior in the field, the biological and environmental factors that influence larval transport, and how these factors influence bonefish populations. This chapter is intended to provide a brief overview of bonefish larval ecology and population recruitment. Updated information on the physical processes that influence the supply of larvae to nearshore nursery habitats is also presented. The significance of this information is discussed with respect to bonefish population dynamics and management.

THE PLANKTONIC LARVAL DURATION OF BONEFISH

At present, only three studies are known to have assessed the time that larval bonefish spend as plankton from hatching of eggs until late-stage larvae arrive in nearshore settlement habitats (Pfeiler et al., 1988; Mojica et al., 1995; Friedlander et al., Chapter 2, this volume). In these studies, ages were determined from late-stage larvae captured prior to settlement using large neuston nets fished in channels or other areas where nearshore settlement occurs (see Shenker et al., 1993 for description of an effective net design). Following capture, age of larvae was calculated by counting daily growth rings on otoliths.

The planktonic larval duration calculated for bonefish is long compared to many tropical marine fish species, but the estimated age at settlement appears to vary somewhat between locations or bonefish species. *Albula* sp. from the Gulf of California is believed to have a larval duration of between 6 and 7 months (Pfeiler et al., 1988). In contrast, bonefish larvae from the Bahamas (presumed to be *Albula vulpes*) had moved onshore at ages ranging from 41 to 71 days with a mean of 56 days (Mojica et al., 1995). A recent study of *A. glossodonta* larvae from Palmyra Atoll in the Pacific have almost identical planktonic larval durations to *A. vulpes* in the Bahamas, with ages of collection from nearshore areas ranging from 48 to 72 days and a mean of 57.2 days (Friedlander et al., Chapter 2, this volume).

Despite the seemingly high similarity between the planktonic larval duration between *A. vulpes* from the Bahamas and *A. glossodonta* from Palmyra Atoll, half way around the world, we should not assume that other bonefish species follow this same pattern. For example, *Albula* sp. collected from the Gulf of California appear to spend more than twice as long in the plankton. Further studies of *A. glossodonta* and *A. vulpes* from other locations and studies of other bonefish species for which we have no information on larval durations is necessary to determine how these patterns may vary. It is also noteworthy that the range in age of individuals varies up to a month in each study. Such high variability in the planktonic larval duration is significant and may indicate that bonefish are capable of delaying metamorphosis and remaining in the plankton longer under certain conditions. Plasticity in the duration of the larval stage may provide individuals with a greater probability of reaching favorable settlement habitats.

The duration that larvae remain in the plankton may influence the distance that larvae are capable of being transported and how populations are connected. Factors such as currents and larval behavior may also influence connectivity; however, reducing the actual distances that populations are effectively connected (reviewed by Cowen, 2002). The duration of the planktonic period of marine species may also have a significant impact on the degree to which populations fluctuate (Eckert, 2003). Gaining a better understanding of the larval durations for all of the eight potential bonefish species (Colborn et al., 2001), and how planktonic larval durations may vary for these species will improve our understanding of how populations are connected and determining management units for bonefish stocks.

TIMING OF ONSHORE LARVAL MIGRATIONS AND SETTLEMENT

Bonefish larvae show distinct annual, monthly, and daily patterns in the timing of larval settlement. While only a handful of studies have addressed these issues for just two to three species of bonefish, they provide a foundation on which to build. These studies examine the processes that influence larval influx on several spatial and temporal time scales.

Annual variability in the influx of late-stage bonefish larvae and their settlement into juvenile nursery areas is based on both the planktonic larval duration (discussed above) and the timing of spawning. Although bonefish spawning has never been documented in the scientific literature, the timing of spawning events can be inferred from examination of bonefish gonads. Anecdotal reports from fishermen can also be useful in estimating spawning times. In the Florida Keys, *A. vulpes* spawning times estimated from gonadal development extend over a 7-month period from November to May (Crabtree et al., 1997). In the Gulf of California, examination of gonads from *Albula* sp. suggests that spawning occurs in the late spring and early summer. Traditional knowledge of bonefish from Kirimati Atoll in Pacific indicates that they may spawn year-round at monthly intervals during the full moon (Friedlander et al.,

Chapter 2, this volume). Based on these estimates of spawning times and the reported planktonic durations of these species, we would expect to see peak annual settlement or population recruitment for *A. vulpes* to occur from December through June in the Florida Keys. For *Albula* sp. in the Gulf of California, peak settlement would be expected to occur in the late autumn and into the winter. Kirimati Atoll bonefish recruitment would be expected to occur throughout the year.

Studies actually measuring settlement rates throughout the year have rarely been conducted for any fish species, but periodic sampling can provide an indication of when peak settlement may occur. In an early study of bonefish, Alexander (1961) found that bonefish larvae were most commonly caught from November through April in the West Indies; however, some larvae were also caught during August. Annual winter channel net sampling of late-stage bonefish larvae at Lee Stocking Island, Bahamas, from the winter of 1990/91 to the winter of 2003/04 resulted in bonefish larvae being caught throughout the December through March sampling periods (Mojica et al., 1995; Dahlgren, unpublished data; see the section on physical processes and bonefish larval influx for presentation of some of these data). Bonefish larvae were also commonly caught in channel nets fished during the summer (June–September, 1992) at the same location, with over 76% of the 1112 bonefish samples taken during the first 12 days of the sampling period beginning in late June (Thorrold et al., 1994; Mojica et al., 1995). All of these studies are in general agreement with the calculated peak of larval ingress based on estimates of peak spawning times and planktonic larval durations. The few fish captured outside the expected recruitment window (Alexander, 1961; Thorrold et al., 1994) may result from long-distance transport from locations where spawning occurs at different times or may reflect variable in planktonic larval durations allowing for delayed recruitment. Since genetic analyses were not conducted on these fish, there is the possibility that these fish were different bonefish species that spawn at different times.

Sampling for *Albula* sp. leptocephali in the Florida Keys show low recruitment compared to the Bahamas and other locations. For example, sampling by Harnden et al. (1999) using two channel nets in Hawk Channel near Long Key in 1993 over 160 nights throughout the year yielded nearly 35,000 larval fishes, but only six *Albula* sp. leptocephali were collected. All six bonefish larvae were captured during summer months. Monthly sampling using channel nets in the Key West National Wildlife Refuge from May to November 1999 yielded no bonefish larvae (Dahlgren, unpublished data). Similarly, towed plankton net sampling at several locations in the upper Florida Keys from July through September 2000 did not yield any bonefish larvae (Sponaugle et al., 2003). Periodic seine netting along beaches throughout the Florida Keys from November 2003 to August 2005, however, yielded occasional leptocephalus larvae from November through May (although the presence of bonefish larvae in monthly samples varied between years), but none were collected during summer months (Adams et al., Chapter 15, this volume). The near absence of bonefish larvae from these samples may be indicative of low recruitment to the Florida Keys or the result of these studies not effectively targeting peak settlement months or settlement areas.

In the Pacific, Friedlander et al. (2004) reported capturing *A. glossodonta* leptocephalus larvae within channels entering Palmyra Atoll's lagoon in March and August, but not in November 2003. This suggests some seasonality to recruitment for this species, but observed patterns may also be an artifact of few sampling times and low bonefish catch rates (e.g., maximum daily capture in channel nets was only three individuals on Palmyra as opposed to several hundred in Bahamian channel net samples).

During periods of larval influx, the number of bonefish larvae moving onshore varies considerably between days and is even based on the time of day (e.g., Mojica et al., 1995). This may be attributed to variable environmental factors and behavioral responses to them. Onshore migrations present bonefish larvae with a suite of new challenges associated with their new environment. Late-stage larvae move from a relatively stable offshore ocean environment lacking much physical structure, to a coastal inshore one subject to greater physical heterogeneity and environmental variability, as well as a new suite of predators. Furthermore, bonefish larvae are also going through morphological and physiological transformations described in Chapter 13. Bonefish larvae have several adaptations that are expected to improve the likelihood of survival through this critical transition.

The first challenge facing bonefish larvae is physically moving from the offshore environment where they develop through various leptocephalus larval stages to the nearshore habitats where they develop into juveniles and then adults. Bonefish larvae, however, are probably not strong swimmers as evident in nearly all of the larvae that are captured in channel nets becoming impinged on the side of the net by the force of the water current and dying (as opposed to other species of leptocephali; C. Dahlgren, personal observation). Thus it is likely that they rely on currents to assist with their onshore transport.

Tidal currents in particular have been shown to influence the timing of settlement (reviewed in Cowen, 2002). In the Bahamas, for example, larvae of almost all fish taxa, including bonefish, are captured in channel nets primarily at night and during incoming tides (Shenker et al., 1993; Thorrold et al., 1994). Similarly, flood tides also bring *Albula* sp. larvae into estuaries in the Gulf of California (Pfeiler, 1984). By timing onshore migrations with tidal cycles, bonefish may improve their chances of reaching shallow water juvenile habitats.

As they move onshore, bonefish larvae must also avoid the gauntlet of predators associated with coral reefs and other nearshore habitats. To facilitate survival at this time, bonefish have adaptations to avoid detection by predators, such as having a transparent body that is difficult for visual predators to detect. The timing of onshore migrations may also be an adaptation to avoid predators. Many late stage larval fish, including bonefish move onshore at night and at the surface of the water (e.g., Shenker et al., 1993; Thorrold et al., 1994). In the Bahamas, 98% of bonefish leptocephalus larvae collected in passes between oceanic environments and nearshore habitats occurred at night (Shenker et al., 1993). In the Gulf of California, nighttime influx of larvae to estuaries was also noted (Pfeiler, 1984). Moreover, 90% of bonefish larvae in the Bahamas were collected in the upper 1 m of the water column (Thorrold et al., 1994, Mojica et al., 1995). Nighttime onshore migrations in surface waters may reduce the risk of detection by visual predators that are most active during the day and live in association with benthic structure.

The timing of settlement within the monthly lunar cycle reflects a similar predator avoidance strategy. Mojica et al. (1995) found bonefish larval influx to be correlated with the period of flood tide under moonless conditions (i.e., the amount of time that the tide was incoming after sunset and before the moon rose or after the moon set and before sunrise), which varied from 0 to 7 h each night. The majority of recruitment occurred on nights with more than 4 h of dark flood tide (Mojica et al., 1995).

BEHAVIORAL ADAPTATIONS OF BONEFISH LARVAE

The discussion so far has focused on the influence that lunar periodicity, tidal currents, and light levels have on the timing of bonefish larval transport to nearshore systems. For these factors to produce observed patterns of bonefish larval influx, bonefish must be capable of detecting them and responding to them in a way that affects their onshore transport. Using an example from the previous section, Mojica et al. (1995) found larval influx to the Exuma Cays, Bahamas, to be greatest during nights with more than 4 h of flood tide under moonless conditions. Plankton tows in Exuma Sound, Bahamas, concurrent with channel net sampling in January and February 1991, however, indicate that bonefish larvae were found from the shelf edge up to 24 km out from shore providing a pool of bonefish larvae available for recruitment throughout the sampling period (Drass, 1992). The abundance of bonefish larvae in nighttime flood tides under moonless conditions and the scarcity of bonefish larvae in daytime or moonlit flood tides, despite the availability of larvae offshore, suggests that larvae are actively positioning themselves for onshore migrations in response to light levels.

Although bonefish larvae do not appear to be capable of swimming against currents to maintain their horizontal position, Mojica et al. (1995) propose that active vertical migrations may provide bonefish with a mechanism for timing onshore migrations. Several studies have shown how fish larvae and other plankton can migrate vertically to maintain their horizontal position or take advantage of favorable conditions for growth, survival, and directional movement (reviewed by Leis and McCormick, 2002). Vertical migrations have not been documented for bonefish, but findings of bonefish larvae concentrated in the upper meter of the water during onshore migrations despite vertical distribution of earlier stage larvae to depths of 25–50 m (Drass, 1992) suggest that they change their vertical distribution as they move onshore, similar to other zooplankton.

Other factors may serve also as cues to influence either the timing or location of settlement (reviewed by Kingsford et al., 2002). Chemical cues or scents from nursery habitats or conspecifics have been shown to influence settlement of a variety of species (Sweatman, 1988; Atema et al., 2002). Other studies have shown that larvae are capable of detecting sound (e.g., breaking waves) from nearshore settlement areas and respond from distances of kilometers away (e.g., Tolimieri et al., 2000). Biological noise from reefs also attracts and induces settlement-stage fishes to settle at specific locations (Simpson et al., 2005). Since postsettlement mortality can be high and may vary between habitats (e.g., Dahlgren and Eggleston, 2000, 2001), the ability of fish to select high-quality habitats may be a great advantage over random onshore transport or sampling different habitats to determine their suitability. Remote detection of cues from settlement habitat may be particularly advantageous when settlement habitat is rare or has a patchy distribution. Whether bonefish can detect cues from suitable settlement habitats and how bonefish behavior influences spatial and temporal settlement patterns are topics for further research.

PHYSICAL TRANSPORT PROCESSES AND BONEFISH LARVAL INFLUX

Based on available evidence, the timing of the average peak in bonefish larval influx is highly influenced by behavioral adaptations that allow bonefish to time their onshore migration based on lunar periodicity. Nevertheless, much variability in the number of bonefish larvae moving onshore on any given night cannot be accounted for by these factors alone. Other environmental variability may play an important role in determining bonefish larval influx.

Analysis of 4 years of wintertime channel net data from the Bahamas did not reveal consistent effects of other potentially important environmental variables and larval influx (Mojica et al., 1995). When significant correlations were occasionally detected, they varied from year to year and were overshadowed by lunar periodicity (Mojica et al., 1995). For example, cross-shelf winds were never correlated with larval influx for bonefish and alongshore winds were significantly correlated with the number of bonefish caught in channel nets nightly in only half of the years sampled, and their affect was inconsistent. Northwest winds were correlated with peaks in larval influx during 1991/92, and peaks in larval influx lagged 3 days behind winds to the southeast the following year (Mojica et al., 1995).

Although Mojica et al. (1995) found no compelling evidence of the importance of wind-driven transport, wind forcing has a significant influence on surface currents in this system (Smith, 2004) and onshore wind events have been shown to impact recruitment of Nassau grouper, *Epinephelus striatus*, which co-occur with bonefish larvae in channel net samples in this system (e.g., Shenker et al., 1993). The apparently conflicting or inconclusive results related to the importance of wind-driven transport for bonefish may be the result of the strength of the lunar/tidal influence overpowering the influence of wind-driven transport over short periods of time (<3 months each year).

To gain a better understanding of how these processes affect bonefish transport and how they vary over time, additional years of sampling are necessary. In this section, we present new data from continued winter channel net sampling of bonefish leptocephalus larvae extending the 4-year period sampled by Mojica et al. (1995) to encompass a period that spans 12 years from 1990/91 to 2001/02, with sampling collected during 9 of those years (no sampling was conducted in the winter of 1998/1999, 1999/2000, and 2000/2001).

The methodology used to sample bonefish larvae is described by Shenker et al. (1993) and Mojica et al. (1995). Briefly, channel nets measuring 2 m across by 1 m deep with a mesh size of 2 mm were fished at the surface nightly from shortly before sunset until shortly after sunrise during all years. Sampling dates varied somewhat annually, but included at least two new moon periods each winter (Table 12.1). While nets were fished at several stations, only those from two stations fished consistently throughout the sampling period were included in analyses.

TABLE 12.1
Channel Net Sampling Dates in the Lee Stocking Island, Bahamas Area during the 1990/91–2001/02 Winter Sampling Seasons

Sampling Season	Date Start	Date End	Days of Sampling
1990/91	December 19	February 21	64
1991/92	December 22	February 22	62
1992/93	December 13	February 24	73
1993/94	December 17	February 23	68
1994/95	December 19	April 2	104
1995/96	January 5	March 20	75
1996/97	December 29	March 11	72
1997/98	January 5	March 6	60
1998/99	No sampling		
1999/2000			
2000/1			
2001/2	January 30	March 18	47

Each morning larval fish samples were collected from nets and all bonefish leptocephalus larvae were identified and counted. All bonefish larvae included in analyses were late stage larvae (late stage Phase 1 or early stage Phase 2; see Chapter 13 for details). For samples collected in the winter of 1994/95 through 2001/2, a subsample of bonefish of different sizes were used in genetic analyses to positively verify their identification as *A. vulpes* (C. Dahlgren, unpublished data). Since no other species of bonefish were detected in subsamples, it is assumed that all bonefish collected were *A. vulpes* and that observed differences in larval size were likely to be due to the capture of fish at different stages of larval development (Chapter 13).

Correlations between the number of bonefish larvae recruiting nightly and various environmental variables were analyzed throughout the sampling period following the approach used by Mojica et al. (1995), which uses cross-correlations between channel net time series and environmental variable time series. To remove the effects of autocorrelations in time series datasets, auto regressive integrated moving average (ARIMA) models were fit to the data and residuals from these models were used in cross-correlation analyses between bonefish larval ingress and environmental variables. Specific variables examined included lunar phase, hours of flood tide under dark conditions, and the speed of both the cross-shelf and alongshore component average nightly winds to account for the effects of any wind-driven surface currents. Cross-shelf and alongshore component of wind was calculated from nightly wind averages (speed and direction) at the weather station on Lee Stocking Island, Bahamas.

Results of these analyses corroborate previous findings that the influx of bonefish larvae is correlated with both lunar phase and the hours of flood tide under moonless conditions each night (Figures 12.1 and 12.2; Table 12.2). Cross-correlation plots of the relationship between the total numbers of bonefish larvae captured each night and percentage of illumination of the moon (Figure 12.1B) and number of hours of dark flood tide (Figure 12.2B) show the periodic nature of bonefish recruitment. For example, there was a negative correlation between bonefish recruitment and lunar phase (percent illumination) at a lag of ±3 days (i.e., low recruitment during full moon) and a positive correlation between recruitment and lunar phase with a lag of 13–15 days (i.e., high recruitment during new moon; Figure 12.1B). Of the 9328 bonefish larvae captured during all sampling periods, 71.5% were caught on nights in which the hours of flood tide under moonless conditions was greater than average (2.83 h), and nearly half of bonefish caught (47.3%) were captured on nights in which there were more than 4 h of flood tide under moonless conditions.

However, some results from this extended dataset differed from the previous analysis of Mojica et al. (1995). When additional years were included in the analyses, there was a change in the total contribution of wind-driven transport to the magnitude of bonefish larval ingress (Table 12.2). The number of larvae collected was correlated with the cross-shelf component of winds only during the night of sampling (Figure 12.3A) when all years were included in the analysis, but no significant correlations were detected between bonefish recruitment and the alongshore component of winds (Figure 12.3B). For nights in which wind data were collected, more than 77% of bonefish captured in nets were on nights in which there was an onshore component of winds and 65% of bonefish sampled were collected on nights in which the onshore component of winds was greater than the average onshore component of 1.65 km/h.

These findings suggest that, although the timing of larval influx may be predominantly based on lunar and tidal periodicity, the magnitude of larval influx is also influenced by wind-driven transport mechanisms. Although statistically significant, the strength of correlations between larval influx and individual environmental parameters are relatively low. The combination of favorable wind and lunar period, however, can account for the majority of bonefish larval influx, despite being only a small fraction of the whole sampling period. Of the 667 nights for which we have data on larval bonefish catches, hours of dark flood tide, and winds, there were only 174 (26%) in which greater-than-average hours of dark flood tide occurred simultaneously with onshore winds, yet these nights accounted for 53% of all bonefish caught during this period.

The importance of the combined affects of onshore winds and hours of dark flood tide can be illustrated graphically using data from 1993/94 in Figure 12.4 showing the number of larvae caught in channel nets as a function of the hours of dark flood tide, cross-shelf winds, and a combination of the two. Every peak in larval influx (e.g., nightly catches of more than 20 bonefish) during this sampling period occurred during nights in which there were greater than average hours of dark flood tide. Five of seven peaks in larval influx, including those of the greatest magnitude for both January and February, occurred during nights of more than 4 h of dark flood tide. Nevertheless, there were also nights of average or below average settlement during nights with more than 2.5 or 4 h of dark flood tide. This suggests that other factors must contribute to above average larval influx. When the influx of larvae is plotted with the cross-shelf component of wind (Figure 12.4B), all but one peak in larval influx occurred during nights in which winds were onshore. Nevertheless, there were also several nights in which wind direction was onshore and recruitment was average or below average.

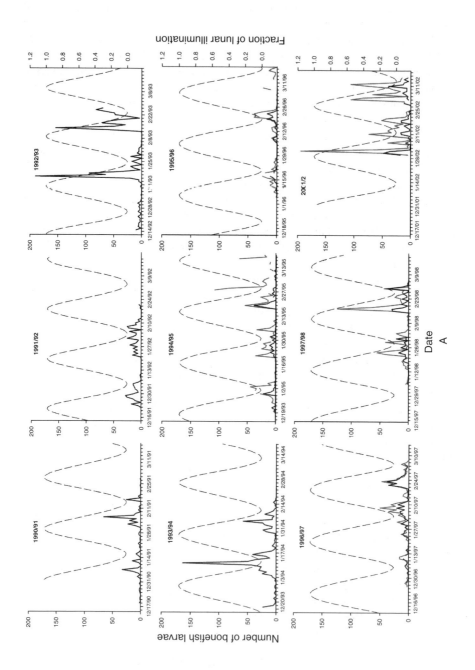

Ecology of Bonefish during the Transition

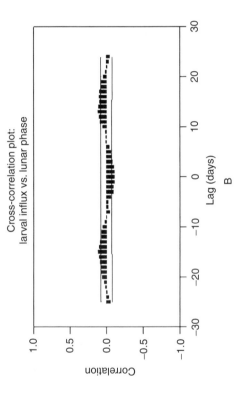

FIGURE 12.1 Number of bonefish larvae captured in channel nets during winter sampling (solid lines) plotted with the lunar phase (dashed line) expressed as a percentage of the full moon (A). Data from 1990/91 to 1993/94 are combined for two sampling stations near Lee Stocking Island, Bahamas. Data from 1994/95 to 2001/2 are presented for the two sampling stations individually. (B) The cross-correlation plot illustrating the relationship between the total number of larvae captured nightly and the lunar phase with a lag time of up to 15 days. A lag of 0 days represents the full moon period (i.e., periods of greatest fraction of moon illuminated). Horizontal lines represent 95% confidence intervals.

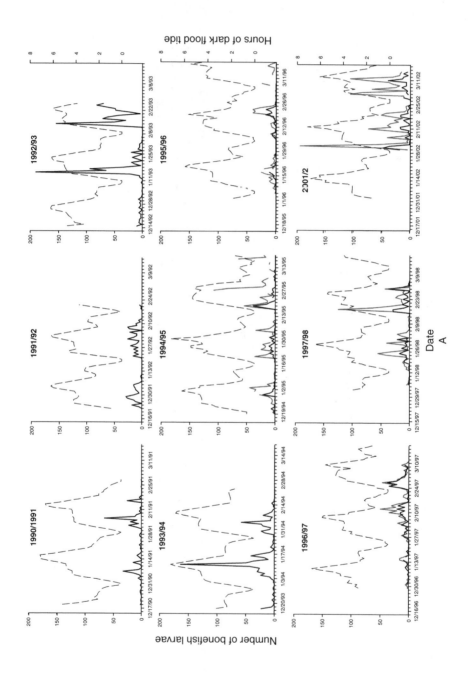

Ecology of Bonefish during the Transition

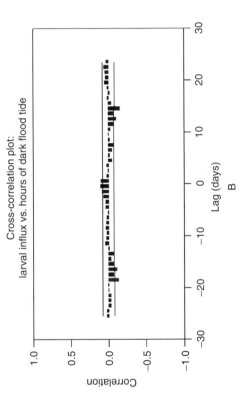

FIGURE 12.2 Number of bonefish larvae captured in channel nets during winter sampling (solid lines) is plotted with the hours of flood tide under dark conditions (dashed line) (A). Data from 1990/91 to 1993/94 are combined for two sampling stations near Lee Stocking Island, Bahamas. Data from 1994/95 to 2001/2 are presented for the two sampling stations individually. (B) The cross-correlation plot illustrating the relationship between the total number of larvae captured nightly and hours of dark flood tide with a lag time of up to 15 days. A lag of 0 days represents periods of the most flood tide under dark conditions. Horizontal lines represent 95% confidence intervals.

TABLE 12.2
Correlations between the Total Number of Bonefish Larvae Sampled Nightly and Nightly Tidal and Wind Conditions

Factor	Correlation Coefficient	Chi-Square	P Value
Lunar phase	−0.107	7.583	0.006
Hours of dark flood tide	0.105	7.44	0.006
Cross-shelf winds	−0.098	6.458	0.011
Alongshore winds	−0.062	2.542	0.111

Note: Statistics reported here are for direct correlations each night with no lag time (see cross-correlation plots, Figures 12.1A and 12.2B for cases in which there were significant correlations involving time lags) for data on time lags in correlations.

When both onshore winds and hours of dark flood tides are combined, however, we see a much tighter relationship between environmental conditions and larval influx. Six of the seven peaks in larval influx occurred when onshore winds coincided with nights with more than 2.5 h of dark flood tide, and the highest monthly peaks occurred when onshore winds exceeded 4 km/h and there were more than 4 h of dark flood tide. In addition, there were few periods of below average larval influx when favorable lunar periods co-occurred with onshore winds. Thus, it appears that larval supply is temporally patchy—when patches co-occur with good conditions, larval influx is high; but if good conditions occur without many larvae nearby or if larvae are available, but conditions are poor, influx is low.

INTERANNUAL VARIABILITY IN LARVAL INFLUX

Another interesting temporal pattern that is evident in the expanded sampling from of bonefish larvae using channel nets in the Bahamas is the variability between years. While each year of sampling shows a pattern of peaks in larval ingress under favorable lunar periods and onshore winds, as discussed above, the total catch across years is not constant. While some variability may be due to the number of nights sampled each year, and the timing of sampling with respect to the months sampled, and the number of new moon cycles included in sampling, these factors alone do not account for interannual variability. When the total catch per lunar month (full moon to full moon) for January and February (the 2 months for which the most data is available) is compared across all years, there is a distinct increase in larval influx over time, with the average of total monthly catches during the winter of 1998 being more than twice that of the first two sampling periods 1991 and 1992 (Figure 12.5). This trend continued throughout the sampling period with monthly catches in 2002 (January only) being double the average monthly catch of 1998.

While the cause of this increase may be driven by a natural fluctuation in bonefish populations, it may also result from a decrease in consumptive fishing for bonefish. The decrease in consumptive fishing for bonefish began with a ban on netting for bonefish in the Bahamas. While this ban was initially implemented prior to the start of larval sampling, lack of enforcement made it ineffective until the 1990s. In addition, the 1990s saw a rise in the importance of bonefish as a resource for catch and release recreational angling in the Bahamas. The combination of the netting ban and the increase in the importance of bonefish as a catch-and-release fishery has reduced the importance of bonefish as a baitfish (primarily for billfish) and the traditional fishery for local consumption. It is possible that such changes in the fishery may be allowing bonefish stocks to increase in the Bahamas, causing greater reproductive output and subsequent larval ingress to nearshore nurseries.

SPATIAL VARIABILITY IN RECRUITMENT

Much less data has been collected to describe spatial variability in larval supply and recruitment to juvenile populations. Channel net sampling by Shenker et al. (1993) at four stations in cuts between islands near Lee Stocking Island, Bahamas, found the total number of larval fish collected between sites to vary nearly fourfold between sites spread less than 5 km apart during a 75-day sampling period. Temporal patterns in larval supply (e.g., data from 1994/95–2001/02, Figures 12.1–12.4) were consistent among sites, however, differing only in the magnitude of larval influx. This can be explained by bonefish larvae being influenced by the same environmental variables across this spatial scale of 5 km or more, but the strength of these variables (e.g., tidal currents, wind exposure) varied somewhat between sites. Consistency in patterns across this spatial scale also suggests that larval bonefish fish patch size is greater than 5 km.

Studies of bonefish larval supply to nearshore settlement habitats have not been effectively conducted on larger spatial scales. Adams et al. (Chapter 15, this volume) and others have collected bonefish larvae as part of large-scale sampling efforts; however, the abundance of larvae collected has been too low for effective comparisons among sites. Comparing results of different studies from different systems is difficult due to differing sampling designs and other confounding factors, such as the use of different gear, sampling protocols, and sampling frequency and timing in different studies. Nevertheless, we can learn something from these comparisons. For example, bonefish larvae are a major component of larval fish moving onshore in the Bahamas (e.g., Shenker et al., 1993), with individual channel nets catching up to several hundred bonefish larvae in a single night, but bonefish larvae are only occasionally captured from other locations (e.g., Friedlander et al., 2004; Adams et al., Chapter 15, this volume). Understanding how bonefish larval recruitment varies from location to location; what causes this spatial variability; and how spatial variability impacts bonefish populations are areas of research that will contribute greatly to our understanding of bonefish populations and our ability to manage bonefish stocks.

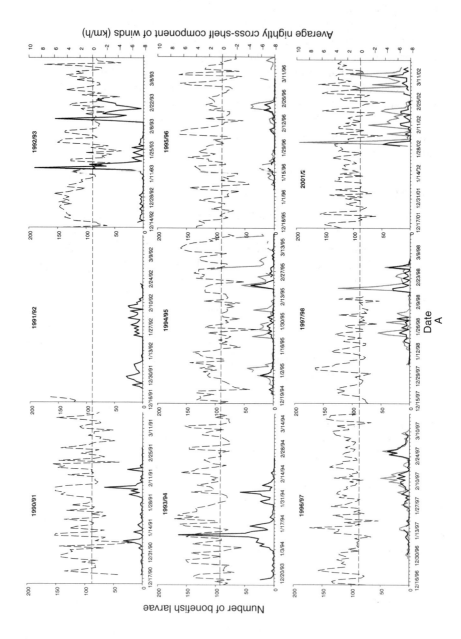

Ecology of Bonefish during the Transition

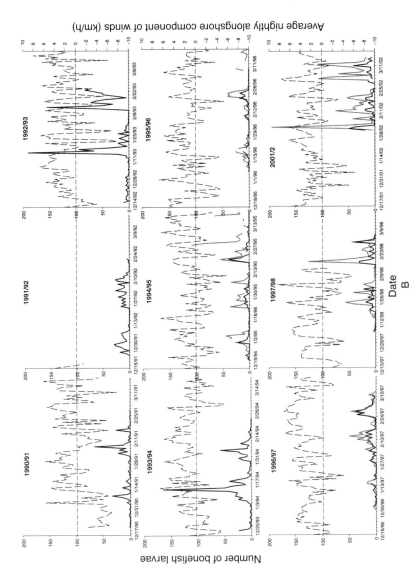

FIGURE 12.3 Number of bonefish larvae captured in channel nets during winter sampling plotted with cross-shelf component of winds (A) and alongshore component of winds (B) shown as dashed lines. Data from 1990/91 to 1993/94 are combined for two sampling stations near Lee Stocking Island, Bahamas. Data from 1994/95 to 2001/2 are presented for the two sampling stations individually.

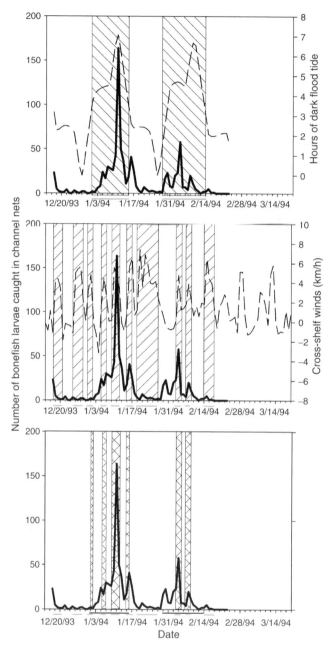

FIGURE 12.4 Abundance of bonefish larvae captured in channel nets during the winter of 1993/94. The top graph shows the number of hours of dark flood tide (dashed line) each night during this sampling period with nights in which there was more than 2.5 h of dark flood tide shaded. The middle graph shows the cross-shelf component of winds each night (dashed line) with all nights averaging onshore winds shaded. The bottom graph shows nights in which there were more than 2.5 h of dark flood tide and onshore winds shaded.

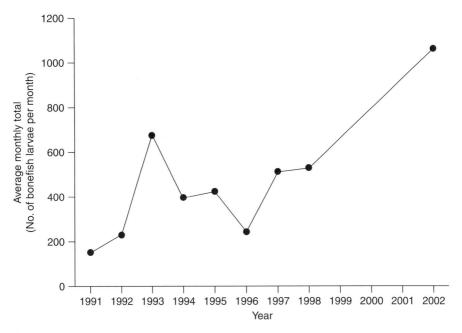

FIGURE 12.5 Average monthly catches of bonefish during winter channel net sampling at Lee Stocking Island, Bahamas for the period of 1990/91 through 2001/2. Monthly catches represent lunar months (full moon to full moon) in channel nets, centered on the January and February new moon periods.

SETTLEMENT AND JUVENILE HABITATS

Habitats used by juvenile bonefish may be quite patchily distributed. Few studies have examined the settlement and metamorphosis of bonefish from leptocephalus larvae to juvenile, and identified the habitats in which this transition occurs (but see Chapter 13 for physiological factors of metamorphosis). Although several studies have reported the habitat in which early juvenile bonefish occur, fish sampled in this study may be at different stages or ages that influence habitat use, or may even be different species. For example, extensive sampling of various windward and leeward shore habitats in the Florida Keys resulted in the capture of nearly 700 juvenile bonefish, with 94% of those identified being *Albula* sp. B (as identified by Colborn et al., 2001) and only 6% were *A. vulpes* (Adams et al., Chapter 15, this volume). This finding is particularly interesting as *A. vulpes* comprise the primary species in the Florida Keys bonefish fishery, yet its juvenile habitats there remain unidentified. It also highlights the facts that the importance of different habitats may vary for juveniles of different bonefish species. In the Florida Keys study, *Albula* sp. B juveniles were found in sand or sand-sparse seagrass mix adjacent to sandy beaches or beachrock shoreline. In contrast, juvenile bonefish were not collected in similar habitats of the Virgin Islands (Mateo and Tobias, 2004), but were found to use mangrove-lined lagoons in St. Croix (Tobias, 1999). Clearly, much research still needs to be done to identify which habitats various species of juvenile bonefish use and which of these habitats are essential for contributing enough juveniles to sustain adult bonefish populations.

RESEARCH PRIORITIES AND APPLICATION FOR MANAGEMENT

As evident in this chapter's review of the factors that influence the early life stages of bonefish, we are only beginning to understand how patterns in distribution and abundance at this stage affect populations as a whole and the bonefish fishery. Based on what is known of the population dynamics of bonefish during their early life history stages, many gaps in our knowledge exist and are opportunities to improve our understanding of bonefish population and our ability to manage human activities that may affect bonefish populations. Based on the fact that nearly all of what we know about the larval and juvenile populations of bonefish come from a few studies in the Bahamas, Gulf of California, and one or two studies from other locations, the most obvious research recommendation would be to conduct more studies on early life stages of bonefish from more locations and for more bonefish species. Studies must also focus on issues that will improve our management of bonefish habitat and fisheries. Specific research priorities are discussed below.

While we have a fundamental understanding of the biophysical factors that affect the timing bonefish larval influx to nearshore areas and temporal variability in abundance, information on spatial variability in these patterns or how spatiotemporal variability in recruitment affects bonefish populations will greatly improve our understanding of bonefish populations. Is temporal variability in recruitment reflected in the age structure of bonefish populations? Are differences in bonefish abundances between sites due to differences in recruitment to those areas, or are recruitment patterns modified by postrecruitment processes? Studies that address these questions will help us understand the structure of bonefish stocks and will help with the design of effective management strategies for bonefish stocks.

Another priority research area involves identifying key characteristics of the habitats that bonefish use from the time of settlement through their early juvenile stages. Which habitats are the most important for settlement and subsequent juvenile development? What characteristics contribute to their relative quality? How do human impacts such as coastal development, dredging, mangrove destruction, or pollution affect these habitats and bonefish use of them? Improving our understanding of juvenile habitat use will greatly improve our ability to identify and protect essential habitats for these species. Adopting a research strategy, such as that outlined by Adams et al. (2006), may greatly improve our understanding of the nursery function of various habitats for bonefish, and address some of these critical questions. This will allow managers to prioritize habitats for protection (e.g., identifying essential habitat) or adopt an ecosystem-based approach for managing bonefish, which includes spatial protection (e.g., marine-protected areas) for critical habitats or specific areas.

Another potentially useful tool for the management of bonefish stocks may be the development of recruitment indices based on the influx of larvae during peak recruitment periods. Such an index may be able to facilitate the detection upward or downward trends in bonefish populations before these trends are otherwise noticeable in the fishery. Such an index would allow marine resource managers to adapt management rapidly to prevent populations from dipping below sustainable levels and evaluate the efficacy of management by examining changes in larval influx in response to adaptive management. In the Bahamas, where we have the greatest amount of larval recruitment data, the increase in total monthly larval influx over an 11-year period (Figure 12.5) may indicate that larval influx may vary in response to increasing spawning population biomass resulting from improved management of the bonefish fishery. This sort of analysis appears to be promising, but more research must be done to determine if the number of larvae recruiting to a system is a reflection on reproductive output of the population or if the number of larvae recruiting to an area influence the number of fish in the fishery in subsequent years before such an index can be implemented.

More questions than answers exist when it comes to understanding the early life stages of bonefish and many management tools are in their early stages of development and have rarely been applied to bonefish. Our current knowledge of the early life history stages of bonefish, however, provides a foundation for future studies and the development of management strategies to effectively ensure the health of bonefish stocks. New research tools and technology have advanced our understanding of bonefish further in the last 5 years than the previous 50 years. The current increased interest in bonefish research by scientists and fishermen alike, is encouraging and should serve as a catalyst for directed research that will improve management of this important species.

ACKNOWLEDGMENTS

The authors thank A. Adams and A. Friedlander for access to unpublished data and manuscripts development during the writing of this chapter. Assistance with channel net collections was provided by D. Eggleston, D. Nadeau, M. Sutyak, L. Etherington, S. Ratchford, T. Wolcott Dahlgren, H. Stewart, M. Blaesbjerg, K. Reinhold, K. Buch, E. Rechisky, C. Greengrass, and M. Bird. The authors also thank J. Ault for his efforts in putting this volume together. The chapter was greatly improved by the comments of J. Ault and two anonymous reviewers. The Perry Institute for Marine Science's Caribbean Marine Research Center provided funding for much of the research presented in this chapter. Additional funding for the preparation of this manuscript was provided by the Curtis and Edith Munson Foundation.

REFERENCES

Adams, A., Dahlgren, C.P., Kellison, G.T., Kendall, M., Layman, C., Ley, J., Nagelkerken, I. and Serafy, J. 2006. The nursery function of back reef systems. Marine Ecology Progress Series 318: 287–301.

Alexander, E.C. 1961. A contribution to the life history, biology and geographical distribution of the bonefish, *Albula vulpes* (Linneaus). Dana-Report, Carlsberg Foundation 53: 1.

Atema, J., Kingsford, M.J. and Gerlach, G. 2002. Larval reef fish could use odour for detection, retention and orientation to reefs. Marine Ecology Progress Series 241: 151–160.

Chambers, R.C. and Trippel, E.A. (eds.). 1997. Early life history and recruitment in fish populations. Chapman & Hall, London.

Colborn, J., Crabtree, R.E., Shaklee, J.B., Pfeiler, E. and Bowen, B.W. 2001. The evolutionary enigma of bonefishes (*Albula* spp.): cryptic species and ancient separations in a globally distributed shorefish. Evolution 55(4): 807–820.

Cowen, R.K. 2002. Larval dispersal and retention and consequences for population connectivity, pp. 149–170. In Sale, P.F. (ed.), Coral reef fishes dynamics and diversity in a complex ecosystem. Academic Press, San Diego, CA.

Crabtree, R.E., Snodgrass, D. and Harnden, C.W. 1997. Maturation and reproductive seasonality in bonefishes, *Albula vulpes*, from the waters of the Florida Keys. Fishery Bulletin 95: 456–465.

Dahlgren, C.P. and Eggleston, D.B. 2000. Ecological processes underlying ontogenetic habitat shifts in a coral reef fish. Ecology 81(8): 2227–2240.

Dahlgren, C.P. and Eggleston, D.B. 2001. Spatiotemporal variability in abundance, distribution and habitat associations of early juvenile Nassau grouper. Marine Ecology Progress Series 217: 145–156.

Doherty, P.J. and Fowler, A.J. 1994. An empirical test of recruitment limitation in a coral reef fish. Science 263: 935–939.

Drass, D.M. 1992. Onshore movements and distribution of leptocephali (Osteichthyes: Elopomorpha) in the Bahamas. MS Thesis, Florida Institute of Technology, Melbourne, FL. 85p.

Eckert, G.L. 2003. Effects of the planktonic period on marine population fluctuations. Ecology 84: 372–383.

Friedlander, A., Beets, J., Caselle, J., Bowen, B., Ogawa, T., Lowe, C., Calitri, T. and Lange, M. 2004. Investigation of the Biology and Monitoring the Stock of Bonefish of Palmyra Atoll, Line Islands, Central Pacific Ocean. Year Two Report to The Nature Conservancy Palmyra Project. 45pp.

Harnden, C.W., Crabtree, R.E. and Shenker, J.M. 1999. Onshore transport of Elopomorph Leptocephali and glass eels (Pisces: Osteichthyes) in the Florida Keys. Gulf of Mexico Science 17(1): 17–26.

Kingsford, M.J., Leis, J.M., Shanks, A., Lindeman, K.C., Morgan, S.G. and Pineda, J. 2002. Sensory environments, larval abilities and local self-recruitment. Bulletin of Marine Science 70(1) Supplement: 309–340.

Leis, J.M. and McCormick, M.I. 2002. The biology, behavior and ecology of the pelagic, larval stage of coral reef fishes, pp. 171–200. In Sale, P.F. (ed.) Coral reef fishes dynamics and diversity in a complex ecosystem. Academic Press, San Diego, CA.

Mateo, I. and Tobias, W.J. 2004. Survey of nearshore fish communities on tropical backreef lagoons on the southeastern coast of St. Croix. Caribbean Journal of Science 40(3): 327–342.

Mojica, R., Jr., Shenker, J.M., Harnden, C.W. and Wagner, D.E. 1995. Recruitment of bonefish, *Albula vulpes*, around Lee Stocking Island, Bahamas. Fishery Bulletin 93: 666–674.

Pfeiler, E. 1984. Inshore migration, seasonal distribution and sizes of larval bonefish, Albula, in the Gulf of California. Environmental Biology of Fishes 10: 117–122.

Pfeiler, E., Mendiza, M.A. and Manrique, F.E. 1988. Premetamorphic bonefish (*Albula* sp.) leptocephali from the Gulf of California with comments on life history. Environmental Biology of Fishes 21: 241–249.

Rothschild, B.J. 1986. The dynamics of marine population. Harvard University Press, Cambridge, MA.

Russ, G.R., Lou, D.C. and Ferreira, B.P. 1996. Temporal tracking of a strong cohort in the population of a coral reef fish, the coral trout, *Plectropomus leopardus* (Serranidae: Epinephelinae), in the central Great Barrier Reef, Australia. Canadian Journal of Fisheries and Aquatic Science 53: 2745–2751.

Shenker, J.M., Maddox, E.D., Wishinski, E., Pearl, A., Thorrold, S.R. and Smith, N. 1993. Onshore transport of settlement-stage Nassau grouper *Epinephelus striatus* and other fishes in Exuma Sound, Bahamas. Marine Ecology Progress Series 98: 31–43.
Simpson, S.D., Meekan, M., Montgomery, J., McCauley, R. and Jeffs, A. 2005. Homeward sound. Science 308(5719): 221.
Smith, N.P. 2004. Transport processes linking shelf and back reef ecosystems in the Exuma Cays, Bahamas. Bulletin of Marine Science 75: 269–280.
Sponaugle, S., Fortuna, J., Grorund, K. and Lee, T. 2003. Dynamics of larval fish assemblages over a shallow coral reef in the Florida Keys. Marine Biology 143: 175–189.
Sweatman, H.P.A. 1988. Field evidence that settling coral reef fish larvae detect resident fishes using dissolved chemical cues. Journal of Experimental Marine Biology and Ecology 124: 163–174.
Thorrold, S.R., Shenker, J.M., Mojica, R., Maddox, E.D. and Wishinski, E. 1994. Temporal patterns in the larval supply of summer-recruitment reef fishes to Lee Stocking Island, Bahamas. Marine Ecology Progress Series 112: 75–86.
Tobias, W.J. 1999. Mangrove habitat as nursery grounds for recreationally important fish species—Great Pond, St. Croix, U.S. Virgin Islands. Proceedings of the Gulf and Caribbean Fisheries Institute 52: 468–487.
Tolimieri, N., Jeffs, A. and Montgomery, J.C. 2000. Ambient sound as a cue for navigation by pelagic larvae of reef fishes. Marine Ecology Progress Series 207: 219–224.

13 Physiological Ecology of Developing Bonefish Larvae

Edward Pfeiler

CONTENTS

Introduction 179
Review of the Early Life History of Bonefishes 180
    Characteristics of the Leptocephalous Larva 180
    Phases of Larval Development 181
Environmental Factors in the Marine Environment and Larval Physiology 182
    Hydrostatic Pressure and Temperature 182
        Depth Distribution of Phase I Leptocephali 182
        Effects of Water Temperature on Larval Physiology and Ecology 183
    Salinity Tolerance and Osmoregulation 186
        Osmotic Considerations in Leptocephali 186
        Salinity Tolerance of Phase II Larvae 186
    Oxygen Availability and Respiration 187
        Metabolic Rates in Phase I Leptocephali 187
        Oxygen Requirements and Survival under Hypoxia
            in Phase II Larvae 187
    Feeding Ecology and Nutrition 188
        Postulated Nutritional Sources of Phase I Larvae 188
        Endogenous Nutrients of Nonfeeding Phase II Larvae 189
        Calcium and Phosphorus Balance in Phase II Larvae 189
Physiological Ecology of Larval Migration and Onset of Metamorphosis 190
    Timing of Inshore Migration 190
    Sensory Reception, Environmental Cues and Hormones 190
Conclusions 190
Acknowledgments 191
References 191

INTRODUCTION

The early life history stages of bonefishes (Albuliformes: Albulidae: *Albula* spp.) inhabit both pelagic and coastal marine environments and thus are subjected to a range of physicochemical conditions that can affect survival, development, and recruitment. In general, pelagic larvae (leptocephali) will not experience the large fluctuations in temperature, salinity, and oxygen levels often seen in coastal environments inhabited by metamorphic leptocephali and juveniles. Metamorphic larvae, for example, have been reported from both hyposaline and hypersaline estuaries,[1,2] as well as from sandy beach habitats (unpublished observations of C.H. Gilbert, cited in Gill;[3] D. Snodgrass and R.E. Crabtree[4]). Gilbert also observed that in the Gulf of California the young fish are very abundant and are "often thrown by the waves on the beach in great masses... ." Metamorphosing bonefish leptocephali, as well as adults, are still very abundant in coastal regions in the Gulf of California, probably owing to fact that in this region the species is not the object of either sport or commercial fishing interests.[5]

Bonefish leptocephali inhabiting the relatively stable pelagic environment would be predicted to show narrower survival limits to a variety of environmental parameters than metamorphic larvae. Testing this hypothesis, however, has proven difficult because bonefish larvae, as with other leptocephali, are often damaged during collecting and handling,[6] and therefore, scant physiological data are available. The information we have has been collected primarily on metamorphic larvae from the Gulf of California (*Albula* sp. A), which, because of their relatively consistent seasonal abundance in the hypersaline mangrove lagoons (*esteros*) during winter and spring,[7] can be easily collected with hand nets resulting in minimum damage and high larval survival. However, much of the physiological ecology of developing bonefishes, especially during the pelagic phase, still must be inferred from oceanographic observations taken at the time of collection and on results obtained on other species with leptocephalous larvae, principally eels.

Another potential problem in understanding adaptations of bonefish leptocephali to the marine environment is the recent discovery from molecular analyses that at least eight valid species of bonefishes, most of which are unnamed, are found worldwide.[8] (See also Bowen et al., Chapter 11, this volume.) Conclusions based on studies of a particular bonefish species inhabiting one geographic area, therefore, may not necessarily apply to all species. For example, the estimated duration of the pelagic larval phase differs between *Albula vulpes* from the Bahamas (~2 months)[9] and *Albula* sp. A from the Gulf of California (~6–7 months).[10] It is probable, however, that most details on the biochemical and physiological adaptations of *Albula* leptocephali described herein will apply to the genus as a whole. In the present chapter, therefore, I have attempted to provide a conceptual framework for understanding the ecology of bonefish larvae and how different environmental parameters might affect the physiology and development of the leptocephalus. A comprehensive review of the developmental physiology of elopomorph leptocephali can be found in Pfeiler.[11]

REVIEW OF THE EARLY LIFE HISTORY OF BONEFISHES

CHARACTERISTICS OF THE LEPTOCEPHALOUS LARVA

The leptocephalus is a specialized larva shared with more than 800 species of fishes grouped into the superorder (or subdivision) Elopomorpha, which includes the bonefishes, spiny eels, and halosaurs (Albuliformes), tarpons and ladyfishes (Elopiformes), true eels (Anguilliformes), and gulpers and bobtail snipe eels (Saccopharyngiformes).[12,13] Although morphology can vary greatly in the different groups of elopomorph fishes, all leptocephali are characterized by having a laterally compressed body resulting in a high surface-to-volume ratio.[11] Most of the thin body is composed of a highly hydrated and transparent extracellular gelatinous matrix containing a variety of acidic glycosaminoglycans (GAGs) linked to protein termed proteoglycans.[14] The leptocephalus lacks a vertebral column, and most other ossified skeletal elements,[6,15] and therefore, the gel matrix provides structural support for the body. The GAG component of the principal proteoglycan in both *A. vulpes* and *Albula* sp. A is a form of keratan sulfate (KS) that possesses a unique sulfation pattern compared to mammalian KS.[16–18] Minor GAGs in bonefish leptocephali include chondroitin sulfate (CS) and hyaluronan,[19] which was previously identified as an undersulfated form of CS.[14,20]

PHASES OF LARVAL DEVELOPMENT

Spawning of adult bonefishes occurs either offshore or in areas where currents carry the fertilized eggs offshore,[21,22] thus the earliest larval stages, termed Phase I leptocephali,[23] are pelagic. The duration of Phase I, although differing in *A. vulpes* and *Albula* sp. A as mentioned before, is relatively long. During Phase I, the leptocephalus increases in size (Figure 13.1), reaching a maximum standard length (SL) of ~70 mm.[7] Most of the increase in larval mass during Phase I is due to synthesis, and the resulting hydration, of the gelatinous matrix.[23]

The metamorphic period (Phase II), during which the larva transforms into a juvenile, is relatively short compared with Phase I, lasting about 10 days at a water temperature of ~22°C.[2,6] Developing bonefish larvae show major changes in morphology and chemical composition during Phase II.[11,15] Phase II larvae lose >60% of their body length (Figure 13.1) and about half of their dry mass as the gelatinous matrix is degraded.[2] The principal organic components that are broken down are lipid (~50% loss) and carbohydrate (~80% loss). Other major changes during Phase II include about an 80% loss of water and an 80–90% loss in NaCl. Some chemical components, however, are conserved during Phase II. These include potassium, phosphorus, calcium, magnesium, and total soluble protein. A complete summary of the changes in organic and inorganic components during Phase II is given in Pfeiler.[11]

Phase II has been further subdivided into Phases IIa and IIb (Figure 13.1).[24] During the initial phase of metamorphosis (Phase IIa), shrinkage rate and loss of wet mass (WM) are relatively rapid as larvae decrease from an average of ~63 to 35–40 mm SL, and from ~0.53 to ~0.25 g WM, in about 3–4 days. During the following 6–7 days (Phase IIb), rates of shrinkage and wet mass loss decrease as larvae reach a size of 20–25 mm SL and ~0.15 g WM.[2,6,24,25] Phases IIa and IIb larvae also differ in their oxygen requirements, hypoxic survival time, and tissue development, as described later.

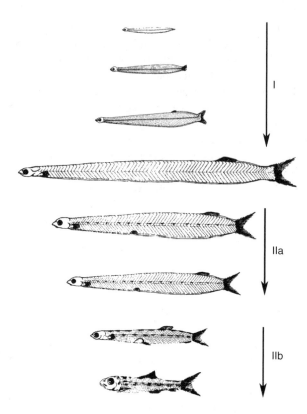

FIGURE 13.1 Drawings of bonefish (*Albula* sp. A) leptocephali from the Gulf of California, Mexico, showing the external morphological changes that take place during the pelagic premetamorphic (Phase I) growth interval and during metamorphosis to a juvenile (Phases IIa and IIb). Drawings of Phase I larvae were adapted from Pfeiler et al.;[10] drawings of Phase II larvae were taken from Pfeiler.[23] All individuals are drawn to scale, with the largest larva (63 mm SL) used as a reference.

ENVIRONMENTAL FACTORS IN THE MARINE ENVIRONMENT AND LARVAL PHYSIOLOGY

HYDROSTATIC PRESSURE AND TEMPERATURE

Depth Distribution of Phase I Leptocephali

Although pressure and temperature tolerances in pelagic Phase I bonefish leptocephali and the effects on these environmental variables on larval physiology are unknown, some tentative conclusions can be drawn from collection data. Plankton tows conducted at night have shown that Phase I bonefish leptocephali are found almost exclusively in the upper 100 m of the water column, with most larvae occurring at depths of <50 m.[10,21,26] These observations suggest that, at least at night, Phase I larvae are not subjected to substantial changes in hydrostatic pressure and temperature. It is not known, however, whether bonefish larvae undergo diurnal vertical migrations. Alexander[21] stated that the few plankton tows made during the day were insufficient to arrive at any conclusions on daily vertical movements. Diurnal vertical migrations, however, have been observed in eel leptocephali, with captures reported at depths of 100–600 m during the daytime,[27–30] and thus the possibility that pelagic bonefish leptocephali undergo vertical migrations cannot be ruled out. The swimbladder is not yet developed in Phase I and early Phase II leptocephali,[11] and therefore it is unlikely that pelagic leptocephali are capable of adjusting their buoyancy during vertical migrations. Although it is apparent that eel leptocephali can withstand moderate hydrostatic pressures, as well as daily fluctuations in pressure, empirical data on pressure effects on the biochemistry and physiology of leptocephali are lacking.

Effects of Water Temperature on Larval Physiology and Ecology

If bonefish leptocephali undergo diurnal vertical migrations, they would also be subjected to substantial fluctuations in water temperature. Although thermal tolerances of Phases I and II leptocephali have not been studied, Alexander[21] suggested that both the horizontal and vertical distribution of bonefish larvae are generally limited by the 20°C isotherm, although a few larvae, referred to as "anomalies," were collected at slightly cooler temperatures. If larvae cannot survive temperatures much below 20°C for short periods, it would obviously restrict them from migrating to deeper waters. But field observations indicate that bonefish leptocephali can withstand temperatures substantially colder than 20°C. Early Phase II bonefish larvae have been collected at a water temperature of 16°C in the Gulf of California[10] and the Florida Keys,[4] and on one occasion were found at a temperature of 12°C in Estero del Soldado in the Gulf of California near Guaymas, Sonora, Mexico.[31] The ability of Phase II bonefish leptocephali to withstand relatively cool temperatures is consistent with findings on Phase I leptocephali of the albuliform *Pterothrissus gissu*, which have been collected at temperatures of 10.2–16.7°C.[32] At the other extreme, Phases I and II bonefish leptocephali have been collected at temperatures approaching 30°C.[4,10,21]

An increase in temperature within the physiological range (15–30°C) increases metabolic rate, as determined by the respiratory electron transport system (ETS) assay, in Phase II leptocephali of *Albula* sp. A[33] and, in addition, increases the activity of two enzymes (β-*N*-acetylglucosaminidase and sulfatase) probably involved in KS degradation during metamorphosis (a third enzyme found in larvae capable of degrading KS, β-galactosidase, shows little temperature sensitivity).[34] With the exception of β-galactosidase, the Q_{10} values for enzyme and ETS activities range from 1.5 to 2.0 (Table 13.1). These results suggest that increased water temperature should increase the rate of metamorphosis, as concluded by Rasquin[6] for *A. vulpes*. Rasquin's conclusion, however, was based on different stages of development reached in two small groups of early larvae ($N = 3$ and 5) that were held at about the same temperature (21–24°C). The second group had been subjected to a higher temperature (27°C) 2 days prior to capture, which was assumed to cause the observed developmental differences between groups. Because rates of metamorphosis can vary

TABLE 13.1
Q_{10} Values for Glycosaminoglycan-Degrading Enzymes (β-N-Acetylglucosaminidase, β-Galactosidase, and Sulfatase), Respiratory Electron Transport System (ETS) Activity, and Metamorphic Rate in Phase II Bonefish Leptocephali (*Albula* sp. A) from the Gulf of California, Mexico

	Q_{10}	Temperature Range (°C)	Reference
β-N-acetylglucosaminidase[a]	1.5	20–30	34
β-Galactosidase[a]	1.1	20–30	34
Sulfatase[a]	2.0	15–24	34
ETS[a]	1.7	17–30	33
Metamorphic rate	2.1	19.7 26.4	This study

[a] Enzyme and ETS activities were determined on early metamorphic larvae (Phase IIa).

between individual larvae held at the same temperature, especially during Phase IIa,[2] the effect of temperature on rates of metamorphosis in two large groups of early Phase II larvae of *Albula* sp. A was investigated. The experimental details are given in Figure 13.2. The results show that leptocephali held at 26.4°C throughout Phase II had a significantly faster rate of metamorphosis, as measured by rate of loss of SL and wet mass, than larvae held at 19.7°C. The high temperature group had essentially completed metamorphosis in about 6 days whereas the control group required about 10 days, which is the expected duration at a temperature of ~20°C.[2] The increased rate of metamorphosis over this temperature range yielded a Q_{10} of 2.1, which was generally concordant with Q_{10} values for enzyme and ETS activities in metamorphic larvae (Table 13.1).

In addition to accelerating the rate of metamorphosis, increased water temperature may alter the normal pattern of inshore migration of leptocephali of *Albula* sp. A in the Gulf of California. Extensive field observations from 1978 to 1987, in addition to systematic sampling conducted during 1981,[7] showed that large numbers of leptocephali of *Albula* sp. A migrated to the *esteros* near Guaymas during the winter and spring of 1978–1982. During the 1982–1983 El Niño–Southern Oscillation (ENSO) event, surface water temperatures were 2–3°C higher than normal in the *esteros* and a notable decrease in larval abundance occurred. Although not investigated systematically, larval abundance increased in subsequent years, and by early 1989, large numbers of leptocephali were present in the *esteros* along the coast of Sonora again. However, no leptocephali were observed in repeated visits to a key collecting site at Estero del Soldado during a very strong ENSO event in January and February 1998, but a few were present at the unusually late date of May 20, 1998.[35] The warm surface waters and nutrient anomalies associated with ENSO events are known to affect fish distributions,[36] but more data will be required to determine if they significantly alter Phase I development and larval migration in bonefishes.

Physiological Ecology of Developing Bonefish Larvae

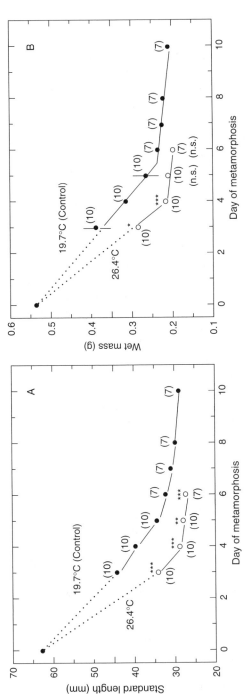

FIGURE 13.2 Effect of water temperature on shrinkage rate (A) and wet-mass loss (B) during metamorphosis of *Albula* sp. A from the Gulf of California. Approximately 100 early Phase II larvae were collected on January 31, 2004 at Estero del Soldado, near Guaymas, Sonora, Mexico (water temperature, 21.1°C; salinity, 36.0‰). Data taken over a 25-year period have consistently shown that recently captured early Phase IIa larvae from Estero del Soldado have a mean standard length (SL) of ~63 mm and a mean wet mass (WM) of ~0.53 g.[2,24] On the day of collection (day 0), one group of about 40 larvae was randomly taken and placed in a plastic container with 60 L of aerated seawater maintained at 26.4 ± 0.2°C with a Visi-Therm submersible aquarium heater (Aquarium Systems, Mentor, Ohio). The remaining larvae were held in 60 L of aerated seawater at room temperature (19.7 ± 1.4°C) and served as the control group. Temperature and salinity were monitored twice a day in both containers using a YSI (Yellow Springs, Ohio) Model 85 digital meter. Salinity was maintained at 35.0–36.0‰ by adding distilled water. Larvae were exposed to a natural photoperiod, and no food was added (metamorphic larvae utilize endogenous nutrients).[15] The number of larvae sampled from each container for each day is shown in parentheses. Sampling was initiated after 3 days because most of the loss in SL and WM occurs during the first 3–4 days of metamorphosis (i.e., during Phase IIa) and any temperature-induced differences in metamorphic rates would be most evident at this time. By day 6, the last seven larvae sampled from the high temperature group had essentially completed metamorphosis. Sampling of control larvae continued until day 10. After blotting each larva, SL (to the nearest millimeter) and WM (to the nearest 0.001g) were determined. Statistical significance of effect of temperature on mean SL and WM for each day was determined with a paired samples *t* test using SYSTAT Version 9. Standard errors for all means fell within the size of the symbol, except those for control WM on days 3 and 5, which are shown as vertical lines. Statistical significance: *, $P < 0.05$; **, $P < 0.01$; ***, $P < 0.001$; n.s., not significant.

SALINITY TOLERANCE AND OSMOREGULATION

Osmotic Considerations in Leptocephali

It has been suggested that the elopomorph leptocephalus represents a developmental strategy that allows the larva to remain in osmotic equilibrium with seawater and thereby delay the development of energy-requiring osmoregulatory mechanisms.[37] The hypothesis is based on the observation that Phase I leptocephali of several different species of eels have serum osmolalities equivalent to ~80–100% seawater. There is also evidence that serum osmolalities decrease in more advanced Phase I larvae of some species, suggesting increased development of osmoregulatory mechanisms during larval growth.[37] Phase II leptocephali of *Albula* sp. A, as well as those of eels, have lower and less variable osmolalities of body fluids,[37,38] supporting the hypothesis, but serum osmolalities have not yet been determined for Phase I bonefish leptocephali.

No data are available on salinity tolerance of Phase I bonefish leptocephali, but because these larvae are pelagic they will not normally be subjected to major salinity fluctuations, with the possible exception of an occasional encounter with a freshwater lens. Reported salinities at collection sites range from ~35 to 37‰.[10,21] During Phase I growth, elopomorph leptocephali take up water and NaCl from the environment.[23] The uptake appears to be directly related to the ion and water-binding properties of GAGs that are synthesized and deposited in the extracellular gelatinous matrix at this time. As described earlier, during Phase II, bonefish leptocephali lose most (~80–90%) of the whole-body water and NaCl that has accumulated as the gelatinous matrix is degraded. A similar percentage decrease in KS occurs,[16] supporting the hypothesis that the developmental cycle of salt and water loading and then unloading during Phases I and II, respectively, is directly related to the corresponding synthesis and breakdown of KS.[23] NaCl efflux in Phase II larvae of *Albula* sp. A may be mediated by chloride-type cells found in the skin.[39]

Salinity Tolerance of Phase II Larvae

In contrast to Phase I larvae, Phase II bonefish leptocephali are found inshore where they may be subjected to a range of salinities depending on locality. For example, salinities can approach 40‰ in the *esteros* in the Gulf of California.[2] In other areas, larvae may encounter dilute conditions from freshwater input when they enter positive estuaries.[1,40,41] Phase II leptocephali of *A. vulpes* have been reported from salinities of 8.8[1] and 10.4‰.[4] Laboratory experiments conducted with *Albula* sp. A have shown that Phase II larvae and early juveniles are euryhaline. The lower and upper incipient lethal salinities of larvae adapted to a salinity of 35‰ and 19°C are 4.2 and 52‰, with corresponding values of 3.3 and 59‰ found for juveniles,[42] suggesting that both larvae and juveniles would be restricted from strictly freshwater habitats. The reported salinities from which leptocephali have been collected fall within the upper and lower lethal salinities determined experimentally.

Not only can Phase II bonefish larvae tolerate a wide range of salinities, but metamorphosis, including the unloading of water and NaCl, appears to be unaffected by salinity extremes.[38] The ability to maintain net efflux rates of NaCl under hypersaline (48‰) conditions, as well as to maintain high rates of water excretion under dilute hyposaline (8‰) conditions, is impressive and is probably related, at least in part, to the breakdown of KS described above.[23]

OXYGEN AVAILABILITY AND RESPIRATION

Metabolic Rates in Phase I Leptocephali

Alexander[21] found that variations in oxygen levels in the epipelagic zone had no effect on the distribution of Phase I bonefish leptocephali. Even though oxygen levels vary, it is unlikely that epipelagic organisms would be exposed to periods of substantial hypoxia or anoxia. Phase I leptocephali lack functional gills, erythrocytes, and hemoglobin, and therefore most gas exchange probably occurs across the thin epithelial layer of the skin.[11] Two characteristics of leptocephali that could compensate for the lack of functional gills are (1) a high surface-to-volume ratio of the larval body, which provides a large respiratory exchange surface and (2) a relatively low wet mass-specific metabolic rate compared with other teleost larvae.[43] Also, eel leptocephali need <50% of the energy required by nonelopomorph larvae of equal dry mass.[44] The relatively low metabolic rates seen in eel leptocephali are a result of the large amount of acellular and nonrespiring gelatinous material that makes up the bulk of the body.[43] Although respiratory rates have not been determined specifically on Phase I bonefish leptocephali, the low values observed in early Phase II bonefish leptocephali,[33] taken together with results from Phase I eel larvae, suggest that they should be low. The low oxygen demand found in Phase I larvae can be considered a survival advantage because energy expenditure also would be relatively low during the extended period that larvae remain in the epipelagic zone.[11] As found for pelagic eel leptocephali,[44] it is probable that Phase I bonefish larvae also devote most of their energy expenditure to metabolism rather than to growth.

Oxygen Requirements and Survival under Hypoxia in Phase II Larvae

Unlike Phase I leptocephali, Phase II bonefish larvae are more likely to encounter hypoxic conditions often found in coastal marine environments.[45] Oxygen requirements, as determined by routine oxygen consumption rates per larva and whole-body ETS activities, approximately double during metamorphosis of *Albula* sp. A.[33] Corresponding to the increase in oxygen demand, survival time in hypoxic seawater decreases about 70% in *Albula* sp. A during Phase II.[24] However, there is an abrupt increase in oxygen consumption, and a corresponding abrupt decrease in hypoxic survival time, when shrinking Phase II larvae reach a size of ~35–40 mm SL (Figure 13.3), which, together with differences in rates of shrinkage described earlier, led to the subdivision of the metamorphic period into early (Phase IIa) and late (Phase IIb) stages.[24] It is during Phase IIb that the developing bonefish larva develops mature erythrocytes and hemoglobin,[6] and changes from mainly cutaneous to branchial respiration.[11] By Phase IIb, the swimbladder also has developed and has become functional.[6,23,38] The effect of hypoxia on ability to successfully

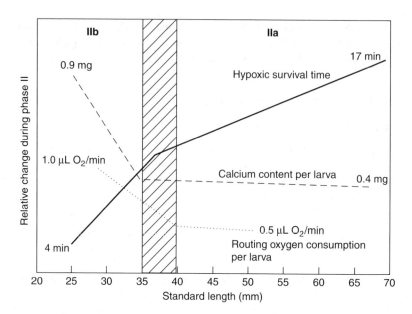

FIGURE 13.3 Changes in hypoxic survival time determined at 21–22°C, routine oxygen consumption rate (20–25°C), and whole-body calcium content during metamorphosis (Phase II) of *Albula* sp. A. The figure was redrawn from published data.[15,24,33] Metamorphosis proceeds from right to left along the *x*-axis. The hatched column represents the transition from Phases IIa to IIb. Initial and final values for each parameter are shown (the mean size of earliest Phase IIa larvae used in the oxygen consumption study[33] was ~50 mm SL).

complete metamorphosis has not been examined, but it is apparent that recently metamorphosed juveniles would be more sensitive to hypoxia than early Phase II leptocephali.

FEEDING ECOLOGY AND NUTRITION

Postulated Nutritional Sources of Phase I Larvae

Although the feeding habits of adult bonefishes have been well documented,[46,47] much less is known of the feeding ecology and nutrition of the pelagic larval stages and early juveniles. After yolk reserves are exhausted about 1–2 weeks after hatching,[11] Phase I leptocephali must rely on exogenous sources of nutrition. The type of food utilized by Phase I larvae is still not completely understood, but the lack of identifiable phytoplankton or zooplankton in the digestive tracts of Phase I bonefish[21] and eel[48,49] leptocephali suggests that they feed primarily at a trophic level below that of most other larval fishes and thereby occupy a specialized niche in the marine ecosystem in which they avoid direct competition with nonelopomorph fish larvae for food resources. Various lines of evidence suggest that the diet of Phase I leptocephali is most likely composed of dissolved (DOM) or particulate (POM) organic matter.[11,23,44,50]

The stable isotope ratio for nitrogen ($\delta^{15}N$) in early Phase II leptocephali of *Albula* sp. A is below that of POM in seawater from the Guaymas area, suggesting that DOM is the more important nutritional source in this species.[51] Analysis of larval energetics and energy content of DOM and POM also suggests that DOM is the primary nutrient source in Phase I eels,[44] but more work is needed to assess the relative roles of DOM and POM in larval nutrition. Uptake of DOM in Phase I leptocephali has been suggested to occur by intestinal absorption[50,52] and by absorption across the thin epithelial layer of the skin.[11,23,48]

Endogenous Nutrients of Nonfeeding Phase II Larvae

In contrast to the requirements for exogenous nutrients during Phase I, Phase II bonefish larvae do not feed throughout most of the metamorphic period.[6,15] Energy requirements during most of Phase II, therefore, derive from catabolism of stored reserves. Energy budget calculations confirm that the energy requirements of metamorphosing larvae can be met by endogenous lipid and carbohydrate (KS) that is broken down, with most (~80%) of the energy obtained from lipid stores.[53]

Calcium and Phosphorus Balance in Phase II Larvae

The loss of the gelatinous body support matrix during metamorphosis is associated with an increase in ossification, especially of the vertebral column and head bones, which is most pronounced during Phase IIb.[15,54] Throughout Phase IIa the leptocephalus of *Albula* sp. A conserves calcium, and then during Phase IIb calcium is taken up from seawater resulting in about a twofold increase in whole-body calcium levels (Figure 13.3).[15] Calcium uptake also occurs at an external calcium concentration of 2.0 mM, or about five times lower than normal seawater, suggesting that calcium uptake is driven by a high-affinity pumping mechanism.[15] Thus, Phase II leptocephali entering hyposaline estuaries with low and fluctuating calcium levels should be able to take up calcium from the environment and undergo normal ossification. High larval mortality, however, was noted at external calcium concentrations of 0.0–1.0 mM, and body deformities and erratic swimming behavior were noted even at 2.0 mM calcium,[15] suggesting a minimum requirement of >2.0 mM calcium for normal development and survival.

Although early Phase II leptocephali of *Albula* sp. A appear to undergo normal metamorphosis in nutrient-free (and phosphate-free) artificial seawater at a normal calcium concentration of 10.1 mM, ossification of the vertebral column and head bones is dramatically reduced.[15] Whole-body phosphorus, required for bone formation along with calcium, is conserved during metamorphosis but, unlike calcium, does not increase during Phase IIb.[51] These results, together with estimates of total larval phosphorus requirements during this period, suggest that larvae are phosphorus limited near the end of Phase IIb and therefore depend on phosphorus in their diet when they resume exogenous feeding in coastal nursery areas. The diet of advanced Phase IIb larvae and early juveniles probably includes small benthic and epibenthic prey such as mollusks, crustaceans, and annelids.

PHYSIOLOGICAL ECOLOGY OF LARVAL MIGRATION AND ONSET OF METAMORPHOSIS

Timing of Inshore Migration

A critical part of the life history of bonefishes is the migration of pelagic Phase I larvae to inshore habitats where they undergo metamorphosis. In the Gulf of California, migration of Phase II larvae of *Albula* sp. A into the *esteros* has been shown to occur at night during the initial phases of flood tide when tidal current velocity is relatively weak.[7] Mojica et al.[9] also found that in the Bahamas leptocephali of *A. vulpes* recruited to inshore habitats mainly at night. In *A. vulpes*, there was also a strong association with lunar cycle, especially with the number of hours of flood tide during the dark cycles, but no correlation was found with other environmental variables such as wind and current patterns. It has been suggested that nighttime onshore movements of leptocephali limits their vulnerability to visual predators.[9] Early Phase II leptocephali of *Albula* sp. A are routinely found <1 m from shore within the *esteros* in the Gulf of California, and although mostly transparent, they are often heavily preyed upon during the daytime by shorebirds.[54]

Sensory Reception, Environmental Cues and Hormones

At present, there is insufficient information to determine how environmental factors and physiological mechanisms might interact to control the onset of inshore migration and the initiation of metamorphosis in bonefish leptocephali. The eyes, olfactory organs, lateral-line system, and pineal gland are well developed in Phase I and early Phase II leptocephali[6,11,21,54] and most probably play important roles during larval migration. There is also evidence that the endocrine system is important in determining the duration of Phase I and in initiating metamorphosis. Studies on Phase I eel (*Conger myriaster*) leptocephali have shown that cortisol levels are high during Phase I, but then decrease at the time of metamorphosis.[55] Levels of thyroxine (T_4) and triiodothyronine (T_3), on the other hand, are low during Phase I,[55] but thyroid gland development increases during metamorphosis in both eels[55] and in *A. vulpes*.[6] The pituitary gland is well developed in metamorphic leptocephali of *A. vulpes*,[6] which also suggests that the hypothalamic–hypophyseal axis plays an important role in the control of metamorphosis.

CONCLUSIONS

Given the relative abundance of bonefishes in tropical and semitropical coastal areas worldwide,[56] it is apparent that leptocephali are successful and well-adapted to the variety of physicochemical and ecological conditions they encounter during development. The relatively low metabolic rate of Phase I leptocephali, together with the abundance of postulated nutritional sources (DOM and POM), are adaptations favorable to a protracted larval phase in the open ocean. Water temperature anomalies associated with ENSO events, however, may play an important role in determining temporal stability and age structure of adult bonefish populations,[5] a possibility that warrants further study. The ability to successfully complete metamorphosis in inshore environments is also a critical factor influencing adult bonefish populations, but with the exception of metamorphic differences in hypoxic survival times described earlier,[24] we know virtually nothing of how environmental contaminants and physical disturbances in coastal marine ecosystems affect survival and development of larval bonefish. Determining the effects of these factors on larval development will be especially important for managing the bonefish fishery, as coastal ecosystems are increasingly being modified by human activities. Other important areas for future research include identification of the environmental and physiological factors involved in triggering inshore migration of Phase I leptocephali and in initiating metamorphosis.

ACKNOWLEDGMENTS

I thank D. Snodgrass and R.E. Crabtree for kindly providing unpublished collection data on bonefish leptocephali from the Florida Keys, and L.T. Findley and T.A. Markow for sharing their field observations on larval bonefish from the Gulf of California. I also thank T. Hernández for help with collecting leptocephali, K.H. Holtschmit for help with the temperature experiments, and W. Moore and J. Egido Villarreal for help with preparing Figure 13.1.

REFERENCES

1. Thompson, B.A. and Deegan, L.A., Distribution of ladyfish (*Elops saurus*) and bonefish (*Albula vulpes*) leptocephali in Louisiana, *Bull. Mar. Sci.*, 32, 936, 1982.
2. Pfeiler, E., Changes in water and salt content during metamorphosis of larval bonefish (*Albula*), *Bull. Mar. Sci.*, 34, 177, 1984.
3. Gill, T., The tarpon and ladyfish and their relatives, *Smithsonian Misc. Coll.*, 48, 31, 1907.
4. Snodgrass, D. and Crabtree, R.E., unpublished data, 1997.
5. Pfeiler, E., Padrón, D., and Crabtree, R.E., Growth rate, age, and size of bonefish from the Gulf of California, *J. Fish Biol.*, 56, 448, 2000.
6. Rasquin, P., Observations on the metamorphosis of the bonefish, *Albula vulpes* (Linnaeus), *J. Morph.*, 97, 77, 1955.
7. Pfeiler, E., Inshore migration, seasonal distribution and sizes of larval bonefish, *Albula*, in the Gulf of California, *Environ. Biol. Fish.*, 10, 117, 1984.
8. Colborn, J. et al., The evolutionary enigma of bonefishes (*Albula* spp.): cryptic species and ancient separations in a globally distributed shorefish, *Evolution*, 55, 807, 2001.
9. Mojica, R., Jr. et al., Recruitment of bonefish, *Albula vulpes*, around Lee Stocking Island, Bahamas, *Fish. Bull.*, 93, 666, 1995.
10. Pfeiler, E., Mendoza, M.A., and Manrique, F.A., Premetamorphic bonefish (*Albula* sp.) leptocephali from the Gulf of California with comments on life history, *Environ. Biol. Fish.*, 21, 241, 1988.
11. Pfeiler, E., Developmental physiology of elopomorph leptocephali, *Comp. Biochem. Physiol. A*, 123, 113, 1999.
12. Greenwood, P.H. et al., Phyletic studies of teleostean fishes, with a provisional classification of living forms, *Bull. Am. Mus. Nat. Hist.*, 131, 339, 1966.
13. Nelson, J.S., *Fishes of the World*, 3rd ed., Wiley, New York, 1994.
14. Pfeiler, E., Acidic glycosaminoglycans in marine teleost larvae: evidence for a relationship between composition and negative charge density in elopomorph leptocephali, *Comp. Biochem. Physiol. B*, 119, 137, 1998.

15. Pfeiler, E., Effect of Ca^{2+} on survival and development of metamorphosing bonefish (*Albula* sp.) leptocephali, *Mar. Biol.*, 127, 571, 1997.
16. Pfeiler, E., Glycosaminoglycan breakdown during metamorphosis of larval bonefish *Albula*, *Mar. Biol. Lett.*, 5, 241, 1984.
17. Pfeiler, E., Isolation and partial characterization of a novel keratan sulfate proteoglycan from metamorphosing bonefish (*Albula*) larvae, *Fish. Physiol. Biochem.*, 4, 175, 1988.
18. Peña, M., Williams, C., and Pfeiler, E., Structure of keratan sulfate from bonefish (*Albula* sp.) larvae deduced from NMR spectroscopy of keratanase-derived oligosaccharides, *Carbohydr. Res.*, 309, 117, 1998.
19. Pfeiler, E., et al., Identification, structural analysis and function of hyaluronan in developing fish larvae (leptocephali), *Comp. Biochem. Physiol. B*, 132, 443, 2002.
20. Pfeiler, E., Characterization and distribution of undersulfated chondroitin sulfate and chondroitin in leptocephalous larvae of elopomorph fishes, *Fish Physiol. Biochem.*, 12, 143, 1993.
21. Alexander, E.C., A contribution to the life history, biology and geographical distribution of the bonefish, *Albula vulpes* (Linnaeus), *Dana-Report, Carlsberg Found.*, 53, 1, 1961.
22. Crabtree, R.E., Snodgrass, D., and Harnden, C.W., Maturation and reproductive seasonality in bonefish, *Albula vulpes*, from the waters of the Florida Keys, *Fish. Bull.*, 95, 456, 1997.
23. Pfeiler, E., Towards an explanation of the developmental strategy in leptocephalous larvae of marine teleost fishes, *Environ. Biol. Fish.*, 15, 3, 1986.
24. Pfeiler, E., Changes in hypoxia tolerance during metamorphosis of bonefish leptocephali, *J. Fish Biol.*, 59, 1677, 2001.
25. Hollister, G., A fish which grows by shrinking, *Bull. N.Y. Zool. Soc.*, 39, 104, 1936.
26. Drass, D.M., Onshore movements and distribution of leptocephali (Osteichthyes: Elopomorpha) in the Bahamas, M.Sc. thesis, Florida Institute of Technology, Melbourne, 1992.
27. Tesch, F.-W., Occurrence of eel *Anguilla anguilla* larvae west of the European continental shelf, 1971–1977, *Environ. Biol. Fish.*, 5, 185, 1980.
28. Schoth, M. and Tesch, F.-W., The vertical distribution of small 0-group *Anguilla* larvae in the Sargasso Sea with reference to other anguilliform leptocephali, *Meeresforschung*, 30, 188, 1984.
29. Castonguay, M. and McCleave, J.D., Vertical distribution, diel and ontogenetic vertical migrations and net avoidance of leptocephali of *Anguilla* and other common species in the Sargasso Sea, *J. Plankton Res.*, 9, 195, 1987.
30. Otake, T., et al., Diel vertical distribution of *Anguilla japonica* leptocephali, *Ichthyol. Res.*, 45, 208, 1998.
31. Pfeiler, E., unpublished data, 1994.
32. Tsukamoto, Y., Leptocephalus larvae of *Pterothrissus gissu* collected from the Kuroshio–Oyashio transition region of the western North Pacific, with comments on its metamorphosis, *Ichthyol. Res.*, 49, 267, 2002.
33. Pfeiler, E. and Govoni, J.J., Metabolic rates in early life history stages of elopomorph fishes, *Biol. Bull.*, 185, 277, 1993.
34. Díaz, R.E. and Pfeiler, E., Glycosidase and sulfatase activities and their possible role in keratan sulfate degradation in metamorphosing bonefish (*Albula* sp.) leptocephali, *Fish Physiol. Biochem.*, 12, 261, 1993.
35. Pfeiler, E., unpublished data, 1998.
36. Simpson, J.J., Response of the southern California current system to the mid-latitude North Pacific coastal warming events of 1982–1983 and 1940–1941, *Fish. Oceanogr.*, 1, 57, 1992.

37. Hulet, W.H. and Robins, C.R., The evolutionary significance of the leptocephalus larva, in *Fishes of the Western North Atlantic*, Memoir No. I, Part 9, Vol. 2, Böhlke, E.B., Ed., Sears Foundation for Marine Research, New Haven, 1989, 669.
38. Pfeiler, E., Effect of salinity on water and salt balance in metamorphosing bonefish (*Albula*) leptocephali, *J. Exp. Mar. Biol. Ecol.*, 82, 183, 1984.
39. Pfeiler, E. and Lindley, V., Chloride-type cells in the skin of the metamorphosing bonefish (*Albula* sp.) leptocephalus, *J. Exp. Zool.*, 250, 11, 1989.
40. Hildebrand, S.F., Family Albulidae, in *Fishes of the Western North Atlantic*, Memoir No. I, Part 3, Bigelow, H.B. et al., Eds., Sears Foundation for Marine Research, New Haven, 1963, 132.
41. Shen, S-C., Notes on the leptocephali and juveniles of *Elops saurus* Linnaeus and *Albula vulpes* (Linnaeus) collected from the estuary of Tam-sui River in Taiwan, *Quart. J. Taiwan Mus.*, 17, 61, 1964.
42. Pfeiler, E., Salinity tolerance of leptocephalous larvae and juveniles of the bonefish (Albulidae: *Albula*) from the Gulf of California, *J. Exp. Mar. Biol. Ecol.*, 52, 37, 1981.
43. Bishop, R.E. and Torres, J.J., Leptocephalus energetics: metabolism and excretion, *J. Exp. Biol.*, 202, 2485, 1999.
44. Bishop, R.E. and Torres, J.J., Leptocephalus energetics: assembly of the energetics equation, *Mar. Biol.*, 138, 1093, 2001.
45. Breitburg, D.L., Episodic hypoxia in Chesapeake Bay: interacting effects of recruitment, behavior, and physical disturbance, *Ecol. Monogr.*, 62, 525, 1992.
46. Colton, D.E. and Alevizon, W.S., Feeding ecology of bonefish in Bahamian waters, *Trans. Am. Fish. Soc.*, 112, 178, 1983.
47. Crabtree, R.E. et al., Feeding habits of bonefish, *Albula vulpes*, from the waters of the Florida Keys, *Fish. Bull.*, 96, 754, 1998.
48. Hulet, W.H., Structure and functional development of the eel leptocephalus *Ariosoma balearicum* (Delaroche, 1809), *Phil. Trans. R. Soc. London (Ser. B)*, 282, 107, 1978.
49. Mochioka, N. and Iwamizu, M., Diet of anguilloid larvae: leptocephali feed selectively on larvacean houses and fecal pellets, *Mar. Biol.*, 125, 447, 1996.
50. Otake, T., Nogami, K., and Maruyama, K., Dissolved and particulate organic matter as possible food sources for eel leptocephali, *Mar. Ecol. Prog. Ser.*, 92, 27, 1993.
51. Pfeiler, E., Lindley, V.A., and Elser, J.J., Elemental (C, N and P) analysis of metamorphosing bonefish (*Albula* sp.) leptocephali: relationship to catabolism of endogenous organic compounds, tissue remodeling, and feeding ecology, *Mar. Biol.*, 132, 21, 1998.
52. Otake, T., et al., Fine structure and function of the gut epithelium of pike eel larvae, *J. Fish Biol.*, 47, 126, 1995.
53. Pfeiler, E., Energetics of metamorphosis in bonefish (*Albula* sp.) leptocephali: Role of keratan sulfate glycosaminoglycan, *Fish Physiol. Biochem.*, 15, 359, 1996.
54. Pfeiler, E., Sensory systems and behavior of premetamorphic and metamorphic leptocephalous larvae, *Brain Behav. Evol.*, 34, 25, 1989.
55. Yamano, K., et al., Changes in whole body concentrations of thyroid hormones and cortisol in metamorphosing conger eel, *J. Comp. Physiol. B*, 161, 371, 1991.
56. Briggs, J.C., Fishes of worldwide (circumtropical) distribution, *Copeia*, 1960, 171, 1960.

14 Reproductive Biology of Atlantic Tarpon *Megalops atlanticus*

John D. Baldwin and Derke Snodgrass

CONTENTS

Introduction 195
Larval History 196
Observations of Tarpon Reproduction 197
Synthesis 200
References 201

INTRODUCTION

There is very little known about the reproductive habits of tarpon in the western north Atlantic. Early studies of tarpon reproduction were typically based on the collection of tarpon larvae from nontargeted sampling (Gehringer, 1959; Wade, 1962; Eldred, 1967, 1968, 1972; Smith, 1980, 1989). In one such study, Smith (1980) collected a series of tarpon larvae in the Gulf of Mexico and the Yucatan Channel. Smith speculated that spawning areas were located off Cozumel Mexico, off the west coast of Florida, and in the southwestern Gulf of Mexico. The most current knowledge has been gained by projects completed and published by Crabtree et al. (1992, 1995, 1997) and Crabtree (1995). These studies ranged in coverage from Florida waters (both Atlantic Ocean and (Florida Straits [FS]), Gulf of Mexico (GOM), to Costa Rica (Caribbean) and involved both larval and adult aspects of life history. The origin of this series of research projects was derived from larval sampling being conducted as a possible tool for monitoring numerous species indices of abundance in the coastal waters of Florida in the GOM. Sampling was conducted between June 1981 and July 1989. Through a keen interest in tarpon biology gained from working as a fishing guide in the Florida Keys, Crabtree delved into the archived collections of the fish biology program at the Florida Fish and Wildlife Research Institute in St. Petersburg, Florida, after taking a research position there in 1990. Crabtree found relatively young (2–25 days old) larvae that had been collected offshore of Florida in the Gulf of Mexico in depths ranging from 90 to 1400 m. The temperatures ranged from 27 to 30°C and 35–36 ppt salinity at the collection sites. The primary focus of the Crabtree et al. (1992) publication was age and growth of these larvae. An aside of this, by taking the youngest of these larvae (3–6 days old) they were able to estimate probable spawning location for these individuals. This was perfect evidence of spawning taking place in the GOM offshore of Florida at distances as far as 250 km. Now, we definitely know that spawning takes place in, at least, June and July in the GOM. Earlier evidence (Smith, 1980) noted larvae and spawning taking place in the western GOM as well.

The next aspect addressed was the relationship of lunar phase with spawning activity in and adjacent to Florida waters. Crabtree (1995) reported on additional larval sampling along with the 1981–1989 samples used previously (Crabtree et al., 1992). The additional sampling was again conducted in mid-July 1990 and again from April through October 1991. Sampling again occurred in the GOM while additional collections were conducted in the FS (Bimini, Bahamas Islands, Palm Beach, Florida, and the Florida Keys). All of the sampling occurred at night in the top 20 m of the water column. No specimens were ever collected off Bimini. Only three larvae were collected off of Palm Beach (bottom depth range of 156–749 m), while 105 were collected off the Florida Keys (Big Pine Key to Long Key) (bottom depth range of 59–230 m). All of these samples had hatching dates between 10 May and 18 July, but additional ancillary samples collected by other projects reported hatching dates as late as 14 August. The hatching dates had distinct peaks coinciding with new and full moon phases. Peak hatching occurred 6.3–7.4 days after the full moon and 3.4–8.7 days after the new moon. Crabtree demonstrated a strong association of spawning with lunar phase. Generally, a peak spawning period in the summer months (late spring to late summer) occurs a week after each major moon phase. This was also supported by tarpon larvae collected by Southeast Florida and Caribbean Recruitment (SEFCAR) sampling conducted offshore Long Key, Florida by Rosenstiel School of Marine and Atmospheric Science (RSMAS) scientists between 1989 and 1993 (Limouzy-Paris et al., 1994).

Based on roughly 1500 specimens, it was estimated that females reach reproductive maturity as small as 1285 mm FL, while males were reproductively active as small as 901 mm FL. All Florida tarpon were mature by 10 years of age although one female was mature by 7 years of age. Of 217 tarpon specimens from Costa Rican waters, the females reach reproductive maturity between 880 and 1126 mm FL while males reach it by 880 mm FL. Florida tarpon spawned between April and July with just remnants occurring as late as mid-August. In Costa Rican waters there was no seasonality observed. Tarpon appeared to spawn year-round in Costa Rican waters on the Caribbean side. Possibly due to this inordinate seasonality, the otoliths of Costa Rican tarpon were difficult to attain accurate estimates of age. Roughly 45% (87) of the total sample size from Costa Rica was readable and a maximum age was estimated at 48 years old. We know that male tarpon live at least 30 years whereas females live at least 50 years in the wild (Andrews et al., 2001). The majority (74%) of Crabtree's fish were estimated between 15 and 30 years of age. Tarpon are broadcast spawners similar to many tropical marine fishes (Peters et al., 1998; Graham and Castellanos, 2005). From oocyte staging in preserved ovaries, it appears that they spawn at least four to five times in a given season (Crabtree et al., 1997).

LARVAL HISTORY

Different from the larvae of most marine teleosts, the leptocephali larvae of eels, bonefish, ladyfish, and tarpon represent a unique developmental strategy.

Leptocephali have a small thin head, for which they are named, and a decidedly laterally compressed, transparent body, which has a leaflike appearance with a high surface to volume ratio. The leptocephali larvae are found in the marine environment and may remain in the plankton for as little as 25 days to several months before moving into estuarine nursery habitats (Tzeng et al., 1998; Zerbi et al., 2001). To sustain this long larval growth period, leptocephali deposit energy reserves in the form of glycosaminoglycans, which also aides in locomotion by forming a firm gelatinous-supporting skeleton for the musculature to work against since the leptocephalus larvae have an unossified bony skeleton (Bishop and Torres, 1999). The leptocephalus was first described from specimens collected in the Mediterranean Sea in 1763 (Smith, 1989). All leptocephali were initially classified as separate species altogether. It was not until 1861 that it was described as a larval form of something else altogether. The larvae of elopiformes (tarpon and ladyfish) were not described until the late 1950s (Gehringer, 1959). These were specimens collected in Florida waters. The leptocephalid larval stage for tarpon lasts for 25–40 days. In the positive growth phase of the larval form, they grow to 20–30 mm SL. At this point, they enter the negative growth phase when they coalesce down to their metamorphic size (~5.5–6.1 mm notochord length). The tarpon leptocephali has a large forked caudal fin, short dorsal and anal fins that are not connected with the caudal fin, pectoral fins that are well developed, and developed but small pelvic fins (Wade, 1962). The gut is a simple straight tube. The shape of the head, vertical position of the fins, and number of myomeres (body sections) are the distinguishing characteristics among elopimorphs. In tarpon, the origin of the anal fin is under the middle of the dorsal fin, the head is not depressed and there are fewer dorsal rays (9–16) than anal rays (16–25). The leptocephali larvae are true oceanic larvae and require high steady salinities for healthy osmoregulation (see Pfeiler, Chapter 13, this volume).

OBSERVATIONS OF TARPON REPRODUCTION

While actual spawning, release of gametes by both sexes during courtship has yet to be documented in the wild; there have been numerous observations and documentations of courtship/prespawning behavior. One such observation occurred 2 days before the new moon of June 2002. Baldwin and Snodgrass conducted surface and underwater observations. Video documentation was made of tarpon, *Megalops atlanticus*, exhibiting courtship–spawning behavior (Figure 14.1). The school of 12–16 tarpon was observed in 1–2.5 m of water, just off the edge of the flat on the oceanside of Tavernier Key, FL. The incident occurred at the start of the incoming tide under cloudy skies with light precipitation. The tightly packed school consisted of a single large (~70 kg) female (presumably) fish that was repeatedly and persistently accosted by 8–10 smaller (10–20 kg) male (presumably) fish. In addition, two to four larger (25–40 kg) fish were following in the school, but were not as attentive to the large female as the smaller males. The small males would consistently bump and rub the female's ventral region (Figure 14.1I–P), at times even pushing her above the surface of the water (Figure 14.1B–D). The males maintained this close contact despite numerous directional changes by the female. If a fish got out of the ideal position of being directly under the female, it would circle back around the female

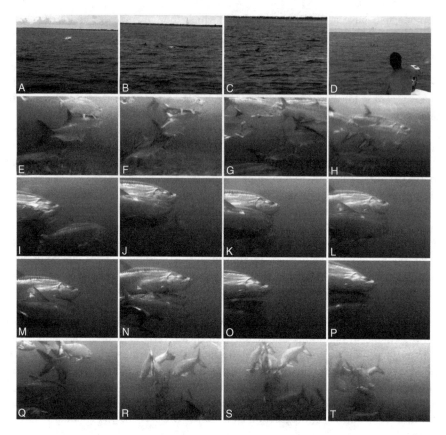

FIGURE 14.1 Still frame captures from video taken on June 8, 2002, 2 days prior to the new moon, highlighting courtship/prespawning behavior of tarpon. These behaviors were observed for over 50 min. A–D shows surface activity of a large female slowly cruising at the surface off the oceanside shoreline of Tavernier Key, FL, with upper portion of her caudal fin exposed. At times she would explode away for a short distance (A and B), be pushed out of the water from beneath exposing her back (C), and be closely followed by other tarpon at the surface, which also exposed upper portion of their caudal fins (D). The tightly packed school of 12–16 tarpon was observed in 1–2.5 m of water (E–H, over a period of 5 s) and consisted of a single large (~70 kg) female (presumably) fish that was repeatedly and persistently accosted by 8–10 smaller (10–20 kg) male (presumably) fish. In addition, two to four larger (25–40 kg) fish were following in the school, but were not as attentive to the large female as the smaller males. The small males would consistently bump and rub the female's ventral region (I–P, over a period of 3 s) and attempt to maintain their position beneath her. If a fish got out of position, it would circle back around the female to reposition itself (I–K). The males maintained this close contact despite numerous directional changes by the female (Q–T, over a period of 3 s).

in order to reposition itself (Figure 14.1I–K). At times the female would be resting/basking at the surface, seemingly motionless with the whole dorsal side of her body and tail out of the water. This typically ended when she would erupt with a short burst of speed and throw a huge boil of water and spray (Figure 14.1A). During the

50 min, which we were able to stay with the school as it paralleled shore, no milt or eggs were observed being released, but certainly could have been. Subsequent to this documented case, the authors have observed this behavior on several other occasions.

From conversations with several guides and anglers, this type of behavior is observed on a relatively common basis. It has been reported from the Loggerhead Keys, Content Keys, Duck Key, Fat Deer Key, Rodriguez Keys, El Radabob Keys, and several areas along Key Largo. It has also been observed just offshore a few locations in Broward County, Florida, such as the observation given by Matt Gardner, a Florida Atlantic University graduate student. "On 29 May 2004 (four days before full moon) we observed a school of 8 tarpon at the Aerojacks beach dive located at John Lloyd State Park, Dania, FL. The habitat consisted of an artificial reef, running perpendicular to shore, made from concrete 'jacks' surrounded by a sandy bottom. The time of the sighting was between 11:45 am and 12:00 pm. We were approximately 200 yards from shore at the time of the sighting. The water temperature recorded on an Oceanic Veo 200 dive computer was 78°F. We were on the bottom at approximately 19 to 15 feet in depth; the fish were about 5 feet above us. The behavior the tarpon exhibited was consistent with that proposed for breeding tarpon. Fish swam in a tight group with the largest individual in the middle of the group. Smaller fish surrounded and nudged the central individual with their heads. Nudges were made to the region just posterior to the opercle on the sides and bottom of the large, central individual. Two of the eight tarpon swam closely behind the group, but did not attempt to nudge the large, central individual during our observations. The group was seen three times on the dive and appeared to be following the reef-line. The behavior described above was consistent among all observations."

From a more historical perspective, we interviewed legendary tarpon fisherman Stu Apte (see Chapter 21, this volume). In the late 1950s and early 1960s, in the Lower Keys, Stu Apte made a living as a backcountry fishing guide in the Florida Keys and kept a daily log of his time on the water.

Stu Apte recalled that as early as mid-April on the moon tide, he would encounter huge schools of tarpon in the Lower, Middle, and Upper Keys in 5 ft of water daisy-chaining. During these years, on numerous occasions he would witness what he assumed to be males, smaller fish "busting" a big female. He would pole into the area (after his client at the time would cast and not get hooked up with a fish) and would find milt in the water. The white milt would stay plainly visible and would be all over the area. He firmly believes that the tarpon were actively spawning during these periods in shallow water. He has also observed hundreds of daisy-chaining tarpon over the years that were punctuated by surface explosions, which is how he would often find schools of fish. These explosions are very similar to those observed by the authors. Daisy-chaining fish are often difficult to catch as they do not appear to actively feed during such times.

Stu Apte has only seen milt in the water in the Florida Keys while he has caught fish that were ripe in other areas of Florida such as Homosassa and Boca Grand. During the spawning season, he frequently encounters ripe males, which release milt upon handling at boatside throughout the state. He has never observed females releasing eggs, however, even with handling. He believes that spawning occurs in shallow nearshore waters on the moon tides, at the top of the tide or the beginning of the outgoing tide, and the spawn gets carried out to deeper offshore waters. One of the reasons why he believes spawning activity is not observed in recent times in shallow nearshore waters is due to the tremendous increase in boat traffic on the water, which disrupts the spawning schools and the fish head for deeper water.

In addition to these sorts of observations, several offshore captains have reported huge schools of tarpon that remain relatively stationary as far as 25–40 km south–southeast of the Keys for several days.

SYNTHESIS

Direct spawning activity of tarpon has never been documented, but is thought to occur offshore in deep water after schools migrate from shallow nearshore staging areas. The above-mentioned behavioral observations and documentation may be of the prelude to or directly after the actual spawning activity. This type of prespawning behavior is not unique to tarpon, but is in fact quite common among broadcast spawning marine fishes (Peters et al., 1998; Graham and Castellanos, 2005). Based on the published information to date, it seems somewhat improbable for spawning to occur at such close proximity to shore, given that the only documented spawning area for Florida tarpon was 250 km offshore in the GOM. However, in contrast to the distant reported spawning areas of Florida, the tarpon spawning grounds in Puerto Rico are estimated to be relatively close to the coast at only 2 km (Zerbi et al., 2001). The shorter distance between putative spawning areas and estuarine nursery areas is also reflected in the young age of tarpon leptocephali (34 days) upon entering the estuarine arrival. Similar deepwater/pelagic environmental conditions are found on the edge of the continental shelf, which in the GOM is located very far from shore, but are found much closer to shore in areas such as Florida's Atlantic coast, the Caribbean coast of Mexico, and the coast of Puerto Rico, such that the environmental conditions probably are a greater factor in determining optimum spawning areas than distance from the coast. As such, although, similar oceanographic conditions (salinity and temperature) to what was documented at the larval capture location occur much closer to shore along the Atlantic side of the Keys than off the GOM coast of Florida. It is quite possible that other oceanographic and physicochemical conditions that draw the GOM fish to that spawning area may occur along a much broader area along the Atlantic coast and closer to the coast. This may lead to a much less centralized spawning location and permit spawning at almost any site of interaction when both genders are in a state of peak reproductive activity. The broad range of locations where such behavior has been observed supports this hypothesis.

There still remain significant gaps in our knowledge about the reproductive biology of tarpon, and more detailed information on all aspects of life history is needed. In particular, further verification of spawning locations throughout the range of Atlantic tarpon are needed for proper management of the species. Research emphasis should be placed on characterizing spawning areas and documenting the environmental conditions under which spawning occurs. The timing and duration of spawning activity also needs further examination, as well as the persistence of spawning locations across years. Embryo and larval ecology and physiology are also areas of critically needed research. By characterizing these sensitive life history stages, our understanding of tarpon biology will be greatly increased and will thus lead to enhanced management practices for the fishery.

REFERENCES

Andrews, A.H., E.J. Burton, K.H. Coale, G.M. Cailliet and R.E. Crabtree. 2001. Radiometric age validation of tarpon, *Megalops atlanticus*. Fish. Bull. 99: 389–398.

Bishop, R.E. and J.J. Torres. 1999. Leptocephalus energetics: metabolism and excretion. J. Exp. Biol. 202: 2485–2493.

Crabtree, R.E. 1995. Relationship between lunar phase and spawning activity of tarpon, *Megalops atlanticus*, with notes on distribution of larvae. Bull. Mar. Sci. 56(3): 895–899.

Crabtree, R.E., E.C. Cyr, D.C. Chaverri, W.O. McLarney and J.M. Dean. 1997. Reproduction of tarpon, *Megalops atlanticus*, from Florida and Costa Rican waters and notes on their age and growth. Bull. Mar. Sci. 61(2): 271–285.

Crabtree, R.E., E.C. Cyr and J.M. Dean. 1995. Age and growth of tarpon, *Megalops atlanticus*, from South Florida waters. Fish. Bull. 93: 619–628.

Crabtree, R.E., E.C. Cyr, R.E. Bishop, L.E. Falkenstein and J.M. Dean. 1992. Age and growth of tarpon, *Megalops atlanticus*, larvae in the eastern Gulf of Mexico, with notes on relative abundance and probable spawning areas. Environ. Biol. Fishes 35: 361–370.

Eldred, B. 1967. Larval tarpon, *Megalops atlanticus*, Valenciennes, (Megalopidae) in Florida waters. Florida Board. Conserv. Mar. Lab. Leafl. Ser. 4(1)4: 1–9.

Eldred, B. 1968. First record of a larval tarpon, *Megalops atlanticus*, Valenciennes, from the Gulf of Mexico. Florida Board. Conserv. Mar. Lab. Leafl. Ser. 4(1)7: 1–2.

Eldred, B. 1972. Note on larval tarpon, *Megalops atlanticus*, (Megalopidae), in the Florida Straits. Florida Dep. Nat. Resour. Mar. Res. Lab. Leafl. Ser. 4(1)22: 1–6.

Gehringer, J.W. 1959. Leptocephalus of the Atlantic tarpon, *Megalops atlanticus* Valenciennes, from offshore waters. Q. J. Flor. Acad. Sci. 21: 235–240.

Graham, R.T. and D.W. Castellanos. 2005. Courtship and spawning behaviors of carangid species in Belize. Fish. Bull. 103: 426–432.

Limouzy-Paris, C., M.F. McGowan, W.J. Richards, J.P. Umaran and S.S. Cha. 1994. Diversity of fish larvae in the Florida Keys: results from SEFCAR. Bull. Mar. Sci. 54(3): 857–870.

Peters, K.M., R.E. Matheson and R.G. Taylor. 1998. Reproduction and early life history of common snook, *Centropomus undecimalis* (Bloch), in Florida. Bull. Mar. Sci. 62(2): 509–529.

Smith, D.G. 1980. Early larvae of the tarpon, *Megalops atlanticus* Valenciennes, (Pisces, Elopidae), with notes on spawning in the Gulf of Mexico and the Yucatan Channel. Bull. Mar. Sci. 30: 136–141.

Smith, D.G. 1989. Introduction to leptocephli. pp. 657–668. *In*: E.B. Bohlke (ed.) Fishes of the Western North Atlantic, part 9, Volume 2, Sears Foundation for Marine Research, New Haven.

Tzeng, W.N., C.E. Wu and Y.T. Wang. 1998. Age of Pacific tarpon, *Megalops cyprinoids*, at estuarine arrival and growth during metamorphosis. Zool. Stud. 37(3): 177–183.

Wade, R.A. 1962. The biology of the tarpon, *Megalops atlanticus*, and the ox-eye, *Megalops cyprinoids*, with emphasis on larval development. Bull. Mar. Sci. 12: 545–603.

Zerbi, A., C. Aliaume and J.C. Joyeux. 2001. Growth of juvenile tarpon in Puerto Rican Estuaries. ICES J. Mar. Sci. 58: 87–95.

15 Rethinking the Status of *Albula* spp. Biology in the Caribbean and Western Atlantic

Aaron J. Adams, R. Kirby Wolfe,
Michael D. Tringali, Elizabeth M. Wallace,
and G. Todd Kellison

CONTENTS

Introduction 203
Materials and Methods 204
    Larval (Leptocephalus) and Juvenile Sampling 204
    Tissue Samples and Genetic Analyses 206
    Otolith Samples 208
Results 209
Discussion 211
Acknowledgments 213
References 214

INTRODUCTION

Worldwide, bonefishes (*Albula* spp.) are ecologically and economically important constituents of tropical, shallow-water systems. Bonefishes support economically important recreational fisheries in numerous locations in the Caribbean (e.g., Bahamas, Belize, Mexico, and Venezuela among the most notable). In southern Florida and the Florida Keys, often credited as the birth place of "flats fishing," bonefishes are an important component of the recreational fishery (Crabtree et al., 1996). Because of their ecological and economic importance, sustainable bonefish fisheries are of particular importance and knowledge of their ecological requirements is essential to successful management.

    Unfortunately for managers, the taxonomy of bonefishes is still being unraveled. Bonefish were once classified as a single circumtropical species (*Albula vulpes* Linnaeus). Robins et al. (1986) noted that two species of bonefish occur on the continental shelf and upper slope of the Atlantic Ocean. Recent genetic research, however, has suggested that at least eight species in the *Albula* genus exist worldwide (Colborn et al., 2001; Bowen et al., Chapter 11, this volume), and revisions of the genus continue. Until recently (Colborn et al., 2001), the western Atlantic bonefish pursued by recreational anglers were assumed to be a single species *A. vulpes*. An ecologically distinct second species (*A. nemoptera*) that reaches about half the maximum size of *A. vulpes* occurs in a limited geographic range and in depths too great for recreational angler interest (Robins et al., 1986). Colborn et al. (2001) identified a third genetically distinct lineage of *Albula* in the Caribbean: a currently undescribed species referred to as *Albula* species B (popularly known as *A. garcia*). Hereafter, *Albula* spp. refers to *A. vulpes* and *A.* sp. B, as addressed in this study.

Little is known about the biology and ecology of these species (but see Ault et al., Chapter 16, this volume), or their relative contributions to recreational fisheries. Published information is on *A. vulpes* from the Florida Keys and Bahamas and mostly limited to the adult life stage: age, growth, and mortality (Crabtree et al., 1996); maturation and reproduction (Crabtree et al., 1997); diet (Colton and Alevizon, 1983; Crabtree et al., 1998); movement (Colton, 1983; Humston et al., 2005); or a combination of these topics (Bruger, 1974; also see Mojica et al., 1995 for larval duration and temporal abundance patterns). However, since these studies were conducted prior to Colborn et al.'s (2001) identification of *Albula* sp. B, verification of these findings may be required. Additionally, timing and location of spawning are not well described for *Albula* spp., and species composition of the fishery has not been quantified. Given the dearth of data on *Albula* spp. in the Caribbean and western Atlantic, studies that contribute additional information on these species are needed.

This chapter presents results of sampling to examine three aspects of *Albula* spp. biology and ecology: (1) postlarval and juvenile spatial and temporal habitat use; (2) species composition of mature *Albula* spp. captured in regional recreational fisheries; and (3) a comparison of age and growth estimates of *A. vulpes* from several Caribbean locations to published results from the Florida Keys. These findings will contribute to better understanding of *Albula* spp. in the Caribbean and western Atlantic and provide direction for future research.

MATERIALS AND METHODS

LARVAL (LEPTOCEPHALUS) AND JUVENILE SAMPLING

Sampling was conducted in the Florida Keys, United States (Figure 15.1), during 2003–2005. Sampling in the Florida Keys occurred every other month from October/November 2003 through January 2005 from Key West to Elliot Key by seine, 23 m × 1.2 m, 3.1-mm mesh center bag seine (for small juveniles) and 45.5 m × 1.8 m, 9.5-mm mesh center bag seine (for larger juveniles). In October/November 2003 and January 2004, six habitat types (i.e., windward and leeward sandy beach; windward and leeward beachrock shorelines; windward and leeward mangroves) were sampled for small juveniles with the 23-m seine. Sampling effort was reduced to only sandy beach and beachrock shoreline for March, May, July, and November 2004, and January 2005 because of zero catches in all habitats except sandy beach and beachrock shoreline in previous samples, and results

Rethinking the Status of *Albula* spp. Biology

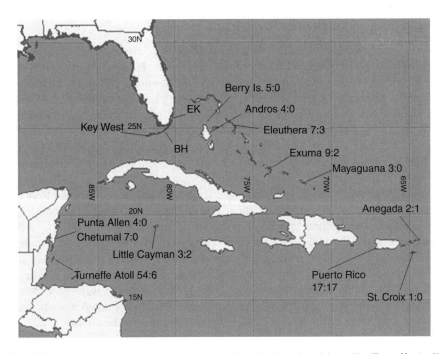

FIGURE 15.1 Map showing locations of sampling for larval and juvenile (Turneffe Atoll, Belize, and Florida Keys, Florida, United States) and adult collections (Turneffe Atoll, Belize; Eleuthera, Exuma, Berry Islands, and Mayaguana, Bahamas; St. Croix, U.S. Virgin Islands; Little Cayman, British West Indies; Anegada, British Virgin Islands; and Chetumal and Punta Allen, Mexico). For Florida Keys locations (Key West = Higgs Beach, Airport Beach; BNP = Biscayne National Park; BH = Bahia Honda State Park). The results of juvenile sampling are listed in Table 15.1. All values reference sampling of adults, all identified as *Albula vulpes*: the first value is the number of adults providing tissue samples; the second value is the number of adults providing otoliths.

of similar sampling conducted in the 1990s (Crabtree et al., 2003). Exploratory sampling was also conducted in shallow (<3 m), sandy-bottom open bays on the Florida Bay side of the Florida Keys. The 23-m (Florida Keys) seine was set perpendicular to shore with one end at or on shore, pulled parallel to shore for 15 m, and either hauled onto shore (sandy beaches – average sample area = 575 m^2) or pursed offshore (all other habitat types – average sample area 166 m^2). For shore sets using the 45.5-m seine, the net was set perpendicular to shore, as above, the outer end pulled in an arc to shore, and the bag end hauled to shore (average sample area = 875 m^2). For offshore sets where the net could not be pulled against shore, the net was pulled for 15 m and pursed (sample area = 166 m^2). All bonefishes were measured (standard length), and tissue samples taken for genetic analysis.

In April 2006, a center bag seine (15 m × 1.2 m, 3.1-mm mesh) was used to sample six sandy beaches at Turneffe Atoll, Belize. Sampling was conducted as for the 23-m seine in Florida (average sample area = 225 m^2).

Tissue Samples and Genetic Analyses

To determine species composition of juvenile *Albula* spp., tissue samples were taken from a subsample of larval and juvenile bonefishes captured in seines. Larval and juvenile bonefishes <80 mm SL were retained whole. For juveniles ≥80 mm SL, a triangle (12 mm × 12 mm × 12 mm) was cut from the soft ray tissue at the rear of the dorsal fin. Whole fish were placed in plastic bags on ice, and then transferred into individual vials containing 95% ethanol. Tissue samples were either placed in individual plastic bags on ice and then transferred to ethanol vials, or placed in ethanol vials on site. Sample location and date were recorded for each collection. Tissue samples from 299 (of 662 total) leptocephalus and juvenile bonefishes captured in the Florida Keys were retained for genetic analysis to identify species. Tissue samples were taken from all months and locations in which leptocephali and juveniles were captured (Table 15.1).

Similarly, to examine species composition of the recreational fishery in the Florida Keys and Caribbean through genetic analysis, tissue samples were obtained from bonefish captured by recreational anglers in 10 locations in the Caribbean and in the Florida Keys. Fin clips were treated as described above. For samples from Puerto Rico (obtained from C. Caldow, NOAA/NOS), otoliths were ground and DNA was isolated from the ground material as described below. DNA was obtained from 17 otoliths, resulting in genetic identification for 48 adult bonefishes.

TABLE 15.1A
Results of Seine Sampling for Juvenile *Albula* spp. in the Florida Keys—Temporal Patterns of Juvenile and Leptocephalus *Albula* Captured in Seine Samples in the Florida Keys

Year		January	March	May	July	August	November
				Month			
2003	No. of samples	–	–	–	–	–	73
	No. of juveniles	–	–	–	–	–	1
	No. of leptocephali	–	–	–	–	–	2
2004	No. of samples	87	128	123	155	–	72
	No. of juveniles	0	149	144	100	–	0
	No. of leptocephali	0	9	12	0	–	0
2005	No. of samples	9	23	53	–	27	–
	No. of juveniles	54	17	161	–	2	–
	No. of leptocephali	5	6	0	–	0	–
	Total samples	96	151	176	155	27	145
	Total bonefish	59	181	317	100	2	3
	No. of bonefish/sample	0.6146	1.1987	1.8011	0.6452	0.0741	0.0207

Note: July 2004 and January and May 2005 are the only months in which *A. vulpes* individuals were captured.

TABLE 15.1B
Results of Seine Sampling for Juvenile *Albula* spp. in the Florida Keys—Genetic Analysis of Juvenile and Leptocephalus *Albula* spp. Captured in Seine Sampling in the Florida Keys in 2003 and 2004

Year/Month	Sample Location	Number Analyzed	Number Identified	
			A. vulpes	*A.* sp. *B*
2003				
November	Elliott Key	3	0	3
2004				
March	Key West	55	0	55
	Bahia Honda	2	0	2
May	Key West	50	0	50
	Bahia Honda	23	0	23
July	Key West	33	0	33
	Bahia Honda	15	4	11
	Elliott Key	11	11	0
October	Bahia Honda	1	0	1
2005				
January	Key West	10	0	10
	Elliott Key[a]	8	1	7
March	Key West	3	0	3
	Elliott Key	10	0	10
May	Key West	2	0	2
	Bahia Honda	71	2	69
August	Key West	1	0	1
	Bahia Honda	1	0	1
	Totals	299	18	281

Note: See Figure 15.1 for sample locations. Only months in which *Albula* were captured and tested are shown.

[a] Elliott Key samples were collected in early February 2005.

Total genomic DNA was isolated from all specimens using the Puregene® DNA Isolation Kit (Gentra Systems, Inc., Minneapolis, Minnesota). Genetic-species-identification (GSI) assays were based on diagnostic nucleotide differences occurring in the mitochondrial DNA (mtDNA) cytochrome-*b* gene (Colborn et al., 2001). Initially, a subsample of bonefishes ($N = 60$) that included both *A. vulpes* and *A.* sp. B specimens was sequenced for the cytochrome-*b* region (with no *a priori* knowledge of individual species identification). Representative voucher specimens are catalogued at the Florida Fish and Wildlife Research Institute. Polymerase chain reaction (PCR) assays were conducted on Hybaid® thermocyclers using ALBA-1, ALBA-2, and ALBA-3 primers (Colborn et al., 2001), Applied Biosystems (ABI)

BigDye® Terminator v1.1, and Taq polymerase (Promega Corp., Madison, Wisconsin). PCRs were run under the following profile: an initial cycle of 94° denaturation for 1 min, 50° annealing for 30 s, 72° extension for 1 min followed by 36 cycles of 94° for 30 s, 55° for 30 s, 72° for 1 min 30 s, and a final 72° extension of 8 min. The PCR products were purified using the Quickstep2 kit (Edge Biosystems, Gaithersburg, Maryland), and the purified products prepared for forward and reverse sequencing with the following thermal profile: 35 cycles of 30 s 95° denaturation, 15 s 55° annealing, and 4 min 60° extension. Cytochrome sequences were aligned in ClustalX (Thompson et al., 1997), and analyzed in MEGA version 2.1 (Kumar et al., 2001) to determine sequence divergence between *A. vulpes* and *A.* sp. B. Example sequences of *A. vulpes* and *A.* sp. B were also compared to the cytochrome sequences submitted by Colborn et al. (2001) in Genbank (http://www.ncbi.nlm.nih.gov/). After confirmation of diagnostic nucleotide sites, new species-specific primers were developed: Avu-CytB F, Avu-CytB R, Aga-CytB F, and Aga-CytB R (Table 15.2). The two forward primers were labeled with dissimilar fluorescent dyes to allow rapid GSI assay of all specimens. Species-specific fragments were amplified under the following conditions: an initial cycle of 94° denaturation for 1 min, 50° annealing for 30 s, and 72° extension for 1 min, followed by 30 cycles of 94° for 30 s, 61° for 30 s, 72° for 1 min 30 s, and a final 72° extension for 8 min. All sequencing and GSI assays were conducted on automated genetic analyzers (ABI models 310 and 3100). The limited amount of DNA obtained from otoliths (17 specimens from Puerto Rico) required that these specimens be fully sequenced in lieu of the diagnostic marker assay.

OTOLITH SAMPLES

Otoliths were extracted from a subsample of adult fish captured in the recreational fishery in the Caribbean for age and growth comparisons with published data from the Florida Keys (Crabtree et al., 1996). Sex was determined during field dissection. Two to four 1–2 mm thick transverse sections containing the otolith core were cut with a Buehler Isomet low speed saw with a diamond blade. The sections were mounted on a microscope slide with thermaplastic glue. Annuli were counted three times by each of the three independent readers with reflected light at magnifications

TABLE 15.2
Summary Data for Cytochrome-*b* mtDNA Markers Developed for Species Identification of Bonefish (*Albula vulpes* and *A.* sp. B)

Marker	Primer Sequence (5'→ 3')	Species	Fragment Size (bp)
Avu-CytB F	CCACTGTACCAATGCATCG	*A. vulpes*	169
Aga-CytB F	ATCCACTGTACTAACGCATCC	*A.* sp. B	171
Avu-CytB R	GTATCTTTACATGGAGACATG	*A. vulpes*	
Aga-CytB R	TTATCTTTACATGGAGACGTG	*A.* sp. B	

of 8–25×. After readers completed reading all otoliths, otoliths with different counts were re-examined. In all cases, differences were reconciled and an age assigned to the otolith.

RESULTS

A total of 750 seine samples were completed at 30 locations in the Florida Keys. A total of 628 juvenile bonefishes were captured along windward sandy beaches (Higgs Beach [$N = 307$] and Airport Beach [$N = 126$], Key West; Bahia Honda State Park [$N = 173$]; Elliot Key [$N = 15$], Biscayne National Park) and along leeward beachrock shorelines (Elliot Key [$N = 7$]). Thirty-four leptocephalus larvae were captured (Table 15.1a), all along windward sandy beaches. Windward sandy beaches are intertidal sand shorelines with subtidal sand bottom immediately adjacent, and seagrass beginning approximately 4–9 m offshore of the beach. Beachrock shorelines are consolidated limestone at the intertidal zone with sand or mixed sand–seagrass bottom immediately adjacent. Juvenile or leptocephalus bonefishes were captured in all months except January and November 2004, and were in greatest abundance in March through July (Table 15.1A). Lengths ranged from 19 to 360 mm SL. No juvenile bonefishes were captured in shallow, sandy-bottom open bays on the Florida Bay side of the Florida Keys.

A total of 35 seine samples were also conducted along six sandy beaches at Turneffe Atoll, and 35 juvenile and three lepteocephalus bonefish ranging from 24 to 56 mm SL were captured at two beaches on the central eastern side of Turneffe (Calabash and Rope Walk).

To identify species, tissue were taken from subsample of the juvenile and leptocephalus captured in the Florida Keys (299 of 662) and Turneffe Atoll (35 of 38). Tissue samples were taken from all months and locations in which leptocephali and/or juveniles were captured (Table 15.1B). During the initial cytochrome sequencing of Florida bonefishes, we found an approximate 9% sequence difference between the two species. This is slightly lower than the 12–15% difference reported in Colborn et al. (2001) and may be attributable to our larger sample sizes. The majority (93.97%) of Florida juveniles and leptocephali assayed with the cytochrome-b diagnostic marker were identified as *A.* sp. B. Only 18 *A. vulpes* were collected from two sites: 14 from Elliot Key (EK) Biscayne National Park (BNP) and 4 from Bahia Honda. The majority of *A. vulpes* juveniles occurred in July 2004 (from 2 seine hauls at EK and 1 seine haul at Bahia Honda). The EK July collection contained only *A. vulpes*, while the Bahia Honda collection was mixed *A. vulpes* and *A.* sp. B. All juveniles and leptochalus larvae captured at Turneffe Atoll were identified as *A.* sp. B. Thus, the findings reported here for juveniles are most applicable to *A.* sp. B.

Fin clips were obtained from 138 adult bonefishes captured at six locations: Turneffe Atoll, Belize ($N = 54$); Eleuthera (7); Exuma (9), Berry Islands (5), and Mayaguana (3), Bahamas; St. Croix, U.S. Virgin Islands (1); Little Cayman, British West Indies (3); Anegada, British Virgin Islands (2); and Chetumal (7), and Punta Allen (4), Mexico (Figure 15.1). Fish ranging from 205 to 711 mm FL were either captured by anglers or by seine nets on flats frequented by guides and anglers. Based upon age estimates from otoliths of a subsample of 31 *A. vulpes* from the 138 above, ages at given lengths differed among locations (Table 15.3; Figure 15.2). Bonefish from Puerto Rico ($n = 17$) were similar in lengths at a given age to those reported for bonefish from the Florida Keys (Crabtree et al., 1996). All *A. vulpes* from all other Caribbean locations ($n = 14$) appeared to exhibit slower growth rates than those from the Florida Keys.

TABLE 15.3
Summary of Otolith-Based Age Estimations for Caribbean Bonefish *Albula vulpes*, Captured in Recreational Fisheries as with Expected Values from the Florida Keys for Comparison

Location	Fork Length (mm)	Age (year)	Sex[a]	Expected Age[b]
Puerto Rico	289	1	M	1
Puerto Rico	292	1	–	1
Puerto Rico	308	1	–	1
Puerto Rico	312	1	M	1
Puerto Rico	316	1	–	1
Puerto Rico	335	1	F	1
Puerto Rico	340	1	M	1
Puerto Rico	341	1	M	1
Puerto Rico	345	1	M	1
Puerto Rico	385	1	M	2
Little Cayman, BWI	279	2	M	1
Puerto Rico	330	2	M	1
Puerto Rico	372	2	M	2
Puerto Rico	397	2	F	2
Puerto Rico	411	2	M	2
Puerto Rico	417	2	M	2
Puerto Rico	352	3	M	2
Puerto Rico	357	3	M	2
Eleuthera, Bahamas	300	4	F	1
Eleuthera, Bahamas	320	5	M	1
Eleuthera, Bahamas	355	5	M	2
Little Cayman, BWI	342	6	F	2
Turneffe Atoll, Belize	410	8	–	2
Turneffe Atoll, Belize	411	8	–	2
Turneffe Atoll, Belize	420	8	–	3
Turneffe Atoll, Belize	426	8	–	3
Exuma, Bahamas	461	8	M	4
Exuma, Bahamas	472	8	F	4
Turneffe Atoll, Belize	423	9	–	3
Turneffe Atoll, Belize	411	10	–	2
Anegada, BVI	560	16	F	6

Note: See Figure 15.1 for sample locations.

[a] M, male; F, female; –, not determined.

[b] Expected ages from Crabtree et al. (1996).

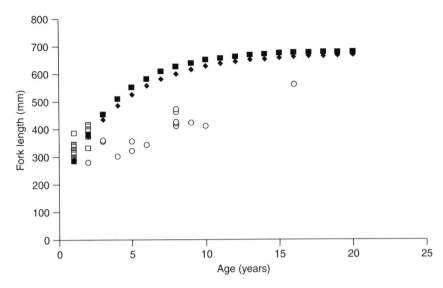

FIGURE 15.2 Observed lengths from Caribbean bonefishes identified as *Albula vulpes* (open symbols: □ = Puerto Rico; ○ = all other Caribbean locations) and predicted lengths of *A. vulpes* from the Florida Keys, Florida, United States (closed symbols: ■ = female; ◆ = male). Florida Keys values are calculated from Crabtree et al. (1996).

DISCUSSION

Our results revealed that (1) most juveniles along sandy beaches appear to be *A.* sp. B; (2) juvenile habitats for *A. vulpes* remain largely unknown; (3) fish captured in the recreational fishery appear to be *A. vulpes*; and (4) *A. vulpes* growth rates appear to differ among locations.

Combined, these findings indicate that additional information is needed to ensure a successful conservation and management strategy for *Albula* spp. in the Caribbean and western Atlantic. The findings reported here raise questions about many aspects of the conventional wisdom of *Albula* in the Caribbean.

That very few juveniles of *A. vulpes*, the species that appears to support the recreational fishery, were captured is disconcerting. Although the declines, or lack of recovery, of adult stocks of many species have been blamed on overfishing, it is becoming increasingly apparent that loss or degradation of habitats may also limit species abundances (e.g., Turner et al., 1999). This generally occurs because essential juvenile habitats or connections between juvenile and adult habitats are lost or severely degraded. Since the extent to which different juvenile habitats contribute fishes to the adult population is essential information for successful conservation of fish populations (Beck et al., 2001), it is imperative that juvenile habitats of *A. vulpes* are determined. Only then can we determine whether juvenile habitat loss has impacted *A. vulpes* populations, and design effective conservation (or even restoration) strategies.

Nonetheless, this research has contributed to knowledge of *A.* sp. B. The temporal occurrence of juveniles suggests that *A.* sp. B spawning occurs primarily in winter. In this study, small *A.* sp. B was in greatest abundance in March and May, and in lesser and roughly equal abundance in late January and July. If it is assumed that larval duration of *A.* sp B is not notably longer than the maximum 72 days for *A. vulpes*, then spawning occurred during fall through early spring, similar to that reported for *A. vulpes* by Crabtree et al. (1997). Concurrent spawning by *A.* sp. B and *A. vulpes* is also supported by the occurrence of juvenile *A.* sp. B in the Florida Keys in this study coinciding with the occurrence of *Albula* leptocephali in the Bahamas in 2004 (C. Dahlgren, personal communication). The Bahamas leptocephali were subsampled, and all were genetically identified as *A. vulpes*.

The occurrence of juvenile *A.* sp. B along sandy beaches suggests a distinct ontogenetic habitat shift. Information from professional fishing captains indicates that adult *A.* sp. B reside in deeper water, whereas this study documents the occurrence of juveniles in shallow shoreline habitats. In contrast, adult *A. vulpes* use a variety of mostly shallow coastal habitats. While it appears that juveniles only rarely used shallow shoreline habitats, the bulk of the resource may be outside the current sampling domain. Sampling in this study and other research suggests that *A. vulpes* juveniles use deeper habitats than were sampled during this study. For example, juvenile bonefishes were not captured in shoreline samples in St. Croix, U.S. Virgin Islands (Adams, unpublished data; Mateo and Tobias, 2004), nor in extensive sampling with multiple gears in the Florida Keys and Florida Bay (D. Snodgrass, NOAA, personal communication). These findings indicate that additional research elucidating species-specific spatial and temporal patterns of juvenile habitat use is necessary to provide sound ecological information that will enable sound bonefish fishery management.

The findings on regional variation in growth rates of adult *A. vulpes* should be treated with great caution because of the low sample size, but do suggest that significant research is needed to specifically address the issue. For example, minimum size at maturity appears to be smaller in the Caribbean than in the Florida Keys: a 342-mm FL female at Little Cayman was sexually mature, whereas 50% sexual maturity is 488 mm FL (95% confidence interval (CI) 472–504 mm) for females in the Florida Keys (Crabtree et al., 1997). Although preliminary, ongoing research in Los Roques, Venezuela, suggests differences in growth rate between Venezuela and the Florida Keys (Posada et al., Chapter 8, this volume). In addition, *A. glossodonta* shows differences in growth and maximum size among locations in the Pacific (Friedlander et al., Chapter 2, this volume).

These findings show a need to reassess conventional wisdom on bonefish biology in the Caribbean, and suggest directions for additional research. The potential for considerable differences in growth rates requires research to verify these findings and to understand the underlying mechanisms. For example, Florida Keys and Puerto Rico coastal habitats receive terrestrial nutrient inputs from rivers that might increase productivity relative to other insular oceanic islands where most other *A. vulpes* were collected in this study. Alternatively, similar latitudinal differences in growth rate have been observed in other species (e.g., Murphy and Taylor, 1990), and may indicate counter-gradient variation (Edwards, 1984; Conover, 1990).

Additionally, the formation of annuli on Caribbean bonefishes otoliths needs to be verified. There was a possibility that bonefish ages in this study were misclassified because otolith increments for young fish tend to be obscured in lower latitudes due to reduced seasonality, making clear interpretation of annuli difficult (Victor, 1982; Caldow and Wellington, 2003). However, annuli were readily apparent for most individuals in this study, and since otoliths in this study were collected in lower latitudes (Belize, British Virgin Islands, Puerto Rico, Cayman Islands) or locations with less temperature variation than the Florida Keys (i.e., the Bahamas), we would expect that the bias would have been to underestimate age.

Within *A. vulpes*, research should also be conducted to verify the apparent regional differences in growth and size at maturity. Higher sample sizes of a wider range of sizes are needed from all locations to test the preliminary conclusions of this study. Data should include length, weight, gonadal stage, meristic information, and genetic identification.

Although so far *A.* sp. B has not been documented in the recreational fishery, considerably higher sample sizes are greatly needed to verify a single-species fishery. Genetic analysis will continue to be a powerful tool in future research of *Albula*, as underscored in this study. This is especially true for differentiation of *A. vulpes* and *A.* sp. B, since *A.* sp. B morphometric information is lacking for all life stages, and limited information indicates significant overlap of morphological characteristics (Crabtree et al., 2003) drawing into question some of the conclusions based on genetic data alone. Genetic analysis may prove useful in examining species geographic ranges and population connectivity, and whether these two sympatric species might hybridize.

Finally, although the documentation of a new *Albula* species in the Caribbean has introduced a new aspect into research and conservation of bonefish in the region, this research also raises questions that are specific to *A. vulpes*. These questions include aspects of both habitat use and regional variations within the species. To the extent that additional research contributes to knowledge on these issues, the research, conservation, and management frameworks for Caribbean bonefish may require modification.

ACKNOWLEDGMENTS

Funding for field sampling was provided by Bonefish & Tarpon Unlimited, National Fish and Wildlife Foundation (Grant No. 2004-0049-000), Turneffe Atoll Conservation Fund (AJA), and Biscayne National Park (GTK). Genetic analysis was funded by Wallop-Breaux Sportfish Restoration Grant #F-69 (MDT). Puerto Rico samples were provided by C. Caldow (NOAA/NOS). Otolith processing was by T. DeBruler. Additional assistance for adult sampling was provided by Turneffe Flats Lodge, Peace and Plenty Bonefish Lodge, and Little Cayman Dive Resort. Thanks to M. Newton, S. Moneysmith, H. Tritt, C. Walter, B. Thornton, P. Madden, and G. Harrison for field sampling assistance. Thanks also to A. Boyer, B. Barbour, B. Frerer, R. Miller, D. Perez, D. Parker, G. Pittard, D. Skok, T. Woodward, A. Murrant, numerous anonymous anglers for fin clip collections, and B. Nauheim, J. Bottcher, and P. Pendergast for arranging some collections.

REFERENCES

Beck, M.W., K.L. Heck Jr., K.W. Able, D.L. Childers, D.B. Eggleston, B.M. Gillanders, B. Halpern, C.G. Hayes, K. Hoshino, T.J. Minello, R.J. Orth, P.F. Sheridan, and M.P. Weinstein. 2001. The identification, conservation, and management of estuarine and marine nurseries for fish and invertebrates. BioScience 51(8): 633–641.

Bruger, G.E. 1974. Age, growth, food habits, and reproduction of bonefishes, *Albula vulpes*, in South Florida waters. FL. Mar. Res. Pub. No. 3. 20pp.

Caldow, C. and G.M. Wellington. 2003. Patterns of annual increment formation in otoliths of Pomacentrids in the tropical western Atlantic: implications for population age structure examination. Mar. Ecol. Prog. Ser. 256: 185–195.

Colborn, J., R.E. Crabtree, J.B. Shaklee, E. Pfeiler, and B.W. Bowen. 2001. The evolutionary enigma of bonefishes (*Albula* spp.): cryptic species and ancient separations in a globally distributed shorefish. Evolution 55(4): 807–820.

Colton, D.E. 1983. Movements of bonefishes, *Albula vulpes*, in Bahamian waters. Fish. Bull. 81: 148–154.

Colton, D.E. and W.S. Alevizon. 1983. Feeding Ecology of Bonefishes in Bahamian Waters. Trans. Am. Fish. Soc. 112: 178–184.

Crabtree, R.E., C.W. Harnden, D. Snodgrass, and C. Stevens. 1996. Age, growth, and mortality of bonefishes, *Albula vulpes*, from the waters of the Florida Keys. Fish. Bull. 94: 442–451.

Crabtree, R.E., D. Snodgrass, and C.W. Harnden. 1997. Maturation and reproductive seasonality in bonefishes, *Albula vulpes*, from the waters of the Florida Keys. Fish. Bull. 95: 456–465.

Crabtree, R.E., D.J. Snodgrass, and F.J. Stengard. 2003. Bonefish species differentiation and delineation of critical juvenile habitat in the Florida Keys. Pp. 7–13, in Investigations into nearshore and estuarine gamefish behavior, ecology, and life history in Florida. Sport Fish Restoration Act Report.

Humston, R., J.S. Ault, M.F. Larkin, and J. Luo. 2005. Movements and site fidelity of bonefish (*Albula vulpes*) in the northern Florida Keys determined by acoustic telemetry. Mar. Ecol. Prog. Ser. 291: 237–248.

Mateo, I. and W.J. Tobias. 2004. Survey of nearshore fish communities on tropical backreef lagoons on the southeastern coast of St. Croix. Carib. J. Sci. 40(3): 327–342.

Mojica, R. Jr., J.M. Shenker, C.W. Harnden, and D.E. Wagner. 1995. Recruitment of bonefishes, *Albula vulpes*, around Lee Stocking Island, Bahamas. Fish. Bull. 93: 666–674.

Murphy, M.D. and R.G. Taylor. 1990. Reproduction, growth, and mortality of red drum *Sciaenops ocellatus* in Florida waters. Fish. Bull. 88: 531–542.

Robins, C.R., G.C. Ray and J. Douglas. 1986. Atlantic Coast Fishes. Houghton Mifflin Company. Boston, MA. 354p.

Turner, S.J., S.F. Thrush, J.E. Hewitt, V.J. Cummings, and G. Funnell. 1999. Fishing impacts and the degradation of or loss of habitat structure. Fish. Manage. Ecol. 6: 401–420.

Victor, B.C. 1982. Daily otolith increments and recruitment in two coral-reef wrasses, *Thalassoma bifasciatum* and *Halichoeres bivittatus*. Mar. Biol. 71: 203–208.

Section III

Population Dynamics and Resource Ecology

16 Population Dynamics and Resource Ecology of Atlantic Tarpon and Bonefish

Jerald S. Ault, Robert Humston, Michael F. Larkin, Eduardo Perusquia, Nicholas A. Farmer, Jiangang Luo, Natalia Zurcher, Steven G. Smith, Luiz R. Barbieri, and Juan M. Posada

CONTENTS

Introduction	218
Atlantic Tarpon (*Megalops atlanticus*)	219
Life Cycle and Resource Ecology	219
Species Distribution and Unit Stock	219
Life Cycle	219
Diet	222
Regional Movements and Migrations	222
Population Dynamics	224
Age and Growth	224
Maximum Age, Maximum Size, and Lifetime Survivorship	232
Maturity and Fecundity	233
Fisheries Exploitation and Human Impacts	234
Atlantic Bonefish (*Albula vulpes*)	235
Life Cycle and Resource Ecology	235
Species Distribution and Unit Stock	235
Life Cycle	236
Diet	239
Regional Movements and Migrations	239
Population Dynamics	242
Age and Growth	242
Maximum Age, Maximum Size, and Lifetime Survivorship	246
Maturity and Fecundity	246

Fisheries Exploitation and Human Impacts 247
Filling Critical Knowledge Gaps for Fishery Management 249
　Resource Ecology 250
　Population Dynamics 251
　Fishery Management 252
References 253

> If he would fish only for sport and the excitement of battle, the mighty tarpon, "Silver King" of finny tribes, often tipping the beam at two hundred pounds, and the agile bonefish, weighing less than ten, but darting with the swiftness of a hawk, and fighting with a hawk's persistent energy, will give him every opportunity for testing his skill and power of endurance against theirs.
>
> —**Kirk Munroe (1909), Florida: a winter playground,
> in Oppel, F. and T. Meisel (1987)**

INTRODUCTION

The tremendous popularity and economic importance of Atlantic tarpon (*Megalops atlanticus* Valenciennes) and bonefish (*Albula vulpes* Linneaus) as gamefishes belies the apparent lack of management-relevant information on population dynamics and resource ecology for the species. Tarpon and bonefish offer a challenge to biologists and fishermen alike. Their complex life history makes their study difficult (Robins, 1977). Tarpon (Elopiformes) and bonefish (Albuliformes) are two of the most primitive assemblages of living bony fishes, an ancient lineage they share with three other orders: Anguilliformes (catadromous and marine eels); Notacanthiformes (spiny eels); and Saccopharyngiformes (gulper eels). The most distinctive commonality of this assemblage (Elopomorpha) is the leptocephalus larval stage, which lives in clear, warm, oceanic waters before metamorphosis to the juvenile stage (Greenwood et al., 1966; Robins, 1977; Smith, 1980; Hulet and Robins, 1989; Shiao and Hwang, 2006; Nelson, 2006).

Rising exploitation pressures, rapid human development, and environmental changes in coastal waters suggest that new information on population dynamics and resource ecology is critically needed to support fishery management strategies to conserve these precious resources. Unfortunately, the body of available scientific information lacks substantive data that are essential to predicting the future course of the fisheries and to making decisions concerning habitat preservation, stock management, and conservation. Also lacking are coherent summaries of knowledge gaps for critical aspects of life history, population dynamics, and fishery impacts. In this paper, we synthesize existing information on tarpon and bonefish population ecology from primary literature and, where appropriate, gray literature sources, and then integrate this information with new data. We concentrate on the data derived principally from the western Atlantic Ocean, particularly studies in Caribbean Sea and Gulf of Mexico waters (Florida, Bahamas, and West Indies).

ATLANTIC TARPON (*Megalops atlanticus*)

LIFE CYCLE AND RESOURCE ECOLOGY

Species Distribution and Unit Stock

Atlantic tarpon are relatively large, highly migratory fish that frequent coastal and inshore waters of the tropical and subtropical central Atlantic Ocean (Robins and Ray, 1986; Crabtree et al., 1995; Ault et al., 2005a; McMillen-Jackson et al., 2005; Luo et al., Chapter 18, this volume). In the western Atlantic, tarpon range from Virginia, Bermuda, Gulf of Mexico, Caribbean Sea to Brazil (Wade, 1962), and infrequently from Nova Scotia to Argentina and the eastern Pacific near the terminus of the Panama Canal (Robins and Ray, 1986). In the eastern Atlantic, they occur primarily along the west coast of Africa from Angola to Senegal (Roux, 1960), and rarely from Portugal, the Azores, southern France to northern Spain (Arronte et al., 2004). Extant distributions of tarpon correspond to those of the tropical and subtropical mangroves (Mendoza-Franco et al., 2004). Genetic studies have shown differentiation between tarpon from Africa and tarpon from the western Atlantic Ocean, suggesting that levels of gene flow between tarpon from these two regions may be low (McMillen-Jackson et al., 2005). Although Blandon et al. (2003) analyzed two African tarpon that possessed the most common western Atlantic mtDNA haplotype, discrepancies may be due to the genetic markers used (McMillen-Jackson et al., 2005). Among the western Atlantic groups, McMillen-Jackson et al. (2005) found similar genetic diversity values suggesting connectivity between these resources, although Costa Rica tarpon could be partially isolated from other populations.

Life Cycle

Tarpon spawning patterns have been inferred from larval distribution patterns and gonadosomatic indices (GSI) of mature adults. Observed larval distribution patterns suggest that spawning of mature tarpon off Florida occurs in offshore waters from April through August (Smith, 1980; Cyr, 1991; Crabtree et al., 1992; Crabtree, 1995); however, Harrington (1966) found tarpon larvae present in the Gulf Stream through November. GSIs peaked for both male and female Florida tarpon in May, with spent females making up less than 25% of the catch in May–July, and more than 90% of the catch in August (Cyr, 1991). Crabtree et al. (1997a) concurred as they found that Florida tarpon spawned during April–July, and by August most fish were either spent or recovering. Smith (1980) back-calculated hatching dates for tarpon in South Florida to coincide with the June–August period. Little is known about the early life history of tarpon, partially because fertilized tarpon eggs have never been observed *in situ* and the specific locations of tarpon spawning sites have never been identified. Indirect evidence of spawning comes from Crabtree et al. (1997a) who observed partially spent female tarpon with ovaries containing postovarian follicles and advanced vitellogenic oocytes in both Florida and Costa Rican waters. Tarpon may be batch spawners, and in Costa Rica they may spawn year-round (Chacon-Chaverri, 1993; Crabtree et al., 1997a) as reproductively active females have been observed in all months. de Menezes and Paiva (1966) examined gonads of tarpon caught off the northeast coast of Brazil and concluded that reproduction probably occurs in October–January (Table 16.1). Tarpon spawning season in Puerto Rico may be year-round but with peaks in March–May and July–September (Zerbi et al., 2001). Schools of gravid tarpon migrate from nearshore and inshore habitats to form large prespawning aggregations approximately 2–5 km offshore (Crabtree et al., 1992), presumably before moving up to 200–250 km offshore for spawning. Crabtree et al. (1992) suggested that mature tarpon enter Florida inshore waters during April–June to feed before moving offshore for spawning. Prespawning aggregations have been referred to as "daisy chains" by anglers, consisting of milling tarpon oriented in a similar direction and swimming in circles. In southern Florida, tarpon have been observed prior to the new moon in groups of several males surrounding one larger female, with the males bumping the female's vents in attempts to stimulate egg release. This may represent premating behavior (see Baldwin and Snodgrass, Chapter 14, this volume). The exact timing, cues, and zones of tarpon spawning have not been described, although Crabtree (1995) suggested that it may be triggered by lunar tidal cycles.

Planktonic leptocephalus larvae of tarpon are widely distributed (Zale and Merrifield, 1989; Crabtree et al., 1992) and common in major western Atlantic Ocean currents (Gehringer, 1959; Eldred, 1967). The 2–3 month phase of larval development occurs up to 250 km offshore in warm, clear, high-salinity waters (Crabtree, 1995). The geographical extent of larval dispersal is unknown, and local eddies and gyres may entrain pelagic larvae and contribute to partial isolation of populations (McMillen-Jackson et al., 2005; Cowen et al., 2006; Steneck, 2006). Tarpon leptocephalus larvae were captured by Crabtree et al. (1992) over depths of 90–1400 m, with sea surface temperatures of 27–30°C and salinities of about 36 ppt. Larval collections suggest that tarpon in Florida waters spawn offshore from May through August, perhaps to October (Smith, 1980; Crabtree et al., 1992; Crabtree, 1995; Crabtree et al., 1995) (Table 16.1). Berrien et al. (1978) collected larval tarpon

TABLE 16.1
Spawning Periodicity of Atlantic Tarpon (*Megalops atlanticus*) in the Western Central Atlantic Ocean

Location	Jan	Feb	Mar	Apr	May	Jun	Jul	Aug	Sep	Oct	Nov	Dec	Source
Brazil	X									X	X	X	de Menezes and Paiva, 1966
Columbia				X	X								Garcia and Solano, 1995
Costa Rica	X	X	X	X	X	X	X	X	X	X	X	X	Crabtree et al., 1997a
Florida				X	X	X	X	X					Smith, 1980; Cyr, 1991; Crabtree et al., 1997a
Mexico				X	X	X	X	X					Perusquia, personal communication
Puerto Rico			X	X	X		X	X	X				Zerbi et al., 2001

off North Carolina, an indication that tarpon spawning may occur along the U.S. south Atlantic coast from Florida to Cape Hatteras (Smith, 1980). Alternatively, this may reflect northward advection from Florida spawning grounds via Gulf Stream currents during the relatively long larval duration (e.g., Cowen et al., 2006). Metamorphic tarpon are found inshore in bay and coastal waters. Recruitment of metamorphic larvae in Costa Rica was highest from December to February and July to October, corresponding with winter storms and the summer hurricane season, respectively (Chacon-Chaverri, 1993). Summer storms, hurricanes, and associated flooding may push metamorphic tarpon into interior streams and pools where the growth into the juvenile stage may be triggered by contact with a freshwater environment (Babcock, 1951; Harrington, 1966; Chacon-Chaverri, 1993). The biophysical processes that interact to transport and ultimately deliver pelagic leptocephalus larvae inshore to juvenile habitats are complex, including significant advection by wind-induced currents associated with hurricanes (Gehringer, 1959; Eldred, 1967, 1968, 1972; Smith, 1980; Cyr, 1991; Crabtree et al., 1992, 1995; Shenker et al., 1995).

Juvenile tarpon occur widely but prefer a warm estuarine or mangrove environment (Robins, 1977). They feed at or near the surface and readily capture insects that fall into the water (Babcock, 1951; Robins, 1977). Small juveniles are restricted to the salt marshes and shallow mangrove-lined estuaries and stagnant pools of varying salinity where predator pressure is low and food supply high (Harrington, 1958, 1966; Erdman, 1960b; Wade, 1962; Mercado and Ciardelli, 1972; Tucker and Hodson, 1976; Chacon-Chaverri and McLarney, 1992; Crabtree et al., 1995; Shenker, personal communication), such as the Everglades and the Big Cypress Swamp in south Florida (Kushlan and Lodge, 1974). Young-of-year (YOY) tarpon have been reported from North Carolina (Hildebrand, 1934), Georgia (Rickards, 1968), Florida (Wade, 1962, 1969), Texas (Simpson, 1954; Marwitz, 1986), including inland reservoirs (via introductions; Howells and Garrett, 1992), the Caribbean islands (Breder, 1933), Mexico, and Central America (Chacon-Chaverri, 1993). As facultative airbreathers, tarpon can tolerate harsh habitats characterized by anoxia, periods of extremely shallow water, and high hydrogen sulfide concentrations (Robins, 1977). Water temperatures below 10°C are lethal to tarpon (Zale and Merrifield, 1989), which helps to explain their constrained northward distribution. Episodic kills of tarpon in cold winters are common in southern Florida, Texas, and northern Mexico where young fish lack access to deeper, warmer water. Interestingly, while adults are common in Costa Rica, Nicaragua, and Panamá, juvenile tarpon are considered rare to absent in these areas (Chacon-Chaverri, 1993).

Late juvenile tarpon are dependent upon deep-water habitats such as canals and sloughs for emigration to coastal bays (Hunt in a personal communication to Kushlan and Lodge, 1974). Juveniles have been found to inhabit mud flats in Puerto Rico during times when connections exist among mud flats and adjacent lagoons (June–February; Zerbi et al., 1999); storms may flush the juveniles into new habitats, providing them with additional food resources (Rickards, 1966). Emigration from juvenile habitats may be mediated by increasing food requirements (Cyr, 1991).

Adult tarpon (>120 cm FL) are primarily coastal fishes that inhabit inshore waters and bays over a wide range of salinities (fresh to hypersaline) and temperatures (17–36°C) (Zale and Merrifield, 1989; Crabtree et al., 1995). Large fish appear tens to hundreds of kilometers offshore and are capable of migrating thousands of kilometers (Ault et al., 2005a; National Marine Fisheries Service tagging database 1960–1999; Luo et al., Chapter 18, this volume). During these offshore migrations they appear to prefer the 26°C isotherm. Reports from temperate waters indicate that migrations of incoming tarpon are generally associated with high seasonal temperatures (Costa Pereira and Saldanha, 1977; Twomey and Byrne, 1985). Recent data from pop-up archival transmitting (PAT) tags deployed in southeastern U.S. coastal waters suggest that tarpon activity may be related to water turbidity (Luo et al., Chapter 18, this volume). The authors illustrate that activity of this particular tarpon (inferred from rapid depth changes) was lowest during periods of high winds associated with passage of frontal systems. Tarpon likely rely on visual perception as predators; increased wind speeds lead to greater wave action and higher turbidity in the water column, decreasing visual acuity.

Diet

Like most large marine fishes, the diet of the tarpon changes according to life stage. There is disagreement over leptocephalus larvae food habits and requirements; some reports conclude that they do not feed (Hollister, 1939), while others suggest that first feeding larvae consume protozoans, rotifers, larvacean houses, and fecal pellets (Dahl, 1971; Mochioka and Iwamizu, 1996). Juvenile tarpon are crepuscular, normally feeding first at sunset and then into the night if sufficient light is present (Robins, 1977). The tarpon's superior mouth and large proportioned jaw allows them to capture food (normally whole prey items) from below via suction (Cataño and Garzón-Ferriera, 1994; Grubich, 2001), particularly with a well-lighted background. Harrington and Harrington (1960), Hildebrand (1963), and Rickards (1968) generally characterized the diet of juvenile tarpon as "carnivorous, but predominantly piscivorous." Harrington and Harrington (1960) reported strong preference of small (1.6–7.5 cm) juvenile tarpon for cyclopoid copepods and fishes (73 and 22%, respectively). The remaining 5% of the diet consisted of mosquito larvae, ostracods, and small shrimps, which fluctuate seasonally with bursts in production (Cataño and Garzón-Ferriera, 1994; Robins, 1977). The size of the prey items consumed by tarpon increases proportional to size (e.g., Rickards, 1968), and as the tarpon grows fishes such as mullets (*Mugil* spp.) and mollies (*Poecilia* spp. and *Gambusia affinis*) become primary prey items. Adult tarpon feed on mullet, silversides, marine catfish, shrimps (pink, brown, and white), blue crabs, ribbonfish, and menhaden, among other fishes.

Regional Movements and Migrations

Tarpon may have resident, migratory, or mixed populations (Robins, 1977). New evidence from PAT tagging suggests that mature tarpon will undertake substantial alongshore and offshore spawning migrations (Ault et al., 2005a; Luo et al., Chapter 18, this volume). Some of these data show that tarpon will travel hundreds to thousands of kilometers in relatively short time periods (<2 mo). In addition, conventional tagging data collected by the National Marine Fisheries Service's Gamefish Tagging Program (Eric Prince, NMFS, personal communication) provide a wealth of information on the long-distance movements of tarpon tagged by anglers. These studies document tarpon transiting vast distances, often in relatively short time periods. These movements can carry tarpon across international borders: tarpon tagged off North Carolina waters have been recaptured off the southern coast of Cuba, while others tagged off Mexico have later appeared in the waters of Texas, Louisiana, and other Gulf states.

There is now little question that tarpon are capable of—and often undertake—long-distance movements. These movements may represent repeated migratory patterns, or there may be significant annual variation in the movement patterns of individuals. The evidence is presently not yet sufficient to address the population significance of long-distance movements in tarpon. Two possibilities are spawning and feeding, and these are not mutually exclusive functions. The Caribbean and Gulf of Mexico has been suggested to be an important spawning zone for tarpon (Smith, 1980). Spawning occurs in the spring and summer in the Yucatan Channel and the Gulf of Mexico, with spawning areas off Cozumel, the west coast of Florida, and in the southwestern Gulf of Mexico (Smith, 1980) (Figure 16.1). Tarpon along the U.S. Atlantic coast and the western Gulf of Mexico appear to undertake seasonal migrations, moving north in the spring–early summer months, then returning south in the

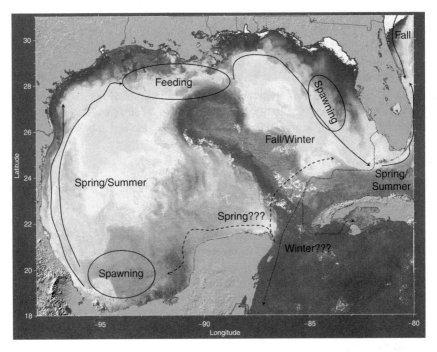

FIGURE 16.1 Map showing spawning and feeding areas for tarpon, *Megalops atlanticus*, and observed migration routes around Florida, the Gulf of Mexico, and the northern Caribbean Sea. Solid lines indicate observed migration pathways from PAT satellite tagging; dotted lines indicate observed routes from NMFS conventional tag study; and dashed lines indicate hypothesized migration routes.

fall and winter. Fishes that comprise the U.S. Atlantic coast run are assumed to originate in Florida; the extent to which these stocks may interact (i.e., mix) is not well understood. On the southward return migrations, it is unknown how many tarpon continue past the Florida Keys and into the Caribbean Sea; however, recreational catches indicate that many tarpon overwinter in south Florida's coastal waters. There are records of fish that originated in Louisiana moving south to Mexico with others moving east to Key West. Tarpon that seasonally move into Texas, Louisiana, and elsewhere in the northern Gulf of Mexico may depend on spawning grounds between Tamaulipas and Veracruz, Mexico (Robins, 1977).

There are no data to assess the potential for trans-Atlantic migrations of tarpon (McMillen-Jackson et al., 2005). Tarpon have occasionally been reported from European waters, presumably after following the warm Gulf Stream across the Atlantic Ocean (Costa Pereira and Saldanha, 1977; Twomey and Byrne, 1985). The European region does not constitute a suitable long-term habitat for tarpon, and the cool, high-salinity waters off North Africa are probably a barrier to north-to-south dispersal (Costa Pereira and Saldanha, 1977; McMillen-Jackson et al., 2005). In tropical waters of the equatorial Atlantic, strong east-to-west currents and advective outflows from the Congo River could assist the dispersal of tarpon larvae and the migratory patterns of adults. This region is a well-known feeding area for tunas and billfish and contains sufficient forage fish to sustain such a migration. These strong currents likely inhibit larval transport in the opposite direction, resulting in asymmetrical exchange. Such a pattern of larval dispersal would essentially isolate African tarpon from an influx of western Atlantic fish, unless there is directed migration by adults between the two continents. Genetic studies in the western Atlantic (McMillen-Jackson et al., 2005) have suggested that tarpon populations are genetically similar, but some regional isolation may occur. However, the western Atlantic Ocean has few absolute barriers for the dispersal of a species such as tarpon with actively migrating adults and a relatively long pelagic larval period.

POPULATION DYNAMICS

In this section we review the population dynamics of Atlantic tarpon. Management-relevant population parameters considered are age and growth, maximum age and survivorship, and reproductive maturity. Parameter definitions are provided in Table 16.2, and specific parameter values are listed in Table 16.3. In this chapter, we have augmented and updated Crabtree et al.'s (1995) database with additional aged samples from follow-on studies conducted at the Florida Fish and Wildlife Research Institute (R. Crabtree and Luiz Barbieri, personal communication), along with new allometric weight-fork length data collected from tarpon tournaments in Coatzacoalcos and Veracruz, Mexico, from 2000 to 2005.

Age and Growth

Age and growth of Atlantic tarpon have been insufficiently documented (Crabtree et al., 1995), and the majority of life history and population dynamics research

TABLE 16.2
Parameters, Definitions, and Units for Population Dynamics Variables Used in This Review of Atlantic Tarpon (*Megalops atlanticus*) and Bonefish (*Albula vulpes*)

Parameter	Definition	Units
t	Age	Years
t_r	Age of recruitment	Years
L_r	Length at recruitment	Centimeters (tarpon) or millimeters (bonefish)
t_m	Age at 50% maturity	Year
L_m	Length at 50% maturity	Centimeters or millimeters FL
t_λ	Oldest age	Years
W_λ	Weight at oldest age	Kilograms
L_λ	Length at oldest age	Centimeters or millimeters FL
W_∞	Ultimate weight	Kilograms
L_∞	Ultimate length	Centimeters or millimeters FL
K	Body growth coefficient	Per year
t_0	Age at which length equals 0	Years
G	Dorsal girth	Centimeters
α	Scalar coefficient of weight–length function	Dimensionless
β	Power coefficient of weight–length function	Dimensionless
$W(t)$	Weight at age a at time t	Kilograms
$L(t)$	Length at age a at time t	Centimeters or millimeters FL
$N(t)$	Numbers at age a at time t	Numbers of fish
M	Instantaneous natural mortality rate	Per year
F	Instantaneous fishing mortality rate	Per year
$S(t)$	Survivorship to age t	Dimensionless
Z	Instantaneous total mortality rate	Per year
$\Theta(t)$	Sex ratio at age t	Dimensionless
$\Gamma(w)$	Fecundity as a function of weight	Ooyctes per female

Note: See Table 16.3 for parameter values.

comes from Florida and the Gulf of Mexico. Size (length) at hatching is 0.2 cm *FL*; Smith (1980) collected a 0.057 cm *FL* larva that retained a portion of the yolk sac. Leptocephalus larvae have extremely high growth rates, yet may remain in the plankton for several months before undergoing metamorphosis into their juvenile form (McCleave, 1993). Crabtree et al. (1992) conducted directed sampling of tarpon larvae to obtain a realistic estimate of distribution and abundance. They examined otoliths (sagittae) of tarpon leptocephali from south Florida and found their ages ranged from 2 to 25 days for sizes ranging from 0.55 to 2.44 cm, respectively. The relationship between fish length and age was estimated for 117 larvae by

$$SL = 2.78 + 0.92t, \tag{16.1}$$

TABLE 16.3
Atlantic Tarpon and Bonefish Population Dynamics Parameters

Species Groups	M	t_λ	L_∞	W_∞	K	t_0	L_m	t_m	L_r	t_r	α	β	L_λ, W_λ from IGFA
Tarpon, female	0.0545	55	181.5	61.8	0.1019	−1.20	128.5	10.9	2.7	0.09	4.568E−05	2.71	238.76 cm, 130 kg
Tarpon, male		44	157.0	41.7	0.1159	−1.51	117.5	10.4	2.7	0.09	4.568E−05	2.71	
	$FL = -1.062607 + 0.896584TL$												
	$FL = 1.08005833 + 1.0423643SL$												
Bonefish	0.1498	20	663.0	4.52	0.3258	−0.29	499.0	4.6	64.0	0.16	1.632E−08	2.99	865.76 mm, 8.62 kg
	$FL = -0.61714 + 0.858092TL$												
	$FL = 1.996345 + 1.032997SL$												

Note: Parameter definitions and units are given in Table 16.2.

where SL is standard length in mm and t is age in days. Approximate hatching dates for these larvae ranged from 12 May to 10 July.

Tarpon larvae do not move inshore until they undergo metamorphosis to the juvenile stage: a critical period that influences whether the pelagic larvae can successfully transit to demersal habitats (Shiao and Hwang, 2006). During this period of metamorphosis, the transparent leptocephali shrink from approximately 2.8 to 1.3 cm SL, perhaps in response to signals associated with inshore waters such as reduced salinity or turbidity (Cyr, 1991). Once metamorphosis is complete, positive growth begins again (Wade, 1962). Thus, size at recruitment L_r to the population is 2.3–3.0 cm, which corresponds to age of recruitment t_r of 30–34 days, respectively. Rickards (1966) found recruited juvenile specimens in Georgia as small as 1.96 cm, and provided the following weight–length equation for tarpon ranging from 19.6 to 273.5 mm SL:

$$\log_{10} W = -5.21753 + 3.18689[\log_{10} SL], \tag{16.2}$$

where W is weight (g) and SL is the standard length (mm). Harrington (1958) described the weight–length relationship for juvenile tarpon 16–45 mm SL as

$$W = -150 + 55.14(1.069^{SL}). \tag{16.3}$$

Crabtree et al. (1995) studied age and growth of juvenile to adult tarpon from Florida. Their specimens were taken from the Florida Keys, Boca Grande Pass, and Indian River Lagoon. The smallest fish in their length samples (sex undetermined) was 6.6 cm FL, while the largest male was 171 cm and the largest female was 204.5 cm FL. The range of lengths in Crabtree et al.'s (1995) samples followed a bimodal distribution, containing many small and some large fish. Generally, the samples lacked fish in the 90–120 cm range with few fish larger than 175 cm. Many of these historical data have been reported in various terms of total length TL, standard length SL, and fork length FL. In this chapter, FL is used as the standard reporting size; using all available Florida data the relationship between FL dependent on TL is estimated as

$$FL = -1.062607 + 0.896584TL, \tag{16.4}$$

where $n = 1074$ and $r^2 = 0.9994$ (Figure 16.2A). The statistical relationship between FL dependent on standard length SL is

$$FL = 1.08005833 + 1.04236437 SL, \tag{16.5}$$

with $n = 1356$ and $r^2 = 0.9994$ (Figure 16.2B).

Another growth function is the allometric relationship

$$W(t) = \alpha L(t)^\beta, \tag{16.6}$$

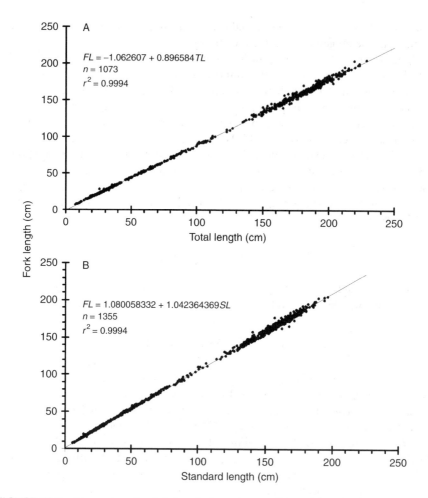

FIGURE 16.2 Regressions for fork length dependent on (A) total length and (B) standard length, for tarpon, *Megalops atlanticus*, from south Florida waters.

where $W(t)$ is the weight at age t; $L(t)$, the length at age t; α, a scalar coefficient of the weight–length function; and β, the power coefficient of the weight–length function. We fit the allometric function to data for 1488 fish from Florida ($n = 1279$), International Game Fish Association (IGFA) world records ($n = 73$), and Mexico ($n = 136$) (Figure 16.3B). Figure 16.3A shows that the function fits the Florida data well.

Crabtree et al. (1997a) described sexually dimorphic growth with females significantly larger than males for tarpon from Florida (confirmed by Andrews et al., 2001) and Costa Rica (cf. Cyr, 1991). Costa Rican tarpon were also examined by Chacon-Chaverri (1993) and it was observed that female tarpon were significantly heavier than males at a given age (i.e., average sizes $\overline{W}♀ = 35.90$ kg; $\overline{W}♂ = 21.36$ kg; and, weights ranged from 7.0 to 74.0 kg, $\overline{W} = 29.68$ kg). Using nonlinear regression methods to analyze *FL* dependent on age data combined from Crabtree et al. (1995)

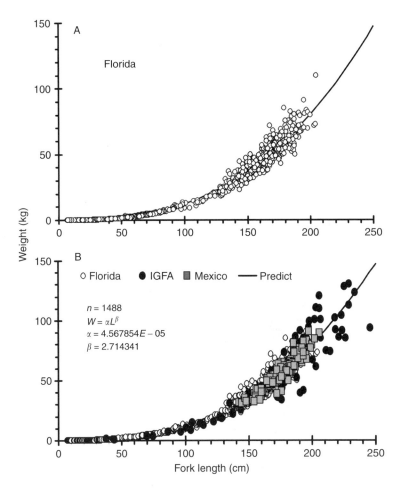

FIGURE 16.3 Plots of observed (symbols) allometric weight dependent on fork length plots and predicted nonlinear regression (solid line) for tarpon, *Megalops atlanticus*, from (A) south Florida and (B) Florida, Mexico, and IGFA world records.

and Ault et al. (2005a), we fit lifetime growth functions to the von Bertalanffy equation,

$$L(t) = L_\infty \left(1 - e^{-K(t-t_0)}\right) \quad (16.7)$$

for females and males. The sample of 316 females had a mean length of 145.6 cm *FL*, while 164 male tarpon had a mean *FL* of 112.6 cm (Figures 16.4A, 16.4C). Weight-at-age was determined by evaluating Equation 16.6 with the results of Equation 16.7 (Figure 16.4B, 16.4D).

Because catch-and-release is rapidly becoming a dominant component of tarpon fishing tournaments, we developed an empirical function to efficiently estimate tarpon weight given readily obtainable measurements of length and girth.

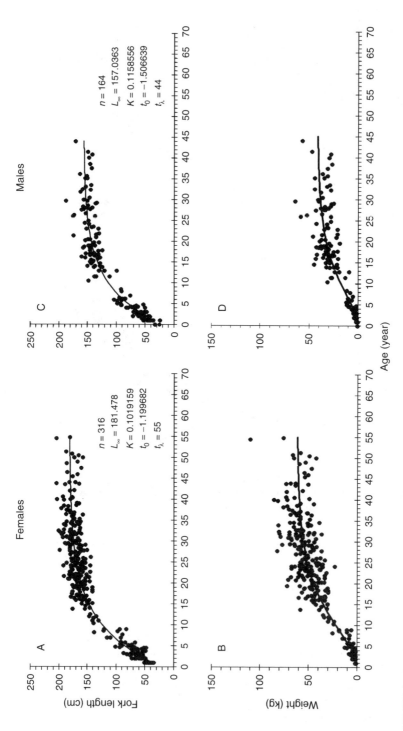

FIGURE 16.4 Observed fork lengths and weights at ages (solid dots) for tarpon, *Megalops atlanticus*, and predicted length or weight on age functions (solid line) from the von Bertalanffy growth model for: (A–B) females ($n = 316$) and (C–D) males ($n = 164$).

Precise measurements of tarpon weight, fork length, and dorsal girth were obtained from several sources: (1) Crabtree et al.'s (1995) study; (2) the IGFA world record database; and (3) regional tarpon kill tournaments (e.g., Veracruz Yacht Club, Coatzacoalcos) held during 2000–2005. These data were used to parameterize a log-linear version of a generalized multivariate linear statistical model,

$$\ln(W_i) = b_0 + b_1 \ln(L_i) + b_2 \ln(G_i) + \ln(\xi_i), \tag{16.8}$$

where $\ln(W_i)$ is the natural logarithm of weight (kg) of the ith observation; L_i is fork length (cm); G_i is dorsal girth (cm); and ξ_i is the error term. Higher-order terms were ignored because partial F-tests revealed that their inclusion into the model did not significantly reduce mean squared error. We took advantage of the high correlation between FL and girth (Figure 16.5A) to produce the fitted model,

$$\ln(W_i) = -10.6027 + 1.8105 \ln(L_i) + 1.1708 \ln(G_i), \tag{16.9}$$

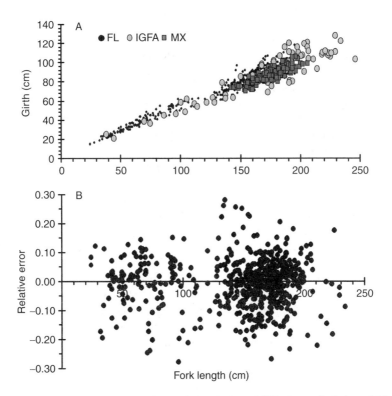

FIGURE 16.5 Plots of south Florida (small black dots), IGFA (open circles), and Mexico (squares) data sources showing: (A) dorsal girth dependent on observed fork length for tarpon, *Megalops atlanticus*, and (B) distribution of relative error of predicted weight dependent on fork length from Equation 16.9.

where $n = 612$ and $r^2 = 0.9954$. The model produces accurate weight estimates shown by the distribution of relative error of predicted weights (i.e., $W_i - \hat{W}/W_i$) given FL and girth, such that more than 92% of all predicted weights are within ±10% of the true weight over the tarpon size range of 25–235 cm FL (Figure 16.5B).

Mortality and growth estimation in tropical fishery populations are normally approached from a size-based perspective because of difficulties in ageing fish. Average size can be converted to mean age by assuming that age t maps directly into, or is a function of, size $L(t)$ and that mean length-at-age from the von Bertalanffy equation can be inverted as

$$t = \frac{-\ln\left[\dfrac{L_\infty - L(t)}{L_\infty}\right]}{K} - t_0. \tag{16.10}$$

Additionally, numbers-at-length can be converted to numbers-at-weight by means of simple allometric relationships.

Maximum Age, Maximum Size, and Lifetime Survivorship

The expected lifetime survivorship of tarpon is a function of all sources of natural mortality, including predation and disease. Beebe (1927) stated that shore birds, ospreys, and eagles were the principal predators of juvenile tarpon, and Harrington (1966) suggested that the rolling habits of the tarpon make them excellent targets for fish-eating birds. As tarpon grow and age, the predator field changes. Sharks, particularly great hammerhead (*Sphyrna mokarran*) and bull (*Carcharhinus leucas*), are known predators of adult tarpon, especially those that have been injured through hooking or catch-and-release by fishermen.

The instantaneous rate of natural mortality rate M for tarpon can be estimated from maximum age. Results from previous analyses of maximum age have not been particularly conclusive. In a detailed age and growth study of south Florida tarpon using otoliths, Crabtree et al. (1995) estimated ages for 164 males and 316 females using otoliths. The oldest two females aged were 55 years old, at lengths of 180 and a 204.5 cm FL, and the oldest male was estimated at 44 years old at 171.0 cm FL (Figure 16.4). In a related study, Crabtree et al. (1997a) found maximum age for 120 sampled Costa Rican tarpon to be at least 48 years, with the majority of ages ranging between 15 and 30 years. However, radiometric analysis by Burton et al. (1999) suggested that female tarpon longevity may exceed 82 years. Similar analyses by Andrews et al. (2001) found that Crabtree et al.'s (1995) 55 years could have exceeded 78 years, and a male estimated at 36 years could have been up to or exceeding 41 years. Andrews et al. (2001) concluded that longevity of female tarpon was at least 50.6 years, but it may exceed 78 years. They noted that a captive tarpon was held alive at the John G. Shedd Aquarium in Chicago for 63 years until it died after jumping out of the tank in 1998. This fish was likely at least 10 years old when first placed in the aquarium (based on its size), supporting the suggestion that maximum age exceeds 70 years.

South Florida samples available for the Crabtree et al. (1995) study were selective toward tarpon smaller than 150 cm (Figure 16.3A). Seasonally, very large tarpon are captured in the tarpon tournaments held at Veracruz and Coatzacoalcos, Mexico. More than 50% of the Mexico tournament-caught fish for the period 2000–2005 were larger than 175 cm FL, whereas only 24% of the Florida recreational catch exceeded this length (Figure 16.3B).

The Florida Fish and Wildlife Conservation Commission (FWC) Saltwater Record Program shows that the state record tarpon of 110.2 kg was landed on conventional gear at Key West, Florida, on February 17, 1975 (FWC Saltwater Record Program, International Game Fish Association, 2006). This catch record suggests that Florida may not be home to the biggest tarpon in recent times, even within the regional ecosystem. The world record Atlantic tarpon of 130 kg was captured off Rubane, Guinea-Bissau, Africa, on March 20, 2003. At least three fish within 2 kg of this size can be found in the IGFA record database, two from Africa and one from Venezuela (International Game Fish Association, 2006; Figure 16.3). Interestingly, McClane (1974) reported an individual over 243 cm FL and 158.8 kg taken from the Hillsborough River Inlet, Florida.

The reported maximum age of fish in the stock (t_λ) allows application of a convenient and consistent method to normalize the annual instantaneous natural mortality rate M to life span following Alagaraga (1984) and Ault et al. (1998). First, we assumed that $S(t_\lambda)$, the fraction of the initial cohort numbers surviving from recruitment t_r to t_λ, can be expressed as

$$\frac{N(t_\lambda)}{N(t_r)} = S(t_\lambda) = e^{-M(t_\lambda - t_r)}. \tag{16.11}$$

Then, assuming an unexploited equilibrium, setting the probability of survivorship of recruits to the maximum age to be 5% (i.e., $S(t_\lambda) = 0.05$), and letting t_r be equal to 0, rearrangement of Equation 16.11 provides an estimate (Table 16.3) of the natural mortality rate,

$$\hat{M} = \frac{-\ln[S(t_\lambda)]}{t_\lambda}. \tag{16.12}$$

Maturity and Fecundity

In Florida, male tarpon reached sexual maturity at approximately 117.5 cm FL, and females were found to be sexually mature by 128.5 cm FL (Table 16.4). Male and female tarpon reach sexual maturity at approximately 10 years of age, but females attain a larger size at maturity due to their more rapid growth. Chacon-Chaverri (1993) and Crabtree et al. (1997a) state that tarpon in Costa Rica reach sexual maturity at smaller sizes than those in Florida. In Costa Rica, males reached sexual maturity at approximately 88 cm FL, and females were found to be sexually mature by 112.6 cm FL. Although Crabtree et al. (1997a) noted that Florida fish were significantly larger than Costa Rican fish of similar age, their sampling in Florida was biased because fish were obtained

TABLE 16.4
Length (cm *FL*) at Sexual Maturity for Atlantic Tarpon (*Megalops atlanticus*) in the Central Atlantic Ocean

Location	Females	Fecundity	Males	Source
Brazil	125.0	Not available	95.0	de Menezes and Paiva, 1966
Florida	128.5	4.5–20.7 million oocytes per fish	117.5	Crabtree et al., 1997a
Costa Rica	112.6	Not available	88.0	Chacon-Chaverri, 1993; Crabtree et al., 1997a

primarily from tournaments and taxidermists and thus selectively harvested for their larger size. Finally, de Menezes and Paiva (1966) found that Brazilian tarpon attained sexual maturity at 95 and 125 cm *FL* for males and females, respectively.

Total fecundity (amount of yolked oocytes) of 32 Florida female tarpon was estimated gravimetrically by Crabtree et al. (1997a) and was positively correlated with body weight. They estimated mean fecundity to range between 4.5 and 20.7 million oocytes per fish, a range that encompassed Babcock's (1951, p. 43) estimate of 12,201,984 eggs for a 64.55 kg–203.2 cm *TL* female tarpon. Cyr (1991) reported minimum fecundity at 1,081,330 oocytes per female, but his maximum of 19,519,400 oocytes per fish fell within Crabtree et al.'s (1997a) reported range.

FISHERIES EXPLOITATION AND HUMAN IMPACTS

Atlantic tarpon are highly valued throughout their range. A big game fish in every sense, Atlantic tarpon are the subject of a host of tournaments, especially in Florida and Mexico (Robins, 1977). As a highly prized sportfish, their large adult body size, strenuous fighting characteristics, and striking silver flash contribute to their appeal (Grey, 1919; Oppel and Meisel, 1987). Tarpon typically attack baits or lures with ferocious intensity and make strong runs, and may make spectacular leaps as high as 10 ft out of the water (McClane, 1974).

Tarpon are fished across the eastern seaboard from Virginia to Florida, across the Gulf of Mexico, and south into Brazilian waters, but the most developed fisheries are found in Florida and Costa Rica (Chacon-Chaverri, 1993). Tarpon are among the most sought-after gamefish off the Caribbean coast of Costa Rica (Chacon-Chaverri, 1993), and Florida anglers spend hundreds of millions of dollars each year sportfishing for tarpon. In Florida, tarpon are targeted by recreational anglers as juveniles and adults (Robins, 1977). The Florida tarpon fishery is seasonal; most tarpon are caught during May to July, although some fish are caught in all months (Crabtree et al., 1995). The fishery is predominately catch-and-release, although large kill tournaments still persist. Tarpon are generally not eaten in the United States (McClane, 1974). Anglers in Florida who wish to harvest a tarpon must first purchase an annual permit. The cost of this permit has remained US$50 per fish since it was introduced in 1989, and there is an annual two-fish limit per person. Since this regulation was established, fewer than 100 tarpon have been harvested per year in Florida (Barbieri et al., Chapter 27, this volume).

Edwards (1998) noted that postrelease mortality is lower for captured tarpon not removed from the water, and that aggressive angling techniques intended to shorten capture (e.g., use of heavy tackle) may reduce release mortality. Recent studies by the FWC (K. Guindon, personal communication) estimated catch-and-release mortality at about 4.1%, with three principal phases of stress to the fish: (1) capture (hooking, angling duration, water temperature, shark attack); (2) handling (hook removal, air exposure, length of retention); and (3) release (revival, shark attack, recovery time).

As tarpon fisheries are predominantly catch-and-release, there is currently a shortage of data necessary for quantitative analysis of the stocks. Fishery catches and efforts are not well documented throughout its range. Unlike the United States, tarpon are highly esteemed for their food value by subsistence fisheries in many Latin America countries such as Mexico, Belize, Cuba, Puerto Rico, Nicaragua, Costa Rica, Colombia, and Trinidad (Hildebrand, 1934; Cataño and Garzón-Ferreira, 1994; Mol et al., 2000; Ramsundar, 2005; Montano et al., 2005). The roe of large females is highly prized in Mexico. Tarpon or "sábalo" soup remains a popular traditional dish along the Caribbean coast of Columbia, which has contributed to localized stock declines (García and Solano, 1995). These declines may also be attributable to habitat destruction, as Restrepo (1968) and Dahl (1971) report that tarpon were dynamited in Columbia for harvesting. Three metric tons of tarpon were produced by aquaculture in Columbia between 1985 and 1987, but this practice appears to have ceased.

ATLANTIC BONEFISH (*Albula vulpes*)

Despite a relatively large and growing body of research investigating the biology of bonefishes (*Albula* spp.), essential data on the life history and population dynamics of *A. vulpes* in the Atlantic Ocean and Caribbean Sea are in short supply. In this section, we review the existing information on bonefish and identify gaps for fishery management.

LIFE CYCLE AND RESOURCE ECOLOGY

Species Distribution and Unit Stock

The Atlantic bonefish, *A. vulpes*, was originally described by Linnaeus in 1758. The bonefish is considered one of the few examples of a cosmopolitan circumtropical distribution in shorefishes. Twenty-three nominal species have been described, all of which were synonymized under *A. vulpes* in 1940 (Colborn et al., 2001). Only two Atlantic species, *A. vulpes* and *A. nemoptera*, have been recognized (Rivas and Warlen, 1967; Robins and Ray, 1986). In this review, we focus on *A. vulpes*. In general, there appears to be little difference between bonefish species in terms of morphology or resource ecology (Colborn et al., 2001). The larvae, juveniles, and adults of all bonefish appear to be similar and are difficult to distinguish morphologically between species. Colborn et al. (2001), from genetic analyses, suggested there may be a deeper water species in the Florida Keys, but recent acoustic telemetry work (Larkin et al., Chapter 19, this volume) shows that supposedly shallow water *A. vulpes* travel to deeper waters, presumably to move offshore to spawn or avoid wintertime weather patterns (i.e., cold fronts) and associated water temperature fluctuations. Bonefish were not previously believed to be ocean migrants as juveniles or adults, but recent results of conventional anchor tagging studies in Florida have documented movement of mature bonefish between the Florida Keys and the Bahamas (Larkin et al., Chapter 19, this volume).

Life Cycle

Studies by Bruger (1974), Mojica et al. (1995), and Crabtree et al. (1997b) provided information on reproductive seasonality of bonefish spawning in south Florida. Bruger (1974) suggested that spawning occurred year-round off the Florida Keys based on finding ripe females in all months. A more extensive study by Crabtree et al. (1997b) concluded that gonadal activity of mature bonefish showed seasonal periodicity. They found vitellogenic oocytes most prevalent during November–May. The seasonal spawning pattern of bonefish in the Bahamas appears similar to the Florida Keys. Mojica et al. (1995) observed ripe adult bonefish from October to May, while Colton and Alevizon (1983b) anecdotally suggested that spawning occurred in October and November. Erdman (1960a) found ripe females only during the period November through January in Puerto Rico. Collectively these data indicate seasonal reproductive activity of *A. vulpes* throughout its Caribbean range, with most activity occurring in fall-winter and early-spring months.

Exceptionally large schools (densely formed and covering an acre or more) are sometimes seen in the Bahamas and Florida from mid-January to April, "milling" in protected shallow bays behind reefs (McClane, 1974; J. Kalman, personal communication). These concentrations of mature-sized animals may be prespawning or spawning bonefish. However, it is widely believed that bonefish do not spawn in the shallow nearshore areas where the fishery exists (Crabtree et al., 1997b). Instead, spawning is presumed to occur in deep water off the coral reef shelf edge, away from these principal foraging grounds (Colton and Alevizon, 1983b; Mojica et al., 1995; Crabtree et al., 1997b). Alexander (1961) suggested that bonefish either spawn offshore or in areas where currents are likely to carry the eggs offshore. First development phase leptocephali are nearly isotonic with sea water, and therefore may require the relatively stable salinity of offshore waters to reduce complications from osmotic variation (Hulet and Robins, 1989).

Mojica et al. (1995) collected metamorphic leptocephali using nets placed 1 m deep in tidal channels near Exuma Sound, Bahamas, in spring and early summer. They reported ages ranging from 41 to 71 days (mean = 56 days) in their sample based on counts of daily increments in otoliths. They concluded that these larvae had been spawned the previous October through January, and also noted one recruitment pulse during June, which prompted them to add that "significant spawning activity" may continue through May. Greatest catch rates coincided with evening hours and flood tides during the new moon. Drass (1992) found bonefish leptocephali to 50 m in Exuma Sound, suggesting that vertical migration likely influences timing of onshore migrations. Erdman (1960a) in Puerto Rico collected metamorphic leptocephali in beach seines during all months except July; abundance peaked during March to May, and suggested a larval duration of 5–6 months. Pfeiler (1984a) suggested that duration of the premetamorphic larval stage of *Albula* sp. in the Gulf of California may be as great as 6–7 months, based on presumed timing of spawning and appearance of metamorphic larvae. However, Mojica et al. (1995) discussed the possibility that bonefish could delay metamorphosis until environmental conditions favored onshore settlement, and Schmidt (1922) stated that leptocephali are capable of living in the oceanic plankton for months. Pfeiler (1984b; see also Chapter 13, this volume) and Bishop and Torres (1999) have discussed the ecological and evolutionary implications of the potentially long larval duration within the subdivision Elopomorpha. Bonefish leptocephali appear infrequently in northern Gulf of Mexico estuaries (Thompson and Deegan, 1982) and Atlantic coast embayments off the central and northeastern United States (Alperin and Schaefer, 1964). These occurrences have been attributed to transport by entrainment in eddies shearing off the Straits of Florida, Gulf Stream, and Loop Current near spawning areas. Transport times from spawning locations in the northern Caribbean Sea, Florida, and the Bahamas, even for a passively drifting larva, could be relatively short, that is, weeks versus months (Cowen et al., 2006; Luckhurst et al., in review).

The location of juvenile bonefish habitats remains an enigma. Erdman (1960a) reported that young postmetamorphic juveniles were found along sandy shorelines with moderate surf off Puerto Rico, and juvenile bonefish (0.11–0.34 kg) inhabited mud bottom, mangrove habitats near shore. Crabtree et al. (1996) conducted limited seine collections in sand and seagrass benthic habitats on the Atlantic coast of Florida between Key West and the Indian River Lagoon that produced 56 YOY bonefish that ranged from 21 to 116 mm *FL*. They did not comment on the spatial distribution of abundance. A comprehensive rollerframe trawl survey to assess fishes and macroinvertebrates of Biscayne Bay revealed that a large number of fish species utilize various benthic habitats as YOY nursery grounds (Ault et al., 1999). However, despite exhaustive sampling over four seasons and 2 years that included areas that adult bonefish frequent, no YOY bonefish were ever observed.

Over the past several years, seining efforts in the Florida Keys have captured some YOY bonefish. Mote Marine (A. Adams, personal communication) sampled along the same beaches where Crabtree et al. (1996) located YOY bonefish. Ault et al. (2005a) reported discovery of YOY bonefish recruiting to the shorelines of Key Biscayne, Florida, from January to June. Some bonefish leptocephalus-stage larvae were collected in January through late April. Despite over 300 YOY bonefish collected by the two sampling programs, captures do not reflect the juvenile density expected to support the south Florida fishery, suggesting that there must be additional recruitment sites outside the sampled areas in the Florida Keys.

Since 1998, a conventional tagging program in south Florida and the Florida Keys (www.bonefishresearch.com; Humston, 2001; Ault et al., 2005a) has caught and released more than 200 tagged bonefish in the 200–350 mm size classes corresponding to ages 0–3 years old along the coastline near Jupiter Inlet, about 150 km north of Miami. The presence of immature fish to the north suggests that primary areas of stock recruitment and nursery grounds could be north of the principal area of the Florida Keys fishery. Growing evidence suggests that bonefish may be migrating south into the fishery after they reach maturity at about 3.5–4 years or 500 mm FL to recruit to the exploited phase of the stock.

Mature adult bonefish in the tropical Atlantic primarily occur in shallow waters, and their habitat for angling purposes are the flats or intertidal areas adjacent to sand and coral islands or mainland beaches. They commonly forage in shallow (<2 m depth) coastal and inshore waters with benthic composition dominated to varying degrees by sand or seagrasses, termed "flats" (Erdman, 1960a; Colton and Alevizon, 1983a; Crabtree et al., 1998b; Colborn et al., 2001); they are reported with less frequency around shallow reefs (Erdman, 1960a; Colton and Alevizon, 1983b; Crabtree et al., 1996). Many of these areas coincide with mangrove-fringed shorelines, and bonefish may utilize the intertidal nexus of mangrove roots for protective cover (Erdman, 1960a; Cooke and Philipp, 2004). Bonefish generally enter the flats on a flood tide and drop back to deeper water on the ebb (Colton and Alevizon, 1983b; Humston et al., 2005). Deep passes or channels between shallow flats are important as conduits of travel and refuges from rapid changes in water temperature in the shallows (Humston et al., 2005).

Colton and Alevizon (1983b) indicated that bonefish in Bahamian waters utilize shallow flats with a variety of habitat characteristics. They tracked tagged bonefish with acoustic telemetry, and primarily observed bonefish in water less than 2 m with temperatures ranging from 24–32°C. Their data showed a strong seasonal trend in abundance of large bonefish (>550 mm FL) that was inverse to water temperatures. They suggested that the periodic exodus of adult bonefish from the flats may be related to spawning activity. They also stated that bonefish were typically observed by anglers and divers in deeper waters near coral reefs during the warm summer months. Erdman (1960a) noted that local commercial fishermen reported that the greatest numbers of adult bonefish were present on the flats in November, coinciding with the onset of cooler weather and the reproductive season. A similar pattern of availability has been suggested by professional guides in south Florida (Ault et al., 2002; Larkin et al., in review).

Recent studies in Florida described distinct seasonal shifts in spatial allocation of charter guide fishing effort targeting bonefish (Humston, 2001; Ault et al., 2002; Larkin et al., in review) that corresponds with bonefish seasonal spatial abundance. They found that bonefish in Biscayne Bay and the Florida Keys were more abundant on the coastal Atlantic "oceanside" flats during cool weather months (November–April), and more abundant in the interior coastal bay waters during warm months (May–October). The shift in spatial distribution of these mature bonefish may likely be related to spawning behavior or forage densities (Ault et al., 2005a). A telemetry study conducted in Biscayne Bay using remote data-logging hydrophones showed patterns in habitat use that may reflect this shift from interior to oceanside flats (Humston et al., 2005). Results from this study also indicated that bonefish spent the majority of their time in shallow and median-depth habitats (<2 m), but appeared to utilize adjacent deep channels (3–5 m) with greater frequency as air temperature increased with the onset of summer months. Data also indicated a potential relationship between ontogeny and site fidelity, roughly in agreement with size-dependent patterns in seasonal habitat selection observed by Colton and Alevizon (1983b).

Diet

The inferior mouth of the bonefish is well suited to bottom feeding on benthic and epibenthic prey (Erdman, 1960a; Colton and Alevizon, 1983a; Crabtree et al., 1998b). The adult bonefish usually digs for food in the bottom with its snout, and sometimes somersaults in the process (McClane, 1974). In the West Indies and Florida, the bonefish may be seen by day along shallow sandbanks and among underwater seagrasses. Bonefish prey preferences and selectivity have been quantified by analysis of stomach contents at several Caribbean locations (Table 16.5). Most of these studies have shown that crustaceans are common prey throughout its range, but regional differences exist between the relative importance of bony fishes and mollusks. In general, bonefish feed on mollusks, shrimps, crabs, marine worms, squids, small fish, and sea urchins.

Regional Movements and Migrations

Movements and migrations of bonefish (*A. vulpes*) have been variously studied in the Caribbean using both conventional mark-recapture and more advanced acoustic telemetry methods. While several mark-recapture studies have been undertaken (e.g., Colton and Alevizon, 1983b), the ongoing bonefish anchor tagging program in south Florida and the Florida Keys led by the University of Miami (www.bonefishresearch.com) and supported by Bonefish & Tarpon Unlimited (www.tarbone.org) spans the period 1998–2006 and represents the only successful study to date (Humston, 2001; Ault et al., 2005a; Larkin et al., Chapter 19, this volume). This study provides relatively high-resolution data that illustrate a range of mostly local movements (≤ 10 km) and evidence of a number of long-range movements. Net movements in excess of 200 km have been recorded, including one fish tagged in south Florida that was recaptured in the Bahamas. The data are replete with substantial variations in time elapsed (days to years) between releases and recaptures, making it difficult to interpret general patterns. However, these data strongly suggest bonefish usually inhabit relatively restricted home ranges (<5 km radius). Long-range movements are predominantly undertaken by mature individuals. The timing of long-range movements appears to have a seasonal component, with many of the longest occurring in conjunction with the passage of cold fronts in late fall and winter. These patterns indicate that long-distance movements by bonefish may serve reproductive purposes.

Recent telemetry results on Florida bonefish suggest a similar pattern, showing that some bonefish exhibit significant site fidelity extending over 60 days or more (Humston et al., 2005). Of 11 bonefish tagged, 7 remained in the area of a single flat (~1 km^2) or returned to the flat daily for significant periods of time after release. Most of these fish left the area after returning for 3–4 days and were not observed again. This would tend to concur with findings reported by Colton and Alevizon (1983b). In contrast, two tagged fish remained within the study area for periods of 40 and 61 days after release. The majority of data support the hypothesis that bonefish generally display short-term site fidelity, shifting the location of their foraging efforts every few days. This nomadic approach to habitat utilization could result in significant long-range displacement over time. The dynamics of individual or group searching and foraging behaviors likely determines the magnitude of displacement

TABLE 16.5
Summary of Diet Studies of Adult and Subadult Bonefish (*Albula vulpes*)

Location	Molluscs		Crustaceans					Worms	Fishes	Source
	Bivalves	Snails	Crabs		Shrimps					
			Portunid	Xanthid	Penaeid	Alpheid				
Bahamas (Grand Bahama)	Rank 1 66% *Tellina* spp.	Rank 3 25.4% unidentified gastropods	Rank 2 40.5% *Callinectes ornatus*	24.8% *Pitho aculeate, Leptodius floridanus*	24.1% *Farfantepenaeus duorarum*					Colton and Alevizon, 1983a
Florida (Lower Florida Keys)	Rank 2 28%		17.1%	Rank 3 18.6% *Panopeus* spp.	17.8% *Farfantepenaeus duorarum*	Rank 1 34.9%			15% no *Opsanus beta*	Bruger, 1974
Florida, primarily Upper Florida Keys (>439 mm *FL*)	Bivalves 24.9%	Gastropods 31.2%	*Callinectes* spp. Majority of nonfishes	Rank 1 57.0%		Rank 2 51.7% primarily *Alpheus normanni*			Rank 3 49.5% primarily *Opsanus beta*	Crabtree et al., 1998b

Location	Rank 1	Rank 2	Rank 3	Other	Reference
Florida, primarily Upper Florida Keys (<440 mm FL)	Rank 1 36.1% primarily *Farfantepenaeus* spp.	Rank 2 23.0%	Rank 3 21.3%	No *Callinectes*; No *Opsanus beta*	Crabtree et al., 1998b
Los Roques, Venezuela	Rank 1-tie 90.4%		Rank 3 44.1% primarily *Codakia orbiculatus*	Rank 1-tie 90.4% primarily *Anchoa* spp.; Polychaetes 22.8%	Weinberger and Posada, 2005
Puerto Rico	Rank 1 40% *Codakia costata*	Rank 2 30% *Cronius tumidulus*	Rank 3 10% *Farfantepenaeus* spp.		Erdman, 1960a; Warmke and Erdman, 1963

Note: The top three food items are ranked 1, 2, and 3. Ranking was based on percent frequency of occurrence, which was included for each food item (%).

over time. If long-range movements function in reproduction, variation in movement behavior with ontogeny is likely to exist (Ault et al., 2002, 2005a; Humston et al., 2005; Larkin et al., Chapter 19, this volume). Humston et al. (2005) indicated that large bonefish appear to display less site fidelity than smaller individuals. This may reflect an increase in home range with increasing body size. It may also suggest that the onset of sexual maturity evokes significant behavioral changes in bonefish, reflecting differences in movement behavior between small and large bonefish. Fine-scale foraging movements are closely tied to tidal cycles (Colton and Alevizon, 1983b; Humston et al., 2005). Cooke and Philipp (2004; see also Chapter 25, this volume) used telemetry and visual float tags to monitor movements of bonefish following capture on hook and line. Most fish monitored in the study moved less than 300 m net distance following release, and those that moved further often returned to the release location soon after. Bonefish were observed "resting" stationary for extended periods, often near mangroves, in the first 30 min after release.

POPULATION DYNAMICS

A summary of population dynamic parameters on bonefish lifespan, natural mortality, allometric growth, reproductive maturity, age–size at recruitment, and maximum size is given in Table 16.2.

Age and Growth

Length–length (i.e., total length TL, standard length SL, and fork length FL) and allometric weight–length relationships for bonefish have been reported by Bruger (1974) and Crabtree et al. (1996). These allow efficient conversion of body length as measured by any of these metrics to be converted in terms of the others. In this chapter, we reanalyzed the data of Crabtree et al. (1996) for Florida plus new data from the Florida Keys bonefish fishery. Lengths are reported in terms of FL. An updated estimate of FL dependent on TL is

$$FL = -0.61714 + 0.858092TL, \tag{16.13}$$

with $n = 869$ and $r^2 = 0.9994$ (Figure 16.6A). Similarly, the relationship between FL dependent on SL was

$$FL = 1.996345 + 1.032997SL, \tag{16.14}$$

with $n = 873$ and $r^2 = 0.9993$ (Figure 16.6B). The nonlinear allometric relationship between weights dependent on FL was

$$W(t) = 1.631709E - 08L(t)^{2.992311}, \tag{16.15}$$

with $n = 870$ (Figure 16.7).

Bruger (1974) used scales to study the age and growth of Florida bonefish up to 680 mm FL, and he determined the oldest bonefish to be 12 years old.

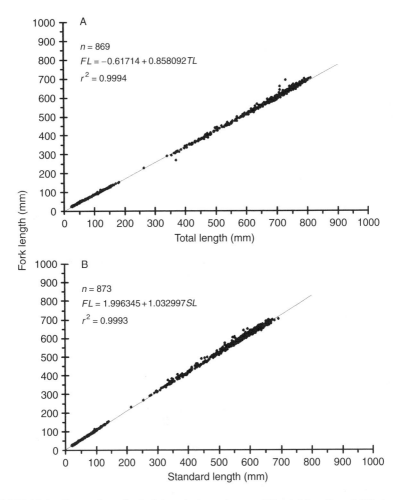

FIGURE 16.6 Regressions for fork length dependent on (A) total length and (B) standard length, for bonefish, *Albula vulpes*, from south Florida waters.

A later analysis by Crabtree et al. (1996), using sectioned otoliths and a substantially larger sample, reported bonefish maximum age to be 19 years. However, only 25 fish in their sample were greater than 12 years of age, and these maximum sizes from Crabtree et al.'s (1996) Florida study were exceeded many times by IGFA world record data (Figure 16.8). In fact, the largest bonefish examined by Crabtree et al. (1996) was not the oldest. Larkin et al. (Chapter 19, this volume) noted that greater than 3% of the more than 4500 fish captured in a long-term bonefish mark-recapture study exceeded the sizes of those used in the Crabtree et al. (1996) study (Figure 16.9). To help resolve this ambiguity, an ongoing study (Ault et al., 2005a) estimated the ages of more than 100 additional bonefish using similar techniques as Crabtree et al. (1996) (Figure 16.10). That work has extended the maximum age to at least 20 years. No differences in sex-specific curves were noted, and the maximum

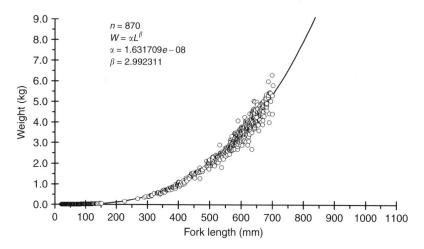

FIGURE 16.7 Plot of observed (open circles) weight dependent on fork length and predicted nonlinear regression (solid line) for Florida bonefish, *Albula vulpes*.

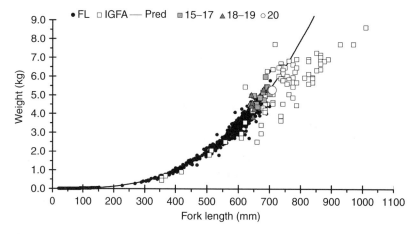

FIGURE 16.8 Plots of observed allometric weight dependent on fork length for bonefish, *Albula vulpes*, in south Florida (dark circles), IGFA world record bonefish (open squares), and predicted nonlinear regression (solid line). Specific data for bonefishes aged 15–20 are overlain for comparison and correspond to predicted ages of weight–length pairs from Crabtree et al.'s (1996) study.

sizes of males and females were similar; thus, the sexes-combined von Bertalanffy growth function was revised to

$$L(t) = 663.0026(1 - e^{-0.3257709(t-(-0.2946266))}). \tag{16.16}$$

Erdman (1960a), with a sample of 446 adult bonefish from Puerto Rico's reefs and shallows, reported an average weight of 0.57 kg and average length of 342.34 mm,

Population Dynamics and Resource Ecology 245

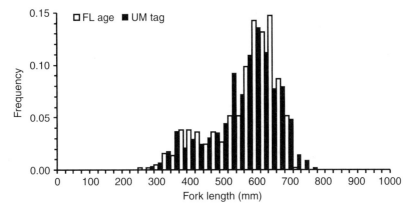

FIGURE 16.9 Comparison of bonefish *Albula vulpes* length–frequency distributions for Crabtree et al. (1996) (open histograms; $n = 448$) and University of Miami tagging program 1998–2006 (dark histograms; $n = 4221$).

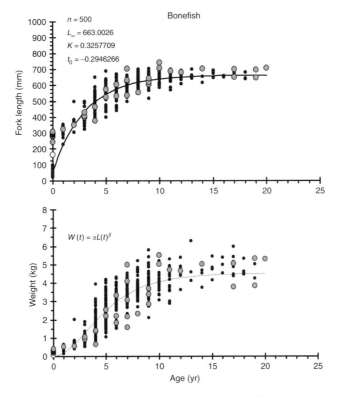

FIGURE 16.10 Observed fork lengths and weights at ages (dark circles) and predicted lengths or weight on age functions (solid lines) from the von Bertalanffy growth model for bonefish, *Albula vulpes*, from Florida waters. Large open circles are new length–weight dependent on age data from this study.

with fish ranging from 0.11 kg and 177.8 mm *FL* to 4.65 kg and 711.2 mm *FL*. While the average size of Erdman's (1960a) sample is smaller than that reported by Crabtree et al. (1996), the overall size range was similar.

Catch-and-release is an important component of bonefish angling and tournaments. To develop an empirical function that would allow precise estimation of the weight of a released fish, we took the measurements of length and girth from Crabtree et al. (1996), and combined them with new data from bonefish tournaments sponsored by the Florida Keys Fishing Guides Association. The final fitted model was

$$\ln(W_i) = -18.45004331 + 2.73712795 \ln(L_{1i}) + 0.3523196 \ln(G_{2i}), \quad (16.17)$$

where $\ln(W_i)$ is the natural logarithm of weight (kg); $\ln(L_{1i})$ is the natural logarithm of fork length (mm); and $\ln(G_{2i})$ is the natural logarithm of dorsal girth (mm), with $n = 98$ and $r^2 = 0.9993$. The function gives an accurate prediction of bonefish weight given estimates of *FL* and girth, such that more than 94% of all predicted weights are within ±10% of the true weight over the size range of bonefish.

Maximum Age, Maximum Size, and Lifetime Survivorship

Predators of bonefish must be fast enough to capture and large enough to consume this wary species. Sharks, groupers, barracuda, and dolphin and porpoise species are known predators of bonefish. Cooke and Philipp (2004 and Chapter 25, this volume) described sharks in the Bahamas as significant predators of bonefish following angling release. In various regions (Bahamas, Belize, wider Caribbean Sea), bonefish are often employed as bait for catching large pelagic or reef fishes (marlin, grouper, etc.). This may indicate that bonefish represent familiar "prey of opportunity" to these deepwater species, perhaps intercepted on occasion as bonefish migrate to offshore spawning grounds.

As noted above, there is disagreement between studies on the potential maximum age of bonefish (Bruger, 1974; Crabtree et al., 1996; Ault et al., 2005a). Discrepancies between the two most recent studies result from differences in size range represented in samples (Ault et al., 2005a), particularly with respect to determining age of very large bonefish. Based on the analysis of data from a bonefish-tagging program in Florida, Ault et al. (2002, 2005a) determined that the maximum age and size estimates reported by Crabtree et al. (1996) might be conservative. Ageing large bonefish revealed that bonefish live to at least to 20 years, which is older than the maximum age reported by Crabtree et al. (1996), corresponding to a natural mortality rate M equal to 0.1498 (Table 16.2). The maximum size bonefish seen in the south Florida conventional tagging study was 810.2 mm *FL*, while the IGFA world record database lists a bonefish of 1009.7 mm *FL* and a weight of 8.62 kg. In a subsequent section of this review, we discuss the implications of extending the maximum age with respect to estimating rates of natural and fishing mortality for the Florida stock.

Maturity and Fecundity

According to Crabtree et al. (1997b), male bonefish reach 50% length of sexual maturity L_m at 418 mm *FL* and an age t_m of 3.6 years, while females reach L_m at

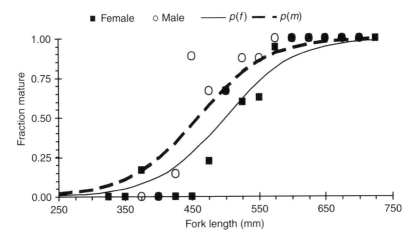

FIGURE 16.11 Observed and predicted fraction mature males (open circles) and females (dark circles) dependent on fork length for Florida bonefish *Albula vulpes*. The $p(f)$ (solid line) and $p(m)$ (dashed line) are the estimated population proportion females and males, respectively.

488 mm *FL* at t_m of 4.2 years (Figure 16.11). All males longer than 477 mm *FL* and all females longer than 594 mm *FL* were mature (Crabtree et al., 1997b). The current legal minimum fish length L_c imposed upon the Florida bonefish fishery is 392 mm *FL*, thus $L_c < L_m$ for both sexes. This situation may be less than favorable for conservation of the resource since bonefish should have at least one chance to spawn during their lifetime. Conservation standards require that each male–female spawning pair must produce at least two offspring that survive to reproductive size–age for the population to be sustainable. Little is known about bonefish maturation from other areas, although Pfeiler et al. (1988) reported 12 *Albula* spp. ranging from 205 to 264 mm *SL* from the Gulf of California that had ripe or ripening gonads.

Crabtree et al. (1997b) estimated bonefish fecundity Γ and found it ranged from 0.4 to 1.7 million oocytes per female and had a significant relation to fish weight as

$$\log_{10} \Gamma = 1.936 + 1.31(\log_{10} W), \tag{16.18}$$

where Γ is the fecundity in oocytes per female and *W* is the mature bonefish weight in grams.

FISHERIES EXPLOITATION AND HUMAN IMPACTS

Bonefish are the basis of economically important recreational fisheries in Florida and throughout the Caribbean. The majority of bonefish angling occurs in the Florida Keys and the Bahamas. The south end of Exuma, the east end of Grand Bahamas, and the middle bight of Andros are well known for numerical abundance of bonefish schools. For convenience and suitable fly-fishing conditions, Biscayne Bay, Key Largo, and Islamorada areas of Florida provide excellent fishing and access to world-record sized bonefish (International Game Fish Association, 2006). Anglers generally target visible fish that are stalked by wading or with a skiff. In Caribbean waters, fishery dynamics (effort patterns, unit stock, catch rates, etc.) have been habitually difficult to assess. The fleet operates over wide areas, and since most of this fishing is catch-and-release, there are no landings data to assess.

In the Florida Keys, fishing for bonefish is a year-round activity and provides an important source of income for professional fishing guides (Larkin et al., in review). The commercial sale of bonefish is prohibited in Florida; the recreational fishery is limited to a bag-limit of one fish per angler per day and a minimum size L_c of 390 mm *FL* (457 mm *TL*). The impacts of the fishery and catch-and-release fishing are not well understood. Based on the data collected by professional fishing guides that was used in subsequent age and growth, reproduction, and food habits studies, Crabtree et al. (1996) suggested that the recreational fishery using hook-and-line gears in the Florida Keys exerts little fishing mortality. They also stated that there was apparently little mortality of bonefish as bycatch in Florida commercial net fisheries. Crabtree et al. (1996) based their conclusions on a catch curve analysis that showed total instantaneous mortality rate Z for bonefish was 0.25 for males and 0.21 for females. They assumed that natural mortality M ranged from 0.2 to 0.3 using the Pauly (1980) method, and concluded that fishing mortality $F (= Z - M)$ was likely very low.

An interesting comparison of the Crabtree et al. (1996) and the University of Miami (UM) tagging data is seen in Figure 16.9. It is important to note that the apparent selectivity patterns of the fleet are essentially identical for the independent investigations and that the selectivity pattern has not changed appreciably over the past two decades. In the Crabtree et al. (1996) study, they indicated that bonefish were not fully recruited to the exploitable phase of the recreational fishery in Florida until they reached a length L_c of 543 mm *FL* (i.e., t_c is 5.0 years). We used the size-based mortality estimation formula of Ault et al. (1998, 2005b) to examine these data, and determined Z for fully recruited bonefish with average size in catch $\bar{L}$ of 612.9 mm *FL*. The Ehrhardt–Ault (1992) method is substantially more robust than catch curve analyses (see Quinn and Deriso, 1999, pp. 364–367). Employing the population dynamics parameters for bonefish given in Table 16.2 resulted in a Z estimate of 0.227, which falls within the range of Z estimated by Crabtree et al. (1996). Given that bonefish live to at least 20 years would suggest that Z exceeds M in the Florida Keys and that F is probably nonzero (see Larkin et al., Chapter 19, this volume, for an extended analysis). In addition, because ~3% of the more than 4000 bonefish tagged in the 1998–2006 UM conventional anchor tag mark-recapture study (max *FL* = 810 mm) exceeded those sizes sampled and aged by Crabtree et al. (1996) (max *FL* = 702 mm), thus it is likely that $M \in [0.12, 0.15]$. Crabtree et al. (1998a) and Cooke and Philipp (2004) provided the only studies on bonefish release mortality from a directed recreational fishery. Given the high potential for a lower estimate of natural mortality, and regarding potential release mortality in the recreational fishery (Cooke and Philipp, Chapter 25, this volume), it is likely that fishing mortality is greater in the Florida bonefish fishery than was reported by Crabtree et al. (1996).

New data are required to resolve these issues. The predominance of catch-and-release recreational fisheries for bonefish stocks has resulted in few quantitative data for analysis of stock response to exploitation. Recent efforts have attempted to tap into the conventional wisdom of the bonefish angling community to aid assessment of the fishery and to integrate user-group expertise into assessments, with success providing new insights into bonefish fishery dynamics in Florida (Ault et al., 2005a; Humston et al., Chapter 30, this volume), and applications of similar methods to assess Bahamian stocks are in the initial stages of development (Danylchuk et al., Chapter 5, this volume). Surveys of Florida Keys charter captains provided initial data on spatiotemporal patterns in effort and release mortality. Volunteer tagging programs have also generated data on size frequency in the recreational catch. In a 2002 survey of the bonefish charter fleet (Larkin et al., in review), captains reported they spent an average of 100 days a year targeting bonefish. Ault et al. (2002) reported a bimodal distribution of guide-based fishing effort targeting bonefish, which peaked in spring and fall. During cool winter months, most fishing effort was expended on the oceanside flats (east of the barrier islands), while in warmer months anglers concentrated their efforts on the interior areas of Florida Bay and Biscayne Bay (Humston, 2001).

New approaches to stock assessment will be required to sustain these important fisheries. Recently, Ault et al. (Chapter 26, this volume) developed and implemented a novel visual census in the Florida Keys to estimate bonefish population size that was conducted during one day each year over three consecutive years. Professional bonefish guides (charter fleet captains) and experienced anglers were recruited to the field effort at specific survey locations between Miami and the Marquesas, and enumerated the numbers of bonefish seen in the sampling area. Statistical sampling techniques were used to estimate adult stock size at approximately 300,000 individuals with relatively good precision. The census has been repeated annually since 2003 and provides a mechanism for quantitative monitoring of status and trends of the bonefish population.

FILLING CRITICAL KNOWLEDGE GAPS FOR FISHERY MANAGEMENT

Beyond Florida waters, there is a paucity of information on bonefish and tarpon resource ecology, population dynamics, and fishery impacts. Islands of the Bahamas and West Indies tout local bonefish angling opportunities as a means of attracting traveling anglers and other tourists. The species are highly regarded sportfishes, and therefore attract attention as potential angling quarry in popular destinations. Where anglers travel to target bonefish, charter fleets of various sizes exist to enhance angling opportunities with specialized vessels and local knowledge. At present, few data are available to quantify the size of these charter fleets, their economic significance to regional communities, or their impact on local stocks. There remain a number of critical knowledge gaps limiting monitoring, assessment, and management of Atlantic tarpon and bonefish stocks. Some of the most critical issues remaining in resource ecology, population dynamics, and fishery management are discussed in the following sections.

Resource Ecology

- *Migrations and dispersal.* It is essential to assess the degree to which population migration or dispersal of tarpon and bonefish might create connections between regional fisheries, particularly as they span international boundaries. This would occur at two key life stages: during the planktonic larval phase when animals are transported by ocean currents via passive or behaviorally mediated dispersal (e.g., Cowen et al., 2006); or, as active adult migrations or dispersal via "ranging" behavior for either feeding or reproductive purposes. Directed tagging studies are the primary means of documenting these behaviors. Genetic analyses may also provide some first-order inferences on connectivity between populations, but these are on much longer time scales and these analyses cannot generally discern contributions between the two life stages. Directed studies on both larvae and adults are therefore essential to precisely quantify exchange. Understanding larval transport and exchange will provide a clearer picture of recruitment teleconnections and interregional contributions. Quantifying adult movement between fisheries will allow management decisions to account for variable fishing mortality rates between regions and viability of conservation targets. Our review suggests that exchange in both stages is likely for tarpon, particularly among the Western Atlantic and Caribbean fisheries but not dismissing potential for dispersal across the southern Atlantic. The smaller bonefish are less likely to undertake large-scale migrations, but emerging data indicate that adult exchange between fisheries is possible on a limited scale. Large-scale studies employing acoustic telemetry arrays and satellite PAT technologies are clearly warranted. Additionally, otolith chemistry investigations may be useful for documenting larval dispersal or ontogenetic movements for both species (e.g., Sandin et al., 2005; Brown and Severin, Chapter 17, this volume).
- *Location of spawning areas.* This particular aspect of population biology represents a significant gap in our understanding of larval production and subsequent population recruitment of tarpon and bonefish. Clearly, identifying locations and essential conditions for spawning has significant implications for studying larval dispersal and recruitment exchange, and allows managers to protect critical spawning and/or recruitment areas. It is further possible that anthropogenic or climatic changes could affect fish movement and orientation behavior along migratory pathways to spawning sites. Technological advances in telemetry tracking and satellite archival tagging technologies will greatly facilitate future study in this area for both species (e.g., Luo et al., Chapter 18, this volume; Larkin et al., Chapter 19, this volume). Increased knowledge can help mitigate potential impacts of habitat destruction, gas and oil development, and other anthropogenic factors.
- *Essential habitats for feeding and nursery functions.* Juvenile biology of bonefish is presently an enigma, and it should be considered a top priority to determine essential nursery habitat for this species. Loss of nursery habitat is a significant concern for conservation of coastal species (e.g., Adams et al., 2006) and can undermine even the most restrictive management

strategies for adult stocks. Quantifying the extent of available nursery habitat supporting tarpon and bonefish stocks, and any spatial variability in their contribution to adult recruitment, would also support conservation efforts for this species. A host of population biology questions remain. For example, why do bonefish recruit to the exploited phase (fishery) so late in life (i.e., at the size/age of sexual maturity) in the Florida Keys? Where are they before they arrive on the fishing grounds, that is, what are the essential habitat(s) for immature bonefish? Otolith chemistry analyses may also hold great potential in these investigations (Dorval et al., 2005; Brown and Severin, Chapter 17, this volume).

- *Dynamics of critical prey species.* Studies of feeding habits have identified the primary prey items for tarpon and bonefish. Continuing research should further assess the status of these prey species populations, the impacts of directed harvest on these species, and define trophodynamic links between predator and prey stocks. Bottom-up forcing in these food webs may determine productivity and community dynamics (e.g., Ware and Thompson, 2005), therefore anthropogenic nutrient inputs, pollution, and habitat destruction in coastal waters could potentially influence stock dynamics of prey and predators. An integrated systems approach to these investigations would be an important shift toward an increasingly ecosystem-based perspective in management of these fisheries (Pikitch et al., 2004; Ault et al., 2005c).

POPULATION DYNAMICS

Key life history and demographic data are needed to assess the risks of increasing and intensifying exploitation effects on tarpon and bonefish resources. Critical information that underlies stock sensitivity to exploitation includes spawning stock biomass, stock and recruitment relationships, and the fecund potential of mature fish among other key aspects of population dynamics.

- *Maximum age–size and life span.* Maximum age defines the lifetime expectation of animal survivorship, but also determines the stock's sensitivity to exploitation. In situations where the stocks have been fished for some time, the probability of seeing these potential maximum sizes–ages is greatly reduced. For robust assessment, these statistics must be known with certainty.
- *Maturity and reproduction.* Maturity and stock reproductive outputs are keystone population-dynamic variables of sustainability. More intensive regional studies of these critical life history parameters are warranted.
- *Lifetime growth.* Outstanding age–growth problems remain for both the youngest and oldest fish of both species. Apparent differences in growth rates around the Caribbean and western Atlantic (e.g., Adams et al., Chapter 15, this volume) suggests the strong need to use standardized methodologies (see Crabtree et al., 1996) as a model to design studies at other locations. In fact, while better studied than most stocks, Florida bonefish still require

additional and intensive focus on very small (<300 mm *FL*) and large (>700 mm *FL*) fish.

FISHERY MANAGEMENT

New and novel types of data and management perspectives will be required to assess whether tarpon and bonefish fisheries are sustainable. What is required is a systems approach to fishery management (Rothschild et al., 1996; Ault et al., 2005c) where strategy is emphasized over tactics in the fishery management institution. The approach would ensure relatively seamless integration of fishery-independent data acquisition with typical data derived from the fishery (i.e., catch, effort, etc.).

- *Unit stock and stock sizes virtually unknown.* Clear definition of unit stock(s) and seasonal spatial distributions and abundance of the resources are needed for development of management strategies.
- *Fishery catches and effort.* Fishery impacts in the region are largely unknown. However, there has been substantial growth in technological capacity and potential fishing power of the fleets over the past several decades. In addition, exponential increases in participants in the fisheries have been observed throughout the region. In many areas there are directed subsistence and commercial fishing operations, which remain largely undocumented. Also required are stock assessments to identify the extent of fishery impacts and to develop management alternatives that build sustainable fisheries.
- *Catch-and-release mortality.* The impacts of catch-and-release fishing, while widely assumed to be negligible, have remained largely unquantified for tarpon and bonefish. However, this must soon be addressed and quantified for fishery management because there is a large and growing number of seasonal catch-and-release tournaments in Florida, the southeast United States, Mexico, Trinidad, etc.
- *Coastal zone development.* Unrestrained growth of human populations in the coastal zones has also accelerated habitat destruction, water quality degradation, and disruption of prey species dynamics.
- *Economic valuation.* To build support and highlight the need for management actions, it is clear that a full evaluation of the economics of recreational and commercial fishing components for these valuable fisheries must be undertaken. Understanding the magnitude of their local and regional economic impacts will help to garner critical public and political support for action.

What is required is a coordinated effort by fishery management to develop the appropriate and precise information on which to base current and future management decisions to build sustainable fisheries for these remarkable species. For highly migratory species like tarpon, it is apparent that a regional management effort that develops essential interstate and international cooperatives will be the most strategic approach to conserve these valuable resources (McMillen-Jackson et al., 2005). But in reality, because both tarpon and bonefish are valued for different reasons in different countries and by different cultures, their effective management will require a unique blend of extensive cooperation among anglers, guides, scientists, and fishery management to ensure their sustainability.

REFERENCES

Adams, A.J., C.P. Dahlgren, G.T. Kellison, M.S. Kendall, C.A. Layman, J.A. Ley, I. Nagelkerken, and J.E. Serafy. 2006. Nursery function of tropical back-reef systems. Mar. Ecol. Prog. Ser. 318: 287–301.

Alagaraga, K. 1984. Simple methods for estimation of parameters for assessing exploited fish stocks. Indian Journal of Fisheries 31: 177–208.

Alexander, E. C. 1961. A contribution to the life history, biology and geographical distribution of bonefish, *Albula vulpes* (Linnaeus). Dana Rept. Carlsberg Found. 53: 1–53.

Alperin, I.M. and R.H. Schaefer. 1964. Juvenile bonefish (*Albula vulpes*) in Great South Bay, New York. New York Fish Game Journal 11(1): 1–12.

Andrews, A.H., E.J. Burton, K.H. Coale, G.M. Cailliet, and R.E. Crabtree. 2001. Radiometric age validation of Atlantic tarpon, *Megalops atlanticus*. Fish. Bull. 99: 389–398.

Arronte, J.C., J.A. Pis-Millan, M.P. Fernandez, and L. Garcia. 2004. First records of the subtropical fish *Megalops atlanticus* (Osteichthyes: Megalopidae) in the Cantabrian Sea, northern Spain. J. Mar. Biol. Assoc. U.K. 84(5): 1091–1092.

Ault, J.S., J.A. Bohnsack, and G.A. Meester. 1998. A retrospective (1979–1996) multispecies assessment of coral reef fish stocks in the Florida Keys. Fish. Bull. 96(3): 395–414.

Ault, J.S., G.A. Diaz, S.G. Smith, J. Luo, and J.E. Serafy. 1999. An efficient sampling survey design to estimate pink shrimp population abundance in Biscayne Bay, Florida. N. Am. J. Fish. Manage. 19(3): 696–712.

Ault., J.S., R. Humston, M.F. Larkin, and J. Luo. 2002. Development of a bonefish conservation program in South Florida. Final report to National Fish and Wildlife Foundation on grant No. 20010078000-SC.

Ault, J.S., M.F. Larkin, J. Luo, N. Zurcher, and D. Debrot. 2005a. Bonefish-tarpon conservation research program. Final Report to the Friends of the Florida Keys National Marine Sanctuary. University of Miami RSMAS, Miami, FL. 91p.

Ault, J.S., S.G. Smith, and J.A. Bohnsack. 2005b. Evaluation of average length as an estimator of exploitation status for the Florida coral reef fish community. ICES J. Mar. Sci. 62: 417–423.

Ault, J.S., J.A. Bohnsack, S.G. Smith, and J. Luo. 2005c. Towards sustainable multispecies fisheries in the Florida USA coral reef ecosystem. Bull. Mar. Sci. 76(2): 595–622.

Babcock, L.L. 1951. The tarpon: a description of the fish with some hints on its capture (fifth edition). Privately printed, Buffalo, NY. 174pp.

Beebe, W. 1927. A tarpon nursery in Haiti. Bull. N.Y. Zool. Soc. 30(5): 141–145.

Berrien, P.L., M.P. Fahay, A.W. Kendall, Jr., and W.G. Smith. 1978. Ichthyoplankton from the RV Dolphin survey of the continental shelf waters between Martha's Vineyard, Massachusetts and Cape Lookout, North Carolina, 1965–66. National Marine Fisheries Service Sandy Hook Lab., Tech. Ser. Rep. 15: 1–152.

Bishop, R.E. and J.J. Torres. 1999. Leptocephalus energetic: metabolism and excretion. J. Exp. Biol. 202(18): 2485–2493.

Blandon, I.R., F.J. Garcia de Leon, R. Ward, R.A. Van den Bussche, and D.S. Needleman. 2003. Studies in conservation genetics of tarpon (*Megalops atlanticus*)—V. Isolation and characterization of microsatellite loci. Mole. Ecol. Notes 3(4): 632–634.

Breder, C.M., Jr. 1933. Young tarpon on Andros Island. Bull. N.Y. Zool. Soc. 36(3): 65–67.

Bruger, G.E. 1974. Age, growth, food habits, and reproduction of bonefish, *Albula vulpes*, in south Florida waters. Florida Marine Research Publications. 3. Florida Department of Natural Resources. St. Petersburg, FL.

Burton, E.J., A.H. Andrews, K.H. Coale, and G.M. Cailliet. 1999. Application of radiometric age determination to three long-lived fishes using ^{210}Pb:^{226}Ra disequilibria in calcified structures: a review. In Musick, J.A. (ed.) Life in the Slow Lane: Ecology and Conservation of Long-lived Marine Animals. American Fisheries Society Symposium 23: 77–87.

Cataño, S. and J. Garzón-Ferreira. 1994. Ecología trófica del sábalo *Megalops atlanticus* (Pisces: Megalopidae) en el area de Ciénaga Grande de Santa Marta, Caribe Colombiano. Revista del Biología Tropical 42(3): 673–684.

Chacon-Chaverri, D. 1993. Aspectos biométricos de una polación de sábalo, *Megalops atlanticus* (Pisces: Megalopidae). Revista del Biología Tropical 41(1): 13–18.

Chacon-Chaverri, D. and W.O. McLarney. 1992. Desarrolo temprano del sabalo, *Megalops atlanticus* (Pisces:Megalopidae). Rev. Biol. Trop. 40:171–177.

Colborn, J., R.E. Crabtree, J.B. Shaklee, E. Pfeiler, and B.W. Bowen. 2001. The evolutionary enigma of bonefishes (*Albula* spp.): cryptic species and ancient separations in a globally distributed shorefish. Evolution 55(4): 807–820.

Colton, D.E. and W.S. Alevizon. 1983a. Feeding ecology of bonefish in Bahamian waters. Trans. Am. Fish. Soc. 112: 178–184.

Colton, D.E. and W.S. Alevizon. 1983b. Movement patterns of bonefish, *Albula vulpes*, in Bahamian waters. Fish. Bull. 81(1): 148–154.

Cooke, S.J. and D.P. Philipp. 2004. Behavior and mortality of caught-and-released bonefish (*Albula* spp.) in Bahamian waters with implications for a sustainable recreational fishery. Biol. Conserv. 118(5): 599–607.

Costa Pereira, N. and L. Saldanha. 1977. Sur la distribution de *Tarpon atlanticus* Valenciennes 1847 (Pisces: Megalopidae) dans l'atlantique oriental. Mem. Mus. Mar. Ser. Zool. 1: 1–15.

Cowen, R.K., C.B. Paris, and A. Srinivasan. 2006. Scaling of connectivity in marine populations. Science 311: 522–527.

Crabtree, R.E. 1995. Relationship between lunar phase and spawning activity of tarpon, *Megalops atlanticus*, with notes on the distribution of larvae. Bull. Mar. Sci. 56: 895–899.

Crabtree, R.E., E.C. Cyr, R.E. Bishop, L.M. Falkenstain, and J.M. Dean. 1992. Age and growth of tarpon, *Megalops atlanticus*, larvae in the eastern Gulf of Mexico, with notes on relative abundance and probable spawning areas. Environ. Biol. Fish. 35(4): 361–370.

Crabtree, R.E., E.C. Cyr, and J.M. Dean. 1995. Age and growth of tarpon, *Megalops atlanticus*, from south Florida waters. Fish. Bull. 93: 619–628.

Crabtree, R.E., C.W. Harnden, D. Snodgrass, and C. Stevens. 1996. Age, growth, and mortality of bonefish, *Albula vulpes*, from the waters of the Florida Keys. Fish. Bull. 94(3): 442–451.

Crabtree, R.E., E.C. Cyr, D.C. Chacon-Chaverri, W.O. McLarney, and J.M. Dean. 1997a. Reproduction of tarpon, *Megalops atlanticus*, from Florida and Costa Rican waters and notes on their age and growth. Bull. Mar. Sci. 61(2): 271–285.

Crabtree, R.E., D. Snodgrass, and C.W. Harnden. 1997b. Maturation and reproductive seasonality in bonefish, *Albula vulpes*, from the waters of the Florida Keys. Fish. Bull. 95(3): 456–465.

Crabtree, R.E., D. Snodgrass, and C. Harnden. 1998a. Survival rates of 665 bonefish, *Albula vulpes*, caught on hook-and-line gear in the Florida Keys. In Investigation into nearshore and estuarine gamefish abundance, ecology and life history in Florida. Technical Report to the US Fish and Wildlife Service, Sport Fish Restoration Project F-59. Florida Fish and Wildlife Conservation Commission.

Crabtree, R.E., C. Stevens, D. Snodgrass, and F.J. Stengard. 1998b. Feeding habits of bonefish, *Albula vulpes*, from the waters of the Florida Keys. Fish. Bull. 96(4): 754–766.

Cyr, E.C. 1991. Aspects of the Life History of the Tarpon, *Megalops atlanticus*, from South Florida. University of South Carolina Press, Columbia, SC. 139p.

Dahl, G. 1971. Los peces del norte de Colombia. Inderena, Bogotá. 391p.

de Menezes, M.F. and M.P. Paiva. 1966. Notes on the biology of tarpon, *Megalops atlanticus* (Cuvier and Valenciennes), from coastal waters of Ceara State, Brazil. Arq. Estac. Biol. Mar. Univ. Fed. Ceara 6: 83–98.

Dorval, E., C.M. Jones, R. Hannigan, and J. vanMontfrans. 2005. Can otolith chemistry be used for identifying essential seagrass habitats for juvenile spotted seatrout, *Cynoscion nebulosus*, in Chesapeake Bay? Mar. Freshwater Res. 56: 645–653.

Drass, D.M. 1992. Onshore movements and distribution of leptocephali (Osteichthyes: Elopomorpha) in the Bahamas. MS thesis, Florida Institute of Technology, Melbourne, FL. 85p.

Edwards, R.E. 1998. Survival and movement of released tarpon. Gulf of Mexico Sci. 1998(1): 1–7.

Ehrhardt, N.M. and J.S. Ault. 1992. Analysis of two length-based mortality models applied to bounded catch length frequencies. Trans. Am. Fish. Soc. 121(1): 115–122.

Eldred, B. 1967. Larval tarpon, *Megalops atlanticus* Valenciennes, (Megalopidae) in Florida Waters. Florida Board of Conservation. Leaflet Ser. 4(4): 1–9.

Eldred, B. 1968. First record of a larval tarpon, *Megalops atlanticus* Valenciennes, from the Gulf of Mexico. Florida Board Conserv. Mar. Res. Lab. Leafl. Ser. 4(7): 1–2.

Eldred, B. 1972. Note on larval tarpon, *Megalops atlanticus* (Megalopidae), in the Florida Straits. Florida Dep. Nat. Resour. Mar. Res. Lab. Leafl. Ser. 4(22): 1–6.

Erdman, D.S. 1960a. Notes on the biology of the bonefish and its sport fishery in Puerto Rico. 5th Inter. Game Fish Conf. 5: 1–11.

Erdman, D.S. 1960b. Larvae of tarpon, *Megalops atlanticus*, from the Añasco River, Puerto Rico. Copeia. 2: 146.

Garcia, C.B. and O.D. Solano. 1995. *Tarpon atlanticus* in Colombia: a big fish in trouble. Naga ICLARM Q. 18(3): 47–49.

Gehringer, J.W. 1959. Leptocephalis of Atlantic tarpon, *Megalops atlanticus* Valenciennes, from offshore waters. Q. J. Fla. Acad. Sci. 21(3): 235–240.

Greenwood, P.H., D.E. Rosen, S.H. Weitzman, and G.S. Myers. 1966. Phyletic studies of teleostean fishes, with a provisional classification of the living forms. Bull. Am. Mus. Nat. Hist. 131: 341–455.

Grey, Z. 1919. Tales of Fishes. The Derrydale Press, Lanham, MD. 267p.

Grubich, J.R. 2001. Prey capture in Actinopterygian fishes: a review of suction feeding motor patterns with new evidence from an elopomorph fish, *Megalops atlanticus*. Am. Zool. 41: 1258–1265.

Harrington, R.W. 1958. Morphometry and ecology of small tarpon, *Megalops atlanticus*, from transitional stage through onset of scale formation. Copeia 1958(1): 1–10.

Harrington, R.W. 1966. Changes through one year in the growth rates of tarpon, *Megalops atlanticus*, reared from mid-metamorphosis. Bull. Mar. Sci. 16(4): 863–883.

Harrington, R.W. and E.S. Harrington. 1960. Food of larval and young tarpon, *Megalops atlanticus*. Copeia 1960(4): 311–319.

Hildebrand, S.F. 1934. The capture of a young tarpon, *Tarpon atlanticus*, at Beaufort, North Carolina. Copeia 1934: 45–46.

Hildebrand, S.F. 1963. Family Elopidae: in fishes of the western North Atlantic. Mem. Sears Found. Mar. Res. 3: 630.

Hollister, G. 1939. Young *Megalops cyprinoides* from Batavia, Dutch East India, including a study of the caudal skeleton and a comparison with the Atlantic species, *Tarpon atlanticus*. Zool. N.Y. 24(4): 449–475.

Howells, R.G. and G.P. Garrett. 1992. Status of some exotic sport fishes in Texas waters. Tex. J. Sci. 44(3): 317–324.

Hulet, W.H. and C.R. Robins. 1989. The evolutionary significance of the leptocephalus larva. In: Bohlke E.B. (ed.) Fishes of the Western North Atlantic. Memoir No. 1. Part 9. Vol. 2. Sears Foundation for Marine Research. New Haven, CT. 669–677.

Humston, R. 2001. Development of movement models to assess the spatial dynamics of marine fish populations. Doctoral dissertation. University of Miami, Miami, FL.
Humston, R., J.S. Ault, M.F. Larkin, and J. Luo. 2005. Movements and site fidelity of bonefish (*Albula vulpes*) in the northern Florida Keys determined by acoustic telemetry. Mar. Ecol. Prog. Ser. 291: 237–248.
International Game Fish Association (IGFA). 2006. World Record Game Fishes: freshwater, saltwater, and flyfishing. Dania Beach, FL. 384p.
Kushlan, J.A. and T.E. Lodge. 1974. Ecological and distributional notes on the freshwater fish of Southern Florida. Fla. Scientist 37(2): 110–127.
Larkin, M.F., J.S. Ault, and R. Humston. In review. Mail survey of south Florida's bonefish charter fleet. Trans. Am. Fish. Soc.
Luckhurst, B.E., E.D. Prince, E.B. Brothers, J.S. Ault, and D.B. Olson. In review. First time occurrence of juvenile Atlantic blue marlin (*Makaira nigricans*) at Bermuda: spawning site verification using ocean transport models. Fish. Bull.
Marwitz, S.R. 1986. Young tarpon in a roadside ditch near Matagorda Bay in Calhoun County, Texas. Tex. Parks Wildl. Dep. Manag. Data Ser. 100: 8.
McClane, A.J. 1974. McClane's New Standard Fishing Encyclopedia. Holt, Rinehart and Winston, New York. 1156p.
McCleave, J. 1993. Physical and behavioral controls on the oceanic distribution and migration of leptocephali. J. Fish Biol. 43: 243–273.
McMillen-Jackson, A.L., T.M. Bert, H. Cruz-Lopez, B. Seyoum, T. Orsoy, and R.E. Crabtree. 2005. Molecular genetic variation in tarpon (*Megalops atlanticus* Valenciennes) in the northern Atlantic Ocean. Mar. Biol. 146: 253–261.
Mendoza-Franco, E.F., D.C. Kritzky, V.M. Vidal-Martinez, T. Scholz, and L. Aguirre-Macedo. 2004. Neotropical Monogenoidea. 45. Revision of *Diplectanoctyla* Yamaguti, 1953 (Diplectanidae) with redescription of *Diplectanocotyla megalopis* Rakotofiringa and Oliver, 1987 on Atlantic tarpon, *Megalops atlanticus* Cuvier and Valenciennes, from Nicaragua and Mexico. Comp. Physiol. 71(2): 158–165.
Mercado, J.E. and A. Ciardelli. 1972. Contribucion a la morfologia y organogenesis de los leptocefalos del sabalo *Megalops atlanticus* (Pisces: Megalopidae). Bull. Mar. Sci. 22: 153–184.
Mochioka, N. and M. Iwamizu. 1996. Diet of anuilloid larvae: leptocephali feed selectively on larvacean houses and fecal pellets. Mar. Biol. 125: 447–452.
Mojica, R. Jr., J.M. Shenker, and C.W. Harnde. 1995. Recruitment of bonefish, *Albula vulpes*, around Lee Stocking Island, Bahamas. Fish. Bull. 93(4): 666–675.
Mol, J.H., Resida, D., Ramial, J.S., and C.R. Becker. 2000. Effects of El Niño-related drought on freshwater and brackish-water fishes in Suriname, South America. Environ. Biol. Fishes 59: 429–440.
Montano, O.J.F., E.D. Dibble, D.C. Jackson, and K.R. Rundle. 2005. Angling assessment of the fisheries of Humacao Natural Reserve lagoon system, Puerto Rico. Fish. Res. 76: 81–90.
Nelson, J.S. 2006. Fishes of the World. John Wiley & Sons, Upper Saddle River, NJ. 624p.
Oppel, F. and T. Meisel. 1987. Tales of Old Florida. Castle Press, Seacaucus, NJ. 477p.
Pauly, D. 1980. On the interrelationships between natural mortality, growth parameters, and mean environmental temperature in 175 fish stocks. J. Cons. Int. Explor. Mer. 39: 175–192.
Pfeiler, E. 1984a. Inshore migration, seasonal distribution and sizes of larval bonefish, *Albula*, in the Gulf of California. Environ. Biol. Fishes 10(1/2): 117–122.
Pfeiler, E. 1984b. Changes in water and salt content during metamophosis of larval bonefish (*Albula*). Bull. Mar. Sci. 34(2): 177–184.
Pfeiler, E. 1988. Isolation and partial characterization of a novel keratin sulfate proteoglycan from metamorphosing bonefish (*Albula*) larvae. Fish Physiol. Biochem. 4: 27–36.

Pikitch, E.K., C. Santora, E.A. Babcock, A. Bakun, R. Bonfil, D.O. Conover, P. Dayton, P. Doukakis, D. Fluharty, B. Heneman, E.D. Houde, J. Link, P.A. Livingston, M. Mangel, M.K. McAllister, J. Pope, and K.J. Sainsbury. 2004. Ecosystem-based fishery management. Science 305(5682): 346–347.

Quinn, T.J. and R.B. Deriso. 1999. Quantitative fish dynamics. Oxford University Press, New York. 542p.

Ramsundar, H. 2005. The distribution and abundance of wetland icthyofauna, and exploitation of the fishes in the Godineau Swamp, Trinidad—case study. Revista de Biologia Tropical 53(Suppl. 1): 11–23.

Restrepo, M. 1968. La pesca en la Ciénaga Grande de Santa Marta. Invest. Pesq. (C.V.M.) 3: 1–69.

Rickards, W.L. 1966. A study of the ecology of first-year tarpon, *Megalops atlanticus* Valenciennes, in a Georgia salt-marsh with laboratory studies of growth rates and ecological growth efficiencies. University of Georgia Press, Athens, GA. 67p.

Rickards, W.L. 1968. Ecology and growth of juvenile tarpon, *Megalops atlanticus*, in a Georgia salt marsh. Bull. Mar. Sci. 18(1): 220–239.

Rivas, L.R. and S.M. Warlen. 1967. Systematics and biology of the bonefish *Albula nemoptera* (Fowler). Fish Bull. 66: 251–258.

Robins, C.R. 1977. The tarpon—unusual biology and man's impact determine its future. In Clepper, H. (ed.), Marine Recreational Fisheries, Volume 2. Sport Fishing Institute, Washington, DC, pp. 105–112.

Robins, C.R. and G.C. Ray. 1986. Atlantic Coast Fishes. Peterson Field Guide. Houghton Mifflin, Boston, MA. 354p.

Rothschild, B.J., J.S. Ault, and S.G. Smith. 1996. A systems science approach to fisheries stock assessment and management. In Gallucci, V.F., Saila, S., Gustafson, D., and B.J. Rothschild (eds.) Stock Assessment: Quantitative Methods and Applications for Small Scale Fisheries. Lewis Publishers (Div. of CRC Press), Chelsea, MI, pp. 473–492.

Roux, C. 1960. Note sur le tarpon (*Megalops atlanticus* C. et V.) des Cotes de la Republique du Congo. Bull. Mus. Nat. Hist. Paris 32(4): 314–319.

Sandin, S.A., J. Regetz, and S.L. Hamilton. 2005. Testing larval fish dispersal hypotheses using maximum likelihood analysis of otolith chemistry data. Mar. Freshwater Res. 56: 725–734.

Schmidt, J. 1922. The breeding places of the eel. Phil. Trans. R. Soc., Ser. B 211: 385.

Shenker, J.M., R. Crabtree, and G. Zarillo. 1995. Recruitment of tarpon and other fishes into the Indian River Lagoon. Bull. Mar. Sci. 57(1): 284.

Shiao, J.-C. and P.-P. Hwang. 2006. Thyroid hormones are necessary for the metamorphosis of tarpon *Megalops cyprinoids* leptocephali. J. Exp. Mar. Biol. Ecol. 331: 121–132.

Simpson, D.G. 1954. Two small tarpon from Texas. Copeia 1: 71–72.

Smith, D.G. 1980. Early larvae of the tarpon, *Megalops atlanticus* Valenciennes (Pisces: Elopidae), with notes on spawning in the Gulf of Mexico and the Yucatan Channel. Bull. Mar. Sci. 30(1): 136–141.

Steneck, R.S. 2006. Staying connected in a turbulent world. Science 311: 480–481.

Thompson, B.A. and L.A. Deegan. 1982. Distribution of ladyfish (*Elops saurus*) and bonefish (*Albula vulpes*) leptocephali in Louisiana. Bull. Mar. Sci. 32(4): 936–939.

Tucker, J.W. and R.G. Hodson. 1976. Early and mid-metamorphic larvae of the tarpon, *Megalops atlanticus*, from the Cape Fear River estuary, North Carolina, 1973–1974. Chesapeake Sci. 17: 123–125.

Twomey, E. and P. Byrne. 1985. A new record for the tarpon, *Tarpon atlanticus* Valenciennes (Osteichthyes-Elopiformes-Elopidae), in the eastern North Atlantic. J. Fish Biol. 26: 359–362.

Wade, R.A. 1962. The biology of the tarpon, *Megalops atlanticus*, and the ox-eye, *Megalops cyprinoides*, with emphasis on larval development. Bull. Mar. Sci. Gulf Caribb. 12(4): 545–622.

Wade, R.A. 1969. Ecology of juvenile tarpon and effects of Dieldrin on two associated species. U.S. Dept. Int. Bur. Sports Fish. Wildl. Tech. Pap. 41: 85.

Ware, D.M. and R.E. Thompson. 2005. Bottom-up ecosystem tropic dynamics determine fish production in the northeast Pacific. Science 308: 1280–1284.

Warmke, G.L. and D.S. Erdman. 1963. Records of marine mollusks eaten by bonefish in Puerto Rican waters. Nautilus 76(4): 115–120.

Weinberger, C.S. and J.M. Posada. 2005. Analysis of the diet of bonefish, *Albula vulpes*, in Los Roques archipelago National Park, Venezuela. Contrib. Mar. Sci. 37: 30–44.

Zale, A.V. and S.G. Merrifield. 1989. Species profiles: life histories and environmental requirements of coastal fishes and invertebrates (South Florida)—ladyfish and tarpon. U.S. Fish and Wildlife Service Biological Report 82(11.104). U.S. Army Corps of Engineers, TR EL-82-4. 17p.

Zerbi, A., C.A. Aliaume, and J.M. Miller. 1999. A comparison between two tagging techniques with notes on juvenile tarpon ecology in Puerto Rico. Bull. Mar. Sci. 64(1): 9–19.

Zerbi, A., C. Aliaume, and J.C. Joyeux. 2001. Growth of juvenile tarpon in Puerto Rican estuaries. ICES J. Mar. Sci. 58(1): 87–95.

17 A Preliminary Otolith Microchemical Examination of the Diadromous Migrations of Atlantic Tarpon *Megalops atlanticus*

Randy J. Brown and Kenneth P. Severin

CONTENTS

Introduction 259
Methods 260
Results 264
Discussion 270
    Nursery Habitats 270
    Adult Habitats 271
Summary 272
Acknowledgments 272
References 272

INTRODUCTION

Atlantic tarpon *Megalops atlanticus* are long-lived members of the Elopomorpha superorder of fishes, which are united by their unique leptocephalus larval form.[15] Tarpon are thought to spawn exclusively in marine waters,[30] but following metamorphosis they are capable of moving freely among marine, brackish, and freshwater systems, indicating a tolerance for a wide range of salinities.[35,36] Movement patterns among these habitats will be recorded in the chemistry of their otoliths and could potentially be described using otolith microchemical techniques.

Otoliths are mineral structures associated with the semicircular canal network near the brain in teleost fish.[22] They are composed primarily of calcium carbonate amid a proteinaceous matrix.[9] They grow throughout a fish's life as material precipitates on the outer surface, and are generally considered to be insoluble.[4] In recent years, fisheries scientists have used trace element distribution within otoliths to describe life history events and patterns of movement of many species. Early laboratory experiments with a variety of fish families indicated that salinity had a strong influence on the chemical composition of their otoliths.[11,21,27] More recent experimental work has shown that the primary factor influencing otolith chemistry is not so much salinity as the molar ratio of trace metal ions to calcium (Ca) ions in ambient water.[2,18,37] Strontium is a 2^+ ion in solution and precipitates in otoliths, replacing Ca ions in the otolith mineral in proportion to the atomic ratio of Sr to Ca in ambient water.[2] Most freshwater systems appear to have a lower molar ratio of Sr to Ca than marine water,[18] and many researchers have evaluated fish movements between the two environments by examining otolith Sr distribution.[1,14,33]

In this study, we present the results of a preliminary examination of Sr distribution in the otoliths from a small number of tarpon. To our knowledge, microchemical analyses of tarpon otoliths have not been previously conducted, so there was some uncertainty as to the utility of the technique for the species. Therefore, the major objective of this work was to determine if the Sr distribution patterns within tarpon otoliths were useful for assessing migration between marine and freshwater habitats. A secondary objective, provided the technique appeared useful, was to examine the lifetime patterns of marine and freshwater habitat use of the sampled tarpon. Of particular interest were the Sr distribution patterns within the first growth increment, as that material precipitated when the fish were in nursery habitats.

METHODS

Tarpon otoliths used in this study were collected from three geographic locations: Lake Nicaragua; the vicinity of the mouth of the Rio Colorado, a distributary of the Rio San Juan; and along the Texas coast in the northwestern Gulf of Mexico (Figure 17.1). Lake Nicaragua is 160 km long by 65 km wide. It is drained by the San Juan River, which flows for 190 km from the lake to the Caribbean Sea.[16] Tarpon from Lake Nicaragua were recovered from the discarded carcasses of fish that had been processed in food fisheries. The only biological data from these fish were that the collectors considered them to be large. Otoliths from the Rio Colorado were collected from fish captured in a sport fishery. Lengths and weights were recorded for these fish and all were larger than the minimum size at maturity.[8,10] Otoliths from the Gulf of Mexico were collected during fisheries surveys in various locations along the Texas coast. Lengths were recorded for these fish, and only one was larger than the minimum size at maturity. Otoliths of four fish from each geographic location were analyzed in this study. Table 17.1 presents the biological data for the fish from each collection area.

Tarpon otoliths were thin-sectioned (sectioned) and polished in preparation for microscopic viewing and microprobe analysis. All otoliths were sectioned in the transverse plane through the core,[26] mounted on a glass slide, and ultimately polished on a lapidary wheel with 1-µm diamond abrasive. Each otolith section was approximately 200 µm thick, and growth increments could be clearly viewed with transmitted light (Figure 17.2). Finally, each otolith section was coated with a thin layer of carbon.

A wavelength-dispersive electron microprobe (WD-EM), an instrument capable of precise and accurate measurement of otolith Sr and Ca concentrations,[5] was used

A Preliminary Otolith Microchemical Examination

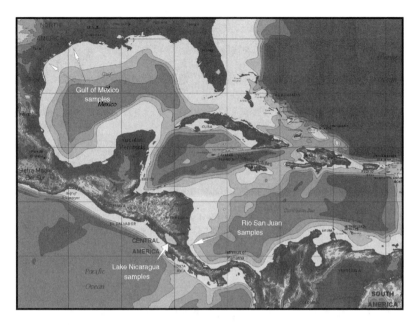

FIGURE 17.1 (Color Figure 17.1 follows p. 262.) Tarpon otolith collection locations within the Caribbean Sea and Gulf of Mexico region.

TABLE 17.1
Details of the Tarpon Sample Collections Used in This Study

Sample Group	Collection Location	Sample	Date	Length (mm TL)	Weight (kg)
Lake Nicaragua	Isla de Zanate	LN-1	2002	—	—
	Isla de Zanate	LN-2	2002	—	—
	Rio Frio	LN-3	2002	—	—
	Isla de Zanate	LN-4	2002	—	—
Rio San Juan	Rio Colorado	RSJ-1	10/21/91	1725	37.0
	Rio Colorado	RSJ-2	10/22/91	1785	47.3
	Rio Colorado	RSJ-3	1/28/92	1820	55.5
	Rio Colorado	RSJ-4	2/22/92	1905	54.5
Gulf of Mexico	Near Sabine Lake	GOM-1	10/19/87	2060	—
	Lower Laguna Madre	GOM-2	10/4/93	885	—
	Matagorda Bay	GOM-3	10/6/93	950	—
	Corpus Christi Bay	GOM-4	10/29/93	937	—

for microchemical analyses of tarpon otoliths in this study. It functions by bombarding points on a sample surface with a focused beam of electrons. Atoms within the material are ionized by the electron beam and emit x-rays unique to each element. Spectrometers are tuned to count the x-rays from elements of interest. The x-ray counts at each sample point are proportional to the elemental concentration in the material.[12,23,24]

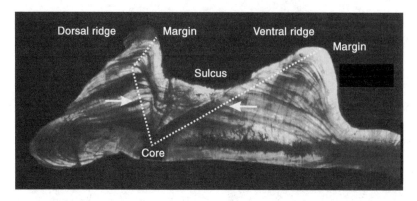

FIGURE 17.2 Optical image of a sectioned tarpon otolith with certain features labeled. The dotted lines illustrate two possible transect paths used in this study: one from the otolith core to the margin up the dorsal ridge that contains a dog-leg, and the other from the otolith core to the margin up the ventral ridge. The white arrows indicate the positions where the transect paths cross the first annulus.

Strontium and Ca x-ray counts (qualitative data) were collected from a series of points along a core (precipitated during the first year of life) to margin (precipitated just prior to the fish's death) transect for each otolith (Figure 17.2). The electron beam was 5 μm in diameter operating at an accelerating voltage of 15 kiloelectron volts (keV), and a nominal current of 20 nA. Center-to-center distance between sequential points was 8 μm. Strontium x-ray counts were collected for 25 s at each point. Total time required for each sample point was approximately 40 s.

Quantitative data collection was more involved and required approximately 260 s for each sample point, 6.5 times longer than qualitative data. Howland et al.[14] created a least-squares linear regression equation relating their qualitative x-ray count data to concentration estimates with a small number of quantitative analyses, and used the equation to convert their x-ray count data to concentration estimates. They found that their raw Sr x-ray counts were highly predictive of Sr concentration ($r^2 > 99\%$). In the interest of efficiency, we followed their approach in this study.

Strontium and Ca concentrations were estimated for 42 sample points across both high and low Sr concentration regions following standard analytical procedures as detailed in Reed[24] and Goldstein et al.[12] Strontianite (USNM R10065) and calcite (USNM 13621) were used as external standards. Detection limits were approximately 340 and 203 ppm for Sr and Ca, respectively. Strontium concentration varied widely in different regions of sample otoliths, from 363 to 4141 ppm, and the estimation error varied with concentration (Figure 17.3). Ca concentration was relatively constant in all sampled regions of otoliths, averaging approximately 40% by weight, or 400,000 ppm, and the estimation error was similarly stable. Molar concentrations of Sr and Ca were derived from absolute concentrations using the equations:

$$\text{Sr moles} \cdot \text{kg}^{-1} = (\text{Sr mg} \cdot \text{kg}^{-1})(1 \text{ mol} \cdot 87.620 \text{ g}^{-1})(1 \text{ g} \cdot 1000 \text{ mg}^{-1})$$

and

$$\text{Ca moles} \cdot \text{kg}^{-1} = (\text{Ca mg} \cdot \text{kg}^{-1})(1 \text{ mol} \cdot 40.078 \text{ g}^{-1})(1 \text{ g} \cdot 1000 \text{ mg}^{-1})$$

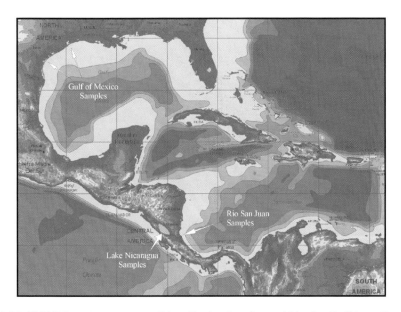

COLOR FIGURE 17.1 Tarpon otolith collection locations within the Caribbean Sea and Gulf of Mexico region.

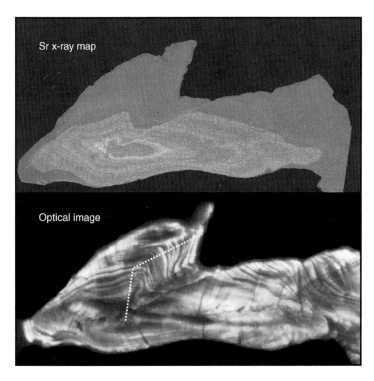

COLOR FIGURE 17.6 Strontium map (top) and optical image (bottom) of otolith from Lake Nicaragua tarpon LN-4. Lighter regions in the Sr map indicate relatively high Sr levels (marine precipitate), and darker regions indicate relatively low Sr levels (freshwater precipitate). The dotted line in the optical image represents the approximate path of the core to margin transect.

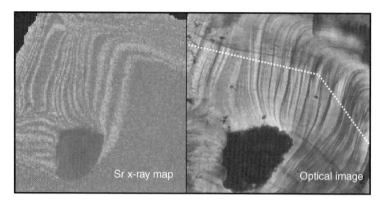

COLOR FIGURE 17.7 Strontium map (left) and optical image (right) of otolith from Rio San Juan tarpon RSJ-1. Light regions in the Sr map indicate relatively high Sr levels (marine precipitate), and dark regions indicate relatively low Sr levels (freshwater precipitate). The dotted line in the optical image represents the approximate path of the core (not visible in this image) to margin transect.

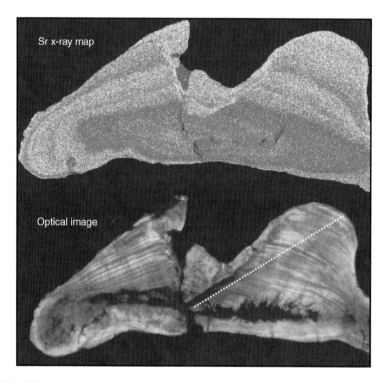

COLOR FIGURE 17.8 Strontium map (top) and optical image (bottom) of otolith from Gulf of Mexico tarpon GOM-1. Light regions in the Sr map indicate relatively high Sr levels (marine precipitate), and dark regions indicate relatively low Sr levels (freshwater precipitate). The dotted line in the optical image represents the approximate path of the core to margin transect.

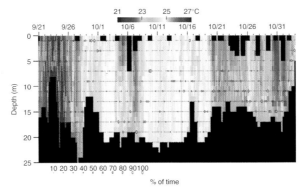

COLOR FIGURE 18.4 Summary map of vertical thermohabitat utilization by a 36.4-kg PAT-tagged tarpon generated with minute-by-minute depth and temperature data from a recovered PAT tag (T-03) for fish that migrated from Savannah, Georgia to Sebastian Inlet, Florida. Temperatures are displayed in color scale ranging from 21 to 27°C. Depth is on the y-axis. Time is on the x-axis running from September 21 to November 3, 2001. Size of open circles indicates percentage of time the tarpon spent at each depth during each 6-h interval. The dashed white line indicates the minute-by-minute depth position of the fish.

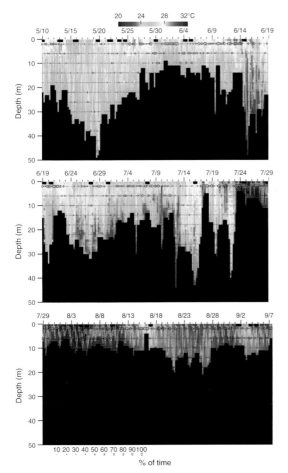

COLOR FIGURE 18.8 A summary map of vertical thermohabitat utilization generated with minute-by-minute depth and temperature data of a recovered PAT tag (T-24). The temperatures are displayed in color scale from 20 to 32°C. The depth is on the y-axis and time is on the x-axis from May 10 to September 8, 2004. The size of open circle indicates the percentage of time the tarpon spent at each depth for each 6-h interval. The dashed white line indicates the minute-by-minute depth distribution of the fish.

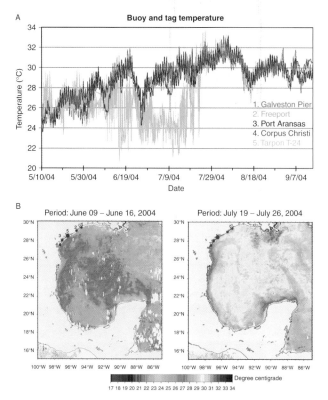

COLOR FIGURE 18.10 (A) Water temperatures recorded at Galveston Pier (red line), Freeport (green line), Port Aransas (black line), and Corpus Christi (blue line) C-MAN stations, and by the recovered PAT tag T-24 (light blue line) from May 10 to September 8, 2004. (B) Sea surface temperature maps for the Gulf of Mexico for the period June 9–June 16, and July 19–July 26, 2004, as determined by MODIS satellite. Note the areas of upwelling (light blue) off south Texas and Campeche Bank, Mexico.

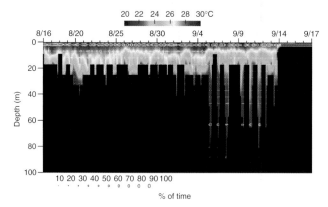

COLOR FIGURE 18.12 Vertical thermohabitat utilization map generated from ARGOS transmitted summary data from a tarpon tagged in Trinidad (T-46). The temperatures are displayed in color scale from 20 to 30°C. The depth is on the y-axis and the date is on the x-axis. The size of open circle indicates the percentage of time the tarpon spent at each depth for each 3-h interval.

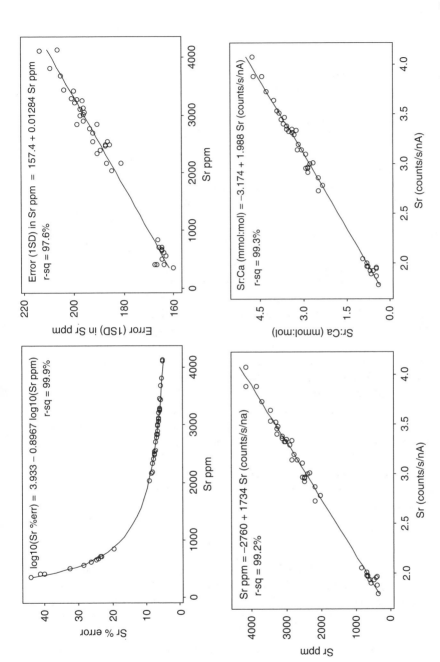

FIGURE 17.3 Least-squared linear regression charts illustrating the results of quantitative assessments of Sr concentration in tarpon otoliths including: error (1SD) in percent of concentration estimate (top left) and in ppm (top right) across the range of Sr concentration encountered in sampled points; and the linear relationships between standardized Sr x-ray count data (counts · s^{-1} · nA^{-1}) and Sr concentration (ppm; bottom left) and the molar ratio of Sr and Ca (mmol · mol^{-1}; bottom right) values.

X-ray count data were standardized to counts per second per nanoampere (counts·$s^{-1} \cdot nA^{-1}$). Least-squares linear regressions of Sr counts·$s^{-1} \cdot nA^{-1}$ on Sr concentration estimates and the molar ratios of Sr and Ca (mmol Sr·mol Ca^{-1}) showed that Sr x-ray count data were highly predictive ($r^2 > 99\%$) of both values (Figure 17.3). Calcium variability was low relative to Sr, and it behaved as a constant in the ratio of Sr:Ca (mmol:mol). Transect data were converted from Sr x-ray counts to Sr concentration (ppm) and the molar ratio of Sr:Ca (mmol:mol) using the regression equations (Figure 17.3) and presented with both units in core to margin line graphs.

In the absence of experimental data or known life history individuals of a fish species, the Sr ppm range to expect in freshwater precipitate, or alternatively in marine precipitate, is unknown. Interpretation of observed Sr ppm distribution in initial explorations must be guided by our general understanding that material precipitated in freshwater is lower in Sr concentration than material precipitated in marine water,[3,28] which may not be true for all freshwater systems.[18] When Sr distribution data are collected from several individuals, the Sr ppm range and distribution common to the species become clear. If it is known that certain individuals inhabited freshwater during their lives, such as the samples in this study that were collected from Lake Nicaragua, then it should be possible to empirically determine a Sr ppm range indicating freshwater precipitate based on low values from their otoliths. Similarly, if it is known that other individuals inhabited fully marine water during their lives, which should include all 12 samples examined in this study, then it should be possible to empirically determine a Sr ppm range indicating marine precipitate based on high values from their otoliths. Histograms of Sr ppm data from the four Lake Nicaragua samples and from all 12 samples tarpon together were prepared to calibrate Sr ppm values to freshwater and marine ranges.

Strontium distribution maps (Sr maps) were produced from 3 of the 12 sample otoliths by collecting Sr x-ray count data from points in a grid pattern across the entire otolith surface. Elemental maps are routinely used in geological applications,[12,24] and have been used in otolith microchemical studies as well.[19,20,34] Sr maps in this study were made with an electron beam 5 µm in diameter, an accelerating voltage of 15 keV, and a nominal current of 100 nA. Center-to-center distance between adjacent points in the grid was approximately 9 µm. X-ray counts were collected for 0.05 s at each point.

RESULTS

Strontium concentration ranged from 511 ppm to 5459 ppm in combined data from Lake Nicaragua tarpon, and from below detectible levels to 5459 ppm in the combined data from all 12 tarpon. A bimodal distribution was evident in both histograms (Figure 17.4), but appeared to be most distinct in the data from Lake Nicaragua tarpon (Figure 17.4A). An extended upper-tail in the distribution of data in both histograms, was the result of high levels of Sr ppm, greater than 3550 ppm, present in a single tarpon otolith from Lake Nicaragua.

The bimodal distribution of Sr ppm suggested that the lower-value mode reflected the Sr ppm range from freshwater precipitate, while the higher-value mode reflected the Sr ppm range from marine precipitate. The lower-value mode had a

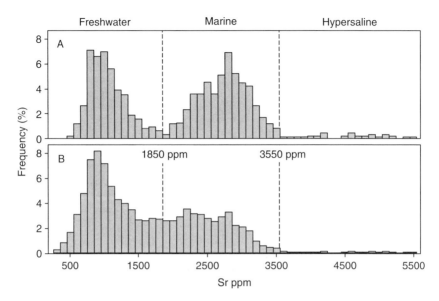

FIGURE 17.4 Histograms of Sr ppm distributions for the four Lake Nicaragua tarpon samples (A) and for all 12 tarpon otolith samples examined in this study combined (B). Proposed Sr ppm ranges for material precipitated in freshwater, marine, and hypersaline environments are indicated.

mean value of approximately 1000 ppm (Sr:Ca ≈ 1.137 mmol:mol), and ranged from the limit of detection (340 ppm or 0.381 mmol:mol) to approximately 1850 ppm (Sr:Ca ≈ 2.111 mmol:mol). The higher-value mode had a mean value of approximately 2800 ppm (Sr:Ca ≈ 3.200 mmol:mol), and ranged from 1850 to 3550 ppm (Sr:Ca ≈ 4.059 mmol:mol). The distinct bimodal distribution in the Sr ppm data from the Lake Nicaragua tarpon (Figure 17.4A) suggests that they inhabited either freshwater or marine water, and did not spend much time in brackish water. The blending of modes in the Sr ppm distribution of all tarpon together (Figure 17.4B) suggests that at least some tarpon in the group spent a substantial amount of time in brackish water, or moved between fresh and fully marine water on a short-enough time scale that a single microprobe point encompassed both fresh and saltwater precipitates. Strontium levels intermediate between the two modes would result in both cases, and these data do not permit distinguishing between the two possibilities.

In their review of tarpon life history literature, Zale and Merrifield[36] reported that juvenile tarpon had been collected in, and apparently tolerated, hypersaline environments with salinities as high as 45 ppt or greater. Since the high Sr ppm values, greater than 3550 ppm, were only seen within the first annual growth increment of one fish (LN-4), they were thought to result from the young fish's presence in a hypersaline environment during much of its first year.

Graphs of the otolith Sr distribution along core to margin transects from each tarpon were evaluated based on the calibration of Sr ppm values among freshwater, marine, and hypersaline environments (Figure 17.5). The Sr distribution graphs

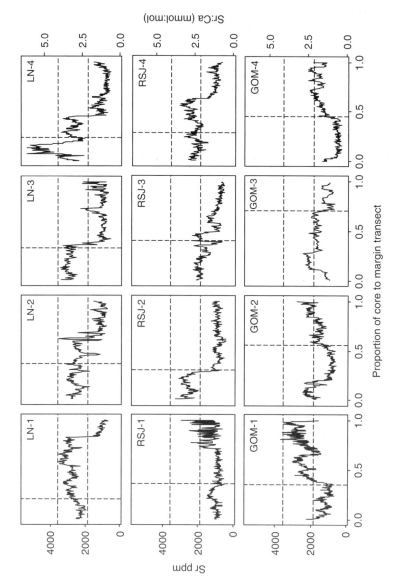

FIGURE 17.5 Strontium graphs from Lake Nicaragua (LN-#), Rio San Juan (RSJ-#), and Gulf of Mexico (GOM-#) tarpon otoliths. The x-axis is in units of proportion of the core (to the left) to margin (to the right) transect for each fish. The vertical dashed line in each graph indicates the approximate position of the first annulus. The lower horizontal dashed line marks the interface between fresh and marine precipitates at approximately 1850 ppm (Sr:Ca ≈ 2.111 mmol:mol), and the upper horizontal dashed line marks the interface between marine and hypersaline precipitates at approximately 3550 ppm (Sr:Ca ≈ 4.059 mmol:mol).

(Sr graphs) from all samples showed evidence of migration between freshwater and marine habitats. Marine levels of Sr ppm predominated in the otolith core regions of Lake Nicaragua tarpon (on the left-hand side of the graphs), and dropped within the freshwater range toward the otolith margins (on the right-hand side of the graphs) (Figure 17.5, top row). An otolith from one Lake Nicaragua tarpon had Sr ppm values to the left of the first annulus (indicated by the vertical dashed line), when the fish was age 0, which rose well beyond 3550 ppm, into the range considered to have precipitated when the fish was in a hypersaline environment (Figure 17.5, LN-4). Strontium ppm values in the freshwater range from all Lake Nicaragua samples were seen to the right of the first annulus, when fish were age 1 or older.

The Sr distribution graphs from three of four Rio San Juan samples were similar to those of tarpon from Lake Nicaragua; marine levels of Sr ppm predominantly to the left of the graphs, when fish were young, and freshwater levels predominantly to the right, when fish were old (Figure 17.5, RSJ-2, 3, and 4). The Sr ppm values from one Rio San Juan sample (Figure 17.5, RSJ-1) contrasted with the others and fell within the freshwater range in the otolith core and rose into the marine range near the margin in a closely spaced, oscillating series of peaks and nadirs that ranged from marine levels at the peaks, to freshwater levels at the nadirs. A similar but less extreme pattern is seen near the otolith margin on the right-hand side of the Sr distribution graph from Lake Nicaragua sample LN-3 (Figure 17.5).

The Sr graphs from three of four Gulf of Mexico samples (Figure 17.5, GOM-1, 2, and 4) contrasted with the other groups in that freshwater levels of Sr ppm predominated to the left of the graphs, when fish were young, and marine levels predominated to the right, when fish were older. The Sr distribution graph from one Gulf of Mexico sample (Figure 17.5, GOM-3) was more similar to Rio San Juan tarpon RSJ-3 and RSJ-4, with higher Sr ppm levels to the left and lower levels to the right. Within the Gulf of Mexico group, only otolith sample GOM-1 represented the Sr distribution record through the lifetime of a mature fish, as samples GOM-2, 3, and 4 were from immature fish thought to be only 3 or 4 years old.

Strontium x-ray maps were created from three of the tarpon otoliths examined in this study, one from each sample group. For each Sr map, the associated otolith optical image with the approximate transect path indicated is presented as well. This allows the Sr distribution apparent in the Sr maps to be compared with visible growth increments in the associated optical images, and with the Sr ppm patterns in the associated Sr graphs.

One Sr map was produced from Lake Nicaragua sample LN-4 (Figure 17.6, top) because of the anomalously high Sr ppm levels within its core region. The growth increments visible in the optical image of this otolith up the ridges on either side of the sulcus, similar to those presented by Crabtree et al.,[7] suggested that this tarpon was more than 15 years old (Figure 17.6, bottom). The transect path from which the Sr data were collected, originated near the core region and extended up the dorsal ridge beside the sulcus (Figure 17.6, bottom, dotted line). The band of high Sr concentration material near the core region was clearly visible as a light band in this Sr map (Figure 17.6, top). Similarly, the outer region of the otolith appeared to be composed of low Sr concentration material (dark area), which agreed with Sr values presented in the Sr graph from this otolith (Figure 17.3, LN-4).

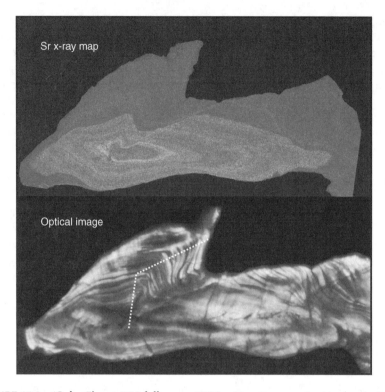

FIGURE 17.6 (Color Figure 17.6 follows p. 262.) Strontium map (top) and optical image (bottom) of otolith from Lake Nicaragua tarpon LN-4. Lighter regions in the Sr map indicate relatively high Sr levels (marine precipitate), and darker regions indicate relatively low Sr levels (freshwater precipitate). The dotted line in the optical image represents the approximate path of the core to margin transect.

A Sr map was produced from Rio San Juan sample RSJ-1 (Figure 17.7, left) because of the rapidly oscillating series of Sr peaks and nadirs apparent in the outer margin of this otolith (Figure 17.4, RSJ-1), which was the focus area of the map. Growth increments were clearly visible in the optical image (Figure 17.7, right) and appeared similar to images of outer otolith annuli presented by Crabtree et al.[7] The Sr map (Figure 17.7, left) revealed that the oscillating series of Sr peaks and nadirs seen in the Sr graph from this fish (Figure 17.4, RSJ-1) were the result of the transect crossing bands of high (light) and low (dark) Sr concentration regions. The image showed 18 or 19 finely spaced Sr bands in the outer margin of this otolith that were associated with visible growth increments.

A Sr map was produced from the Gulf of Mexico sample GOM-1 (Figure 17.8, top) because it was the only sample examined from a mature tarpon in which Sr values in the saltwater range predominated all the way to the otolith margin on the right-hand side of the Sr graph (Figure 17.5, GOM-1), which is from material precipitated during later life. The transect from which the Sr data were collected extended from the core region (low center) to the margin (upper right) along the ventral ridge beside the sulcus (Figure 17.8, bottom, dotted line). Considering annuli criteria discussed and illustrated

A Preliminary Otolith Microchemical Examination 269

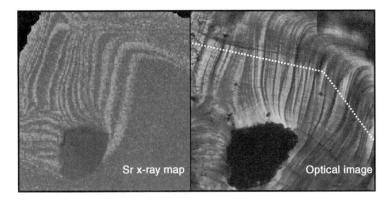

FIGURE 17.7 (**Color Figure 17.7 follows p. 262.**) Strontium map (left) and optical image (right) of otolith from Rio San Juan tarpon RSJ-1. Light regions in the Sr map indicate relatively high Sr levels (marine precipitate), and dark regions indicate relatively low Sr levels (freshwater precipitate). The dotted line in the optical image represents the approximate path of the core (not visible in this image) to margin transect.

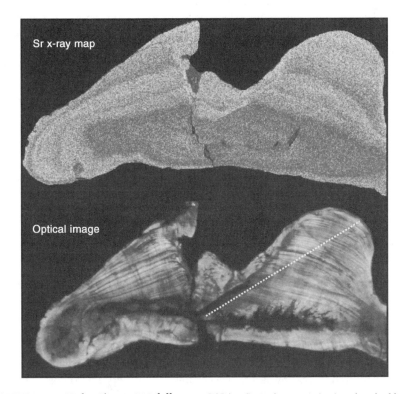

FIGURE 17.8 (**Color Figure 17.8 follows p. 262.**) Strontium map (top) and optical image (bottom) of otolith from Gulf of Mexico tarpon GOM-1. Light regions in the Sr map indicate relatively high Sr levels (marine precipitate), and dark regions indicate relatively low Sr levels (freshwater precipitate). The dotted line in the optical image represents the approximate path of the core to margin transect.

by Crabtree et al.,[7] the age of this tarpon would be estimated at well over 20 years. The Sr map showed that the Sr concentration in this otolith was generally low (dark) in the core region, and predominantly high (light) in the outer region (Figure 17.8, top), just as the Sr graph indicated (Figure 17.5, GOM-1).

DISCUSSION

Calibration of otolith Sr levels with the salinity or molar ratios of Sr and Ca of ambient water in which the fish lived has been a major focus of otolith microchemical studies of diadromous species. The experimental work of Secor et al.[27] with striped bass *Morone saxatilis*, an anadromous species, Tzeng[32] with Japanese eel *Anguilla japonica*, a catadromous species, and Zimmerman[37] with anadromous salmonid species, showed that one could use otolith chemistry to sort between freshwater and marine influence. The work of Bath et al.[2] and Kraus and Secor,[18] however, highlighted the need to understand regional freshwater chemistry, as well as salinity to avoid erroneous interpretation of otolith chemistry data. Other factors that warrant consideration when interpreting otolith chemistry data include: the spatial resolution of the microprobe point,[13] which relates directly to the time period covered within the sampled material; temperature effects on Sr incorporation, which have not proven to have a ubiquitous influence;[3] and growth rate of fish.[25] Despite these interpretive limitations and complications, the otolith microchemical technology can provide tremendous insights into major lifetime migration patterns between freshwater and marine habitats that could not be known otherwise. The results of this study indicate that otolith microchemical techniques can be effectively applied to tarpon life history research.

Considering the small sample sizes examined in this study, it would be imprudent to make population-level inferences based on the results. However, data from each of the fish examined provided unique insights into its habitat use and migratory patterns through life, information not easily available through other technologies (e.g., conventional anchor tagging, acoustic telemetry, and satellite pop-up archival transmitting (PAT) tagging).

NURSERY HABITATS

Young tarpon are generally thought to rear in coastal marsh and stream habitats.[36] Of the 12 tarpon examined in this study 4 appeared to live primarily in freshwater during their first year of life (Figure 17.5; RSJ-1, GOM-1, 2, and 4), while the other 8 fish appeared to live primarily in marine waters (Figure 17.5; LN-1, 2, 3, and 4, RSJ-2, 3, and 4, and GOM-3). All of these fish could have been in coastal salt marsh habitats, but the relatively stable high Sr counts taken from material in the first growth increment of several tarpon (Figure 17.5, LN-2 and 3, and RSJ-2) suggest that marine habitats may have been utilized for rearing as well. Determining the nursery environment, freshwater vs. marine, utilized by an individual tarpon appears to be possible. Describing the primary nursery environment of tarpon populations will require a systematic approach to sampling and analysis. This knowledge could eventually lead to the identification of actual nursery locations required to sustain specific adult populations.

Scientists are exploring otolith microchemical techniques to identify nursery areas of certain fish species by linking chemical properties within otolith core material to specific river systems or water bodies. Severin et al.[29] examined the utility of electron microprobe techniques to identify multiple stocks of walleye pollock *Theragra chalcogramma* based on otolith core region chemistry. Thorrold et al.[31] dissolved otoliths of weakfish *Cynoscion regalis* in nitric acid and conducted chemical and isotopic analyses with an inductively coupled plasma mass spectrometer, an instrument capable of detecting very small concentrations of elements, as well as isotopes. Kennedy et al.[17] used micromilling techniques to obtain small volumes of otolith material from Atlantic salmon *Salmo salar* that were subsequently dissolved in nitric acid and the Sr isotope ratios were analyzed in a mass spectrometer. Campbell et al.[6] used a proton microprobe to link absolute Sr concentration in core regions of anadromous Dolly Varden *Salvalinus malma* otoliths with the chemistry of their rearing streams. None of these attempts at chemical signature identification of rearing locations or stock identification have been as simple or utilitarian as the identification of freshwater or marine habitation–based otolith Sr distribution, or Sr:Ca molar ratios. The challenges to identifying the specific rearing areas of tarpon based on otolith chemistry would be great, but it may eventually be possible with careful study.

Adult Habitats

Tarpon sampled in the geographic region associated with the Lake Nicaragua drainage appeared to spend extended periods of their adult lives in freshwater. Tarpon RSJ-2 (Figure 17.5), for example, appeared to enter freshwater after the first year of life and remained there from that point on. This fish was measured at 157 cm fork length and was assumed to have been sexually mature for several years. The Sr values from two other fish from the Lake Nicaragua and Rio San Juan sample groups showed similar patterns of marine life early and what appeared to be exclusively freshwater life later (Figure 17.5, LN-1 and RSJ-3). The question then arises as to where these fish spawned? Is it possible that spawning for some tarpon occurs in freshwater even if 100% of the eggs fail? Or did these fish make such brief spawning migrations into marine water that the band of Sr enhanced material was too narrow to significantly influence the Sr values for individual microprobe points? If mature fish living predominantly in the Rio San Juan drainage were making brief trips to marine water to spawn, a sampling study using acoustic telemetry near the mouth of the river could be designed to detect their outward migrations as prespawning fish, and return migrations as postspawners.

Only one tarpon, GOM-1, from the Gulf of Mexico sample group was of a length that would indicate maturity (Table 17.1). This fish was associated with freshwater during its first year, but it appeared to spend its entire adult life in marine water (Figure 17.5, GOM-1; Figure 17.8). This life history scenario contrasts with the life histories of tarpon examined from the Rio San Juan drainage, in which most fish were associated with marine water early in life and freshwater later (Figure 17.5). If the life history of tarpon GOM-1 was representative of Gulf of Mexico tarpon in general, it would lend support to the localized adult populations hypothesis suggested by Crabtree et al.[7] Carefully designed sampling studies of adult tarpon populations combined with otolith microchemical analyses would be required to determine if regional life history patterns really exist.

SUMMARY

Otolith microchemical techniques appear useful for examining the diadromous migrations of tarpon. With current technology, it is possible to determine whether a tarpon lived in a freshwater or marine environment during its first year of life. Lifetime patterns of migration between freshwater and marine environments appear to be distinct. Knowledge of regional freshwater chemistries will be required to accurately interpret tarpon habitat use based on otolith chemistry. Known life history fish would be extremely useful for verification of Sr concentration patterns in various environments, as identified in this study. Time resolution is obviously greater in younger years than in older years because of the decline in growth increment width with age, resulting in fewer data points per year of growth. Population level deductions will require systematic sampling designs, coupled with otolith microchemical analyses. Identifying nursery areas based on links between the chemistry of otolith core material and the chemistry of specific bodies of water will require further investigation.

ACKNOWLEDGMENTS

This was a cooperative project involving the U.S. Fish and Wildlife Service, Fairbanks Fish and Wildlife Field Office, and the University of Alaska Fairbanks, Advanced Instrumentation Laboratory. We received assistance in one form or another from many other organizations and individuals, including the late Monty Millard of the Co-Ni Foundation; Connie Stevens, Luiz Barbieri, Roy Crabtree, Stu Kennedy, and Bob Muller of the Florida Marine Research Institute; Ivonne Blandon, Rocky Ward, and Britt Bumguardner of the Texas Parks and Wildlife Department; and Gene Huntsman and Allen Andrews. Helpful comments by two anonymous reviewers and by Karin Limburg improved an earlier version of the manuscript. Their assistance was greatly appreciated.

REFERENCES

1. Babaluk, J.A. et al., Evidence for non-anadromous behaviour of Arctic charr (*Salvelinus alpinus*) from Lake Hazen, Ellesmere Island, Northwest Territories, Canada, based on scanning proton microprobe analysis of otolith strontium distribution, *Arctic*, 50, 224–233, 1997.
2. Bath, G.E. et al., Strontium and barium uptake in aragonitic otoliths of marine fish, *Geochim. Cosmochim. Acta*, 64, 1705–1714, 2000.
3. Campana, S.E., Chemistry and composition of fish otoliths: pathways, mechanisms and applications, *Mar. Ecol. Prog. Ser.*, 188, 263–297, 1999.
4. Campana, S.E. and Neilson, J.D., Microstructure of fish otoliths, *Can. J. Fish. Aquat. Sci.*, 42, 1014–1032, 1985.

5. Campana, S.E. et al., Comparison of accuracy, precision, and sensitivity in elemental assays of fish otoliths using the electron microprobe, proton-induced x-ray emission, and laser ablation inductively coupled plasma mass spectrometry, *Can. J. Fish. Aquat. Sci.*, 54, 2068–2079, 1997.
6. Campbell, J.L. et al., Strontium distribution in young-of-the-year Dolly Varden otoliths: potential for stock discrimination, *Nucl. Instrum. Methods Phys. Res. Sect. B*, 189, 185–189, 2002.
7. Crabtree, R.E., Cyr, E.C., and Dean, J.M., Age and growth of tarpon, *Megalops atlanticus*, from South Florida waters, *Fish. Bull.*, 93, 619–628, 1995.
8. Crabtree, R.E. et al., Reproduction of tarpon, *Megalops atlanticus*, from Florida and Costa Rican waters and notes on their age and growth, *Bull. Mar. Sci.*, 61, 271–285, 1997.
9. Degens, E.T., Deuser, W.G., and Haedrich, R.L., Molecular structure and composition of fish otoliths, *Mar. Biol.*, 2(2), 105–113, 1969.
10. de Menezes, M.F. and Paiva, M.P., Notes on the biology of tarpon, *Tarpon atlanticus* (Cuvier & Valenciennes), from coastal waters of Ceara State, Brazil, *Arq. Estac. Biol. Mar. Univ. Fed. Ceara*, 6, 83–98, 1966.
11. Farrell, J. and Campana, S.E., Regulation of calcium and strontium deposition on the otoliths of juvenile tilapia, *Oreochromis niloticus*, *Comp. Biochem. Physiol. A*, 115(2), 103–109, 1996.
12. Goldstein, J.I. et al., *Scanning electron microscopy and x-ray microanalysis*, 3rd ed., Kluwer Academic/Plenum Publishers, New York, 2003.
13. Gunn, J.S. et al., Electron probe microanalysis of fish otoliths-evaluation of techniques for studying age and stock discrimination, *J. Exp. Mar. Biol. Ecol.*, 158, 1–36, 1992.
14. Howland, K.L. et al., Identification of freshwater and anadromous inconnu in the Mackenzie River system by analysis of otolith strontium, *Trans. Am. Fish. Soc.*, 130, 725–741, 2001.
15. Hulet, W.H. and Robins, C.R., The evolutionary significance of the leptocephalus larva, in *Fishes of the Western North Atlantic*, Bohlke, E.B., Ed., Sears Foundation for Marine Research, Yale University, New Haven, 1989, 669–677.
16. Incer, J., Geography of Lake Nicaragua, in *Investigations of the Ichthyofauna of Nicaraguan Lakes*, Thorson, T.B., Ed., University of Nebraska, Lincoln, 1976, 3–7.
17. Kennedy, B.P. et al., Reconstructing the lives of fish using Sr isotopes in otoliths, *Can. J. Fish. Aquat. Sci.*, 59, 925–929, 2002.
18. Kraus, R.T. and Secor, D.H., Incorporation of strontium into otoliths of an estuarine fish, *J. Exp. Mar. Biol. Ecol.*, 302, 85–106, 2004.
19. Limburg, K.E. et al., Flexible modes of anadromy in Baltic sea trout: making the most of marginal spawning streams, *J. Fish Biol.*, 59, 682–695, 2001.
20. Limburg, K.E. et al., Do stocked freshwater eels migrate? Evidence from the Baltic suggests "yes," *Am. Fish. Soc. Symp.*, 33, 275–284, 2003.
21. Mugiya, Y. and Tanaka, S., Incorporation of water-borne strontium into otoliths and its turnover in the goldfish *Carassius auratus*: effects of strontium concentrations, temperature, and 17b-estradiol, *Fish. Sci.*, 61(1), 29–35, 1995.
22. Platt, C. and Popper, A.N., Fine structure and function of the ear, in *Hearing and Sound Communication in Fishes*, Tavolga, W.N., Popper, A.N., and Fay, R.R., Eds., Springer-Verlag, New York, 1981, 3–36.
23. Potts, P.J., *A handbook of silicate rock analysis*, Chapman & Hall, London, 1987.
24. Reed, S.J.B., *Electron microprobe analysis*, 2nd ed., Cambridge University Press, Cambridge, 1997.
25. Sadovy, Y. and Severin, K.P., Elemental patterns in red hind (*Epinephelus guttaltus*) otoliths from Bermuda and Puerto Rico reflect growth rate, not temperature, *Can. J. Fish. Aquat. Sci.*, 51, 133–141, 1994.

26. Secor, D.H., Dean, J.M., and Laban, E.H., Otolith removal and preparation for microstructural examination, in *Otolith Microstructure Examination and Analysis*, Stevenson, D.K. and Campana, S.E., Eds., *Can. Spec. Publ. Fish. Aquat. Sci.*, 117, 19–57, 1992.
27. Secor, D.H., Henderson-Arzapalo, A., and Piccoli, P.M., Can otolith microchemistry chart patterns of migration and habitat utilization in anadromous fishes? *J. Exp. Mar. Biol. Ecol.*, 192, 15–33, 1995.
28. Secor, D.H. and Rooker, J.R., Is otolith strontium a useful scalar of life cycles in estuarine fishes? *Fish. Res.*, 46, 359–371, 2000.
29. Severin, K.P., Carrol, J., and Norcross, B.L., Electron microprobe analysis of juvenile walleye pollock, *Theragra chalcogramma*, otoliths from Alaska: a pilot stock separation study, *Environ. Biol. Fish.*, 43, 269–283, 1995.
30. Smith, D.G., Introduction to leptocephali, in *Fishes of the Western North Atlantic*, Bohlke, E.B., Ed., Sears Foundation for Marine Research, Yale University, New Haven, 1989, 657–668.
31. Thorrold, S.R. et al., Accurate classification of juvenile weakfish *Cynoscion regalis* to estuarine nursery areas based on chemical signatures in otoliths, *Mar. Ecol. Prog. Ser.*, 173, 253–265, 1998.
32. Tzeng, W., Effects of salinity and ontogenetic movements on strontium:calcium ratios in the otoliths of the Japanese eel, *Anguilla japonica* Temminck and Schlegel, *J. Exp. Mar. Biol. Ecol.*, 199, 111–122, 1996.
33. Tzeng, W. et al., Strontium bands in relation to age marks in otoliths of European eel *Anguilla anguilla*, *Zool. Stud.*, 38, 452–457, 1999.
34. Tzeng, W., Severin, K.P., and Wickstrom, H., Use of otolith microchemistry to investigate the environmental history of European eel *Anguilla Anguilla*, *Mar. Ecol. Prog. Ser.*, 149, 73–81, 1997.
35. Wade, R.A., The biology of the tarpon, *Megalops atlanticus*, and the ox-eye, *Megalops cyprinoides*, with emphasis on larval development, *Bull. Mar. Sci. Gulf Caribb.*, 12, 545–622, 1962.
36. Zale, A.V. and Merrifield, S.G., Species profiles: life histories and environmental requirements of coastal fishes and invertebrates (South Florida)—ladyfish and tarpon, *U.S. Fish and Wildl. Serv. Biol. Rep.*, 82(11.104), TR EL-82-4, 1989.
37. Zimmerman, C.E., Relationship of otolith strontium-to-calcium ratios and salinity: experimental validation for juvenile salmonids, *Can. J. Fish. Aquat. Sci.*, 62, 88–97, 2005.

18 Seasonal Migratory Patterns and Vertical Habitat Utilization of Atlantic Tarpon (*Megalops atlanticus*) from Satellite PAT Tags

Jiangang Luo, Jerald S. Ault, Michael F. Larkin, Robert Humston, and Donald B. Olson

CONTENTS

Introduction 275
Methods 277
    Pop-Up Archival Transmitting Tags 277
Results 278
    Regional Deployments of Pop-Up Archival Transmitting Tags 278
    Pop-Up Locations and Data Retrieval 279
    Summary of Seasonal Migration Patterns 281
    Vertical Thermohabitat Utilization 286
        Tarpon PAT T-03 286
        Tarpon PAT T-24 290
    Insights from New Pop-Up Archival Transmitting Technology 293
Discussion 296
Acknowledgments 298
References 298

INTRODUCTION

Anglers have flocked to the U.S. southeast Atlantic and Gulf of Mexico coastlines since the mid-1800s to catch Atlantic tarpon (*Megalops atlanticus*), the "silver kings" of the sea (Oppel and Meisel, 1987). The U.S. tarpon fishery is a fundamental component of the multibillion dollar regional recreational marine fishing and tourism industry. For many a veteran angler, the pinnacle of a fishing career is achieved with the explosion of chromed muscle that signifies his first tarpon hook-up. The experience was once common on the Texas coast, so common, in fact, that through the 1950s, tarpon tournaments were widespread and presidents and potentates made the journey to the Gulf coast to catch a "silver king." Then the Texas tarpon stock seemed to disappear. By the early 1970s, the much sought-after sportfish were rarely encountered off Texas. There has been serious speculation about reasons for the declines, and whether they could occur elsewhere.

Over the past 40 years, many experienced anglers throughout the southeast region have noted progressive declines in the quality of tarpon fishing at other favored fishing spots. Rapidly growing human populations are placing tarpon under increasingly intense exploitation pressures, and associated coastal development has negatively impacted their sensitive nursery habitats. While U.S. fisheries for tarpon are predominantly catch and release, sources of mortality such as capture for human and animal consumption are believed to be substantial south of U.S. borders. The tarpon's migratory range documented in this study places the resource out of U.S. waters for certain periods of the year, and thus the stock is subjected to unregulated harvest during this period of each year. As such, tarpon are prized for their roe in Mexico, and Mexican waters may be the spawning source of juvenile tarpon found in Texas and Louisiana. Because tarpon movements and migrations are poorly understood, there is concern that the stocks U.S. anglers target during the early summer to fall seasons are the same fish that inhabit Caribbean and central American waters during the remainder of the year, where tarpon are harvested for commercial markets or subsistence.

From 1960 to 1999, more than 10,000 tarpon were fitted with anchor tags and released. However, less than 200 recaptures of these tagged fish have been reported (unpublished NMFS data). This relatively low reporting rate can be attributed in part to few reports from neighboring foreign waters where anglers are often unaware of mark-recapture experiments initiated in the United States, from capture by foreign nationals who cannot read tag information imprinted in English, or by those who are simply unable to report a recaptured tag. Those recaptures reported have indicated limited exchange of tarpon between U.S. coastal areas and Caribbean waters, but in reality, these data allow only limited inferences on movement behavior and migrations, restricted to the start and end points of the time at liberty. To improve the basis of fishery-management decision making for tarpon, there is clearly a great need to conduct focused research concerning tarpon migration paths, timing of movement events related to feeding and reproduction, and the degree of stock connectivity, particularly as related to the distribution of regional fishing pressures.

The unit stock for tarpon remains unknown, but is perhaps the most critical piece of information needed for sound fishery management and conservation decision making. Satellite-based pop-up archival transmitting (PAT) tags provide a unique method for tracking large-scale movement and migration patterns of tagged fish and defining the unit stock when fishery-dependent recaptures or tag-reporting rates are suspected to be low. PAT tags do not rely on labor-intensive individual tracking methods. Archived data from PAT tags are returned by transmissions to orbiting satellites, thus they do not require specific recapture of a tagged fish. PAT tag technologies have facilitated an improved understanding of migrations, behavior patterns, and environmental preferences (i.e., ocean "habitat" utilization) for a number of large marine fishes, including tunas (Block et al., 1998a,b,c, 1999; Lutcavage et al., 1999; Block et al., 2001, 2003, 2005); billfish (Sedberry and Loefer, 2001; Graves et al., 2002; Kerstetter et al., 2003; Luo et al., 2006; Prince and Goodyear, 2006); sharks (Boustany et al., 2002; Loefer et al., 2005); and the ocean sunfish (*Mola mola*) (Seitz et al., 2002). In 2001, we began a research program using PAT technology to assess movements and habitat-use patterns of Atlantic tarpon. The goal of this research was to monitor and assess seasonal migration patterns and stock connectivity, and to define vertical thermohabitat utilization of tarpon in the western Atlantic Ocean and Gulf of Mexico.

METHODS

POP-UP ARCHIVAL TRANSMITTING TAGS

The PAT tags used in this study were manufactured by Wildlife Computers (www.wildlifecomputers.com). Electronic components of the tags are fully cast in a ceramic tube measuring 21 mm in diameter, and the added float measured approximately 40 mm in diameter at its widest point. The cast tube and float are tested and confirmed to withstand 2000 m of pressure. Overall length of the tag (not including the antenna) is 175 mm and total weight in air is 75 g (Figure 18.1).

All tarpon were angled on the heaviest hook-and-line tackle feasible to reduce time from hook-up to release. At boatside, tarpon were guided into a specially designed sling, which remained in the water to prevent injury to the tarpon (Figure 18.1). Fork length and dorsal girth were measured to the nearest centimeter. The PAT tag was attached to the animal via a tether and dart-type anchor in the dorsal-lateral area. During deployment, tags archived time-referenced data on depth, temperature, and light level. Since one of our principal objectives was to characterize migratory patterns of tarpon contributing to the U.S. recreational fishery, we directed tagging efforts within the putative range of this stock by tagging tarpon in coastal waters off Florida, Texas, and Mexico. However, we also tagged tarpon opportunistically over a broad geographic area to maximize spatial distribution of tag deployments.

To draw inferences on the behavior of the tarpon, an "activity index" was calculated based on depth data obtained from recovered PAT tags. Daily values of the activity index were calculated as a function of the number of instances in a day's record when successive depth measurements collected at 1-min intervals differed by more than 2 m (tag accuracy is ±1 m). We converted depth data into a binary series that identified depth changes between successive time points as either 0 (<2 m) and 1 (≥2 m). Daily activity index was then calculated as the percent of active time (fraction of ones) during each day of deployment.

We also analyzed variations in depth and light intensity to draw inferences on ambient turbidity and light attenuation. The coefficient of light attenuation k was calculated as

$$k = -\frac{(\ln(L_2) - \ln(L_1))}{(d_2 - d_1)}, \tag{18.1}$$

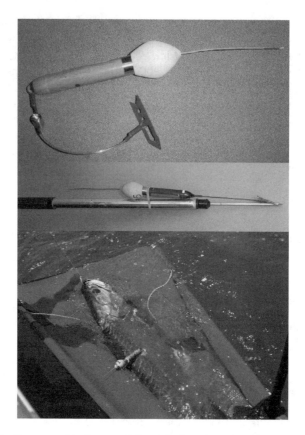

FIGURE 18.1 Pop-up archival transmitting (PAT) technology showing the PAT tag and the process of fitting a tarpon with the tag prior to release. Note the blue "tarpon sling" we specifically designed to reduce trauma to the fish during the tagging process.

where L_1 and L_2 are the light levels at depths d_1 and d_2 for each 1 min interval when depth changes were greater than 2 m. We calculated the k for each daytime data interval and then calculated an average daily value of light attenuation. Given an estimate of light attenuation, we calculated visibility distance based on the empirical equation of Man'kowski (1978),

$$Z_{vis} = \frac{33.5}{10^{\wedge}k}, \qquad (18.2)$$

where Z_{vis} is the visibility distance in meters.

RESULTS

REGIONAL DEPLOYMENTS OF POP-UP ARCHIVAL TRANSMITTING TAGS

Since September 2001, 48 Atlantic tarpon have been tagged with satellite PAT tags at locations ranging from North Carolina, South Carolina, Georgia, Florida,

Louisiana, and Texas, and at international locations such as Veracruz and Coatzacoalcos, Mexico, and Trinidad, British West Indies (Table 18.1; Figure 18.2). The research began in fall 2001 along the southeastern coast of the United States, placing four PAT tags on tarpon at Oriental, North Carolina; Hilton Head, South Carolina; Savannah, Georgia; and Stuart, Florida. In 2002, five tags were deployed in Florida with the help of sportsmen along the west coast from Tampa Bay to the Florida Keys. In 2003, seven tags were deployed in the western Gulf of Mexico: two at Veracruz, Mexico; three at Venice, Louisiana; and two in the eastern Gulf of Mexico at Boca Grande, Florida. To further refine the migratory pathways for tarpon out of Mexico and into the U.S. waters, we returned to Veracruz in May 2004 and 2005, and tagged seven and six tarpon, respectively. The timing of these trips coincided with the May full moon when large tarpon were abundant. In 2006, 18 tags were deployed. These included four tags in the Florida Keys; two tags at Galveston, Texas; one tag at Port O'Connor, Texas; five tags at Veracruz, Mexico; three tags at Coatzacoalcos, Mexico; and three tags at Port of Spain, Trinidad, British West Indies.

POP-UP LOCATIONS AND DATA RETRIEVAL

Over the 5 years of study, we have had mixed success in tag reporting via ARGOS satellites (Table 18.2; Figure 18.2). Overall, 25 PAT tags have successfully popped off and transmitted data that were received by the ARGOS satellite network and transmitted to us. Eleven of the fish tagged since May 2006 are scheduled to pop off (October 2006 to May 2007) beyond the date of finishing this chapter. Twelve of 48 tags (25%) we have deployed were never heard from after their scheduled pop-off dates. However, this situation has improved markedly over time due to innovations in tagging and sling construction. From 2001 to 2002, only 4 out of 10 tags (40%) were heard. From 2003 to 2005, 14 out of 20 tags (70%) were heard. During this time, we also physically recovered six tags. We recovered one of the tags when we searched a beach area indicated by the coordinate position of the tag from ARGOS. The other five tags were found and returned to us by beachcombers.

The minimum distances (net displacements) moved by the tagged tarpon between tagging and pop-off locations were estimated as the along coastline distance, shown in Table 18.2. In general, the distance traveled between the tagging and the pop off increased as the number of days that the PAT tag remained on the fish increased (Figure 18.3A). In a few cases, the tagged fish did not travel very far (≤ 100 km). By and large, these were relatively small and immature tarpon that weighed less than 50 kg (Figure 18.3B). We believe that these resulted from tag detachment from the fish shortly after tagging possibly due to predation or tagging induced mortality (such as T-02, T-17). An exception was tarpon T-26 for which we received sufficient data indicating the fish was alive on the pop-off date, and which could be explained by unseasonably warm temperatures in the region for the period of time the tag was on the fish (see next section for more details). The average speed of movement over the duration of deployments ranged from 0 to 52.6 km day^{-1} (Table 18.2), with a mean of 12.6 km day^{-1}. Net displacements between point of deployment and tag jettison location ranged from 5 to 1730 km, with a mean of 540 km.

TABLE 18.1
Atlantic Tarpon PAT Tag Deployment in the United States, Mexico, and Trinidad during 2001–2006

PAT Name	Tag Location	W (kg)	Tag Date	Lat (dd)	Lon (dd)	Setup Pop-Off Date
T-01	Hilton Head, South Carolina	38.6	09/06/01	32.3313	−80.7486	01/03/02
T-02	Oriental, North Carolina	45.5	09/03/01	35.0489	−76.5328	03/04/02
T-03	Savannah, Georgia	36.4	09/21/01	31.7023	−81.0799	11/04/01
T-04	Stuart, Florida	34.1	09/28/01	27.0012	−80.1399	02/15/02
T-05	Islamorada, Florida	40.9	05/18/02	24.8421	−80.0709	08/05/02
T-06	Long Key, Florida	34.1	05/28/02	24.7976	−80.8675	12/02/02
T-07	Long Key, Florida	34.1	05/28/02	24.7976	−80.8675	10/24/02
T-08	Boca Grande, Florida	61.4	06/13/02	26.6991	−82.2581	12/15/02
T-09	Tampa Bay, Florida	79.5	06/25/02	27.6069	−82.7650	12/15/02
T-11	Tampa Bay, Florida	40.9	06/27/02	27.6054	−82.7643	06/15/03
T-12	Boca Grande, Florida	36.4	06/06/03	26.6992	−82.2697	09/01/03
T-13	Boca Grande, Florida	45.5	06/07/03	26.6992	−82.2697	10/01/03
T-14	Venice, Louisiana	38.6	09/04/03	28.9621	−89.1998	11/10/03
T-15	Veracruz, Mexico	90.0	05/11/04	19.3657	−96.2744	07/14/04
T-16	Veracruz, Mexico	90.0	05/10/04	19.3407	−96.2867	08/11/04
T-17	Veracruz, Mexico	85.0	05/10/04	19.3767	−96.2890	07/21/04
T-18	Veracruz, Mexico	65.9	06/12/03	19.3200	−96.2706	08/15/03
T-19	Veracruz, Mexico	61.4	06/15/03	19.3175	−96.2556	12/15/03
T-20	Veracruz, Mexico	78.0	05/10/04	19.3767	−96.2890	07/28/04
T-21	Veracruz, Mexico	72.2	05/28/05	19.3280	−96.2943	07/15/05
T-22	Veracruz, Mexico	60.0	05/29/05	19.3598	−96.2896	07/31/05
T-23	Veracruz, Mexico	84.1	05/29/05	19.3598	−96.2896	08/15/05
T-24	Veracruz, Mexico	55.0	05/10/04	19.3767	−96.2890	09/08/04
T-25	Veracruz, Mexico	78.0	05/10/04	19.3767	−96.2890	08/18/04
T-26	Venice, Louisiana	34.1	09/04/03	28.9603	−89.2042	11/10/03
T-27	Venice, Louisiana	27.3	09/04/03	28.9452	−89.2000	02/10/04
T-28	Veracruz, Mexico	85.5	05/29/05	19.3598	−96.2896	08/31/05
T-30	Veracruz, Mexico	80.0	05/11/04	19.3657	−96.2744	09/01/04
T-31	Ocean Reef, FL	31.8	05/10/06	25.3333	−80.2667	10/30/06
T-32	Veracruz, Mexico	78.1	05/28/05	19.3228	−96.2771	09/15/05
T-33	Veracruz, Mexico	48.3	05/29/05	19.3598	−96.2896	09/30/05
T-35	Bahia Honda Bridge, Florida	39.6	05/15/06	24.6583	−81.2633	10/01/06
T-36	Bahia Honda Bridge, Florida	21.8	05/15/06	24.6583	−81.2633	10/01/06
T-37	Galveston, Texas	43.1	08/05/06	29.1760	−94.9127	12/25/06
T-38	Key Largo, Florida	29.5	06/01/06	25.2765	−80.2899	10/04/06
T-39	Veracruz, Mexico	78.1	05/27/06	19.3471	−96.2909	10/20/06
T-40	Galveston, Texas	48.8	08/05/06	29.1760	−94.9127	01/08/07
T-41	Veracruz, Mexico	42.0	05/28/06	19.3715	−96.2950	11/03/06
T-42	Veracruz, Mexico	63.2	05/28/06	19.3440	−96.2963	11/10/06
T-43	Veracruz, Mexico	79.4	05/28/06	19.3695	−96.2945	11/06/06
T-44	Veracruz, Mexico	56.5	05/28/06	19.3687	−96.2954	11/15/06

(Continued)

TABLE 18.1 (continued)
Atlantic Tarpon PAT Tag Deployment in the United States, Mexico, and Trinidad during 2001–2006

PAT Name	Tag Location	W (kg)	Tag Date	Lat (dd)	Lon (dd)	Setup Pop-Off Date
T-45	Trinidad, British West Indies	35.8	08/15/06	10.6863	−61.7140	04/02/07
T-46	Trinidad, British West Indies	54.1	08/15/06	10.6776	−61.7213	02/05/07
T-47	Trinidad, British West Indies	41.1	08/16/06	10.6888	−61.7106	05/02/07
T-48	Port O'Connor, Texas	53.3	09/10/06	28.4058	−96.3946	01/08/07
T-49	Coatzacoalcos, Mexico	42.8	09/30/06	18.1866	−94.4356	01/22/07
T-50	Coatzacoalcos, Mexico	41.7	09/30/06	18.1866	−94.4311	02/12/07
T-51	Coatzacoalcos, Mexico	43.6	09/30/06	18.1822	−94.4358	02/26/07

Note: Tag T-29 and T-34 were not deployed due to battery failures; T-10 was used in a tank geolocation experiment. A total of 48 tags have been deployed to date.

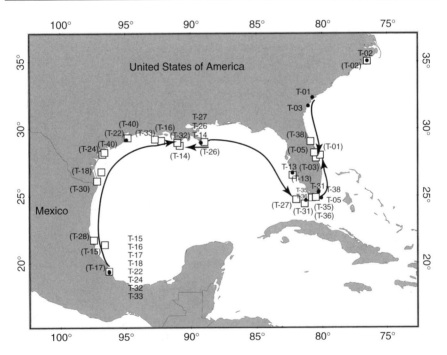

FIGURE 18.2 Locations of tarpon PAT deployments (solid dots) and pop-offs (open squares) in the southeast United States and Gulf of Mexico. Diagrammatic migratory paths are indicated by the arrowed curves.

SUMMARY OF SEASONAL MIGRATION PATTERNS

Conventional anchor tags recaptures of tarpon (unpublished NMFS data) discussed previously provided limited information on movement behaviors and migration patterns of tarpon between the U.S. coastal areas and Caribbean waters. The general

TABLE 18.2
Results of Atlantic Tarpon PAT Tag Deployments during 2001–2006

PAT Name	Actual Pop-Off Date	Pop-Off Lat (dd)	Pop-Off Lon (dd)	Pop-Off Location	Days on Fish	Distance along Shoreline (km)	Mean Speed (km/d)	Min–Max Temp	Max Depth (m)
T-01	01/03/02	28.006	−80.174	Melbourne, Florida	119	500	4.2	22–27	56
T-02	03/04/02	35.001	−76.551	Oriental, North Carolina	182	5	0.0	d	d
T-03[a]	11/04/01	27.877	−80.478	Sebastian Inlet, Florida	44	460	10.5	21–27	25
T-04	Never heard								
T-05[a]	08/05/02	28.180	−80.630	Merritt Island, Florida	79	440	5.6	25–33	16
T-06[a]	Never heard			Melbourne, Florida					
T-07	Never heard								
T-08	Never heard								
T-09	Never heard								
T-11	Never heard								
T-12	Never heard								
T-13[a]	07/01/03[b]	26.529	−82.215	Captiva, Florida	24	20	0.8	26–33	11
T-14	09/22/03[b]	28.720	−90.840	Timbalier Island, Louisiana	18	160	8.9	27–30	40
T-15[a]	06/08/04[b]	21.350	−96.617	Tampico, Mexico	28	230	8.2	21–33	79
T-16	08/11/04	29.087	−92.246	Marsh Island, Florida	93	1580	17.0	d	d
T-17	07/10/04[b]	19.375	−96.289	Veracruz, Mexico	61	15	0.2	d	d
T-18	07/09/03[b]	26.789	−96.873	Corpus Christi, Texas	27	900	33.3	d	d
T-19	Never heard								
T-20	Never heard								
T-21	Never heard								
T-22	07/08/05[b]	29.303	−94.818	Galveston, Texas	40	1290	32.3	d	d
T-23	Never heard								
T-24[a]	09/08/04	28.092	−96.788	Port Aransas, Texas	121	1070	8.8	20–34	48

Tag	Pop-off date	Latitude	Longitude	Location					
T-25	Never heard								
T-26	11/10/03	28.991	−88.939	Venice, Louisiana	67	15	0.2	22–31	40
T-27	02/10/04	24.698	−81.996	Key West, Florida	159	1160	7.3	20–29	28
T-28	07/07/05[b]	21.693	−97.463	Tampico, Mexico	39	310	7.9	22–31	44
T-30	06/06/04[b]	26.100	−97.200	Port Isabel, Texas	26	820	31.5	[d]	[d]
T-31	06/05/06[b]	24.839	−80.802	Long Key, Florida	26	85	3.3	24–31	4
T-32	09/07/05[b]	28.931	−91.014	Timbalier Island, Louisiana	102	1730	17.0	23–33	24
T-33	08/28/05[b]	29.196	−92.733	Oak Grove, Louisiana	91	1530	16.8	22–34	28
T-35	05/20/06[b]	24.392	−81.370	Big Pine Key, Florida	1	30	30.0	26–28	0
T-36	06/07/06[b]	24.864	−80.504	Islamorada, Florida	18	50	2.8	26–32	8
T-37	12/25/06	[c]							
T-38	07/01/06[b]	29.020	−80.911	Edgewater, Florida	26	440	16.9	16–31	16
T-39	10/20/06	[c]							
T-40	08/13/06[b]	28.225	−96.634	Port O'Conner, Texas	4	210	52.5	31–32	16
T-41	11/03/06	[c]							
T-42	11/10/06	[c]							
T-43	11/06/06	[c]							
T-44	11/15/06	[c]							
T-45	09/26/06[b]	13.207	−63.269	St. Vincent	37	290	7.8	21–31	48
T-46	09/17/06[b]	12.696	−63.529	Grenada	29	165	5.7	20–31	88
T-47	05/02/07	[c]							
T-48	01/08/07	[c]							
T-49	01/22/07	[c]							
T-50	02/12/07	[c]							
T-51	02/26/07	[c]							

[a] Tag was recovered.
[b] Premature release.
[c] Scheduled future pop-off date.
[d] No temperature and depth data available.

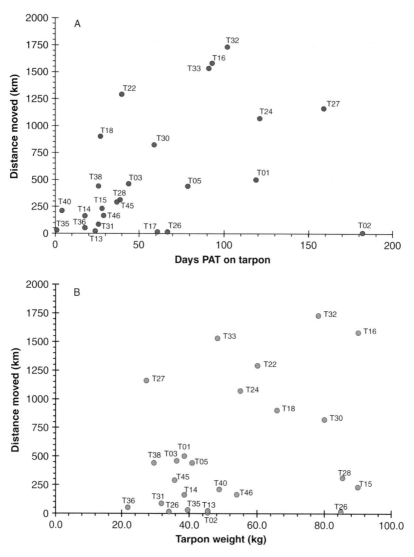

FIGURE 18.3 (A) Minimum distances traveled by PAT-tagged tarpon as function of number of days tag was on the fish; (B) Minimum distance traveled as a function of size (weight in kilograms).

hypothesis that emerged was that tarpon migrated north in spring, and then south in fall and winter in both Atlantic and Gulf waters. Our results from the 25 successful PAT tag deployments support the hypothesis with four, and perhaps five, detailed seasonal migratory patterns (Figure 18.2; Table 18.2).

First, in the fall along the southeast U.S. Atlantic coast, southerly migratory behavior of tarpon was indicated by two tags (T-01, T-03). Tag T-03 was placed on a 36.4-kg tarpon on September 21, 2001, at Savannah, Georgia. The tag popped off on schedule on November 4, 2001, about 460 km south near Sebastian Inlet,

Florida. Once the tag popped off, we found that the ARGOS system provided relatively precise geographic latitudinal location data on the tag (www.argosi-system.org) that allowed us to find the tag on the beach within only a few 100 m (usually the best ARGOS position is ±100 m) of the location provided by ARGOS. This first recovered tag provided over 64,000 individual minute-by-minute records on water temperature, depth, and light levels over the entire 43.5 days deployment. We found that on its daily migration route this fish generally remained in 26°C water (78°F; range: 71–81°F) over a depth range 0–25 m (mean 10 m) as it migrated south at about 11 km day^{-1}. Southerly movements were accelerated by the passage of cold fronts. Similarly, tag T-01 released at Hilton Head, South Carolina, on September 6, 2001, also popped off at Sebastian Inlet, Florida, on January 3, 2002, within a few miles of where we located tag T-03.

The second seasonal pattern, a spring to summer northward migration along Florida's Atlantic coast, was indicated by one tarpon tagged (T-05) at Islamorada, Florida, on May 18, 2002, that popped off on August 4, 2002, near Cape Canaveral, Florida. This pattern was recapitulated by another tarpon (T-38) tagged in Key Largo, Florida, on June 1, 2006, that popped-off prematurely on July 1, 2006, at Edgewater, Florida, about 50 mi north of Cape Canaveral (Figure 18.2; Table 18.2). Both T-05 and T-38 tags popped off in the Intercoastal Waterway. T-05 was found by a resident of Merritt Island, Florida, in the seawall behind her home. This tarpon had traveled some 440 km north from the point of release in 79 days. This recovered tag provided 112,761 data points (time, temperature, depth, and light level) during the 2.5 months (79 days). T-38 also traveled 440 km, but did so in only 26 days, for an average speed of 16.9 km day^{-1}.

The third, a winter migration from Venice, Louisiana, to Key West, Florida, in the eastern Gulf of Mexico, was observed for one fish (Figure 18.2; Table 18.2). On September 4, 2003, we tagged three tarpon (T-27, T-26, and T-14) near Venice, Louisiana. Subsequently, T-14 popped off on October 10 about 160 km southwest of the release on the continental shelf; and, a second tag (T-26) popped off on November 10, within 15 km of the release point. We believe that the first two fish had not yet begun their seasonal southward migrations because ocean water temperatures recorded by a National Oceanic and Atmospheric Administration (NOAA) buoy remained unseasonably warm (>24°C) through early December 2003. However, on February 10, 2004, tag T-27 popped off near Key West, Florida. If we assume the tarpon left Venice on December 10, based on when SSTs dipped below 24°C (NOAA, Buoy data), then this tarpon traveled 1160 km in 60 days for an average speed of about 20 km day^{-1}.

The fourth, a summer northward migratory pattern, was observed for nine tagged fish in the western Gulf of Mexico, originating from Veracruz, Mexico, and extending to at least Texas and Louisiana, United States (Figures 18.2; Table 18.2). In May 2003, June 2004, and June 2005 we tagged 15 large tarpon about 3–5 km offshore of the Antigua River just north of Veracruz. Five of these tags never transmitted any data. One tag (T-17) popped off after 61 days near where it was deployed. The other nine tags popped off at various distances along the western Gulf of Mexico coast ranging from 230 (T-15) to 1730 km (T-32) north of Veracruz (Figure 18.2). Migration speeds ranged from 8 to 33 km day^{-1}, with an average of 17.2 km day^{-1}. Detailed results are presented in Table 18.2.

More recently, of several PAT tags deployed near the Island of Trinidad, British West Indies, in August 2006, two popped off prematurely from 165 (Grenada) to 290 (St. Vincent) km north of the Trinidad release points after only 1 month on fish. While it is far too early to make any broad or general statements concerning the migratory patterns of tarpon in the southern Caribbean Sea region, the general northward movement along the southern Lesser Antilles Island chain may suggest some level of stock connectivity with the northern Caribbean Sea.

VERTICAL THERMOHABITAT UTILIZATION

PAT tags transmitted via ARGOS satellites summarized depth and temperature histograms and depth and temperature profiles. In the early years of our study, the amounts of transmitted data were greatly limited due to tag battery failures and other factors. From 2001 to 2004, of the 15 PAT tags that transmitted geolocations, 5 of these did not transmit any depth or temperature data (Table 18.2). Since 2005, the PAT technology has improved significantly. From 2005 to 2006, of the 10 tags that transmitted geolocations, only one of these failed to transmit any depth and temperature data. The overall temperature range recorded by tags was 16–34°C, and the maximum recorded depth was 88 m (Table 18.2).

We have been fortunate to have physically recovered six PAT tags. The high-resolution depth, temperature, and light-level data obtained in these cases indicated that five out of six tags stayed on the respective fish for time periods ranging from 24 to 121 days. One tag (T-06) stayed on the fish for less than 1 day, as indicated by constant depths for rest of the duration, which may be an instance of predation mortality resulting from a shark attack postrelease. Results from other five recovered tags revealed detailed vertical and thermohabitat utilization patterns of tarpon along their seasonal migratory journey. Here, results from two tags are summarized.

Tarpon PAT T-03

PAT tag T-03 was placed on a 36.4-kg tarpon on September 21, 2001, at Savannah, Georgia, and subsequently popped off on November 4, 2001, a duration of 44 days. The vertical thermohabitat utilization of that tarpon is visualized in Figure 18.4, where the minute-by-minute depth distribution is indicated by the dashed white line. A 6-h summary depth histogram indicated by open circles shows percent of time the tarpon spent at each depth bin. The maximum depth during the deployment period was 25 m, with an average of 10.2 m. Distinct diel vertical movement patterns were observed on some days, but not on others (Figure 18.4). These patterns are much clearer in close-up time-restricted plots (Figure 18.5). The tarpon's general diel vertical movement pattern was to remain shallow during the day and then venture to depths at night (Figure 18.5A). However, during full moon periods the tarpon was active in the water column during both day and night (Figure 18.5B). The background color of Figure 18.4 shows vertical thermohabitat preferences of the tarpon, generated from the minute-by-minute temperatures recorded by the PAT tag. Water temperatures recorded by the tag ranged from 21.5 to 27.1°C (Figure 18.6A) with a mean of 24.8°C over the duration of the deployment. However, examining the time series in detail indicates that the fish experienced three distinct temperature regimes

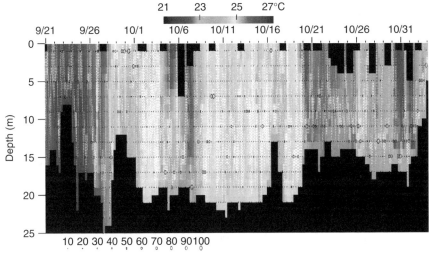

FIGURE 18.4 (Color Figure 18.4 follows p. 262.) Summary map of vertical thermohabitat utilization by a 36.4-kg PAT-tagged tarpon generated with minute-by-minute depth and temperature data from a recovered PAT tag (T-03) for fish that migrated from Savannah, Georgia to Sebastian Inlet, Florida. Temperatures are displayed in color scale ranging from 21 to 27°C. Depth is on the y-axis. Time is on the x-axis running from September 21 to November 3, 2001. Size of open circles indicates percentage of time the tarpon spent at each depth during each 6-h interval. The dashed white line indicates the minute-by-minute depth position of the fish.

during different time periods. From September 21 to 29, the modal water temperature recorded was 26°C with a ±1°C range. From September 30 to October 26, modal temperature was 25°C with a ±1°C range. Finally, from October 27 to November 4, modal temperature shifted to 23°C with a similar ±1°C range.

Comparison of water temperatures recorded at NOAA buoys and C-MAN stations along the coast from Georgia to Florida (Figures 18.6B and 18.6C), and temperature data recorded by PAT T-03 revealed greater detail on environmental factors influencing the timing of the tarpon's southward migration. Water temperatures at the Gray's Reef buoy (B41008), and the St. Augustine, Florida, C-MAN station (SAUF1) were very similar during the period of interest (Figure 18.6B), despite that SAUF1 is 180 km south of B41008. From September 21 to October 6, temperatures on the PAT tag were similar to those recorded at SAUF1 and B41008 (Figure 18.6B), indicating that the tarpon was likely in the area between these two stations at that time. However, after October 6, temperatures from the tag were higher than those recorded at SAUF1 and B41008, indicating that the tarpon had moved south of SAUF1. Between October 6 and 25, temperatures on the tag were between those recorded at SAUF1 and the Cape Canaveral, Florida buoy (B41009) (Figure 18.6C), indicating that the tarpon was located between these two stations. On October 26, temperatures on the tag were greater than temperatures recorded at the Cape Canaveral buoy and less than those recorded at the Lake Worth, Florida C-MAN

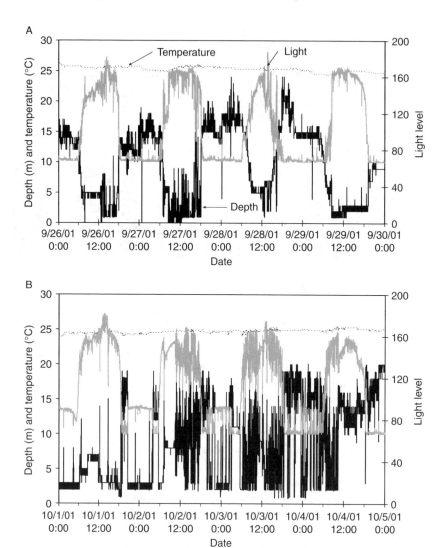

FIGURE 18.5 Examples of minute-by-minute depth (solid black line), temperature (dashed line) and light level (light gray) data recorded by the PAT tag (T-03) for the period of September 26–30, 2001 (A), and for the period October 1–5, 2001 (B). The light level is indicated by the gray line with the second y-axis in relative logarithm unit.

station (LKWF1), indicating that the tarpon had likely moved south of Cape Canaveral, Florida. One day later (October 27), in a 2-h period from approximately 19:00 to 21:00 p.m., the temperature recorded by the tag decreased sharply (below temperatures at B41009). During this time the tarpon stayed mainly shallow (<5 m; Figure 18.4). At 21:30 p.m., the fish dove to depths >10 m, and this event was accompanied by a sharp increase in temperature. The low temperatures of the surface waters suggest the presence of a cool freshwater layer atop the warmer saline waters

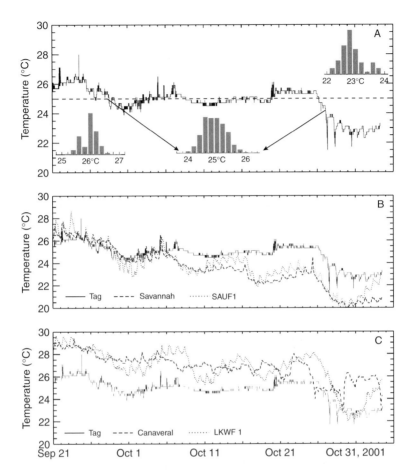

FIGURE 18.6 Water temperatures recorded by tarpon (T-03) with inserts of histogram distribution for three time periods (A). Comparison of temperature recorded on the tag and at Gray's Reef buoy (B42008) of Savannah, Georgia and C-MAN station (SAUF1) at St. Augustine, Florida (B), and buoy (B41009) at Cape Canaveral, Florida and C-MAN station (LKWF1) Lake Worth, Florida (C).

below. This would indicate that the tarpon might be in the outfall plume of the Indian River Lagoon inlets (e.g., Sebastian or St. Lucie) during this event.

From the minute-by-minute depth and light data obtained from T-03, we were also able to calculate the tarpon's daily activity index and apparent water visibility (Figure 18.7). The activity index ranged from 2 to 25% during the deployment period. Baseline values of the daily activity index were generally around 5% with frequent increases to 15% or greater (Figure 18.7A). Peaks in daily activity generally occurred when visibility was at or near 15 m (Figure 18.7B). The data indicate that tarpon activity was strongly influenced by the passage of cold fronts, as revealed by wave height data recorded at Savannah and Cape Canaveral (Figure 18.7C). Lowest activity occurred during the passing of cold fronts, which were marked by increased wave energy and lowered water visibility (Figures 18.7B and 18.7C).

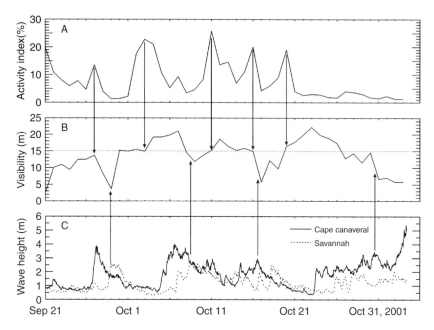

FIGURE 18.7 Fish activity (A), water visibility (B) calculated from light levels and depth profiles recorded on the tag, and wave height (C) at two buoys (B41008, B41009). All high activities occurred around 15-m visibility, and all low visibility and low activities followed cold fronts (high waves).

Tarpon PAT T-24

A 55-kg (121-lb) tarpon tagged on May 10, 2004, at Veracruz, Mexico, popped off as scheduled on September 8, 2004—a duration of 121 days (T-24). The tag was found on Port Aransas Beach, Texas, that same day and subsequently returned to us. Tag T-24 produced over 177,000 individual records of depth, water temperature, light level, and time during the 4-month deployment. The vertical and thermohabitat utilization of the tarpon is visualized in Figure 18.8. The maximum depth during the deployment was 48 m (157 ft), with an average of 6.3 m. Diel vertical movement patterns were observed. For example, from June 15 to 19, at first light the tarpon began its dive into deeper waters and remained down during all the sunlit hours until dusk, when it moved back into surface waters and stayed there all night long, with only a few deep excursions (Figure 18.9). This behavior pattern was exactly the opposite of that observed for tarpon T-03 from the southeastern U.S. coast.

Tarpon T-24's vertical thermohabitat map indicates that the fish had experienced great temperature variations during its migration from Veracruz, Mexico, to Port Aransas, Texas (Figure 18.8). Temperatures ranged from 20.7°C (71.8°F) to 33.7°C (88.2°F) with of mean water temperature of 27.7°C (79°F), and a maximum daily range over 5°C. The cool water temperatures experienced by this tarpon from June 19 to July 10 (Figure 18.10A) suggests that the fish was in the cool upwelling area of the

Seasonal Migratory Patterns and Vertical Habitat Utilization 291

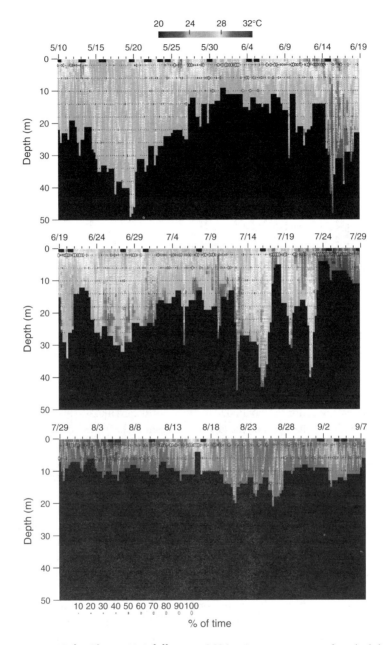

FIGURE 18.8 (Color Figure 18.8 follows p. 262.) A summary map of vertical thermohabitat utilization generated with minute-by-minute depth and temperature data of a recovered PAT tag (T-24). The temperatures are displayed in color scale from 20 to 32°C. The depth is on the y-axis and time is on the x-axis from May 10 to September 8, 2004. The size of open circle indicates the percentage of time the tarpon spent at each depth for each 6-h interval. The dashed white line indicates the minute-by-minute depth distribution of the fish.

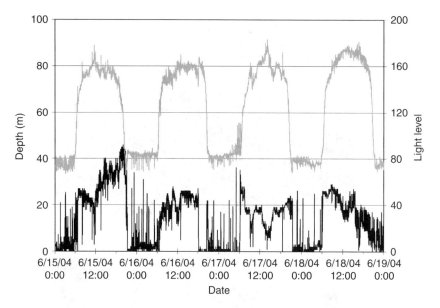

FIGURE 18.9 Minute-by-minute depth (solid black line) and light level (light gray) data recorded by the PAT tag (T-24) for the period of June 15–June 19, 2004. The light level is indicated by the gray line with the second y-axis in relative logarithm unit.

coastline proximal to the U.S.–Mexico border during this period (Figure 18.10B). As Gulf waters warmed, this fish continued to move northward into warmer coastal waters, so that by late July it had reached the lagoonal system of the Texas coastline. High temperatures, such as those recorded on the tag (Figure 18.10A), were also recorded by NOAA C-MAN stations along the Texas coast at Corpus Christi, Port Aransas, Free Port, and the Galveston Pier. From July 25 to September 8, temperatures on the tag were very closely matched by all four C-MAN stations.

During 2004, another fish tagged at Veracruz (T-15) popped off at Tampico, Mexico, but this PAT tag was subsequently found on the beach at Galveston, Texas. This find allowed us to compare the extensive data obtained from the two recovered tags for fish that were tagged a few days apart and within 100 km² of each other. Of interest are the strikingly similar environmental preferences between PAT tags T-15 and T-24 in terms of mean temperatures and ranges (Table 18.3). Note also that the 8 days temperature means shown in Table 18.3 are strikingly similar for both fish, and reflect an uncanny preference for 26°C (79°F) waters over the 40-day period of comparable data. However, the smaller fish (T-24) did not go as deep as T-15. Another interesting point is the similarity or coherence between the temperature profiles of tarpon T-15 and T-24 (Figure 18.11) over the period May 11–May 31. The correlations are striking. Note that both animals, although tagged in locations totally independent of each other, may have schooled and generated similar temperature records, first a variable temperature record with the animal moving up and down in the temperature-stratified (24–29°C) water column during May 12–15, then spent May 15–20 in a water column of uniform temperature (~26°C).

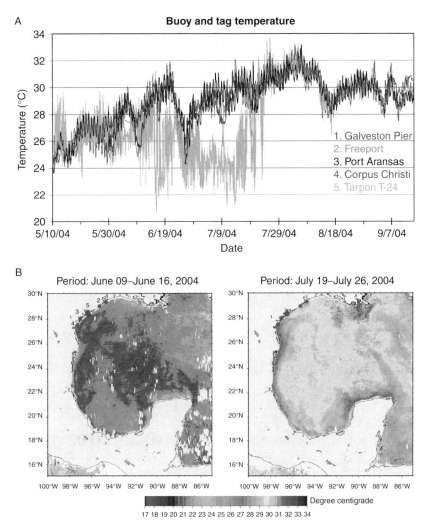

FIGURE 18.10 (Color Figure 18.10 follows p. 262.) (A) Water temperatures recorded at Galveston Pier (red line), Freeport (green line), Port Aransas (black line), and Corpus Christi (blue line) C-MAN stations, and by the recovered PAT tag T-24 (light blue line) from May 10 to September 8, 2004. (B) Sea surface temperature maps for the Gulf of Mexico for the period June 9–June 16, and July 19–July 26, 2004, as determined by MODIS satellite. Note the areas of upwelling (light blue) off south Texas and Campeche Bank, Mexico.

INSIGHTS FROM NEW POP-UP ARCHIVAL TRANSMITTING TECHNOLOGY

With the improvements in new PAT tag technologies available in 2006, we were able to receive over 95% of summarized depth, temperature, and profile of depth and temperature data via Argos satellites for tags stayed on fish over 30 days. This enabled us to generate similar vertical thermohabitat maps without having physically recovered the tags (Figure 18.12). Tarpon (T-46) was tagged in the late afternoon of

TABLE 18.3
Eight-Day Means, Minimums, Maximums, and Standard Deviations of Depth, Water Temperature, and Light Levels from PAT Tags for 198-lb Tarpon and 121-lb Tarpon for Four Corresponding Time Periods between May 11 and June 8, 2004

198-lb Tarpon (T-15)

		Depth	Temp	Light	T (°F)	Depth (ft)
May 11–15	Mean	8.15	26.42	118.27	79.55	26.49
	Min	0	24.10	67.00	75.38	0.00
	Max	29.00	29.90	187.00	85.82	94.25
	Stdev	6.78	1.42	39.18	2.55	22.04
May 16–23	Mean	11.85	26.21	116.84	79.18	38.52
	Min	0	24.50	68.00	76.10	0.00
	Max	33.00	29.30	181.00	84.74	107.25
	Stdev	8.26	0.53	39.31	0.95	26.84
May 24–31	Mean	11.53	25.94	119.55	78.69	37.48
	Min	0	22.10	66.00	71.78	0.00
	Max	79.00	30.10	185.00	86.18	256.75
	Stdev	11.09	1.09	37.82	1.96	36.04
June 1–8	Mean	5.27	25.79	125.02	78.41	17.14
	Min	0	20.70	66.00	69.26	0.00
	Max	45.00	28.90	197.00	84.02	146.25
	Stdev	8.71	1.74	41.02	3.13	28.32

121-lb Tarpon (T-24)

		Depth	Temp	Light	T (°F)	Depth (ft)
May 10–15	Mean	6.94	25.56	125.06	79.81	22.55
	Min	0	24.10	67.00	75.38	0.00
	Max	28.00	29.70	192.00	85.46	91.00
	Stdev	6.89	1.38	40.39	2.48	22.38
May 16–23	Mean	14.55	26.05	126.60	78.90	47.29
	Min	0	23.30	67.00	73.94	0.00
	Max	46.00	28.90	194.00	84.02	149.50
	Stdev	11.33	0.75	42.87	1.34	36.81
May 24–31	Mean	4.81	26.87	132.36	80.36	15.64
	Min	0	24.10	68.00	75.38	0.00
	Max	24.00	29.70	195.00	85.46	78.00
	Stdev	4.93	1.13	43.57	2.03	16.03
June 1–8	Mean	1.96	26.59	132.87	79.86	6.36
	Min	0	24.50	69.00	76.10	0.00
	Max	15.00	28.90	196.00	84.02	48.75
	Stdev	2.54	0.64	40.95	1.15	8.26

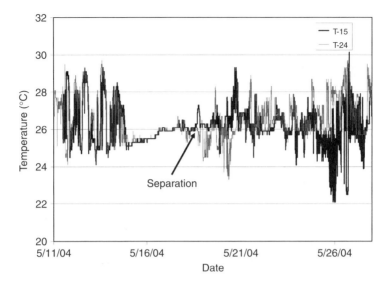

FIGURE 18.11 Water temperatures recorded by two recovered PAT tags T-15 (black) and T-24 (red) for the period May 11–May 28, 2004. Note the striking similarity in temperatures before the separation on May 19.

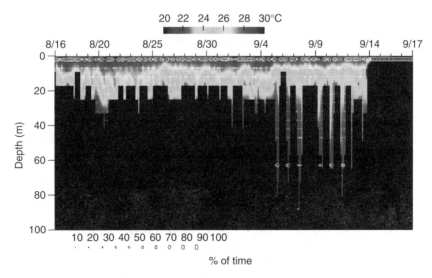

FIGURE 18.12 (Color Figure 18.12 follows p. 262.) Vertical thermohabitat utilization map generated from ARGOS transmitted summary data from a tarpon tagged in Trinidad (T-46). The temperatures are displayed in color scale from 20 to 30°C. The depth is on the y-axis and the date is on the x-axis. The size of open circle indicates the percentage of time the tarpon spent at each depth for each 3-h interval.

August 15, 2006, off Trinidad, near the coast of Venezuela. This tag prematurely popped off about 100 mi north of Trinidad on September 17, 2006, near Grenada. The west coast of Trinidad is surrounded by tannic-stained estuarine waters that show sharp stratification around 15–20 m depth. Below these depths dissolved oxygen levels are near zero. This environmental condition is also clearly reflected by the shallow depth distribution of tarpon T-46 between August 16 and September 4 (Figure 18.12). During this period, the tarpon spent 90% of its time in shallow depths (<10 m), and only occasionally (10% of time) ventured to depths of 40 m. Notably, between Trinidad and Grenada there is a very deep passage, which separates the South American continent and the Lesser Antilles. As the tarpon traveled from Trinidad to Grenada, it crossed this passage, which mostly likely occurred after September 5, when the tarpon PAT tag recorded the deepest depths (88 m) reached among all the PAT tags that we deployed to date (Figure 18.12). On September 5, the tarpon dove into deep water (>50 m) at sunrise and stayed deep until late afternoon. This behavior was repeated for the next 5 days, except on September 8, which was the day after the night of full moon when the tarpon stayed at surface day and night. We believe that it is possible that the tarpon was spawning in the passage between Trinidad and Grenada on this day.

DISCUSSION

Our results have made it clear that many of the tarpon that cruise the southwestern Gulf of Mexico coast off Mexico are apparently part of the stock supporting the U.S. recreational fisheries during the late spring to early fall each year. Defining the unit stock is the most critical piece of information needed for sound fishery management and conservation decision making. An incorrect unit stock range will certainly lead to a failure for any U.S.-centric management and conservation policies. While our study did not address the entire range of tarpon, it has demonstrated that the unit stock range for tarpon that seasonally frequent the U.S. waters extends beyond the U.S. territorial borders. Thus, sound management and conservation strategies for Atlantic tarpon must involve all the states and countries through which tarpon pass on their annual migrations and to which they are subjected to exploitation and loss.

The patterns of seasonal migration and vertical thermohabitat utilization observed in this study are most likely driven by the combination of their environmental preferences, feeding needs, and reproductive behavior. This study shows that the most preferred temperatures of tarpon are from 24 to 26°C in spring and fall, 28–30°C in summer. Seasonal migrations would appear to be most likely cued to the changes in water temperature in combination with the movement and distribution of prey. In the spring, as the temperature increases the tarpon move northward, and in the fall as the temperature decreases the tarpon move south. Different diel vertical movement patterns observed in this study might be a result of different feeding strategies for different prey species encountered at different geographical locations on its migration route. Similarly, on most of the days when no clear diel vertical movement patterns were observed, it could also be a strategy for feeding on nondiel distributed prey.

This study showed that PAT technology has promise for studying the biology of coastal marine species such as the Atlantic tarpon, despite potential obstacles to successful deployment and data retrieval. Most PAT studies to date have reported a frustratingly low success rate of tags with respect to data communications via satellite (Block et al., 1998b; Lutcavage et al., 1999; Block et al., 1999; Sedberry and Loefer, 2001). However, tags deployed on coastal species likely have a higher probability of being recovered by scientists due to higher angler density and likelihood of tags washing ashore. From 2001 to 2005, 6 out of 30 tags were recovered and returned to us, a recovery rate of 20%. As our results showed, the recovered tags provided high-resolution time series of activity and habitat characteristics.

Variation in weather conditions can cause rapid changes in turbidity in coastal waters, thereby confounding accuracy of normal geolocating algorithms based on diel patterns of light intensity (e.g., Hill and Braun, 2001). Given the already low resolution of these location methods, these traditional algorithms may have little utility for identifying movement patterns over the relatively short ranges of interest for many coastal species compared with pelagics. However, unique features of coastal waters appear to provide mechanisms for identifying individual positions with greater accuracy. Temperature and bathymetry data in particular can be used to interpret tag records for location purposes. The large number of monitoring stations along the coast providing direct measurement of water temperature and weather conditions can be very useful in this regard, and provide an advantage over the inconsistent availability of remotely sensed data.

There is some concern that natural and urban high-relief structures (shoreline vegetation canopy, buildings, bridges, etc.) may interfere with effective transmission of data via satellite. If tags wash ashore, the lack of immersion can also interfere with transmission strength, but the newer generation tags are capable to transmit while on beach as indicated by our 2006 tags. In general, the size of tags is probably the greatest impediment to applying this technology to a broad range of coastal species. Many coastal fisheries species do not attain an appropriate size for carrying PAT tags (e.g., bluefish, striped bass, and reef species). In this sense, tarpon represent an ideal candidate species for present PAT equipment due to their large size and the long-range movements as revealed by this study. Our results suggest that PAT tags have great potential for characterizing tarpon movement behavior and habitat preferences, and as the technology continues to improve they can effectively be used on other coastal species of interest.

Our goal of future PAT tagging research is to fill critical information gaps in understanding regional and far-field connectivity of these important game fish, and to identify spawning locations in the Gulf of Mexico and the Caribbean Sea. We believe that seasonal placement of PAT tags at about 20 strategically chosen sites and dates over several years could provide unique insights into the range of the tarpon's movements and migrations, and spawning sites along Florida, Gulf, Central America, and Mexican coastlines. Knowledge of the range and duration of migrations and specific spawning locations is essential to assuring the proper management of tarpon fisheries and the future conservation of this ancient species.

ACKNOWLEDGMENTS

This research would not have been possible without the help of many captains and anglers who assisted us in providing vessels, catching fish, and tagging the tarpon. We are grateful for the help and technical assistance of Scott Alford, Stu Apte, Mark Badzinski, Youssef Barquet, Capt. George Beckett, Curtis and Andrew Bostick, Maria Jose Calderon Caracas, Felipe Fernandez Ceballos, Roy Crabtree, Bill Curtis, Alberto Pavon David, Tom Davidson, Fuzzy Davis, Dave Denkert, Jorge Diaz, Paul Dixon, Jim Farley, Frank Fowler, John Frazier, Gary Fungui, Heliodoro Garza, Tom Gibson, Doug "The Bass Professor" Hannon, Jeff Harkavy, Dan Jacobs, Capt. Derek Jordan, Doug Kelly, Julian Lajornade, Alberto Maderia, Iain Nicolson, Carlos Partida, Billy Pate, Eduardo Perusquia, James Plaag, Jesus Quijano, Gerardo Fernandez Quijano, Fernandez Quijano, Angel Requejo, Jose Manual Parada Rey, Danny Romino, Gabriel Romo, Geoffrey Samuels, George Santoy, Lance "Coon" Schouest, Joe Skrumbellos, Pedro Sors, Jason Schratwieser, Derke Snodgruss, Monty Trim, Bruce Ungar, Felix Juan Malpica Valverde, Steve Venini, Rufus Wakeman, in addition to sportfishermen of Veracruz Yacht Club, Coatzacoalcos Fishing Club, and La Soufriere Marina. This research was supported by funding provided by Bonefish & Tarpon Unlimited, Sanctuary Friends Foundation of the Florida Keys, National Fish and Wildlife Foundation, Florida Fish and Wildlife Conservation Commission, Texas Tarpon ProAm, Texas Coastal Conservation Association, and Texas Parks and Wildlife Department.

REFERENCES

Block, B., Boustany, A., Teo, S.L.H., Walli, A., Farwell, C., Williams, T., Prince, E.D., Stokesbury, M.J., Dewar, H., Seitz, A., and Weng, K.C. 2003. Distribution of western tagged Atlantic bluefin tuna determined from archival and pop-up satellite tags. Col. Vol. Sci. Pap. ICCAT 55(3): 1127–1139.

Block, B.A., Dewar, H., Farwell, C., and Prince, E.D. 1998a. A new satellite technology for tracking the movements of Atlantic bluefin tuna. Proc. Natl. Acad. Sci. U.S.A. 95(16): 9384–9389.

Block, B.A., Dewar, H., Williams, T., Prince, E.D., Farwell, C., and Fudge, D. 1998b. Archival and pop-up satellite tagging of bluefin tuna. FASEB J. 12(5): A676.

Block, B.A., Dewar, H., Williams, T., Prince, E.D., Farwell, C., and Fudge, D. 1998c. Archival tagging of Atlantic bluefin tuna (*Thunnus thynnus thynnus*). Mar. Technol. Soc. 32(1): 37–46.

Block, B.A., Dewar, H., Blackwell, S.B., and Boustany, A. 1999. Archival and pop-off satellite tags reveal new information about Atlantic bluefin tuna thermal biology and feeding behaviors. Comp. Biochem. Physiol. Part A Mol. Integrative Physiol. 124: S21.

Block, B.A., Dewar, H., Blackwell, S., Williams, T., Farwell, C.J., Prince, E.D., Boustany, A., Teo, S.L.H., Seitz, A., Fudge, D., and Walli, A. 2001. Electronic tags reveal migratory movements, depth preferences and thermal biology of Atlantic bluefin tuna. Science 293: 1310–1314.

Block, B.A., Teo, S.L.H., Walli, A., Boustany, A., Stokesbury, M.J., Farwell, C.J., Weng, K.C., Dewar, H., and Williams, T.D. 2005. Electronic tagging and population structure of Atlantic bluefin tuna. Nature 434: 1053–1164.

Boustany, A.M., Davis, S.F., Pyle, P., Anderson, S.D., Le Boeuf, B.J., and Block, B.A. 2002. Satellite tagging: expanded niche for white sharks. Nature 412: 35–36.

Graves, J.E., Luckhurst, B.E., and Prince, E.D. 2002. An evaluation of pop-up satellite tags for estimating postrelease survival of blue marlin (*Makaira nigricans*) from a recreational fishery. Fish. Bull. 100: 134–142.

Hill, R.D. and Braun M. J. 2001. Geolocation by light level, the next step: latitude. In: J.R. Sibert and J.L. Nielsen (eds.). Electronic tagging and tracking in marine fisheries. Kluwer Academic, Dordrecht, pp. 315–330.

Kerstetter, D.W., Luckhurst, B.E., Prince, E.D., and Graves, J.E. 2003. Use of pop-up satellite archival tags to demonstrate survival of blue marlin (*Makaira nigricans*) released from pelagic longline gear. Fish. Bull. 101: 939–948.

Loefer, J.K., Sedberry, G.R., and McGovern, J.C. 2005. Vertical movements of a shortfin mako in the western North Atlantic as determined by pop-up satellite tagging. Southeast. Nat. 4(2): 237–246.

Luo, J., Prince, E.D., Goodyear, C.P., Luckhurst, B.E., and Serafy, J.E. 2006. Vertical habitat utilization by large pelagic animals: A quantitative framework and numerical method for use with pop-up satellite tag data. Fish. Oceanogr. 15(3): 208–229.

Lutcavage, M., Brill, R., Skomal, G., Chase, B., and Howey, P. 1999. Results of pop-up satellite tagging of spawning size class fish in the Gulf of Maine: do North Atlantic bluefin tuna spawn in the mid-Atlantic? Can. J. Fish. Aquat. Sci. 56(2): 173–177.

Man'kowski, V.I. 1978. Empirical formula for estimating the light attenuation coefficient in sea water from secchi disk readings. Okeanologiya 18(4): 750–753.

Oppel, F. and Meisel, T. 1987. Tales of Old Florida. Castle Press, Seacaucus, NJ, 477 pp.

Prince, E.D. and Goodyear, C.P. 2006. Hypoxia-based habitat compression of tropical pelagic fishes. Fish. Oceanogr. 15(4): 451–464.

Sedberry, G.R. and Loefer, J.K. 2001. Satellite telemetry tracking of swordfish, *Xiphias gladius*, off the eastern United States. Mar. Biol. 139(2): 355–360.

Seitz, A.C., Weng, K.C., Boustany, A.M., and Block, B.A. 2002. Behaviour of a sharptail *mola* in the Gulf of Mexico. J. Fish Biol. 60(6): 1597–1602.

19 Tagging of Bonefish in South Florida to Study Population Movements and Stock Dynamics

Michael F. Larkin, Jerald S. Ault, Robert Humston, Jiangang Luo, and Natalia Zurcher

CONTENTS

Introduction 301
Methods 301
    Anchor Tagging 302
    Acoustic Telemetry 302
Results 306
    Anchor Tagging 306
    Acoustic Telemetry 308
Discussion 316
Acknowledgments 319
References 319

INTRODUCTION

As documented in this volume, little is known about bonefish population and fishery dynamics, stock spatial distribution, spawning migrations, or movements between fishing areas, despite the importance of this species as a premier game fish (e.g., Ault et al., Chapter 16, this volume). In addition, information on the spatial ecology of bonefish is particularly sparse, and could help fill important knowledge gaps for management. A lack of this kind of vital information hinders development of management practices to ensure sustainability of their fisheries. In this chapter, we discuss progress on a research program in south Florida using anchor tag and acoustic telemetry methods to evaluate and quantify bonefish movements and aspects of population dynamics.

METHODS

We used two complementary methods, anchor tags and acoustic telemetry, to study movements, migrations, stock structure, and population dynamics because each has its own advantages and disadvantages. Deployment of anchor tags broadly over the range of the fishery requires little training on how to tag bonefish and record necessary information. However, such mark-recapture data can only provide information on the time and location of tagging/release and, with luck, subsequent recapture. No inference is gained on the whereabouts or behavior of fish during the intervening period at liberty. Use of acoustic telemetry methods for monitoring bonefish movements—while labor intensive and requiring substantial technical expertise to deploy—provides high-resolution spatial and temporal data on fish movements and apparent responses to environmental cues.

Anchor Tagging

Initiated in 1998, in consultation with professional fishing captains and concerned anglers who ply the waters of the Florida Keys year-round, we developed a bonefish anchor tagging program that used volunteered expert efforts and well-tested quantitative methods in fishery science (Humston, 2001; Ault et al., 2002; Ault et al., Chapter 26, this volume). Volunteers received a tagging kit consisting of a Mark II pistol grip gun, Floy FD-94 T-bars tags (www.floytag.com), a document containing tagging instructions and guidelines, mechanical pencils, and a waterproof data sheet. After volunteers caught a bonefish using hook-and-line gear, they used the tag-gun to insert a relatively thin, high-grade polymer plastic T-bar tag into an "anchored" position between the dorsal musculature and bone structures of the fish (Figure 19.1A). At the time of each tag deployment, the date, location, fork length (FL), and weight (W) were recorded. In some cases, only FL was recorded and in these situations W was determined using an allometric growth equation of the form $W = \alpha L^B$ described by Ricker (1975) estimated from fishery data from Crabtree et al. (1996) and this study (Ault et al., Chapter 16, this volume) using nonlinear regression methods.

Locations of the anchor-tagged releases were grouped into two regions: (1) Florida Keys—extending from Key Biscayne south to the Marquesas and (2) Palm Beaches—extending from the St. Lucie Inlet to Miami Beach (Figure 19.2). Average sizes in the population for each region, computed between length at recruitment (L_r) and maximum length in the stock (L_λ) in mm FL and their standard errors (SE), were compared using a Student's t-test (Sokal and Rolf, 1994). In addition, total instantaneous mortality rate (Z) was computed by year for the Florida Keys region by evaluating the average length of tagged fish computed between length at full recruitment to the exploitable phase (L_c) and (L_λ) using the length-based mortality methods of Ault et al. (2005). Because of partial availability of immature bonefish in the Florida Keys fishery, we assumed that the length at full recruitment to the exploitable phase (i.e., fishery) was 545 mm FL (or about 5 years), following Crabtree et al. (1996).

Acoustic Telemetry

We implemented an automated acoustic telemetry (AT) system to continuously monitor movements of the bonefish among the shallow flats and channels alongshore

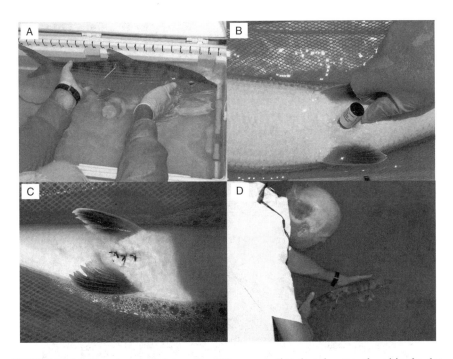

FIGURE 19.1 Steps in bonefish tagging: (A) a conventional anchor tag placed in the dorsal musculature for a bonefish being measured; (B) an acoustic telemetry transmitter being inserted into the abdomen through a small incision in the peritoneal cavity; (C) closure of the incision with three sutures; and (D) a tagged bonefish being gently released after complete resuscitation.

of a series of barrier islands extending from the Ragged Keys to the southern end of Biscayne Bay at Caesars Creek (Figure 19.3). We placed 40 VEMCO (www.vemco.com) VR2 hydrophone receivers and data logging stations in an array that maximized the likelihood of tag detection whenever bonefish accessed or entered the flats, passed through telemetered channels, or cruised along the bayside or oceanside receiver array. To facilitate the evaluation of potential spawning migrations, we arranged two transects of receivers that were oriented cross-shelf extending from the barrier island to the reef tract (roughly east–west) to increase the likelihood of capturing north–south migrations along the "outside" flats, and presented the possibility of documenting offshore movements if they occurred near either transect. The array had a spatial coverage of about 20 km running north to south and 8 km from east to west. We employed VEMCO Model V16 coded acoustic transmitting pinger tags operating on 69 kHz frequency with 158 dB (1 µPa at 1 m) power output to maximize transmission power and receiver detection range. Tag transmissions were separated by 30–79 s (random) delays, providing battery life of at least 62 days postactivation. Acoustic tags were 58 mm long and 16 mm in diameter and were surgically implanted into the peritoneal cavity of bonefish using field procedures described in Humston et al. (2005) (Figure 19.1). All AT-tagged fish were also identified with conventional

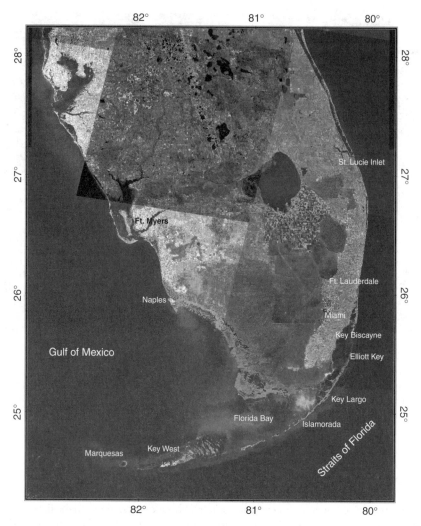

FIGURE 19.2 Map of south Florida showing the bonefish tagging study region running from the St. Lucie Inlet to the Marquesas.

T-bar anchor tags (Floy Tag) prior to release into the array of hydrophone receivers. We focused our tagging efforts around the spawning season, defined as November to May by Crabtree et al. (1997), to capture migrations and movements relative to frontal passages.

Owing to the difficulty of obtaining bonefish for our study, we employed a professional bonefish captain to help catch bonefish and then bring them to holding pens with recirculating seawater at the University of Miami's Rosenstiel School near Key Biscayne. At the facility, AT-tagging procedures were conducted and fish were then released back into the ocean within 1–3 days. Most tagged bonefish placed in holding pens were released close to or within the VR2 array. Bonefish tagged in the

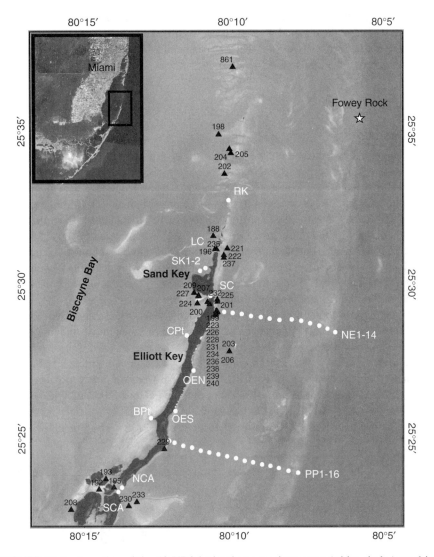

FIGURE 19.3 Location of the 40 VR2 hydrophone receiver array (white circles) used in the bonefish acoustic telemetry study extending from the Ragged Keys (RK) to southern Biscayne Bay at Caesars Creek. Release points of specific AT-tagged bonefish are shown as triangles with the tag identification numbers.

field were released near their capture location, which was typically ≤1 km from one of our hydrophone receivers.

Assessment of hydrophone reception ranges was conducted following the procedure described in Humston et al. (2005). Results of range testing of acoustic telemetry tags by hydrophone receivers revealed that detections were limited to 200–450 m at relatively shallow (<2 m) sites; while at deep (≥2 m) receivers, AT-tag detections ranged from 500 to 700 m. Tag range-testing results were used to guide configuration of the receiver array in a longshore manner that monitored shallow barrier island passes, and placement of two offset cross-shelf receiver lines that extended to the deep barrier coral reef was designed to intercept both southward and northward moving bonefish (Figure 19.3).

RESULTS

ANCHOR TAGGING

From January 1998 through July 2006, a total of 4617 bonefish were captured by about 100 volunteer captains and experienced anglers, fitted with anchor tags and then released into south Florida coastal waters ranging from St. Lucie Inlet down through the Florida Keys and west to the Marquesas (Figure 19.2). Both the number of bonefish released with tags, and those subsequently recaptured, have increased throughout the duration of the program (Figure 19.4). An average of about 45 fish per month ($\bar{T}$) were tagged during the 9-year program. The number of bonefish tagged per month peaked twice during a given year, during March to May and from November to December (Figure 19.5). These months, by and large, corresponded to prime fishing seasons of optimum water temperatures and favorable tides.

To date, more than 160 (3.5%) tagged bonefish have been recaptured. Patterns documented from recaptured anchor-tagged bonefish revealed individual movements and stock mixing throughout the Florida Keys. The majority (87.6%) of observed bonefish movements were <20 km, with 51% of those fish being recaptured ≤2 km of where they were tagged. In contrast, 17 bonefish were observed to have moved >20 km, and some for substantial distances that ranged up to 200 km. One bonefish tagged in February 2005 off Key Biscayne was recaptured 321 days later off Andros Island, Bahamas, traveling a least-linear distance of >300 km (Figure 19.6).

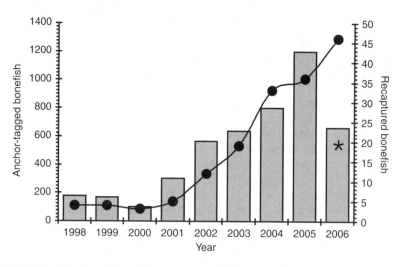

FIGURE 19.4 Numbers of bonefish anchor tagged (histograms) and recaptured (dark circles and black line) by year from 1998 to 2006. The symbol * indicates a partial year of information through July 2006.

Tagging of Bonefish in South Florida

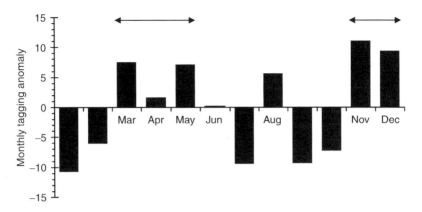

FIGURE 19.5 Anomaly of average number ($\bar{T} = 45$) of anchor-tagged bonefish by month for all years (1998–2006) combined. The horizontal arrows indicate prime fishing seasons that occur during periods of optimum water temperatures and favorable tides.

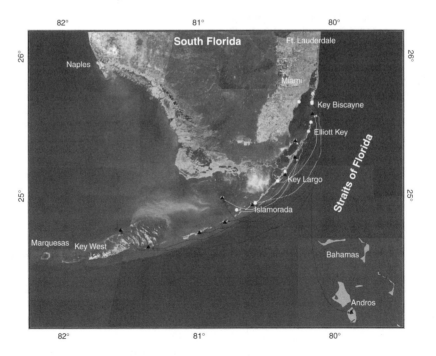

FIGURE 19.6 Summary of long-range (>20 km) bonefish movements from recaptured anchor-tagged bonefish for which both release and recapture locations were known. Dots indicate tagging locations and triangles recapture locations. Black lines indicate southbound movements and white lines northbound movements.

In terms of the distances moved and days at liberty for recaptured bonefish, no apparent trends were observed (Figure 19.7A). Bonefish were observed to have moved substantial distances (>100 km) in periods that ranged from about 15 to >100 days. Some bonefish appeared to have only moved a few kilometers, even after almost 2 years at liberty. Distance moved was not necessarily time-dependent for the size range of recaptured bonefish, and the size at release and number of days at liberty were not correlated (Figure 19.7B), suggesting no differential mortality by size due to tagging. All bonefish that had moved >20 km were recaptured during spring to early summer (March–June) and fall (October and December). Large, sexually mature (>488 mm FL from Crabtree et al., 1997) bonefish were responsible for all but one of the long-distance movements >50 km (Figure 19.7C).

Lengths were available for 4527 tagged bonefish ranging from 223 to 810 mm FL. From these, both lengths and weights available for 3299 bonefish were used to estimate a mean allometric growth function for weight dependent on length

$$W = 1.632 \times 10^{-8} L^{2.992311},$$

where W is the weight in kilograms and L is the FL in millimeters. This was very similar to that reported by Ault et al. (Chapter 16, this volume). Length frequency data obtained from tagging strongly suggested that bonefish are not fully recruited to the Florida Keys fishery until they exceeded the minimum size of sexual maturity (i.e., 488 mm FL) (Figure 19.8). Comparison of available length data for fish tagged in Florida Keys relative to the Palm Beaches region is shown in Table 19.1A. Generally, fish tagged in the Palm Beaches were significantly smaller ($P \leq 0.001$) than those from the Florida Keys region. Annual estimates of average length of bonefish in the exploitable phase (≥ 545 mm FL) for the Florida Keys region were fairly constant throughout the project years (Table 19.1B). The length frequency distribution from our conventional anchor tag study was virtually identical to that obtained in Crabtree et al.'s (1996) age and growth study (Ault et al., Chapter 16, this volume). Length-based mortality estimation procedures using average lengths from the tagging study following the procedures of Ault et al. (2005) resulted in estimates of total annual instantaneous mortality rate that averaged 0.208, and ranged from 0.156 to 0.368 (Table 19.1B).

ACOUSTIC TELEMETRY

From March 2004 to July 2005, a total of 40 bonefish were implanted with AT-transmitter tags and released. Individual data on fish sizes, dates of release, days in the study area, field-tagged or holding-pen bonefish, and day of last recorded transmission are provided in Table 19.2. The size range of bonefish tagged with AT transmitters was similar to those from the anchor tag study (Figure 19.8). Of the 40 fish AT tagged and released, 31 (78%) of these were subsequently detected at least once, producing 57,070 unique tag detections within the 40 receiver array (Table 19.3). Nine AT tags were never detected by any of the receivers; 5 of these were released north of the array. Field-tagged bonefish had higher tag detection rates than lab-tagged fish; however, by and large, lab-tagged fish were captured in

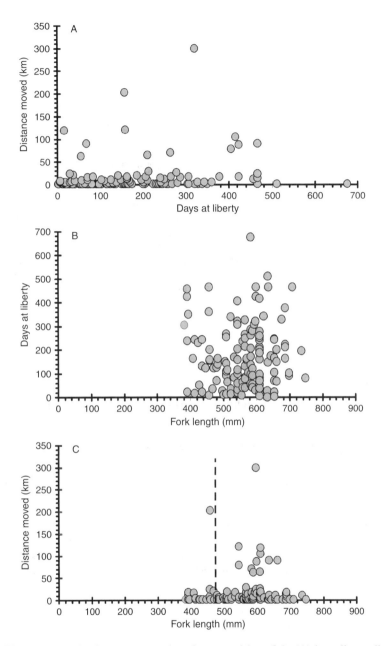

FIGURE 19.7 Results from recaptured anchor-tagged bonefish: (A) least-linear distance moved dependent on days at liberty; (B) days at liberty dependent on size (mm FL) at release; and (C) least-linear distance moved dependent on size at release. Dashed vertical line shows minimum size at 50% sexual maturity (488 mm FL from Crabtree et al., 1997).

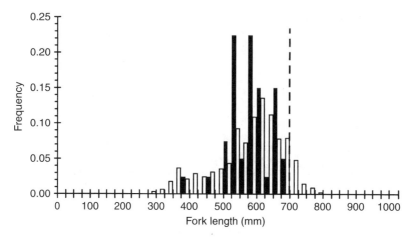

FIGURE 19.8 Comparison of length frequency distributions for anchor-tagged bonefish (open columns; $n = 4527$) from 1998 to 2006 and acoustic telemetry–tagged bonefish (black columns; $n = 40$). Dashed vertical line shows the largest (679 mm FL) bonefish aged by Crabtree et al. (1996).

TABLE 19.1
Estimates of Average Size from the South Florida Bonefish Conventional Tagging Program

Average Size in the Population $\bar{L}$ (mm FL) and Standard Error for All Bonefish Tagged from 1998 to 2006

Region	$\int_{L_r}^{L_\lambda}\bar{L}$	n	SE	Size Range $[L_r, L_\lambda]$
Florida Keys	563	4063	1.38	[261, 810]
Palm Beaches	338	281	2.22	[223, 533]

Annual Average Size in the Exploitable Phase $\bar{L}$, Sample Size, and Estimated Total Instantaneous Annual Mortality Rate $\hat{Z}$ of Tagged Bonefish (where $L_c = 545$, $L_\infty = 663.0$, and $K = 0.3258$)

Year	$\int_{L_c}^{L_\lambda}\bar{L}$	n	$\hat{Z}$
1998	600.1	108	0.368
1999	612.5	100	0.230
2000	620.7	50	0.156
2001	614.9	178	0.207
2002	622.5	346	0.140
2003	620.5	371	0.158
2004	617.6	486	0.183
2005	613.16	636	0.224

TABLE 19.2
Summary of Acoustic Telemetry Data for 40 Tagged Bonefish between March 2004 and July 2005

Tag ID	Type	W (kg)	FL mm	Release Date	Release Location	Last Transmission	Days Detected
188	F	2.95	594	07/14/2005	Boca Chita	08/16/2005	29
192	F	1.81	489	06/12/2005	Caesars Creek	ND	0
193	F	2.04	511	06/12/2005	Caesars Creek	09/17/2005	96
195	F	4.42	654	02/01/2005	Caesars Creek	05/06/2005	94
196	R	1.13	445	01/22/2005	Lewis Cut	01/23/2005	1
197	F	1.47	478	02/03/2005	Elliott Key	05/04/2005	90
198	R	2.72	566	02/09/2005	Soldier Key	NR	0
199	F	1.81	500	02/01/2005	Elliott Key	04/28/2005	86
200	F	2.04	533	03/24/2005	Sands Cut (B)	06/24/2005	84
201	F	3.18	588	05/09/2005	Sands Cut (B)	06/07/2005	19
202	R	2.72	566	05/05/2005	Soldier Key	ND	0
203	R	1.81	522	05/18/2005	Elliott Key	ND	0
204	R	4.65	643	05/02/2005	Soldier Key	ND	0
205	R	3.18	599	05/13/2005	Soldier Key	05/14/2005	1
206	R	2.83	522	05/18/2005	Elliott Key	05/18/2005	1
207	F	4.20	588	03/03/2005	Sands Cut (B)	04/06/2005	34
208	F	2.61	572	05/14/2005	Cutter Bank	05/21/2005	6
209	F	3.97	632	04/05/2005	Sands Cut (B)	05/29/2005	54
210	F	2.72	566	03/24/2005	Sands Cut (B)	06/14/2005	79
221	R	5.33	643	03/06/2004	Sands Key (O)	ND	0
222	R	3.52	588	03/06/2004	Sands Key (O)	03/06/2004	1
223	R	2.04	500	10/05/2004	North Elliott Key	10/10/2004	5
224	R	3.29	610	03/27/2004	Sands Cut (B)	03/27/2004	1
225	R	2.95	566	05/21/2004	Elliott Key	05/21/2004	1
226	R	2.15	522	05/21/2004	Elliott Key	05/27/2004	6
227	R	4.31	632	03/27/2004	Sands Cut (B)	04/02/2004	6
228	F	1.81	511	03/17/2004	Elliott Key	04/14/2004	28
229	F	2.61	566	03/17/2004	Elliott Key	03/28/2004	10
230	F	3.63	594	04/21/2004	Caesars Creek	05/04/2004	13
231	R	2.27	544	05/21/2004	Elliott Key	06/15/2004	25
232	R	4.54	632	05/20/2004	Elliott Key	ND	0
233	F	2.72	555	12/14/2004	Caesar's Creek	12/26/2004	12
234	R	2.38	522	12/14/2004	Elliott Key	12/14/2004	1
235	R	2.95	561	10/26/2004	Lewis Cut	ND	0
236	R	1.81	478	12/14/2004	Elliott Key	12/18/2004	4
237	F	4.65	632	06/04/2004	Lewis Cut	08/22/2004	78
238	R	4.88	654	04/24/2004	Elliott Key	04/24/2004	1
239	F	0.79	357	12/19/2004	Elliott Key	01/05/2005	17
240	R	3.40	555	06/04/2004	Elliott Key	06/07/2004	3
861	R	1.36	522	05/20/2005	Soldier Key	ND	0

Note: Size of sexual maturity is 488 mm FL (Crabtree et al., 1997). ND indicates that there were no tag detections by the array. Days detected is the time (days) between the first and last tag detection. R is the type for fish tagged at University of Miami facility and F for field tagged. B is bayside and O is oceanside of the specific barrier island.

TABLE 19.3
Tag Detections at Specific Receivers for Individually AT-Tagged Bonefish

Receiver Location	Tag Identification Number														
	188	193	195	196	197	199	200	201	205	206	207	208	209	210	222
Bayside															
SN2								71			6		41	144	
SN1								103			164		134	101	
CPt															
BPt															
Oceanside															
RK	4			1	228	6					6			4	
LC	225	1143			1455	2309		123					89	63	1
SC		8	107		540	244	20				14			24	
NE1		67	498		4029	3064	36	55		2					
NE2			16		218	94									
NE3															
NE4															
NE5															
NE6															
NE7															
NE8															
NE9															
NE10															
NE11															
NE12															
NE13															
NE14															
OEN		30	9		13	38	9	23							
OES		236	17		27	2003	733	430							
PP1		1663	10			3856	279	226							
PP2		318				166	28	33							
PP3		2				3									
PP4						1									
PP5															
PP6															
PP7															
PP8															
PP9															
PP10															
PP11															
PP12															
PP13															
PP14										8					
PP15															
PP16															
NCA		744	58			5						1			
SCA			30	301		2						35			
Total	229	4241	1016	1	6510	11791	1105	1064	8	2	190	36	264	336	1

Tagging of Bonefish in South Florida

223	224	225	226	227	228	229	230	231	233	234	236	237	238	239	240	Total
															6	268
															10	512
																0
																0
			7	7								90				353
142			89				65	53				2233				7990
	153		29		1774	29	37						21	66	3	3069
488		79	12	98	7130	35	368	241		279	141		15	594	29	17260
5			29	49	1016		3			5	55			1		1491
			3	17											1	21
			25							52	28				3	108
				2						208	107					317
			24	38						39	4					105
				21							6					27
			5				12			72						89
			28	6			307			226	9					576
			10	1			37			71						119
										25						25
										7						7
																0
																0
				1	1614	229	50	11			17					2044
											11					3457
						105	39	8098								14276
				14			2	616			2					1179
				12				8								25
				11												12
				7												7
				1												1
				47							15					62
				80							3					83
				92												92
				1												1
																0
				18						114						132
				23						491						514
				2						489						499
										90						90
																0
						30	733	10	103							1684
							13	6	188							575
635	153	79	451	358	11534	428	1666	9043	291	2168	398	2323	36	661	52	57070

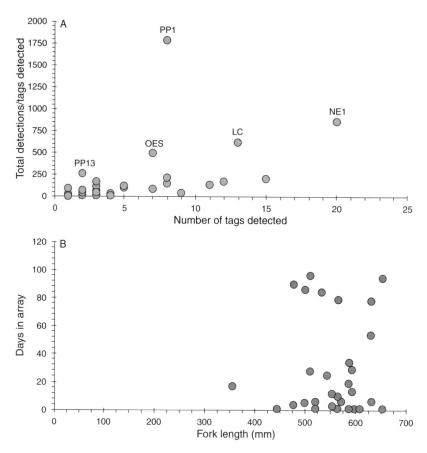

FIGURE 19.9 Bonefish acoustic telemetry results: (A) total detections at a given receiver divided by individual tags detected dependent on individual tags detected at receiver. Data points referring to receivers with greatest numbers of total tag detections divided by unique tags detected are labeled and (B) days detected in array dependent on bonefish size.

foraging habitats north of the array and released to the south within the array. Lower tag-detection rates by lab-tagged fish may have been biased as these fish likely returned to their original capture locations after release. Receivers with the greatest number of different tag detections were those placed at unique geographic features like bay-to-ocean passes and oceanside points (i.e., NE1, PP1, LC, and OES) (Figure 19.9A). There were no apparent trends in receiver detections by the size of fish (Figure 19.9B).

Bonefish exhibited substantial variation in patterns of movement (Figure 19.10), at times moving rapidly alongshore either north to south or vice versa. Five bonefish covered the entire eastern shore of Elliott Key, a distance of approximately 13 km in ≤3.5 h. In several other cases, bonefish moved seaward in apparent response to the drops in barometric pressure (Figure 19.10B). A total of five large mature bonefish were detected at receivers NE2-NE10 and PP2-PP15 located in water depths to 20 m

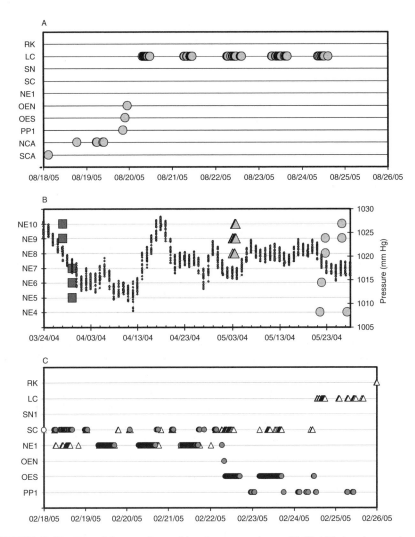

FIGURE 19.10 Bonefish acoustic tag detections at receivers: (A) ID 193 showing northward movement in the array during August 2005; (B) IDs 227 (squares), 230 (triangles), and 226 (circles) from March to May 2004 showing movements east of the barrier islands in deepwater habitats near the barrier coral reef from in relationship to barometric pressure (line); and (C) apparent aggregation behaviors of IDs 197 (triangles) and 199 (circles) at receiver locations during February 2005.

and distances ranging from >0.5 to 6 km east of the barrier island flats of Elliott Key (Table 19.3; Figure 19.3). Often fish returned to the same areas they had previously frequented. Not surprising for fish that are known to school, several AT-tagged bonefish were simultaneously detected at the same receivers (Figure 19.10C). Owing to the relatively large detection ranges of receivers, we can only conclude that these fish displayed similar spatiotemporal patterns in movements and habitat selection, but we cannot rule out aggregation behaviors. This occurred among similar-sized bonefish (i.e., IDs 197 and 199), and for bonefish of different sizes (IDs 195 and 199). Bonefish IDs 197 and 199 stayed in the same proximal area over several days before departure.

DISCUSSION

Prior to this study, conventional wisdom in south Florida among knowledgeable fishing captains and experienced anglers was that bonefish inhabited a limited home range, often frequenting the same foraging areas, and that bonefish rarely, if ever, moved even more than a few kilometers. These perceptions and the noted regional resource declines (e.g., Curtis, 2004) motivated the need for additional study. However, there was little empirical or scientific guidance concerning the most effective methods to track bonefish movements. Two previous conventional anchor tagging studies, one in Florida and one in the Bahamas, were largely unsuccessful. Bruger (Florida Marine Research Institute, personal communication) initiated an anchor-tagging program in the Florida Keys in the 1970s, but in that era anchor-tag technology was crude and none of his tagged fish was ever recaptured, despite the tag-and-release of several hundred bonefish. Colton and Alevizon (1983) anchor-tagged 214 bonefish in the Bahamas, but had only one recapture of a fish that remained at liberty for only a few days postrelease. These studies led to much skepticism by fishing guides concerning potential mortality from tagging, and a general lack of confidence in anchor-tag methodologies. Colton and Alevizon (1983) tracked a small number (i.e., 13) of AT-tagged fish using a boat-mounted directional hydrophone. They stated that bonefish movements ranged widely through available habitats; however, their conclusions were equivocal as they were only able to detect and track three fish after 24 h, and two of these for a 5-day maximum.

The dearth of successful anchor-tagging studies made our unprecedented number of tagged releases (4617) and subsequent recaptures (162) even more remarkable. This research has revealed that bonefish do move, and frequently for substantially long distances. While most of our anchor-tagged bonefish were recaptured within 20 km of their release locations, about 12% of these moved large distances that ranged up to 200 km. All but one of the long-distance (>20 km) movements were made by sexually mature bonefish, but in fact, the one presumed immature bonefish was larger than Crabtree et al.'s (1997) minimum size of maturity for males. Tagging studies with other fishes have documented similar results, that is, immature fish favor small-scale movements, while large, sexually mature fish undertake large-scale movements (Pollock, 1982).

Some bonefish moved great distances (>100 km) in relatively short periods (15 days); these were typically fish of sexually mature sizes during the putative reproductive season (Crabtree et al., 1997). But about 51% of the recaptured bonefish were caught ≤2 km from where they were originally tagged. Some of these fish may have been immature, and therefore did not undertake spawning-related movements but instead continued to frequent the same foraging areas (e.g., Humston et al., 2005). However, a large proportion of recaptured fish exceeded the average size at maturity (Figure 19.7C). Time-at-liberty for many of the mature fish spanned or coincided with the spawning season, and evidence suggests that bonefish do not spawn on or near the inshore flats (Crabtree et al., 1997). This suggests that either (1) these mature bonefish had migrated to spawning sites and then returned to the same foraging areas or (2) some of these mature bonefish may not have participated in spawning migrations. Reasons for lack of participation in spawning migrations for mature fish are generally unknown, but such behavior has been documented in other fishes (Harden Jones, 1968; Pollock, 1982).

One anchor-tagged bonefish was documented to have migrated >300 km across the Straits of Florida to the inshore flats of Andros Island, Bahamas. The cross Florida Straits movement to Bahamian waters raises the question of the appropriate scale of unit stock that defines the Florida bonefish fishery. It is possible that exchange of mature adult bonefish between Florida and the Bahamas could be an important factor for consideration by an international management regime if this kind of observed movement is a common occurrence. It also suggests that new attention should be given to the degree of genetic differentiation between Florida and Bahamas bonefish, with inference for stock mixing based on gene flow (e.g., Gharett and Zhivotovsky, 2003). In addition, since bonefish are schooling fish (Robins et al., 1986; Crabtree et al., 1996), the recaptured fish likely traveled with others.

Anchor-tagging data have also provided some important preliminary improvements in population dynamics information for Florida bonefish. Despite considerable variability in the statistics reported by volunteer captains and anglers, the mean allometric growth model generated from the tag database compared favorably with estimates derived from data collected by scientists. Length-based mortality estimates from tagging program data fell within the 95% confidence interval for male bonefish reported by Crabtree et al. (1996) and provided unique insights into the status of the Florida's bonefish stock.

Size distribution data from our study suggested that bonefish live longer than previously estimated. Crabtree et al. (1996) aged 451 bonefish from the Florida Keys and found a maximum age of 19 years; their largest fish sampled was 679 mm FL. Our tagging database contained 245 bonefish (or about 5% of the bonefish tagged) larger than Crabtree et al.'s (1996) largest fish. The likelihood that bonefish live longer has been already confirmed by a 709-mm FL bonefish that has been aged at 20 years. Maximum age has significant implications concerning stock sustainability with respect to expected response to exploitation and environmental changes. Fish stocks with greater maximum ages tend to be more susceptible to declines from relatively small reductions in survivorship (Ault et al., 2005); thus, additional ageing of large fish from the fishery is warranted. However, this research must be conducted in a judicious manner since there is an understandably high sensitivity among fishing guides concerning sacrificing the largest fish for scientific purposes.

Because we lacked information concerning how far offshore bonefish would travel, in the design phase of our acoustic telemetry study we took a risk-averse position and set two cross-shelf lines of receivers that extended several kilometers eastward of the shallow inshore flats where the fishery is prosecuted. We found that AT-tagged bonefish were detected at substantial distances offshore and at times close to the barrier coral reefs. The observed offshore movements may have been bonefish possibly on spawning migrations because those fish were generally well above the minimum size of sexual maturity, and moved during peak reproductive months according to gonadosomatic indices reported by Crabtree et al. (1997). Johannes (1978) reported an apparent bonefish spawning migration to sandy areas adjacent to Micronesian reefs, and also suggested a strong lunar connection with bonefish spawning behavior. In our study, offshore movements of some bonefish corresponded directly with a full moon or to drops in barometric pressure, at times coinciding with cold front passage. Cold fronts are seasonally intensive in south Florida during periods of peak bonefish reproductive activity, and may be natural physical cues for offshore movements. This type of physical-biological behavior has been documented for striped mullet, *Mugil cephalus* (Behzad Mahmoudi, Florida Marine Research Institute, personal communication), in the region. Bonefish spawning may coincide with spring ebb tides that maximize tidal transport of eggs and pelagic leptocephalus larvae, and presumably favor survivorship. The relatively close proximity of the Gulf Stream current in the Straits of Florida to the deep seaward edges of the barrier coral reefs could greatly facilitate retention or northward transport (Stommel, 1976). Northward dispersal of bonefish larvae is suggested by the presence of substantially smaller bonefish in the Palm Beaches region than the Florida Keys. The Palm Beaches, particularly the southern end of the Indian River Lagoon at St. Lucie Inlet, may be an important recruitment and natal development area for postlarval bonefish. Small immature bonefish from the Palm Beaches region may ultimately recruit to the exploited phase of the stock by southward migration to the Florida Keys after reaching sexual maturity. It is of interest to note that 108 young-of-the-year bonefish were collected in a seine net in Great South Bay, New York (Alperin and Schaefer, 1964).

Potential mortality from either the anchor- or AT-tagging process was a concern amplified by the fishing community. We feel that this concern has been generally assuaged by our findings that indicated no size-selective mortality. In our study, we used the same type of T-bar tags as those by Baeza and Basurto (1999), who reported no mortality associated with tagging. Further evidence was the fact that bonefish in our research carried external anchor tags for significant periods of time (i.e., >1.8 years). The acoustic telemetry study also supported the notion of minimal mortality associated with tagging. One bonefish fitted with a transmitter was caught by a fisherman several months after release, and while the fisherman dutifully noted and reported the external anchor tag, he was unaware that the same bonefish also carried an internal AT transmitter. Humston et al. (2005), using the same tagging procedure also had an AT-tagged bonefish recaptured by a fisherman several months after release.

These results have been encouraging, but much work is still required to better integrate the capabilities of the two tagging methodologies, and to facilitate better understanding of the timing and location of spawning migrations, behavioral responses to environmental cues, and key aspects of population dynamics. Such integrated studies will require higher-intensity tagging efforts and expansion of acoustic arrays to cover more inshore and cross-shelf habitats. Greater anchor tagging efforts need to be spread proportionally across the range of the unit stock, and be linked to a systematic coverage of captain- and tournament-based catch-and-effort logbook data to facilitate improved estimates of stock size, growth, and survivorship. Equipment costs and manpower requirements will always be limiting steps.

Anchor-tagging studies bear minimal fiscal costs to supply participants with the necessary equipment (i.e., tag guns, anchor tags, and data sheets), but they require significant human costs in volunteer participation to meet program goals. In contrast, AT tagging requires substantial fiscal and manpower costs to deploy and maintain receivers and tag fish, and to conduct computer-intensive data analyses. However, costs were minimal relative to several important benefits achieved by this tagging research: it provided a means to address gaps in understanding basic bonefish biology and movement patterns needed for better management practices; it provided a partnership between scientists and the recreational angling community to work together for the goal of more fish in the water; and it was a means to educate the angling public about the importance of catch-and-release fishing.

The methodology of this study could be easily replicated for other bonefish fisheries in a way that would allow effective comparison of results to those obtained from Florida, and such studies could serve as important mechanisms to build sustainable bonefish fisheries.

ACKNOWLEDGMENTS

We thank the many volunteer captains and anglers who tagged more than 4000 bonefish for the project. Captains Joel Kalman, Joe Gonzalez, Bob Branham, and Steve Venini provided critical technical assistance for the acoustic telemetry study. We greatly appreciate the field assistance provided by Nick Farmer, Bobby Gibson, Patrick Dorsy, Brian Teare, Michael Feeley, Ashley McCrea, and Mark Fitchett. Fiscal support for this research was graciously provided by Bonefish & Tarpon Unlimited, Sanctuary Friends Foundation of the Florida Keys, International Women's Fishing Association, and National Fish and Wildlife Foundation.

REFERENCES

Alperin, I.M. and R.H. Schaefer. 1964. Juvenile bonefish (*Albula vulpes*) in Great South Bay, New York. NY Fish Game J. 11(1): 1–12.

Ault, J.S., R. Humston, M.F. Larkin, and J. Luo. 2002. Development of a bonefish conservation program in south Florida. Final Report to the National Fish and Wildlife Foundation on Challenge Grant No. 2001007800-SC. Washington, DC.

Ault, J.S., S.G. Smith, and J.A. Bohnsack. 2005. Evaluation of average length as an indicator of exploitation status for the Florida coral reef fish community. ICES J. Mar. Sci. 62: 417–423.

Baeza, J.A. and M. Basurto. 1999. Effect of two types of external tags in the survival of *Albula vulpes* (Linnaeus), of the Biosphere Sian Ka'An Reserve, Mexico. Rev. Invest. Mar. 20: 4–7.

Colton, D.E. and W.S. Alevizon. 1983. Movement patterns of bonefish, *Albula vulpes*, in Bahamian waters. Fish. Bull. 81: 148–154.

Crabtree, R.E., C.W. Harnden, D. Snodgrass, and C. Stevens. 1996. Age, growth, and mortality of bonefish, *Albula vulpes*, from the waters of the Florida Keys. Fish. Bull. 94: 442–451.

Crabtree, R.E., D. Snodgrass, and C.W. Harnden. 1997. Maturation and reproductive seasonality in bonefish, *Albula vulpes*, from the waters of the Florida Keys. Fish. Bull. 95(3): 455–466.

Curtis, B. 2004. Not exactly fishing (alligator fishing). Chapter 11, pp. 167–171, in Bonefish B.S. and Other Good Fish Stories, T.N. Davidson (ed.). Hudson Books, Canada.

Gharrett, A.J. and L.A. Zhivotovsky. 2003. Migration, pp. 141–174, in Population Genetics: Principles and Applications for Fisheries Scientists, E.M. Hallerman (ed.). American Fisheries Society, Bethesda, MD.

Harden Jones, F.R. 1968. Fish Migration. Edward Arnold, London.

Humston, R. 2001. Development of movement models to assess the spatial dynamics of marine fish populations. Doctoral dissertation. University of Miami, Miami, FL.

Humston, R., J.S. Ault, M.F. Larkin, and J. Luo. 2005. Movements and site fidelity of bonefish (*Albula vulpes*) in the northern Florida Keys determined by acoustic telemetry. Mar. Ecol. Prog. Ser. 291: 237–248.

Johannes, R.E. 1978. Reproductive strategies of coastal marine fishes in the tropics. Env. Biol. Fish. 3(1): 65–84.

Pollock, B.R. 1982. Movements and migrations of yellowfin bream, *Acanthopagrus australis* Gunther, in Moreton Bay, Queensland, as determined by tag recoveries. J. Fish Biol. 20: 245–252.

Ricker, W.E. 1975. Computation and interpretation of biological statistics of fish populations. Bull. Fish. Res. Bd. Can. 191: 382.

Robins, C.R., G.C. Ray, and J. Douglas. 1986. A field guide to Atlantic coast fishes of North America. Houghton Mifflin Company, Boston. 354p.

Sokal, R.R. and F.J. Rohlf. 1994. Biometry. 3rd edition. W.H. Freeman and Company, New York.

Stommel, H. 1976. The Gulf Stream. University of California Press, Berkeley, CA. 248p.

Section IV

Lore and Appeal of Fishing for Tarpon and Bonefish

20 Bonefish Are without Question My Favorite Fly Rod Quarry

Sandy Moret

CONTENTS

Introduction 323
The Florida Keys 324
The Bahamas Archipelago 325
A Sea of Change for Bonefish 326

Sandy Moret and his wife, Sue, have owned and operated the Florida Keys Fly Fishing School and Florida Keys Outfitters since the early 1990s. They live at Tarpon Flats in Islamorada.

Sandy has been grand champion of the Florida Keys' most prestigious fly tournaments, three times in The Gold Cup Tarpon Tournament and five times in The Islamorada Invitational Bonefish Fly Championship. He is often seen as a guest angler on The Walker's Cay Chronicles, The Reel Guys, *and* Andy Mill's Sportsman's Adventures. *Sandy has written numerous articles on saltwater fly-fishing. He has fished and explored extensively throughout the Bahamas, Central and South America, the Seychelles, Christmas Island, and Palau. He helped pioneer Russian Atlantic salmon fishing on the Kola Peninsula.*

Sandy is a past president of The Everglades Protection Association, the founding organization for The Florida Conservation Association. He also served at the appointment of Governor Bob Graham on the East Everglades/Everglades National Park Advisory Council and other elected and appointed positions for Everglades restoration. He is a founding member and sits on the advisory board of Bonefish & Tarpon Unlimited.

INTRODUCTION

I was raised in Atlanta and spent my early years quail hunting and bass and trout fishing in Georgia with spinning and casting rods, as well as the fly rod. The fly rod was fun and considered a novel way to fish back in the 1960s, but it got a lot of strange looks from the bass gang.

After moving to Miami in 1972, I was invited for an afternoon tide's bonefishing on Biscayne Bay with my soon to be bonefish mentor, Flip Pallot. Flip had spent his formative years fishing the saltwater flats of the Florida Keys and is one of the recognized masters of the art. That trip changed my life.

FIGURE 20.1 Author Sandy Moret holding his favorite fly rod quarry, the bonefish.

To learn that you could actually stalk like a wading bird through the shallows and sight fish was unbelievable. To see bonefish cruise water so shallow that their backs broke the surface while searching for prey amazed and intrigued me. I watched as Flip silently laid a tiny imitation shrimp fly 3 feet in front of the bonefish, then when the fish was within 1 foot, make a tiny twitch with the rod tip. To see the bonefish hump up on the fly and pin it to the bottom as the line came tight and then the rooster tail as the bonefish burned out 80 yards of line blew me away. I became a bonefish junkie at that instant (Figure 20.1)!

More importantly, I have had the opportunity to fish places on this watery planet that most anglers only dream about. I am certainly not a scientist or marine biologist; however, I have had the good fortune to be an observer in the field. My lab has been the flats of the Florida Keys, The Bahamas, Seychelles, Christmas Island, Palau, Yucatan, and Belize. Following are some random observations on bonefishing, where bonefish live, and what the future holds.

THE FLORIDA KEYS

The Florida Keys consist of a chain of mangrove and rock islands extending south from Key Biscayne to Key Largo and Islamorada. Starting below Islamorada, the Keys bend westward from Long Key past Key West and on to the Marquesas. This hook running south then west makes up the border between Florida Bay and the Atlantic Ocean. The northern border of Florida Bay is the mainland of Florida, and to the west is the open Gulf of Mexico. We refer to the flats on the east and south side of the Keys as Oceanside flats, and on the Florida Bay side they are called the "back country." The northern third of the Florida Keys separates Biscayne Bay from the Atlantic.

This unique feature of the Florida Keys flats allows anglers to fish a low-rising tide on the Oceanside flats until the water becomes too deep for tailing bonefish, then with a 10- or 15-minute boat ride into Florida or Biscayne Bay, low tide is just beginning again. This allows anglers much longer prime tides for fishing, and I do not know of any other place where this occurs so prominently.

A lot of current is created as the tide squeezes through the openings between the Keys on the tides and it seems bonefish feed more aggressively in stronger currents. Maybe because the speed of the water carrying its prey is faster, the fish must react faster when it has the opportunity.

The Florida Keys flats also seem to have more lush turtle grass than other locations I have fished. I think the grass must hold considerably more food for the bonefish than sandy, rocky, or sea fan bottoms, and I suspect this may have something to do with the size of the Florida Keys bonefish. I also think the proximity to the Everglades estuary system must play a big part in the Keys bonefish size, but I do not know what it is.

Most of the world records come from the Keys. In a typical bonefish tournament with 25 angler teams with emphasis on large fish, the catch will usually include 8–10 fish over 10 pounds, half a dozen fish over 12 pounds, and a 13- or 14-pound-plus fish. This is absolutely amazing.

While I have heard of large bonefish and seen photos of fish over 10 lb caught on the flats of The Bahamas, Seychelles, Cuba and New Caledonia, I have never caught a bonefish over 9 pounds in any locale other than the Keys. I have spent hundreds of days fishing many of those locations and probably caught well over a thousand bonefish in those places. I am sure some big fish must be there and occasionally I see a photo of one, but I think they are few and far between.

THE BAHAMAS ARCHIPELAGO

I remember a day in Middle Bight of Andros Island when my wife, Sue, and I had caught a dozen bonefish in the 3- to 5-pound range. Just as I stepped up to the bow for another turn, a fish cruised up onto the flat. Sue, me, and our guide all freaked out over how gigantic the monster bonefish looked in the water. I caught the fish, thinking it was 12 pounds and we weighed it in one of those net/scale contraptions at 9 pounds. All three of us knew the fish must have been much larger and suspected the scale to be way off. When we returned to the lodge, I filled a gallon jug with water and weighed on the net/scale. The jug of water weighed in at 8 pounds on the money. My guess is that when so many bonefish in the 3- to 5-pound range are seen and caught, a nine pounder, which is two to three times larger, just looks like a giant.

I have fished a fair amount at Grand Bahama, Exuma, Bimini, Chubb, Great Sail, Abaco, and Andros in the Bahamas. I find flats with lots of white sandy bottoms, areas with rock and sea fans, and sloughs that run through low-lying mangroves that open into shallow bays. Few places have the strong current of the water passing through the bridges of the Keys.

The expansive white sand flats of the Bahamas and the Seychelles may not hold as much prey for the bonefish, but they make for great wade fishing. Stalking on foot is a classic way to bonefish and it is rarely possible in the Florida Keys because of the soft marl bottom usually found where the bonefish feed.

Many bonefish anglers carry a tremendous assortment of fly patterns to cast at bonefish. There are zillions of baitfish, crab, shrimp, worm, and eel imitations that fill books and bins in fly shops, all of which makes for lots of fun and great discussion topics over libation at the end of the day. That is all fine, but just give me a small merkin crab with a body from the size of a nickel to a quarter. Tan over sand, brown, and tan over grass. Use bead chain eyes in shallow water and lead eyes as the water gets deeper. I think there are crabs everywhere bonefish come on the flats, and day in and day out, I have found crabs to be the best fly pattern for bonefish.

A SEA OF CHANGE FOR BONEFISH

Bonefish are one of the most valuable fish per pound on the planet. Most bonefish anglers must travel, hire a guide, arrange lodging, and buy tackle and appropriate clothing. It is not an inexpensive pastime or passion for some. A top of the line fly rod and reel runs $1300, and guide fees run $400 to $500 per day. Catching two bonefish on fly in the Keys is pretty good. If you do that for a week, you will catch one over 10 pounds. Catching 6–12 is pretty good most other places. I have mixed feelings as to what the future holds for bonefish and bonefish anglers. These valuable fish must run a gauntlet of obstacles to continue to lure anglers to them.

Virtually no one kills bonefish in the Keys. My concern here is the loss of habitat and changes in water quality. Agricultural and urban runoff and inconsistent freshwater flows into the Everglades have wreaked havoc on this ecosystem for almost 100 years.

Boating pressure, especially by uninformed and just plain incompetent boaters running over grass flats, will increase dramatically over the coming years and geographic positioning system (GPS) is not a good substitute for "reading the water." Look at the fact that one major housing development in Florida City has been approved for more than 5000 units. If 20% of those homes have boats, and half of the boaters go out on a weekend, there could be an additional 500 boats on the waters between Key Largo, Flamingo, and South Biscayne Bay.

The government of the Bahamas has done an excellent job in protecting bonefish. They seem to realize the value of bonefishing to their economy, especially in the out islands. The sparse human population of the out islands may also ensure healthy fisheries for years to come. I understand bonefish can no longer legally be sold in the Bahamas, and I have also seen a change for the good in guide and lodge policies over the years.

Up through the late 1970s, it was very common for the guides to keep a bonefish for dinner after a day of fishing at The Deep Water Cay Club on Grand Bahama. They preferred the larger ones especially. What is the impact to the local fishery if a lodge with 12 boats times 100 fishing days takes the largest of the fish from the waters year after year? Just during the last 15 years, bonefish have been rarely kept by the guides and my impression is that the average size of bonefish is increasing with more 4- to 7-pound fish than years ago. Actually, I rarely see a guide anywhere in the Bahamas take a bonefish anymore and the guides on Andros have been real leaders for the release ethic. The Seychelles has a strong catch and release ethic for bonefishing on several of the islands as well.

I am told that bonefish are netted heavily for food at Christmas Island and many other Pacific locations. Poverty and the need for protein at sustenance level conditions vs. recognition of the economic benefit of a sport fishery will become an even bigger issue in years to come. In 1994, the local government and the nature conservancy arranged for me to explore the potential of a viable bonefish population on Palau in Micronesia. I waded and drifted fantastic looking flats that looked like a combination of Turneffe Atoll, the Bahamas, and the Seychelles. I spoke with locals and had a local "old timer" show me bonefishing his way near Pelelieu. That was with a hand line in 10 feet of water using sections of a sea worm that was two and a half feet long and a half inch in diameter. We caught a wrasse or two and had a sea snake swim up behind the boat. When a dugong swam by, the old timer mimicked shooting it with a rifle and pantomimed eating it. In a week of searching prime habitat, we never saw a bonefish, but we heard plenty of stories of how many there once were. I believe an entire bonefish population was harvested beyond the recovery potential of such a limited and fragile environment as exists on Palau. In the long run, the future of bonefishing around the planet is going to be determined region by region by local governmental authorities who understand the value of a viable fishery. Areas with access and infrastructure for anglers to pay money for the sport will likely flourish as bonefish resorts. As human populations grow on remote Pacific and Indian Ocean islands and bonefish are netted for food, the outlook is not good.

21 Record Tarpon on a Fly Rod

Stuart C. Apte

CONTENTS

Introduction 329
Background 330
Tarpon Fishing Records 331
A Record of My Own 333

> Stuart "Stu" C. Apte, of Tavernier, Florida, is an all-around angler considered a pioneer in fly-fishing for tarpon. In the 1960s, he developed a huge following as a Florida Keys guide. He is also a writer, photographer, and holder of more than 40 IGFA world records. Stu has invented fly patterns and knots and helped design rods and reels. A U.S. postage stamp commemorates his tarpon fly. Stu had the distinction of joining Ted Williams, Ernest Hemingway, and Zane Grey when he was enshrined in the International Game Fish Association Hall of Fame in 2005.

INTRODUCTION

Although a fresh wind from the south had spread a glistening chop across the water's surface, through Polaroid glasses I spotted a school of about 40 fish traveling in our direction. At first they were only faint wavering shadows, but instinct told me they were tarpon, probably of good size. "Be ready," I instructed Guy Valdene who stood in the stern of the boat, "to cast at 10 o'clock." "Ready," he answered. As soon as the lead fish were in range, my companion false cast once and placed the orange and yellow streamer fly a little off target. Just the same, a very big fish turned out of the school and came after it. I held my breath; waiting for the strike, which I knew would be more like an explosion. But in his excitement, and from inexperience also, I suppose, he struck too soon and took the fly away from the tarpon's big open mouth.

"Cast it back again," I shouted. This time the fly fell closer to the tail end of the school and instantly a small fish had it. In the same split second it was up and out of the water in the kind of jump that never fails to make the bristles stand up on the back of my neck. "Bow," I shouted, "bow from your waist and thrust the rod toward the fish, giving it some slack every time he jumps." I was pretty sure it was a male tarpon because of its size and shape. What followed was a typically wild and watery contest, which only the silver king can provide. The fish seemed to jump two places at once and I had a busy time poling the boat to keep Guy in the fight. He did a pretty good job because 30 minutes later he had maneuvered the tarpon into position for gaffing. I jammed the push pole into the bottom, lashed the boat rope to the pole, and grabbed my lip gaff. On the first attempt I put the release gaff into its lower jaw and that is when all hell broke loose.

For several seconds the tarpon shook its head crazily while I just held on. That was enough to sheer off the metal part of the gaff which had rusted nearly through, unseen, inside the wooden handle. So I had to boat, unhook and release the fish, which was a 60-pounder, with my bare hands alone. When I was finished, my fingers were raw and "bleeding." But what a happy scene it was. "Congratulations," I said to my companion, over and over, because it was his first tarpon on any tackle and the largest fish he had ever taken in a short career of flyrodding. It was also the first fish in an uncommonly unsuccessful fishing trip to the Lower Florida Keys, which had started weeks before.

"Now the jinx is broken," Guy laughed, "and the pressure is off. Now you take the rod and cast to the next tarpon we see. What I need is a chance to rest and unwind." That was an unusual offer since Guy was the client and I was the guide. After splitting a cold beer, I picked up the rod, checked the leader for fraying and stood ready to cast. At that moment, I had no idea that the greatest fishing experience of my life was soon to begin. Out of the corner of my eye I saw tailing tarpon coming our way. But that is getting ahead of my story.

BACKGROUND

Fishing has always been the greatest thing in my life. In fact it is my life because I am a professional fisherman. I am never happier than when I am prospecting the Florida Keys flats for tarpon, flyrod in hand. Developing new and better tackle is another special interest of mine. Luckily, I live in Miami, Florida, my birthplace, where fishing is always excellent and nearby. It is also an area where light tackle angling for big fish had produced a whole new cult of outdoorsmen. Curiously enough, I do remember my first childhood fishing trip, in my next-door neighbor's Goldfish Pond, and I also recall getting up at 4 o'clock on many mornings and bicycling to Biscayne Bay near the present site of Rickenbacker Causeway. There I squeezed in a few hours of fishing before school. Then at 36, I often did the same thing before or after one of my flights as a copilot with Pan Am. In 1942, at the age of 12, I became infected with the light tackle bug when I acquired an old bamboo casting rod and a freshwater plug–casting reel. With it I landed my first tarpon, a 15-pounder, but broke the tip of the rod when it was caught in the spokes of my bike one morning. That was a blow. Not much later I earned enough money for my first used flyrod and I was hooked for keeps. Homer Rhode Jr., then a game warden and a well-known fisherman, taught me how to tie flies for saltwater fishes. In 1949, I caught my first tarpon on a fly—a 20-pounder. A few months later I landed my first really big tarpon, a 96.5-pounder, but it was taken on a plug near Marco Island, on the southwest cost of Florida. It seems like I always had a rod of some kind in my hands in those days.

But fishing had to wait for the Korean Conflict, when I enlisted in the Navy as an aviation cadet and went into flight training. After graduation I flew F9F Panther Jets, FJ-3 Furies, and the first all Delta Wing, F7U-3 Cutlass from the decks of various aircraft carriers. All the while, I dreamed of fishing at home and only a few days after release from active duty I caught my first flyrod tarpon, which weighed over 100 pounds. It was the greatest possible way to celebrate being a civilian again.

TARPON FISHING RECORDS

With all the flying experience, it was only natural to seek and obtain a pilot's job with Pan-American World Airways. But lacking seniority in the early days with the airline, there were frequent, long layoffs, and during these periods I began to guide winter tourists from a headquarters at Little Torch Key, 28 miles north of Key West. The truth is that I flew very little for several years and became a full-time guide. It was a very meager beginning, however, because my first guide boat was borrowed from a friend, Bill Curtis, who became an immensely successful guide. Because I enjoyed both the sport and the work so much, I had considerable success. I would willingly fish from daybreak to dusk if the customer were willing. As a result, I built up a clientele of expert, serious fishermen who fished with me as friends. Luckily, I guess, my customers began catching more than their share of prizes and citations in the Metropolitan Miami and other Florida fishing tournaments. At the same time, gradually I became more of a specialist in flyrodding and I concentrated more and more on fly-fishing for tarpon. I consider them just about the most exciting and most unpredictable of all our game species. They are big, they are strong, and they are highly acrobatic. In addition, the big ones were very available where I was guiding. Some good breaks came my way—such as guiding jobs in motion pictures and in a 2-day tarpon fishing contest for experts, which was filmed as a 1962 *ABC Wide World of Sports* fly-fishing spectacular. In that production, I handled the boat for Joe Brooks, angling writer, authority, and friend who easily won the fly rod tarpon competition. Two years earlier, I was guiding Joe when he caught the world record tarpon on a flyrod. It weighed 148.5 pounds and pulled me into the water twice when I gaffed it. That catch was the most thrilling fishing experience I had ever known.

I had a hand in other records. I guided Kay Brodney, a librarian in our Library of Congress, when she caught the 137.5-pounder that remains to date the women's flyrod tarpon record. Twice Ray Donnersberger, a Chicago businessman, was fishing with me when he captured the best flyrod tarpon in the Metropolitan Miami Fishing Tournament. And I guided Russ Ball of Bryn Mawr, Pennsylvania when he caught the world record tarpon of 170.5 pounds on spinning tackle using manufactured stated 8-pound test line. Although I had little opportunity to fish myself during those busy years of full-time guiding, I did catch the largest tarpon at that time ever caught by a guide on his own. It weighed 132.5 pounds. Mark Sosin of New Jersey was handling the boat for me that day.

To qualify for a saltwater flyrod record, the catch must be made within certain rules and regulations (see www.igfa.org). These regulations are observed by all major fishing clubs and tournaments and have recently been given official continent-wide status by the Saltwater Fly Rodders of America, the international custodian of the records. The flyrod can be no shorter than 6 feet and no longer than 10 feet. The fly must be cast in the orthodox manner of fly-fishing, not trolled or drifted. The fish must be fairly hooked, fought, and brought to gaff or net without the aid of another person, except to handle the boat or the landing device. No doubt the most important regulation for a record is that the leader cannot test more than 12 pounds, although a shock tippet of any strength may be used because of the sharp teeth and gill plates of certain saltwater species. Very recently the Saltwater Fly Rodders have added other tippet classes—6-, 10-, and 15-pound tests—but we are not concerned with those here.

The outfit that I presently use includes a 9-foot medium action 5-1/3-ounce flyrod with oversize snake guides, a carbon alloy tip-top, and a thin foregrip above the main grip to help hold the rod when playing big fish. The large snake guides permit better shooting of the large (WF11F) tapered lines I use. The most unique feature of the rod is a 4.5-foot insert, which I helped to develop. This fits into (or inside) the rod's butt section after the fish is hooked and thereby gives extra backbone against the struggle of a really big fish. However, it would interfere with casting if inserted before the fish is hooked. My reel is a single action, positive retrieve with capacity for 200 yards of 27-pound test backing, plus the 30 yards of WF11F fly line. The butt section of my leader is 6 feet of 30- or 40-pound test monofilament joined to the flyline with a nail knot and covered with Plyobond, a plastic coating. To meet flyrodding regulations, my middle section of leader is at least 18 inches of 12-pound test monofilament. Beyond that, I use 12 inches (the maximum permitted) of 100-pound test monofilament as a shock tippet. The three sections are joined by the Stu Apte improved blood knot, which I developed when testing lines for the Stren division of the DuPont Corporation. This is an excellent all-around outfit for flyrodding the salt.

With it, or with something very similar, I have set world records for dolphin (58 pounds), jack crevelle (24 pounds), Pacific sailfish (136 pounds), and yellow-fin tuna (28 pounds). All these were set during two trips to Panama.

It was only natural that I should start thinking seriously, perhaps too seriously, about catching some records right in my own backyard. The flyrod tarpon record, for instance. For one thing, where I live, the intense spirit of competition is always present. Expert fishermen and members of the prestigious Miami Beach Rod and Reel Club, such as Luke Gorham, Al Pfleuger, Jr., Lee Cuddy, and Bart Foth, are constantly out to set new records in South Florida waters. But far more important than that, I knew that there were many tarpon in the Lower Keys much larger than any ever taken before. I had seen them many times and on a number of occasions I had seen them hooked. So have most of the other tarpon guides. Just last year, for example, a customer of guide George Hommel hooked a tarpon larger than 200 pounds on fly rod; Hommel actually had it on the gaff, but it broke free. "I have no doubt whatsoever," the veteran guide remarked later, "that soon someone will land a 200-plus-pounder on a fly."

In 1962, I had come close to it myself. The camera crew of a national TV network was filming a tarpon movie near Big Pine Key and I spent a couple of days casting for tarpon to give then some fill-in jump shots. One of the fish I hooked was 185 pounds, absolute minimum, and I held onto it for several jumps. But the lint wrapped around the reel handle and the 12-pound leader snapped.

I remember other fish of similar size being hooked—and lost. The day before Joe Brooks got his 148.5-pound flyrod record, I was poling for Dave Newell when he hooked and nearly boated a fish in the 170- to 180-pound class. There must have been an unusual invasion of huge tarpon at that time because only a few days later I was guiding Lee Cuddy, a Miami tackle dealer, when he also tied into an extraordinary fish of at least 180 pounds and probably much more.

Record Tarpon on a Fly Rod

It would be impossible to forget that occasion. The fish struck an orange and yellow streamer near Coupon Bight and, after a 2-hour fight, joined up with a passing school of tarpon and actually led the school out to deep water. First, it passed through Newfound Harbor Channel, and from there we followed it through Hawk Channel to the edge of the Gulf Stream between Looe Key and American Reef Light, a total distance of about 7 miles. Once a hammerhead shark made a pass at the tarpon but unaccountably turned away. At times we were close enough to the fish to see the orange and yellow fly in its jaw. When the fish finally broke off after 5 hours and 15 minutes of pressure, Lee had tears in his eyes. I did not blame him because so did I.

In 1964, I came within an eyelash of gaffing a 200-pounder, for sure. The fish was hooked by Ray Donnersberger in a location that I would rather not identify, but which I now call Monster Point. After taking Ray's fly, the fish ran into very shallow rocky water where it began to jump crazily. It seemed to go berserk. This was one time I believed the fish was badly hurt by jumping and falling back on the rocks because it began spewing blood in all directions until the water was amber colored. After one wild jump almost beside the boat, I reached for the gaff. But then the 12-pound test leader slipped under the gill plate and was cut. The fish swam away, slowly and weakly.

Two years later, in 1966, I was guiding Mark Sosin and Leon Martush (a fly line manufacturer from Michigan) around Sugarloaf Key. As luck would have it, the first tarpon he had ever hooked—in fact the first he had ever seen—was in the 175-pound category. And he did an excellent job of playing it. After an hour or so, the fish broke off. That day was abnormally rough and windy, otherwise I believe we could have boated that fish.

Perhaps I should not go any further because these next items may seem impossible. Anyhow, the Russ Ball 170.5-pounder, which is still the spinning record, was the smallest fish in a school of about 20.

And in 1963 I saw a tarpon that was much closer to 300 than to 250 pounds. It was by far the biggest I have ever seen in or out of the water anywhere. That morning I was guiding Sam Clark, a Washington, DC attorney and an experienced salmon fisherman. We were fishing the ocean side of Sugarloaf Key when the monster inhaled Sam's fly. It made just one jump, close enough at 30 feet so that we had a good clear look at it. Sam was so shaken that he froze on the reel and of course the 12-pound leader snapped.

A RECORD OF MY OWN

So it is not any wonder that I have seriously started thinking about a record of my own. For the last couple of years I have reserved the best periods, which I call the prime tarpon tides, for this purpose. These come during the new and full moons of March, April, May, and sometimes June in the Lower Florida Keys, as long as the water is warm enough (74° or more) and the wind is less than 10 knots, I base this on my own experience of the past 15 years. But let me explain that these are not the only times when tarpon fishing is productive.

In February of 1966, with a few days off from flying, I flew over to Deep Water Cay in the Bahamas with my wife, Bernice, mostly to relax and for a change of pace.

One evening we were fly-fishing a shallow flat for mutton snappers when we met another angler doing the same thing with unusually good skill and coordination. He introduced himself as Guy Valdene of Palm Beach and we began a fortunate friendship. That evening, we spent several hours discussing flyrodding in the salt—and especially for tarpon. The upshot was that I agreed to guide him for tarpon during the full moon tides of March and April. It did not work out very well, at least not at first during March. The winds, which were seldom less than 25 knots, nearly blew us off the flats. And it was very cold. It reminded me exactly of the spring tide periods, a year before which I spent record hunting with guide Russ Gray and old friend Erwin Bauer, who is an adventure writer and editor of the *Saltwater Fisherman's Bible*. In 1966, we did hook some fish in spite of the weather, but nothing of any consequence. It takes far more than unfavorable weather to discourage me. Luckily Guy was a kindred spirit and we were back in the Keys to fish the new moon tides in early April as planned. For the first couple of days the weather was considerably better, but tarpon had not yet moved onto the flats in mass. Still Guy had about 50 chances to present flies to tarpon and briefly hooked several. I believe that two of them would have been good enough to break the existing flyrod record. But no fish of any size were landed by the time the weather deteriorated.

Once more the fishing ranged from unpleasant to almost impossible. "I'm still game," I said while driving back to Miami. "Count me in," Guy replied. That was how that April full moon found us back in the Lower Keys. On the second day of fishing, we began to encounter large numbers of tarpon and it was on a flat that I code-named The Bullfighters. Guy broke the ice when I boated and released his first tarpon—the 60-pounder described in the opening of this chapter. A few moments later I stood in the stern, flyrod in hand, hoping to catch another—only much, much larger. I did not have to wait long. The boat drifted very slowly for a short distance when I spotted a school of 40 or 50 fish cruising just out of casting range. The water was calm and I knew they would be spooky, so I crouched down low to cut the size of my silhouette against the sky. From that position I waited for them to come within casting distance. The fish leading the school was a large one, but there was a much bigger tarpon near the center of the school. Still I decided to make my presentation to the first fish, rather than risk having my line fall across other tarpon to reach the largest and thereby spook all of them. Suddenly, I felt the same surge of excitement and suspense I always feel when big tarpon are the targets.

I cast my yellow and orange saddle hackle streamer in the path of the pack, allowed it to settle. As the fish approached, I began the retrieve. The leader charged after it for about 10 feet and when it started to turn away, I stopped the fly, and I had butterflies in my stomach.

The tarpon came back, I twitched the fly, and the fish had it. I set the hook instinctively. And I could feel the tarpon react all the way down to my heels. The first run was short, interrupted by two jumps, and did not even strip off the 30 yards of flyline and get into my backing. I figured the fish to be between 115 and 120 pounds—good trophy size, but not the record fish I am always thinking about. After that initial run, I even turned to Guy and suggested that he take the rod and fight the fish just for the practice. "No, thanks," he answered, "I'll just relax and watch the pro." (See Figure 21.1.)

Record Tarpon on a Fly Rod

FIGURE 21.1 Tarpon are well known for their (A) powerful runs and head-shaking tactics and (B) acrobatic leaps.

Although I did not know it at the time, Guy's reply was the second biggest break of the whole trip for me. His earlier suggestion that I cast while he poled the boat was the first. And after that first short run, my companion had to pole in earnest. The second run was longer and included two more jumps. But thanks to the combination of Guy's poling and my pumping for all the tackle would stand, I worked the tarpon in close to the boat where it made a head-shaking half-lurch out of the water. Mentally, I revised its weight upward to about 130 pounds or slightly more.

At this point, the tarpon had been hooked 10 minutes and if Guy had been fishing, I would have made my move with the gaff. Probably I would have connected because my record at boating fish is good. The leader was through the tip-top of the flyrod. But something happened. The fish spooked and began a fantastic run. No bonefish I have ever hooked ran any faster. There was no stopping it or even slowing it down. First, Guy tried to keep up with the mad dash by poling, but no soap, so we decided to start the outboard. To do so quickly, suddenly became the most urgent thing on earth because for the first time since I had filled the fly reel 4 years before, I could see the metal spool through the last windings of backing. Somehow I got the outboard tilted down into position and the drag adjusted on the reel. But the sudden acceleration of motor almost tossed me overboard. And about the same time, the boat was aimed in the right direction, there was 20 feet of slack line in the water and the fish had dead-ended its run into a tiny mangrove island. "Neutral, put the motor in neutral," I shouted. At the same time, I reeled frantically to regain lost line. Since the strong tidal current was flowing past the island, I was not entirely certain if I was hooked to the tarpon or to the mangroves. And since I was still not thinking of this as a record fish, I remember turning to Guy and telling him how important it always is to rapidly regain line in a situation like this.

"Now turn off the motor," I instructed. "Jam push the pole into the bottom and hold onto it." From this position I gradually worked the tarpon away from the island and onto the flat. Halfway to the boat it careened out in another head shaker. "At least 130 pounds," I mumbled out loud. Now the fish began to behave in a very strange, erratic manner. I thought possibly that a shark might be chasing it but saw no sign of any other fish nearby. The tarpon darted back and forth in figure eights several times, as if trying to catch its tail. I thought I felt the leader rub across the fin and it was a sickening thought. Next it made another short run and at the end of it a magnificent gill-rattling leap, putting more butterflies in my stomach. But that last run and wild leap seemed to have drained the fight from the fish. Slowly I pumped it in toward the boat and told Guy to grab the gaff and try to hook the fish right from beneath the middle of its belly. You may remember that my rusted lip gaff had broken earlier that day. Now all I had on board was a short-handled hand gaff and Guy stood ready to use it. As soon as I worked the tarpon close enough, he made his move and connected. As I knew it would, the tarpon exploded.

Struggling with immense power, the tarpon rapped Guy across the head with the gaff handle. Then it tossed the gaff free into the water about 20 feet away. That should have ended the fight right there. My chance of ever boating the fish would have been small, almost nonexistent, except that the tide carried the boat directly over the gaff lying in very shallow water.

With one free hand, I reached over the side and retrieved the gaff. A moment later I sunk it into the tarpon and just held on; there was nothing else to do. When it finally stopped pummeling me and the gunwale of the boat, Guy helped me wrestle it aboard. Then I sat down, wet and weak from the excitement. Only 18 minutes had elapsed since the fish had been hooked, but what an action-packed 18 minutes it had been.

Record Tarpon on a Fly Rod

All at once, here in the boat, the fish looked bigger than it had ever appeared in the water. "Probably will go 140," I said hopefully, "and maybe as much as 150." But to tell the truth, it did not look as big to me as some other tarpon I have had aboard my boat *Mom's Worry*.

Instead of taking it in to be weighed immediately, I suggested that we go looking for a bigger one. "Things are working our way," I said to Guy, "so let's not waste time." During the rest of the afternoon, with me on the push pole, Guy hooked three more tarpon and landed one, an 80-pounder. We released it.

It was not until 4 h later that we pulled into the Sea Center dock at Big Pine Key, where we hoisted the fish onto the scale operated by Herb Pontin, an official weighmaster for the Metropolitan Miami Fishing Tournament. For a minute I could not believe my eyes when the scales read 151 pounds even (Figure 21.2). All at once, I was being pounded on the back and being congratulated. Then it slowly sank in that I had a new flyrod tarpon world record, by 2.5 pounds. I had caught the largest game fish of any kind ever taken on a flyrod with a 12-pound test leader. How do you describe a moment like that? I cannot. A witnessed affidavit of the catch, a photo, plus my leader and the

FIGURE 21.2 Among saltwater anglers, the most sought-after record is that of landing a tarpon on fly. Stu Apte is shown with his long-standing 68.3-kilogram (151-pound) world record, caught in 1966 at Big Pine Key in the Lower Florida Keys.

fly had to be submitted to Saltwater Fly Rodders of America for official recognition, which is normally withheld for 6 months. The record-catching fly, incidentally, had been tied for me by Larry Kreh, the 14-year-old son of fishing and flycasting expert Lefty Kreh from Miami. But I am not resting on any laurels. Perhaps even while you are reading this story, I will be drifting the Florida flats, flyrod ready to cast. As I said before, there are some 200-pounders waiting to be caught. And I want to be the first flyrodder to catch one.

22 Learning from History

Bob Stearns

CONTENTS

Introduction 339
Where Have All the Bonefish Gone? 340
Decline of the Silver King 342

Bob Stearns holds a B.S. in meteorology (1958) from Florida State University. He is currently an editor-at-large for Saltwater Sportsman Magazine. *He was a research associate at the University of Miami School of Marine and Atmospheric Sciences from the mid-1960s to 1973, participating first in hurricane research projects (aboard hurricane hunter aircraft) and later in biological studies of Biscayne Bay. He began writing for fishing and boating magazines regularly in 1969, and became a full-time magazine writer/photographer and consultant to the fishing/boating industry in 1973. He has published more than 1500 feature articles and columns in many popular national magazines, such as* Boating, Sea, Yachting, Motor Boating & Sailing, Rudder, Motorboat, Field & Stream, Outdoor Life, Sports Afield, Fishing World, Saltwater Sportsman, Small Boat Journal, Saltwater Fly Fishing *(Boating Columnist), and many others. His first book,* The Fisherman's Boating Book, *was published in 1984. In 1991, he completely revised the* Saltwater Fisherman's Bible *with Erwin A. Bauer.*

During the past 30 years, Bob has appeared on numerous television fishing programs, such as The Fisherman *series (Glen Lau Productions, for Brunswick Corp.), the* Outdoor Life *series (by the magazine of the same name), and* Sportsman's Adventures. *In 1987 he appeared as the host in a video production titled* The Alaskan Angler *(produced by VideoLore). His consulting activities have included companies such as 3M, Cruisers Inc., Aquasport, Carver, DuPont, Mon Ark, and others. As a consultant he has worked with boat and boating materials manufacturers toward the goal of product improvement, especially in the areas of design utilization, safety, and fuel efficiency. He has also worked with several fishing tackle manufacturers on product development.*

INTRODUCTION

I caught my first bonefish in Marathon, Florida, on a sunny day in March, 1951. It was my second trip to the Keys from my home in North Carolina. I was hooked for life! And in the process of learning about the fine art of bonefishing from that late great master of the flats, Capt. Harry Snow, Sr., I also came face-to-face with tarpon, permit, and other gamefish that share the same shallow water environments.

I finally caught my first big tarpon in June 1952, and once again I was hopelessly addicted forever. There is no known cure for these afflictions; I simply got caught up in an endless cycle that sometimes took me to far-flung corners of the planet in search of relief. Thus, for almost five decades now, I have fished for bonefish throughout much of the tropical Atlantic, from the Florida Keys, Bahamas, Bermuda, and Caribbean, southward to the northern coast of South America, and even a few places in the tropical Pacific. As for tarpon, I have caught them on both sides of the Atlantic, the Gulf of Mexico, the western Caribbean all the way down to the north coast of South America, and even in some rivers on the Pacific side of Central America, thanks to the Panama Canal.

Misery loves company, especially when it involves serious angling addiction. Along the way I have been fortunate enough to have shared many casting decks with such legends in the world of bonefish and tarpon as Capt. Bill Curtis and Capt. Lee Baker, Al Pflueger, Stu Apte, Lefty Kreh, and the late Harold LeMaster, developer of the immensely popular Mirrolure.

WHERE HAVE ALL THE BONEFISH GONE?

Now that over a half century has passed since my first encounters with bonefish and tarpon, I cannot help but wonder if my reflections on those early days are really as accurate as I would like to believe. Or are they clouded by powerful memories of very successful outings and a natural tendency to overlook those days when the action was slow or nonexistent? So I recently turned to two long-time friends who were there in those exciting early days of the 1950s and are still actively pursuing these same gamefish today.

I first fished with Capt. Bill Curtis in March 1972. I had an assignment from *Saltwater Sportsman Magazine* to do a how-to article about catching bonefish on a fly ("How to fly a bone" appeared in the August 1972 issue). Bill was unquestionably the most popular bonefish guide on Biscayne Bay in those days, and booked almost every fishable day. The only way we could find time to get together on the flats to do the story was in the late afternoon, after the end of his day's charter. Bill put me into so many bonefish on our first outing, which lasted only for the last 2 hours of daylight, that I had all of the information and photos I really needed. But, to tell the truth, I wanted so much to fish with him again that I suggested yet one more late afternoon session "just to tie up some loose ends." I doubt that I fooled Bill for even 1 second, but he graciously agreed. Day two, a week or so later, was a wild repeat of day one.

"There were so many bonefish around back then," he recalls, "that I never had to go more than a few miles from the dock to find all that I needed to keep my customers happy."

On one occasion in 1980, when I was doing a bonefish story for *Outdoor Life* magazine (this time on Bill's highly successful career as a bonefish guide—the story appeared in the June 1981 issue as "The bonefish master"), he provided me with some interesting numbers indeed. In essence, Bill was guiding approximately 300 days per year, and his average take was 1500–2000 bonefish per year for the previous 20 years (over 95% released). That is typically five or more fish per day over a wide range of weather conditions and skill levels for his angling customers (Figure 22.1), all caught from Key Biscayne southward to northern Key Largo.

"That just isn't possible today," he told me one evening in 2005. He estimates the current numbers of bonefish on the flats are only 10–15%, compared to his early

FIGURE 22.1 Bonefish (*Albula vulpes*). (A) Cruising the flats and (B) "tailing."

days as a guide. In addition, angling pressure on these fish has increased by more than tenfold since then. This begs the obvious question: Has the bonefish population decreased that substantially, or has the angling pressure forced these fish to change their habits and spend less time on the flats where they can be fished by sight? The latter has already been clearly demonstrated to occur with tarpon fished visually in shallow water.

Nowadays, for instance, I find more Florida flats empty of bonefish than I do with fish on them. And I rarely see those massive schools of spawning bonefish that I so commonly found during the 1960s, 1970s, and early 1980s, or the big schools of traveling bonefish that I frequently encountered on the flats. The fish I do find today are widely scattered and far more wary, a clear sign of frequent angler interaction. This is even becoming truer on some of the more popular bonefish spots in the Bahamas. But the change in the U.S. fishery over the past 25 years is thus far much greater.

It is certainly possible that bonefish may not have decreased in numbers *that* much, but have instead significantly changed their feeding habits to correspond with the continued increase in angling pressure and boat traffic in shallow water. They have also long been known to feed after sundown; Zane Grey wrote enthusiastically about fishing for them with bait on moonlit nights. This leads to another question: Do they nowadays feed more on the flats at night, when they can do so undisturbed by anglers, and in deeper water during daylight hours?

DECLINE OF THE SILVER KING

The numbers of tarpon available today are still fairly good, at least in some areas, even though the overall population definitely appears to be down significantly from the halcyon days of the 1950s and 1960s when they were greatly abundant almost everywhere. Back then it was rare indeed that any well-placed lure or fly, which looked remotely like something edible was passed up (Figure 22.2). This was even typically true in gin-clear water; it was a common sight to see a 6-foot tarpon leave a school and travel many feet to gulp a live bait or a lure.

FIGURE 22.2 Leaping tarpon (*Megalops atlanticus*) close to the boat.

As veteran tarpon angler Al Pflueger told me recently, "Back then they were everywhere, always happy, and always willing to eat anything you threw at them. Today there are far, far less of them, they are definitely no longer happy, and they rarely eat even the most well-placed bait or lure." Pflueger feels their numbers have decreased by "at least" 75% since the mid-1960s.

Even though tarpon have been protected for many years in Florida by the requirement of a $50 "kill tag" (per fish) that has to be purchased in advance from the state before a dead tarpon can be brought back to the dock, their availability has nevertheless steadily declined throughout a large part of their historical range of greatest abundance. Some anglers blame this on loss of habitat, others on less available food. And still others point to increased angler pressure that they feel has apparently altered both the tarpon's behavior and annual migration pattern.

For example, tarpon traveling along the oceanside of the Florida Keys and in other areas where the water is clear have become very difficult to deceive. They even sometimes pass up live bait, and are frequently observed to either ignore or actually turn and flee from an artificial lure or fly. But, in those areas where the water is more turbid, and at times even in clearer water when the light levels are low (early and late in the day, heavy overcast, etc.), they sometimes become more willing to strike both artificial and natural baits. A growing number of anglers now prefer to fish at night for two reasons: this predator is well equipped for feeding after dark, thanks to its exceptionally large eyes, and does so regularly. And obviously under these conditions, they are not particularly alarmed by the wire or heavy monofilament leader attached to the bait or lure.

Inshore tarpon movements in water less than 20 feet deep have also changed drastically in some areas. Many feel this is also the direct result of unrelenting angling pressure. Along much of the Gulf Coast of Florida, for instance, where big schools of large migrating tarpon formerly congregated near shore with precise seasonal regularity—where they often went through what appeared to be a spawning ritual by swimming several fish across in a large head-to-tail circle (anglers call this "daisy-chaining")—today they are more likely to be found scattered in deeper water a mile or more offshore. And even if they are daisy-chaining, they are more easily disturbed by lures, flies, or even live bait.

Tarpon like to feed in deeper channels and inlets, where the tide brings baitfish, crabs, and shrimp within easy reach. A good example is the 50-foot-deep ship channel, called Government Cut, which leads from the open Atlantic right up to downtown Miami. There are at least some large tarpon (e.g., 50–100 pounds) in this channel at all times, as long as the water temperature is over 70°F. During the late fall, winter, and early spring months these numbers at times frequently increased into the thousands.

When I first started fishing the "cut" back in the late 1960s, hookups were easy both day and night with deep-swimming plugs, jigs, and of course any live bait. I fished for them almost entirely during daylight hours because they were so eager to strike and it was so much fun to watch their incredible silvery aerial display in the bright, early morning sunlight. Hooking as many as 10–15 between daybreak and noon was not uncommon; on a really good day, two of us would put 40 or more in the air before lunch. But as the years passed, their numbers slowly decreased and they became more and more difficult to catch with "pure" artificial lures. By the early 1980s, I had to resort to tipping a jig with a large live shrimp to get their attention. For a few years this was almost as effective as plugs were 15 years earlier, but by 1990 the number of strikes continued to decrease until hooking one or two fish in the morning became a real challenge. These fish were definitely getting more and more "educated." And also every year there were less and less of them.

This steady downward trend from those early years of great abundance is being repeated all along both coasts of Florida, where angling pressure continues to increase with every passing year. Therefore, are we seeing an actual decline in the tarpon population, or are we simply pushing them farther and farther offshore?

My own experience and instincts, as well as others I've discussed this with, tell me it is both. And that the same situation almost certainly applies to bonefish as well.

23 Memories of the Florida Keys: Tarpon and Bonefish Like It Used to Be!

Mark Sosin

CONTENTS

Introduction 345
Fishing for Tarpon and Bonefish in the 1940s–1960s 346

An award-winning writer, photographer, radio personality, and television producer, Mark Sosin has an impressive list of credits that span virtually all phases of outdoor communications. He is the executive producer and on-camera host of Mark Sosin's Saltwater Journal, *broadcast to all 50 states and several foreign countries on The Outdoor Channel. More than 3000 of his articles have been published in major magazines and he is currently writing his 30th book.*

Considered a leading educator and one of America's most knowledgeable fishing authorities, Mark teaches outdoor techniques through a series of seminars and clinics, serves as a consultant to national companies, is an advisor to the International Game Fish Association, and shares his expertise with government agencies and conservation groups. He is a director emeritus of The Billfish Foundation and a former trustee of the University of Florida's Whitney Laboratory.

A past president of the Outdoor Writer's Association of America and recipient of its coveted Excellence In Craft Award as well as its prestigious Ham Brown Award, Mark is a member of the American Society of Journalists and Authors, Society of Professional Journalists, Southeast Outdoor Press Association, and Metropolitan Outdoor Press Association. He has been enshrined in both the IGFA Fishing Hall of Fame and the Freshwater Fishing Hall of Fame.

INTRODUCTION

The Overseas Highway from Miami to Key West officially opened on July 4, 1938. The two-lane highway with bridges so narrow that two trucks could barely pass each other followed the basic route of Flagler's railroad, which was destroyed in the Hurricane of 1935. Few people realize that it was originally a toll road with a collector's booth at the east end of Long Key Bridge.

Seven months after the opening, my father drove to Key West and took me with him. Marathon, which had been the camp from which Flagler built his railroad, stood out as the only community. Islamorada and Key Largo barely existed. I remember seeing signs for live shrimp along the road for miles, only to find the little shack closed when we passed it. To say that there was very little in the Keys was an understatement.

With the exception of Seven Mile Bridge and Bahia Honda Bridge, I have walked and fished every foot of all the other Keys bridges. There were no catwalks and no areas designated for fishing or vehicle parking. You simply walked along the bridge railing and fished anywhere you felt would be productive. When a bus or big truck would approach, the prudent thing was to climb over the bridge railing and hold on tightly with your body hanging over the water below. Otherwise, you ran the risk of being hit by the vehicle's rearview mirror. At night, you never fished near any of the signs on a bridge. Cars would come by and people would shoot at the signs, which usually were punctured with bullet holes.

FISHING FOR TARPON AND BONEFISH IN THE 1940s–1960s

Through the 1940s, 1950s, and 1960s, fishing pressure was a fraction of what it is today and loss of habitat was minor at best. Even as a teenager before I had the luxury of a boat, the Keys bridges proved to be a virtual fishing paradise for tarpon and bonefish. I can recall seeing a school of at least 1000 bonefish finning in the current and on the surface on the north side of one of the bridges.

Whenever we wanted to catch bonefish, we would walk up on the bridge and study the flats that ran along the shoreline on either end of the bridge. Usually, you could spot a few feeding bonefish, walk down the wingwall of the bridge, wade quietly out on the flat, and catch a couple. In those days, a well-placed bait or lure seldom resulted in a refusal.

Jumping tarpon from the bridges became a favorite evening pastime. I would save old plugs, particularly topwater, with rusted hooks and broken parts. The tarpon usually held on the uptide side of the bridge right along the shadow line. On some nights, casts were made upcurrent and the lures given life on their way back to the bridge. Invariably, a tarpon would crash that lure, jump and throw the plug, or dive under the span and cut the line on the barnacle encrusted concrete.

An alternative method centered on what we termed "bridge trolling." That required a rather stiff rod and stout line. The idea was to hook a live or dead mullet in front of the dorsal fin (or any other bait), lower it to the surface of the water, and then start walking across the bridge while towing the bait. Artificial lures could also be used. The strike was sudden and ferocious, but the battle seldom lasted long. In the rare instances where a tarpon could be contained, the procedure was to walk it down to the end of the bridge, scamper down the wingwall, and then release the fish.

In the late 1940s and very early 1950s, I began to have the occasional use of a boat, and it made a significant difference. Push poles were handmade from closet doweling in those days and outboards were not very powerful. You really did not need a big engine in the Keys, because the fish were reasonably close. There were very few guides and only a handful of fishermen. Even in the late 1950s, you could start at Upper Harbor Key and pole the long flat toward the Contents on the first of the incoming tide. It was not unusual to see seemingly endless schools of 20–50 bonefish with an occasional larger school. Eyeballing 1000 bonefish on that flat alone was relatively common.

At the same time, one could fish Biscayne Bay in and around Virginia Key among the mangrove shoots that existed then and see countless bonefish. The Cape flat, Stiltsville flats, and on down to Soldier Key held a wealth of bonefish with virtually no pressure.

The flats within sight of the Overseas Highway in Islamorada and down toward Lower Matecumbe were a haven for bonefish and particularly big bonefish. Back when Jack Kertz owned Bud and Mary's, he and I would cross the narrow channel right in front of the marina and fish the flat on the other side. Local guides usually passed it up because it was too close to the dock, but that flat held plenty of gray ghosts. You could also walk along the beaches and later the resorts in Islamorada on the oceanside and see numbers of bonefish.

Tarpon were everywhere in those days. Most of the channels in the Keys would be loaded with the silver king every spring and early summer. These fish also prowled the flats in huge schools. If a fly rod angler (and there were very few in those days) could not jump more than 20 tarpon in a day's outing, it was a very poor day. Usually, all you had to do was get the fly in front of the fish without spooking them and one would certainly eat. Today, you can cast to tarpon after tarpon and they refuse the fly.

There were days when we played a little game with a fly rod. The object was to toss a fly in a school of tarpon and see how many strikes you could get on a single cast. When a fish ate the fly, you exerted no pressure until the fish dropped the fly. Then, you started the retrieve again until another one ate. My all-time record was four fish on one cast. That should give you an idea of the abundance of big fish and their willingness to eat.

Coupon Bight in the Lower Keys was an area that held large quantities of tarpon, but seldom received much publicity. Only a handful of guides fished it back then, but the tarpon were there. If you were tired of battling those overweight silver kings, you could work in the shallower flats for bonefish and do extremely well.

It is absolutely amazing how fabulous the Keys were as a destination for bonefish and tarpon back in the days when fishing pressure was minimal and habitat encroachment was minor. One could see so many fish on any given day that presentation errors were of little consequence. More fish would swim into view in a matter of minutes. And, as the season progressed, tarpon would leave the backcountry and swing over to the oceanside in numbers too great to even estimate. I remember one angler looking at a school of tarpon approaching with the traditional bigger fish in the lead and exclaiming, "Here comes Grandma and the whole string band."

One might claim that memory tends to magnify things, but not in this case. Those of us who were there will tell you without hesitation of the huge schools of bonefish and tarpon that seemed to be everywhere in the Florida Keys and were willing to eat almost anything tossed at them.

Section V

Ecosystem-Based Management and Sustainable Fisheries

24 National Parks and the Conservation of Bonefish and Tarpon Fisheries

James T. Tilmant

CONTENTS

Introduction 351
Concern for Conservation within Parks 355
Opportunities for Increased Conservation and Protection 356
References 358

INTRODUCTION

As discussed in several chapters of this volume, when it comes to the management of bonefish and tarpon throughout the world, there is cause for concern. These concerns are really twofold: (1) declines in the abundance of fish and (2) loss of quality-fishing opportunities. Many places that used to be home to healthy populations of bonefish and tarpon have seen declines in their abundance and, in some cases, no longer sustain the recreational fisheries that existed there (Texas Parks and Wildlife 1988; Reiger, 1992; Larmouth, 2002). In some cases, the fish populations may still be present, but the level of recreational use, overall amount of boating activity, and local shoreline developments have taken away the opportunity for solitude and a quality-fishing experience. Most often, both of these concerns are related to growing human populations within the area and an associated increase in fishing and boating activity. With an increase in human population usually come increased fishing activity, increased incidental fish mortality, increased unreported harvest, habitat degradation, loss of key spawning and nursery areas, and increased direct and indirect ecosystem impacts. Conservation areas that can provide protected habitat for tarpon and bonefish and a place where populations of these species can continue to exist are limited. Those areas that do exist should be carefully evaluated and managed to maximize their benefits.

Our nation's marine national parks (NPs) offer one of the most promising opportunities to contribute to not only the conservation of bonefish and tarpon species and their habitat, but also the preservation of opportunities to have a quality flats–fishing experience also. NPs are set aside to conserve our nation's most prized and significant natural, historic, and cultural resources, and to provide for their recreational enjoyment. Over the nearly 100 years since the establishment of the NP system, some of the most fantastic natural resources of our nation, including marine resources, have been placed under the stewardship of the National Park Service (NPS). But these resources are not just locked up for preservation; they are set aside for the public to enjoy. The 1916 Organic Act creating the NPS states that:

> The service thus established shall promote and regulate the use of the Federal areas known as national parks, monuments, and reservations ... to conserve the scenery and the natural and historic objects and the wild life therein and to provide for the enjoyment of the same in such manner and by such means as will leave them unimpaired for the enjoyment of future generations.

The dual mandate for NPS managers is to ensure the conservation of park resources while providing opportunity for recreational enjoyment of such resources.

In carrying out this mandate, the NPS has recognized from its very beginning that recreational fishing, when properly managed, is an appropriate and compatible use of our NPs. When the NP system was created, many decisions had to be made about what the appropriate visitor activities and use are. It was recognized that to allow use meant that there would have to be some impact on the resources. Such impacts should be minimized, but are acceptable if they do not impair the future existence and conservation of the resources overall. To allow public enjoyment of the parks, roads were built, campgrounds created, hiking trails developed, lodges and visitor centers constructed, all resulting in some impact to the resources. It is the balance between acceptable levels of impact and adequate conservation of the resources that park managers must continually address.

To enable public enjoyment of the aquatic resources, the NPS recognized that fishing was appropriate and acceptable, and fishing has become a long and continued traditional use of our NPs. However, to put the purpose of fishing in the right context for our NPs, the NPS has slowly evolved a fisheries management policy that emphasizes the quality of the fishing experience, under natural conditions, for recreational enjoyment, and it has not put an emphasis on the take of fish. NPS fishing management policies (National Park Service, 2001) state:

> Recreational fishing will be allowed in parks when it is authorized, or not specifically prohibited by federal law, provided it does not jeopardize natural aquatic ecosystems or riparian zones. When fishing is allowed, it will be conducted in accordance with applicable federal laws and treaty rights, and state laws and regulations. The Service may restrict fishing activities whenever necessary to achieve management objectives outlined in a park's resource management plan or to otherwise protect park resources or public safety, unless such restrictions would violate a federal law or treaty.

Based on this policy, the NPS promotes catch-and-release fishing wherever and whenever it makes sense, encourages conservative regulations be adopted, and emphasizes the opportunity to fish in a natural undisturbed environment for native species.

This is great news for bonefish and tarpon fishermen. When it comes to bonefish and tarpon fishing, there could not be a better match with the NPS fishing policies.

The goals of most bonefish and tarpon fishermen are very similar to those of the NPS management objectives:

- Protected flats where fish can range uninhibited by fast-moving, high-powered boat traffic.
- Maintenance of solitude to be able to pole quietly across flats or along channel edges seeking to find fish naturally feeding.
- Catch and release regulations so those fish will be there for others to experience again.
- Long-term recognition of the importance of these fish resources, protection of their habitat, and enforcement of regulations to ensure that unscrupulous fishermen do not destroy the opportunity for others to enjoy catching them.

We are fortunate that within south Florida a large amount of the important tarpon and bonefish habitat is protected within NP units. The expansive mangrove and estuarine marshes of Everglades NP and Big Cypress National Preserve provide over a half million acres of juvenile tarpon habitat (Figure 24.1), while those portions of Florida Bay within Everglades NP, along with the waters of Dry Tortugas NP, are world famous for the mid-to-adult-sized tarpon that visit these areas in search of the abundant pilchards, sardines, and mullet that occur there (Sosin and Kreh, 1983; Cole, 1991). In Everglades NP, fishermen interviews have revealed that

FIGURE 24.1 Expansive mangrove and estuarine habitat in Everglades National Park and Big Cypress National Preserve provide over a half million acres of juvenile tarpon habitat. (Photo courtesy of Dr. Kevin Whelan, National Park Service.)

FIGURE 24.2 One of the many tidal flats within Biscayne National Park, which are well known for their large bonefish.

an estimated 3500 tarpon and 1500 bonefish are caught annually by more than 2000 anglers targeting these species within park waters (Schmidt et al., 2002). The tidal flats within Biscayne NP are also well known for their large bonefish (Figure 24.2). A recent keys-wide survey of flats fishing guides revealed that 25% of all bonefishing by south Florida guides occurs within Biscayne NP (Ault et al., 2003). Based on park fisheries survey data reported by Lockwood and Perry (1998), it is estimated that the flats within Biscayne NP are used for more than 3250 fishing trips per year by individual and guided recreational fishermen. Biscayne NP also supports a good tarpon fishery around the cuts and channels between the bay and offshore reefs.

But south Florida is not the only place where NPs provide important habitat for bonefish and tarpons species. Within the Caribbean, Virgin Islands NP, Virgin Islands Coral Reef National Monument (NM), Buck Island Reef NM, and Salt River Bay National Preserve, all contain shallow water flats and provide habitat that is at least occasionally used by bonefish and tarpon. These parks are not yet overrun with flats fishing skiffs and offer relatively pristine fishing opportunities, but use is expected to increase (R. Boulon, Virgin Islands NP, personal communication). Within the Pacific, two NP units in Hawaii, Kaloko-Honokohau and Pu'ukohola Heiau National Historic Sites, include adjacent coral reefs and sand flats, while the NPs of American Samoa and War-in-the-Pacific National Historic Park in Guam both contain extensive reef, bay, and flats habitat (Figure 24.3) that may be supportive of Pacific bonefish species (U.S. Army Corps of Engineers, 1980; Amesbury et al., 1999).

FIGURE 24.3 An example of extensive reef and flats habitat that may support Pacific bonefish at the National Park of American Samoa.

The extent of important bonefish and tarpon habitat within our NPs and the growing popularity of recreational flats fishing make tarpon and bonefish conservation and the sustainability of these fisheries important to the NPS. But the important role that parks may play in the overall conservation of these species needs to be emphasized by all who have an interest in these species.

CONCERN FOR CONSERVATION WITHIN PARKS

Most public would likely agree that it is reasonable to expect that the populations of fish within NPs would be in better condition than populations in similar habitats elsewhere. If they are not, one could reasonably ask, what is it that the park is providing and how can we consider it successful? However, parks are not isolated and with the extensive coastal development that has occurred in essentially all marine areas, coupled with continually greater recreational fishing activity, the NPS is very concerned about maintaining the integrity of park resources within its marine parks.

Recent studies suggest that fish populations within Biscayne NP are not substantially different than elsewhere in south Florida and most harvested species are in very poor shape (Ault et al., 2001). Within the U.S. Virgin Islands, similar studies found that fishery resources within the park were not significantly different from those at several comparable locations outside of the park (Beets and Rogers, 2002).

In addition to fish population declines, recreational boating activity continues to increase, particularly in high human population areas like south Florida or popular recreation destinations like the Virgin Islands (Milon and Thunberg, 1993; Ault, 2001; NPS Office of Public Use Statistics). Opportunities for a quality flats–fishing experience, free from general boater disturbance and noise, is decreasing in parks like Biscayne and Everglades. During 2004, over 60,000 boater visits are estimated to have occurred within Biscayne NP and over 50,000 boat visits were estimated to have occurred within Everglades NP (NPS Office of Public Use Statistics). In general, recreational boater use in south Florida has been increasing by about 2–3% per year during the past 10 years (Ault et al., 2001). Coupled with this increasing use has been an increase in recreational use conflicts (R. Clark, Biscayne NP, personal communication).

Increases in boater use have also resulted in increased boat damage to shallow seagrass beds and flats, which are prime bonefish habitat (National Parks and Conservation Association, 2005; NPS, 2006). Vessel grounding events at Biscayne and Everglades NP are common, with over 200 vessel groundings reported each year at Biscayne (NPS, 2005). These reported grounding events are estimated to represent only a small portion of the actual grounding incidents that occur.

OPPORTUNITIES FOR INCREASED CONSERVATION AND PROTECTION

Because of these findings concerning the condition of fish stocks and levels of recreational use in south Florida, the NPS has approached the State of Florida and agreed to work with the State Fish and Wildlife Commission staff to further address what may be needed for adequate conservation of our fishery resources, including bonefish and tarpon. The NPS and State have agreed to develop a long-range fisheries management plan at Biscayne NP and consider implementing new management regulations as necessary to achieve a set of desired future conditions reflective of the park's overarching conservation mandates. This process will include public input and discussion of existing data and information, as well as the formulation of the target-desired conditions. Several public information sessions have already been held in conjunction with this process, and a draft fisheries management plan was released for additional public comment in the spring of 2006 (Figure 24.4). The fisheries management planning process at Biscayne allows all of those with a specific interest in bonefish and tarpon (flats fishing in general) to provide input.

In addition to specific fisheries management planning efforts, all NP units periodically undergo a review and revision, as necessary, of their general management plans (GMPs). GMPs lay out the overarching long-term goals and philosophies for management of that particular park and outlines how and for what specific purposes the park will be managed. It must address visitor activities, park developments, and resource protection. It must show how the park will be managed to ensure that a wide variety of visitor activities will be provided for, while ensuring that resources remain unimpaired. One of the key elements that is addressed in GMPs is any park zoning of recreational activities. Typically, if there are going to be areas of a park set aside for certain types of activities, the zoning necessary to achieve this is a GMP issue. For example, the idea of restricting the use of motors over high-quality fishing flats where elements of solitude and low fish disturbance

FIGURE 24.4 Opportunity for public input at a meeting of the stakeholder advisory committee for Biscayne National Park's fisheries management plan.

are needed could be addressed in the park GMP. The GMP identifies management zones (areas with differing management objectives and degrees of conservation), significant management issues, and sets the direction for management within the foreseeable future.

Since 2004, several of the marine parks within south Florida and the Caribbean have been undergoing such review and revision of their GMPs. Biscayne and Everglades NPs are currently developing revised GMPs that are considering zoning some of the flats for no-motor use. A revision of Dry Tortugas NP's GMP was completed in 2003, but new plans will soon be developed at both Virgin Islands NP and Buck Island Reef National Monument, and at American Samoa in the South Pacific. Recent park expansions at each of these areas have made their GMPs obsolete. All of these planning efforts represent "open door" periods for consideration of changes necessary to achieve adequate conservation of park resources and improvement in recreational use, including fishing. They are an excellent time to raise the awareness of habitats essential to the conservation of tarpon and bonefish species that may be within these parks and to consider special protection needs.

To become involved in the review and revision of a park's GMP, contact the park directly and request to be put on its mailing list of interested parties. If fishermen who care about the future of flats fishing and the protection of bonefish and tarpon resources do not take advantage of these opportunities for stating their concerns to park managers, these fisheries will undoubtedly slowly erode, both in quality and availability, in the future.

REFERENCES

Amesbury, S.S., D. Ginsburg, T. Rongo, L. Kirkendale, and J. Starmer. 1999. War-in-the-Pacific National Historic Park Marine Biological Survey. University of Guam Marine Laboratory, Technical Report to U.S. National Park Service.

Ault, J.S., M.F. Larkin, and A.A. Barranco. 2003. Access-Intercept Survey of Biscayne National Park Marine Resource Users. Final Report on Contract No. NPS H500000B494, Biscayne N.P. University of Miami, Rosenstiel School of Marine and Atmospheric Sciences, Division of Marine Biology and Fisheries, 4600 Rickenbacker Causeway, Miami, FL 33149. 31p.

Ault, J.S., S.G. Smith, G.A. Meester, J. Luo, and J.A. Bohnsack. 2001. Site Characterization for Biscayne National Park: Assessment of Fishery Resources and Habitats. NOAA Technical Memorandum NMFS-SEFC-468. 185p.

Beets, J. and C. Rogers. 2002. Changes in fishery resources and reef fish assemblages in a marine protected area in the U.S. Virgin Islands: the need for a no-take marine reserve. In: M.K. Kasim Moosa, et al. (Eds.). Proceedings of the Ninth International Coral Reef Symposium, Bali, Indonesia, October 23–27, 2000. Allen Press, Lawrence, KS. pp. 449–454.

Cole, J. 1991. Tarpon Quest. Lyons and Burford Press, NY. 106p.

Larmouth, D. 2002. Tarpon on Fly. Amato Publications, Portland, OR. 95p.

Lockwood, B. and W.B. Perry. 1998. Biscayne National Park Annual Fisheries Report, 1997. NPS Unpublished Report. Biscayne National Park, Homestead, FL.

Milon, J.W. and E.M. Thunberg. 1993. A regional analysis of current and future Florida resident participation in marine recreational fishing. Florida Sea Grant College Program Project No. R/FDNR-3D. University of Florida, Gainesville, FL. Sea Grant Report No. 112.

National Parks and Conservation Association. 2005. Florida Bay: a resource assessment. NPCA State of the Parks Report. NPCA, Washington, DC. 9p.

National Park Service. 2001. Management Policies 2001. U.S. Department of the Interior, National Park Service. Washington, DC. NPS D1416/December 2000. 137p.

National Park Service. 2006. Seagrass Restoration Plan/Programmatic Environmental Impact Statement. Biscayne National Park, Homestead, FL. 106p.

Reiger, G. 1992. The Silver King. Meadow Run Press, Stone Harbor, N.J. 227p.

Schmidt, T.W., J. Osborne, J. Kalafarski, and C. Greene. 2002. Annual Fisheries Report; Everglades National Park. Unpublished Report by South Florida Natural Resources Center, Everglades National Park, Homestead, FL.

Sosin, M. and L. Kreh. 1983. Fishing the Flats. Winchester Press, Piscataway, NJ. 160p.

Texas Parks and Wildlife. 1988. Saltwater Finfish Research and Management in Texas: A Report to the Governor and the 71st Legislature. Texas Parks and Wildlife Department, Austin, TX. 67p.

U.S. Army Corps of Engineers. 1980. American Samoa Coral Reef Inventory. U.S. Army Corps of Engineers, Pacific Ocean Division Tech Report on Contract No. DACW84-79-C-0022 for Development and Planning Office, Government of American Samoa. 44p.

25 Improving the Sustainability of Catch-and-Release Bonefish (*Albula* spp.) Fisheries: Insights for Anglers, Guides, and Fisheries Managers

Steven J. Cooke and David P. Philipp

CONTENTS

Introduction 360
Bonefish Catch-and-Release Considerations 361
Environmental Factors 362
    Water Temperature 362
    Oxygen Concentrations 363
The Angling Event 364
    Degree of Exhaustion 364
    Handling and Air Exposure 365
Predation Issues 368
Mortality Estimates 369
Hooking Injury 371
Emerging Issues in Bonefish Catch-and-Release 372
    Novel Hook Designs 372
    Facilitating Recovery 373
Guidelines for Catch-and-Release Angling of Bonefish 374
    Guidelines for Anglers 374
    Guidelines for Guides 374
    Guidelines for Fisheries Managers 375
Research Recommendations 376

Conclusions 377
Acknowledgments 378
References 378

INTRODUCTION

Bonefish are one of the most sought after, but elusive gamefish in the world. Until recently, they were regarded as a single species, but more recently, scientists have recognized at least eight different species that are genetically distinct, yet morphologically indistinguishable (Colborn et al., 2001). At present, guides, anglers, and fisheries managers functionally consider these species as a single grouping, the bonefish (*Albula* spp.); we take the same approach in this chapter. Recreational fisheries that target bonefish are characterized as having highly specialized and skilled anglers that often fish with guides or outfitters (Policansky, 2002). Interestingly, bonefish fisheries are also somewhat unique in that almost all of the bonefish captured by anglers are released upon capture (Humston, 2001; Policansky, 2002). The small fraction of those individual fish not released are used for subsistence food, or more commonly, to generate taxidermy mounts of trophy catches. The popularity of bonefish recreational fisheries and the wide circumtropical distribution of the species make them important elements of many local economies (McIntosh, 1983; Humston, 2001; Ault et al., 2002). Bonefishing lodges and guide service industries have been developed in remote regions of the South Pacific (e.g., Christmas Island), the Indo-Pacific seas (e.g., the Seychelles), the Caribbean (e.g., Mexico, Bahamas), and extensively in the United States (e.g., Florida Keys, Hawaii). In addition to the actual lodging and guide fees, transportation costs to reach these destinations and the specialized equipment required to catch these fish can result in substantial economic benefit even in locales where bonefish do not reside.

Although the bonefish is clearly an important icon of the recreational fishing industry, little is known about the effects of different angling practices on these fish (Ault et al., 2002). At present, there are only two studies that explicitly examine issues associated with the effectiveness of catch-and-release strategies for bonefish (Crabtree et al., 1998b; Cooke and Philipp, 2004). Both of these studies provide information on hooking mortality, and one study (i.e., Cooke and Philipp, 2004) provides some information on post-release behavior. Neither study, however, provides an assessment of the sublethal physiological effects of catch-and-release angling. Although knowing the number of fish that die as a result of catch-and-release angling is essential for basic fisheries management activities (Wydoski, 1977), other sublethal effects can reduce the biological fitness of angled individuals (Cooke et al., 2002a). There is clearly a need to understand how bonefish respond to catch-and-release strategies. Evidence suggests that some local populations are experiencing declines in abundance and shifts in size structure (e.g., Bruger and Haddad, 1986; Anon., 2001; Ault et al., 2002). Conservation-minded anglers and guides are looking to fisheries managers and scientists for catch-and-release guidelines to use with bonefish. Although the lack of information on this species precludes a simple summary of existing literature on bonefish catch-and-release, there are characteristics of bonefish, bonefish fisheries, and bonefish angling techniques that are similar to other recreational fisheries that have been better studied with regard to catch-and-release.

In this chapter, we combine information from catch-and-release research on many different species of fish with our specific data from bonefish catch-and-release experiments to assess the potential range of disturbances arising from angling bonefish. We use that information to develop a series of recommendations for improving the effectiveness of catch-and-release techniques for bonefish. Where appropriate, we incorporate insights from guides and anglers engaged in bonefish catch-and-release on a daily basis. We also outline a research agenda for assessing and improving bonefish catch-and-release strategies. As additional information regarding the effects of catch-and-release angling on bonefish is made available, the guidelines can be refined to reflect new research findings. This conservative approach to developing species-specific catch-and-release guidelines is a risk-averse strategy that will help ensure that bonefish catch-and-release fisheries are sustainable (Cooke and Suski, 2005).

BONEFISH CATCH-AND-RELEASE CONSIDERATIONS

When a fish is hooked by an angler, there are many factors that can affect the outcome of the event for the fish (Figure 25.1). At best, the fish will survive the event, recover quickly, and experience no long-term sublethal impairments. At worst, the fish will not survive. Although catch-and-release anglers strive for the former outcome, it is often more probable that the outcome will be either intermediate to these two extremes, or in some cases, even skewed toward risk of death. Some of the factors that may affect the outcome are intrinsic such as fish age, sex, previous exposure to stressors, maturity, condition, size, and degree of satiation. These intrinsic factors are largely out of the realm of factors that an angler can control or alter to benefit the fish, and indeed, few of these factors have been studied with sufficient rigor to provide any conclusive statements on any species of fish. The environment where the fish is angled and released can also affect the outcome. Pertinent environmental conditions include abiotic factors such as water temperature, oxygen concentration, depth, or habitat complexity, as well as biotic factors such as predator burden and presence of disease. Although these factors cannot be controlled by anglers, most of them can be readily assessed by the angler, and if deemed to be detrimental, the angler could relocate to an alternative location. The remainder of the factors that typically influence the outcome of an angling event are generally controlled by the angler, including choice of fishing equipment (terminal tackle and gear, e.g., bait/lure/fly type, hook type, rod, reel, and line) and behavior of the angler (e.g., during the fight, when the fish is landed, and how it is handled and released). The factors identified here most likely manifest themselves as a series of cumulative stressors, rarely acting independently (Wood et al., 1983; Cooke et al., 2002a). Below, we discuss the factors that are most likely to be relevant to catch-and-release angling for bonefish. It is clear that catch-and-release angling has the opportunity to result in a negative outcome due to many factors, and it is the responsibility of the angler and the guide to conduct themselves in a manner that puts the fish on the best trajectory for a positive outcome.

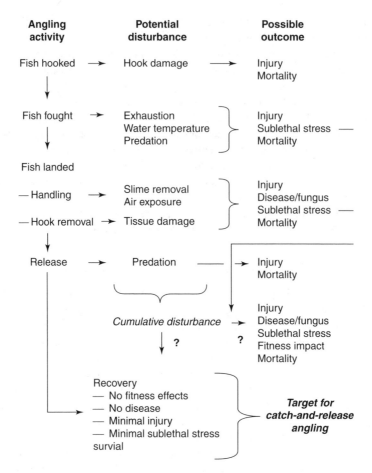

FIGURE 25.1 Schematic illustration of the potential effects of different angling activities on bonefish and the possible outcome of the effects on individual bonefish.

ENVIRONMENTAL FACTORS

WATER TEMPERATURE

Bonefish occupy shallow environments in subtropical and tropical regions that are characterized by relatively high water temperatures (i.e., 20–35°C). Although the water temperatures in nearshore tropical marine environments do not vary seasonally to the same extent as temperate freshwater environments (Mann and Lazier, 1996), the maximal water temperatures experienced by bonefish in late summer and early fall could lead to an increased risk of post-release mortality relative to other seasons. At present, there is only anecdotal evidence from guides and anglers suggesting that bonefish angled to exhaustion may not survive when angled at high water temperatures. Water temperatures experienced by bonefish in the summer may reach 32°C in Florida (Crabtree et al., 1996; Ault et al., 2002) or 34°C in the Bahamas (Colton and Alevizon, 1983a,b). In a recent study on bonefish in the Bahamas, there were no clear links between water temperature and mortality, but because the study was conducted in March, the maximal water temperature observed was less than 28°C (Cooke and Philipp, 2004).

The general principle that beyond some thermal optima fish performance is constrained (e.g., Farrell et al., 1996; Schreer et al., 2001; Farrell, 2002) is ubiquitous for fishes. Although it is accepted that water temperature has a profound effect on cellular processes and metabolism (e.g., Prosser, 1991), there have been few assessments of how tropical marine fishes captured by angling are affected by water temperature. Of these, few were conducted at the highest seasonal water temperatures, that is, where one would expect that temperature would be detrimental (e.g., no effect of water temperature on hooking mortality of snook (*Centropomus undecimalis*; Taylor et al., 2001). There are, however, many examples in temperate recreational fisheries where temperature has been consistently identified as an important determinant of the degree of sublethal disturbance and mortality (see Muoneke and Childress, 1994). For example, Atlantic salmon (*Salmo salar*) exhibited high levels of mortality when water temperatures exceeded ~18°C (Thorstad et al., 2003). Below that water temperature, however, mortality was negligible. Similar patterns have been observed for largemouth bass captured in fishing tournaments; there was a strong positive correlation between water temperature and mortality (Wilde, 1998).

Underlying the apparent association between high water temperatures and mortality are a series of physiological disturbances. Beyond a species-specific thermal threshold, fish approach their maximal metabolic rates (Anderson et al., 1998) and experience limitations in maximal cardiovascular performance (Farrell, 2002). Fish under thermal stress also face extreme biochemical alterations (Wilkie et al., 1996). Wilkie et al. (1997) determined that while warmer water may facilitate postexercise recovery of white muscle metabolic and acid–base status in Atlantic salmon, extremely high temperatures increased mortality rates. Greater oxygen debt may also be correlated with higher water temperatures (McKenzie et al., 1996).

Oxygen Concentrations

Temperature also influences oxygen availability, with high water temperatures resulting in marked reductions in dissolved oxygen. If hypoxic (species-specific threshold), these conditions are known to cause physiological disturbance to fish as they attempt to maximize oxygen transport to essential tissues and prevent mortality (Wu, 2002). The shallow, nearshore environments often frequented by bonefish can exhibit substantial fluctuations in dissolved oxygen (Diaz, 2001), suggesting that bonefish may experience periods of localized hypoxia. Although fish will often alter their movements patterns and distribution in response to hypoxia (e.g., Davis, 1975), if fish are inhabiting regions that are still within their range of tolerance, exposure to additive stress such as angling and higher water temperatures, may result in severe cardiorespiratory disturbances or even death. Interestingly, although the hypoxia sensitivity of metamorphosing bonefish increases threefold as the bonefish transits between leptocephali and juveniles (Pfeiler, 2001), at present there is no information on the sensitivity of adult bonefish to either environmental hypoxia or high water temperature.

THE ANGLING EVENT

Degree of Exhaustion

The degree of exhaustion has been identified in several studies as being a contributor to the magnitude of physiological disturbance from and the duration of recovery to an angling event (see Kieffer, 2000). For example, Gustaveson et al. (1991) determined that the length of angling duration (i.e., length of fight) in largemouth bass (varying between 1 and 5 min) was correlated with the degree of physiological disturbance measured by hematological parameters such as cortisol and plasma lactate. Similarly, in red drum (*Sciaenops ocellatus*), a coastal marine fish, plasma glucose, cortisol, lactate, and osmolality all increased with increased duration of the fight (varying between 10 and 350 s; Gallman et al., 1999). In addition, striped bass (*Morone saxatilis*) in Maryland angled for long durations also had more severe physiological disturbance (plasma pH, pO_2, and pCO_2) relative to briefly angled individuals (Thompson et al., 2002). Marine pelagic fishes including bluefin tuna (*Thunnus thynnus*), yellowfin tuna (*Thunnus albacares*), blue shark (*Prionace glauca*), and white marlin (*Tetrapturus albidus*) are frequently fought for long durations (up to 1 h), and usually experience pronounced acedemia and high plasma lactate that increase with the duration of angling (Skomal and Chase, 2002). In addition to the magnitude of disturbance, the time needed for recovery can also be prolonged with longer angling durations. For example, Schreer et al. (2001) reported that smallmouth bass (*Micropterus dolomieu*) exposed to brief simulated angling in a swim tunnel recovered consistently more rapidly than those fish exercised until exhaustion; heart rate and cardiac output returned to resting values twice as rapidly for briefly angled smallmouth bass relative to exhaustively angled individuals. Extended angling duration can also result in death through mechanisms outlined in Black (1958) and Wood et al. (1983). Indeed, Thompson et al. (2002) noted that mortality of striped bass increased threefold when angling duration increased from 1 to 3 min at 26°C. This water temperature is near their upper thermal tolerance so that exercise can cause significant mortality. Interestingly, at 8°C, no mortality was observed when fish were angled for similar durations.

The duration of the angling event depends primarily on the type of tackle used, the test strength of the line, and size of fish angled, but it can also be affected by water temperature and habitat (especially depth). Larger individuals within a species may require longer periods of time to land, such as observed for Atlantic salmon (Thorstad et al., 2003). In their study, the duration of the angling events ranged from 1 to 49 min with fish undertaking between 0 and 10 runs (mean of 3.7 runs). Plasma lactate increased and plasma pH decreased with increased angling duration in Atlantic salmon (Thorstad et al., 2003). Similar to Atlantic salmon, bonefish also engage in multiple high intensity runs during the fight, although this has not been quantified. Preliminary evidence from bonefish research in the Bahamas indicates that larger fish require longer to get exhausted when angled, but this varied on a site-specific manner, perhaps as a result of different fishing gear (i.e., line and rod strength; Cooke and Philipp 2004; Figure 25.2). To our knowledge, there are no published studies that have explicitly contrasted the duration of time required to land fish using different tackle. In some cases, fish that were landed rapidly (<20 s) have been used as "unangled controls" in physiological studies (Kieffer et al., 1995).

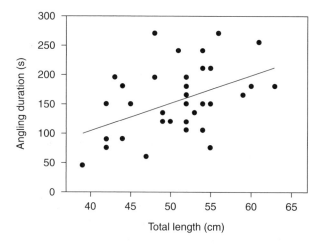

FIGURE 25.2 Relationship between angling duration for bonefish (i.e., degree of exhaustion) and fish size (total length in cm). (Data from Cooke and Philipp, 2004.)

Collectively, the trends in the freshwater and marine literature point toward increased physiological disturbance and risk of mortality as fish are fought for longer durations. These effects appear to be pronounced when combined with high water temperatures.

HANDLING AND AIR EXPOSURE

The processes of landing the fish and removing the hook entail several opportunities for fish to experience injury and sublethal physiological disturbance. The challenge is to land and handle the fish carefully without them being fully exhausted from the angling fight. Landing the fish is accomplished either by hand or with the aid of a net. For bonefish, we urge anglers to use their wetted hands to land bonefish and to avoid using nets. Barthel et al. (2003) determined that the use of a landing net for freshwater fish can result in physical injury and increased risk of mortality relative to fish landed by hand. In addition, the degree of injury (including dermal disturbance and fin fraying) varies with the type of landing net mesh, with knotless nylon being the least injurious and knotted, large mesh being the most damaging (Barthel et al., 2003). Discussions with bonefish guides revealed that most relied on using wetted hands to land and handle bonefish. Several guides that we have conversed with, as well as those who participated in a forum in the Bahamas (Anon., 1999), expressed concern over handling and its role in removal of slime. Although there are no studies that explicitly examine the role of slime removal in causing increased mortality, it is clear that increased abrasion and slime removal magnify the risk of pathogenic infections, in particular those associated with fungus (Barthel et al., 2003). It has also been suggested that excessive handling and slime removal can increase the ease with which sharks can detect the bonefish scent (Anon., 1999). There has also been concern that sunscreen on the hands of anglers can also harm bonefish with anecdotal reports of stained handprints noted on recaptured individuals (Anon., 1999). Although some guides indicated that wet cloths can be used for handling bonefish, our experience suggests that the cloths have the potential to remove excessive slime. An alternative means to handle bonefish is by using the Boca Grip or a similar device that clamps the mouth tissue allowing the angler to restrain the fish without gripping the entire body. However, it may be advisable to first net the fish and hold in the water because it would be difficult to use a Boca Grip on a nonexhausted fish. Although this type of tool could minimize physical contact with the fish it may still require that the fish be exposed to air. Although we advocate minimizing handling of fish, unless the angler is using specialized hook designs that facilitate release, landing and handling the fish is required. In muskellunge (*Esox masquinongy*) fisheries, fish that were handled with care by experienced anglers had higher survival rates than fish that were excessively handled by inexperienced anglers (Newman and Storck, 1986).

Hook removal can be facilitated through the use of specialized tools or tackle. For example, an obvious way of reducing injury and handling is to use barbless hooks. The general concensus in the literature for fish including rock bass (*Ambloplites rupestris*; Cooke et al., 2001), rainbow trout (*Oncorhynchus mykiss*; Taylor and White, 1992), lake trout (*Salvelinus namaycush*; Falk et al., 1974), and nearshore marine fishes (Schaeffer and Hoffman, 2002), is that barbless hooks result in reduced tissue damage at the point of hook insertion. Furthermore, due to the absence of the barb, the hook can be removed easily, thus reducing handing time and air exposure as evidenced in rock bass (Cooke et al., 2001) and a number of tropical marine fishes (Diggles and Ernst, 1997; Schaeffer and Hoffman, 2002). The increased duration of air exposure from using barbed hooks can result in sublethal physiological disturbances that are greater than when fish are captured on barbless hooks (Cooke et al., 2001).

In addition to difficulty in removing the hook, air exposure can also be the result of poor handling skills, taking photographs, lack of knowledge, or poor conservation ethic. In a study on Atlantic salmon, fish exposed to air postangling were more frequently characterized as being in poor condition relative to those not exposed to air (Thorstad et al., 2003). In general, air exposure results in the collapse and subsequent adhesion of gill filaments, compromises respiration, and leads to ion disturbances and metabolic acidoses. To date, the most thorough study of air exposure was on rainbow trout, where blood oxygen tension and the amount of oxygen bound to hemoglobin both fell by over 80% during brief air exposure (Ferguson and Tufts, 1992), causing severe anoxia. Furthermore, fish exposed to air typically experience greater acid/base disturbance than fish that were exercised but not exposed to air (Ferguson and Tufts, 1992). This type of physiological disturbance has also been documented for largemouth bass (*M. salmoides*) in simulated fishing tournaments (Suski et al., 2004) and in Atlantic salmon in the River Alta, Norway (Thorstad et al., 2003). Similar studies on smallmouth bass (Cooke et al., 2002b) determined that the time required for cardiovascular variables to recover was correlated with the duration of air exposure. Extended air exposure beyond some time threshold results in permanent tissue damage, which may lead to heightened rates of mortality. Short-term mortality (12 h) of rainbow trout was negligible for control rainbow trout and low for trout that were exercised to exhaustion but not exposed to air (12%; Ferguson and Tufts, 1992). When trout were exposed to air for either 30 or 60 s following exhaustive exercise, mortality increased to 38 and 72%, respectively.

Improving Sustainability of Catch-and-Release Bonefish 367

We know very little about the effects of handling and air exposure on bonefish. Cooke and Philipp (2004) indicated that almost 80% of bonefish exposed to lengthy handling and air exposure lost equilibrium upon release and tended to exhibit patterns of erratic swimming characterized by alternating periods of high intensity swimming with frequent resting (see Figure 25.3). Bonefish handled and released

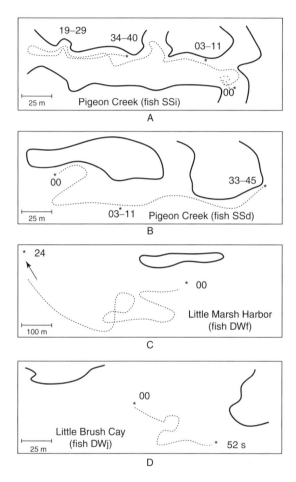

FIGURE 25.3 Representative traces of bonefish behavior after being captured by angling, affixed with visual tags, and released (in Cooke and Philipp, 2004). Here we visualized two fish from each of their study sites. The asterisk (*) adjacent to time 00 is the location where the fish was released. Where two time periods are noted indicates periods (time in seconds) during which the fish were stationary. Traces (A) and (B) were completed in Pigeon Creek, San Salvador for fish SSi and SSd, respectively. Both of these fish had lost equilibrium when first returned to the water. When they swam away, they did so rapidly at first and then spent several periods resting. Traces (C) and (D) were completed in the waters adjacent to Deep Water Cay for fish DWf and DWj, respectively. Neither of these fish lost equilibrium upon return to the water. Fish (C) exhibited slow and steady swimming. Fish (D) was attacked by a shark 52 s after release and thus exhibited erratic behavior while being chased. The three other bonefish (A, B, C) survived the monitoring period.

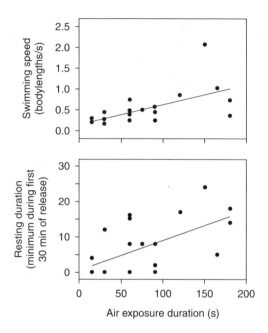

FIGURE 25.4 Effects of air exposure duration on the post-release behavior of bonefish in the Bahamas (San Salvador and Deepwater Cay). (Data from both transmitter and visually tagged fish from Cooke and Philipp, 2004.)

with minimal air exposure generally did not lose equilibrium and exhibited slow and steady swimming post-release (Figure 25.3). Combining data from the two study sites used by Cooke and Philipp (2004) results in a general positive relationship between the duration of air exposure and the response variables post-release swimming speed and amount of time resting (Figure 25.4). The available data on bonefish and other species make it abundantly clear that handling and air exposure should be minimized.

PREDATION ISSUES

In the popular fishing literature, the presence of predators in bonefish habitats is noted with numerous references to personal safety while engaged in bonefishing (e.g., Kaufmann, 2000). These popular sources also report incidences of predation while the fish is on the line and immediately post-release. There is no doubt that bonefish occupy habitats that are shared by many predatory fishes including bull (*Carcharhinus leucas*) and lemon (*Negaprion brevirostris*) sharks, and barracuda (*Sphyraena barracuda*). Unlike marine fisheries, most angled and released fish in freshwater systems and estuaries have relatively low risk of predation by other fishes, and only in few specific situations is predation from other organisms an issue (e.g., estuarine crocodiles, *Crocodylus porosus*, attacking barramundi, *Lates calcarifer*; raptors attacking Pacific salmonids, *Oncorhynchus* spp.).

Only one study has examined the rates of shark predation on released bonefish. In that study, Cooke and Philipp (2004) noted that several fish were attacked while on the line, a well-known phenomenon among bonefish anglers and guides. In other instances, fish were chased by predators while on the line, but aggressive angling and creation of a disturbance by splashing with the wading pole (to distract the sharks) resulted in fish landing safely. Predation by hammerhead sharks (*Sphyrna mokarran*) on caught-and-released tarpon (*Megalops cyprinoides*) has also been observed. Although Edwards (1998) noted that only one of the 27 tarpon was attacked after catch-and-release, that concern increases when anglers use light tackle that results in severe exhaustion, thus making fish more susceptible to predation. A group from the Florida Marine Research Institute (K. Guindon, personal communication) is currently assessing predation in tarpon post-release after capture on light- vs. heavy-fishing gear. Even fish much larger than bonefish or tarpon can be attacked and killed by predators after catch-and-release angling. Jolley and Irby (1979) noted that one of the eight Atlantic sailfish (*Istiophorus albicans*) released after angling, which had an eye injury from the hook, was attacked by a shark about 6 h after release.

There is no doubt that predation during capture and after release of bonefish can be substantial and appears to be correlated to the relative abundance of sharks. Cooke and Philipp (2004) noted that the more sharks that were seen by anglers during the time when the fish was being angled, the greater the chance that fish were attacked by predators after release. Thus, anglers could be able to gauge the relative abundance of sharks and potential for mortality from predation. Fish killed upon release were usually attacked very soon after release. To date, almost all predation events involving juvenile lemon sharks that we have witnessed involve attacks from behind, initially severing the caudal fin and then returning to consume the rest of the fish (Cooke, unpublished data). In some cases, attacks were characterized by splashing and disturbance of substrate followed by a period of relative calm, although in other instances, the entire attack was very subtle. Anglers, however, should also be able to identify when predators are attacking released bonefish and take action to prevent the predation.

MORTALITY ESTIMATES

Until recently, there have been no data quantifying the mortality associated with catch-and-release angling for bonefish. Current estimates from various studies are presented in Table 25.1. As discussed above, mortality can occur due to extreme physiological disturbance, hooking injury, or predation. Crabtree et al. (1998b) excluded predators to focus on catch-and-release mortality arising from physiological disturbances and injury. These authors held 10 bonefish in a small pond and repeatedly angled them over several years. Mortality in the absence of predators was quite low (4.1%), especially considering that some individuals were captured more than 10 times. The restricted nature of the pond environment, however, may have resulted in unrealistically brief angling durations that minimized sublethal disturbances. More recently, Cooke and Philipp (2004) assessed the hooking mortality of bonefish in different field sites in the Bahamas. At a site with low predator abundance, no fish died during the monitoring period (24 h) despite long angling durations, extended

TABLE 25.1
Compilation of Existing Mortality Estimates for Bonefish

Citation	Study Method	Mortality (%)
Colton and Alevizon, 1983a	Telemetry in the Bahamas	Some
Cooke and Philipp, 2004	Telemetry in low shark predation region of the Bahamas	0
Cooke and Philipp, 2004	Float tags in low shark predation region of the Bahamas	0
Cooke and Philipp, 2004	Telemetry in high shark predation region of the Bahamas	39.2
Cooke and Philipp, 2004	Float tags in high shark predation region of the Bahamas	42.4
Crabtree et al., 1998b	Repeated capture of 10 individuals in a pond in Florida	4.1
Danylchuk, unpublished data	Telemetry in Bahamas	Unlikely— not in the short term
Humston et al., 2005	Telemetry in Florida	Possible

air exposure, and loss of equilibrium at release. Conversely, in a region with higher predator abundance, mortality rates were approximately 40% despite the fact that fish were landed rapidly and exposed to air less than at the low predator site. Collectively, the data from Cooke and Philipp (2004) support the suggestion that mortality from hooking injury and physiological disturbance is usually low. In the presence of predators, however, mortality can be extremely high, suggesting that fish need to be released in the best possible condition. A recent study on Pacific threadfin (*Polydactylus sexfilis*) reported that stressed individuals were preyed upon by sharks preferentially over nonstressed controls as part of an aquaculture enhancement project (Masuda and Ziemann, 2003). Interestingly, all of the mortality occurred in the first hour after release. This is consistent with the research by Cooke and Philipp (2004), where all mortality observed occurred within about 30 min of release (e.g., Figure 25.3D).

Critical analysis of research focused on elucidating movement patterns of bonefish that can provide some additional insights into possible catch-and-release mortality rates in bonefish. For example, Colton and Alevizon (1983a) ultrasonically tagged 13 bonefish in the Bahamas that were captured by a combination of angling and netting. The authors noted that fish survived the implantation and recovered with no noticeable effect, provided that predators were not in the immediate vicinity at time of release. Although there is no discussion of what led to that conclusion, it suggests that the authors encountered some level of post-release predation in preliminary studies. In addition, although the authors implanted 13 fish, only 3 were located 24 h later. The authors concluded that they had no evidence that bonefish experienced mortality associated with capture, handling, or predation. Instead, the authors assumed that the fish left the general area where the study was being conducted. More recently, researchers working in the same region of the Bahamas (i.e., Deep Water

Cay; Cooke and Philipp, 2004) used the ability to locate ultrasonically implanted fish 24-h post-release as an indicator of bonefish survival. In all instances, researchers were able to locate four of the six tagged bonefish. The remaining two individuals were attacked by sharks shortly after release during immediate post-release tracking. Because all of the fish that were monitored for 48 h also remained in the general vicinity (within ~1 km) of release, it is equally plausible that the low relocation rates of released bonefish by Colton and Alvizon (1983a) may be attributable to post-release predation, not just emigration from the study area.

An ongoing telemetry study in Biscayne Bay, Florida may also provide some preliminary information about catch-and-release mortality of bonefish (Humston et al., 2005; Ault et al., 2005). Researchers using ultrasonic telemetry during a 60-day monitoring period (Humston et al., 2005) consistently located 7 of 11 tagged fish on the same flat where they were tagged. The fate of the other four individuals could have most likely included movement, but failed transmitters, natural mortality, harvest, or catch-and-release mortality (including predation) are also possibilities that affected their fate.

In a survey of bonefish guides and anglers in Florida, Ault et al. (2002) asked respondents to identify the percentage of bonefish that they released that fell into the categories of (1) excellent condition, (2) partially impaired, or (3) will not survive. Respondents reported that on average 87.9% of fish were released in excellent condition (median of 90%). Of the remaining fish, a mean of 9.8% were classified as partially impaired (median of 5%) and 2.5% were classified as will not survive (median of 0%). These data are consistent with our findings in the absence of predators. Bonefish guides and anglers in Florida identified predator abundance as being the largest contributor to mortality of bonefish (Ault et al. 2002). Other factors identified as being important include length of the fight, hook location, water temperature, type of bait/lure, and water depth (in decreasing order of perceived importance; Ault et al., 2002). An additional 18% of respondents wrote in "other" factors of which the time of handling and air exposure were noted as important (Ault et al., 2002). Interestingly, all of the factors identified by the guides as important are also those for which we have highlighted potential opportunity in the development of catch-and-release strategies.

HOOKING INJURY

Catch-and-release angling for bonefish in the Bahamas revealed that injuries resulting from hooking were minimal (Cooke and Philipp, 2004). The hooks penetrated in the jaw region almost exclusively, despite using a variety of flies, lures, and dead shrimp. Studies on other species of fish have documented differences in anatomical hooking locations depending on bait types; deeper, potentially more lethal hooking sites for organic bait, shallower, less injurious sites from flies and lures (e.g., walleye, *Sander vitreus*, Payer et al., 1989; bluegill, *Lepomis macrochirus*, Siewert and Cave, 1990; rock bass, Cooke et al., 2001). A likely reason for shallow hooking with organic bait in the Cooke and Philipp (2004) study was the large size of the hook and the shrimp that we used. Several studies of other species have documented inverse relationships between hook size and injury/mortality (bluegill, Burdick and

Wydoski, 1989; striped bass, Diodati, 1991). Despite giving the bonefish ample time to ingest the bait, no fish were hooked deeply. If bonefish were to be deeply hooked or if the fish were to break the line, there may be an advantage to avoiding stainless steel hooks to facilitate hook degradation and expulsion. Apparently, some guides already advocate for the use of hooks made from substances that degrade more rapidly than stainless steel (Anon., 1999). Another factor potentially contributing to shallow hooking is the mouth morphology of bonefish. Bonefish have massive pharyngeal tooth plates used for crushing hard structures (Crabtree et al., 1998a) that may make deep hook penetration difficult. Consistent with our shallow hooking locations (i.e., anterior relative to caudal), we observed few incidences of bleeding. Those fish that did bleed exhibited very localized, minor bleeding due to penetration of perfused tissue, not from penetration of vital organs (e.g., esophageal region, gill arches, heart) that can often prove fatal (e.g., Pelzman, 1978; Skomal et al., 2002). Although minor, this amount of bleeding may prove lethal if sharks are in the vicinity. If a captured bonefish was bleeding, we recommend that it be held in a livewell/cooler for 2 min to allow clotting before release. Deeply ingested hooks should be left in place with the line cut.

EMERGING ISSUES IN BONEFISH CATCH-AND-RELEASE

Several recent developments are relevant to bonefish fisheries. The first deals with the development of terminal tackle (i.e., fishing hooks) intended to maintain high catch rates while reducing injury and mortality. The second development is the discovery that there are actions that anglers can take after landing a fish that may actually facilitate the recovery of individuals that are to be released.

NOVEL HOOK DESIGNS

Recent developments in hook technology show promise for reducing gear-induced injury and mortality. Circle hooks have been touted for several years by outdoor media and hook manufacturers as a means to reduce catch-and-release mortality (see Figure 25.5A). Circle hooks tend to cause reduced overall mortality and reduced hooking depth compared with conventional J-style hooks (Cooke and Suski, 2004). Although there are no studies that examine the effectiveness of circle hooks for bonefish, based upon the strong assumption that circle hooks will reduce injury and mortality, several recent bonefish tournaments have included provisions for bonus points for bonefish captured on circle hooks (Cooke and Suski, 2004). Furthermore, in a survey of bonefish guides and anglers in Florida, some respondents stated that circle hooks might be contributing to a reduction in bonefish mortality (Ault et al., 2002). Some anglers are attempting to use circle hooks for targeting bonefish with flies (Cooke, pers. obs.). The existing literature, derived primarily from salmonids, suggests that circle hook fly designs are ineffective at hooking fish and result in similar levels of injury to J-style fly hooks (Julie Meka, personal communication). Although we are unaware of any anglers using circle hooks while angling with bait (e.g., shrimp, conch), we suspect that circle hooks would reduce injury while maintaining capture efficiency for those types of fisheries, if not for fly-fishing.

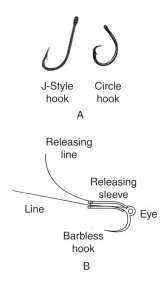

FIGURE 25.5 (A) Schematic of J-style hooks and the recently popular circle hook. (B) An even newer hook design is the Shelton self-releasing hook (see Jenkins, 2003 for full description). On landing the fish, the angler grasps the releasing line while the fish is in the water causing the hook to rotate out of the mouth of the fish without actually touching the fish. Such a hook design may provide substantial benefit for bonefish.

Circle hooks still require handling the fish to remove the hook. Some anecdotal observations suggest that circle hooks are more difficult to remove than J hooks when penetrating similar areas (Cooke and Suski, 2004). A new hook design that shows promise for reducing or eliminating handling is the "self-releasing" Shelton hook (Jenkins, 2003; see Figure 25.5B). In a study of rainbow trout, mortality rates of fish caught on barbless circle hooks that were removed had four times higher mortality rates than fish captured on the barbless Shelton self-releasing hook (Jenkins, 2003). Fish captured on Shelton hooks can be removed without handling the fish when the angler pulls on a tag line that activates a release mechanism. The hook reverses direction by 180° and exits the fish when gentle pressure is applied to the main line. We are currently unaware of anyone using these hooks for bonefish.

FACILITATING RECOVERY

Recent research primarily focusing on salmonids indicates that slow-speed swimming postexercise can facilitate recovery (Milligan et al., 2000). This knowledge is being applied to reduce bycatch mortality of commercial troll-caught salmonids (e.g., Farrell et al., 2001). In addition, this information is also being applied to facilitate recovery of tournament-caught largemouth bass (Cory Suski, unpublished data). Although we have no direct evidence that this would work on bonefish, it is worthy of future examination. In fact, holding fish in large coolers or aerated livewells commonly found on flats fishing vessels for short periods prior to release may provide an opportunity for fish to restore enough energy reserves to enable them to escape potential predators. The livewells used in freshwater fishing tournaments were once regarded as stressful, but if provided with adequate water quality and if fish are kept at low density, some fish can actually recover while retained in livewells (Cooke et al., 2002b; Suski et al., 2004).

GUIDELINES FOR CATCH-AND-RELEASE ANGLING OF BONEFISH

At present, there are three groups of stakeholders that require information on how to minimize disturbance and injury, and maximize survival of bonefish. Each group has unique characteristics and roles, necessitating some level of specialized guidelines. Although the delineation between guide and angler may seem trivial, it must be clear exactly what each group must do for their role in bonefish conservation. The guidelines developed below are based on the best available knowledge at this time and are somewhat conservative (Cooke and Suski, 2005). As additional information becomes available, particularly from empirical bonefish catch-and-release research, these guidelines can be refined (either relaxed or intensified) to reflect the improved state of knowledge.

GUIDELINES FOR ANGLERS

The fishery manager must provide anglers with knowledge and direction. The fishing guide must recommend locations and promote practices to the angler that result in minimal injury and mortality. The ultimate responsibility for the handling and care of fish captured by recreational angling, however, lies with the angler. There are many actions or choices that can be made by an informed angler that will minimize injury, reduce sublethal disturbance, and avoid mortality:

- Use heavy tackle in regions with many predators. Tire fish only to the point where they can be landed safely; do not fight fish to exhaustion. As many bonefish anglers use fly-fishing gear, it is important that the gear be sufficiently robust to enable anglers to land fish quickly without leading to complete exhaustion.
- Minimize or eliminate air exposure. Fish should be handled in the water to minimize loss of slime and the physiological consequences of air exposure. When handling angled bonefish, they should be gripped firmly posterior to the operculum, and held supine in the water.
- In certain locations, many of the fish released could succumb to predation. Post-release predation of bonefish usually occurs rather rapidly (seconds to minutes) and is quite evident; when this is occurring, anglers need to alter fishing location, even when fishing is "good."
- Share your experiences with fisheries managers who can use that information to guide future investigations and to identify "emerging issues."

GUIDELINES FOR GUIDES

Guides and lodge operators can play an important role in directing angler behavior. Guides need to make destination decisions that could sometimes produce reduced

Improving Sustainability of Catch-and-Release Bonefish

bonefish catch rates in an effort to reduce predation rates. Another important role of the guide is to educate and develop angler ethics so that when anglers fish in the absence of guides, they continue to adjust their behavior to benefit bonefish.

- Base decisions regarding where to fish on the presence of predators recognizing that this may not be practical given the spatial relationship between predators and prey. When encountering high predator density, relocate the client to a region with less predation pressure.
- Encourage anglers to use heavy tackle (i.e., stiff fly rods, heavy-spinning gear) that facilitates the rapid landing of fish, particularly when predators are present.
- Require the use of barbless hooks and encourage anglers to try newer hook technologies that appear to reduce injury.
- Handle fish quickly and efficiently, without the use of landing nets, cloths, or any excessive handling that results in slime removal, using pliers to rapidly remove hooks.
- Minimize air exposure by limiting the number of photographs and by holding fish in the water (either in cooler or ocean).
- Pay close attention to fish behavior upon release, particularly during the warmest periods of the year when water temperatures and low oxygen conditions may impart added stress on fish. If released fish appear to be in distress, consider relocating to areas closer to deeper, cooler water.
- In the presence of predators, translocate fish in coolers or aerated livewells within short distances for release into complex habitats (e.g., dense mangroves) where bonefish may be protected from predators.

GUIDELINES FOR FISHERIES MANAGERS

The paucity of information on bonefish and bonefish fisheries makes it difficult for fisheries managers to make informed management and conservation decisions. Fishery managers must serve as a link between the research scientists and guides and anglers to ensure that they are provided with the most pertinent information that is consistent with regional fisheries management and conservation plans.

- Develop outreach and educational materials that promote proper bonefish handling techniques and disseminate that information to all stakeholders.
- Incorporate training on effective catch-and-release techniques into professional guide or captain certification processes, including formal assessment instruments.
- Consult local client groups to determine what issues are specifically applicable to the types of fisheries and ecological characteristics of the fishery and respond to the possible biological implications of the issue. These issues may be site specific or fishery specific.
- Consider limiting bonefishing in regions with high levels of predation. Otherwise, encourage guides and anglers to alter locations frequently and to vary temporal and spatial patterns from day to day so that predators do not become accustomed to angler behavior.

- Fisheries managers must realize that there is much latitude between a fish that is live and one that is dead. Fisheries management has historically been concerned with knowing rates of fishing mortality. However, sublethal impacts can retard growth, alter fitness, and even result in fishing-induced selection, all of which can indirectly contribute to fishing mortality.

RESEARCH RECOMMENDATIONS

It is evident from this synthesis that we do not know a great deal about the effects of catch-and-release angling on bonefish. Underpinning this deficiency is a fundamental lack of information on the basic ecology, behavior, and physiology of bonefish. This theme will undoubtedly be echoed throughout the other chapters in this volume. Here, we identify research needs that we feel are most pressing for conserving bonefish:

Construct baseline information on blood and muscle biochemistry and determine how these parameters are affected by angling. Prior to initiating controlled laboratory assessments on the sublethal effects of catch-and-release angling, it would be useful to document the level of disturbance experienced by fish in the field post-capture (e.g., Thorstad et al., 2003; Suski et al., 2004). Complete blood and muscle biochemistry profiles would also help to identify whether bonefish respond to stress in ways similar to other fishes. Factors worthy of investigation include the type of gear, duration of angling event, and water temperature.

Conduct controlled experiments to document the disturbance and recovery trends of blood and muscle biochemistry, hormones, and the cardiorespiratory system. Controlled laboratory assessments can be used to manipulate factors such as the duration of air exposure, degree of exhaustion, and water temperature to determine how these factors may contribute to sublethal disturbances or mortality, and how they alter recovery duration. Laboratory assessments would most likely involve cannulation to collect serial plasma samples or cardiovascular monitoring devices to record cardiorespiratory activity.

Determine what characteristics of a released bonefish attract sharks and other predators. When a bonefish is captured and released, sharks are able to cue in on that specific individual when released, even in the presence of other bonefish. Knowledge of the cues that identify the released bonefish to the sharks would possibly permit altering bonefish release in some manner to reduce the ability of the predators to locate the released fish.

Assess the effects of different strategies for facilitating recovery of angled bonefish. As discussed above, there has been recent interest in trying to develop strategies that actually facilitate recovery of commercial bycatch and caught-and-released fish. It would be useful to know if short-term retention in a live-well type device could provide captured bonefish adequate time to recover such that they would be able to evade predators upon release.

Assess the swimming performance of bonefish before and after exercise and compare it to the swimming ability of different predator species. Current literature suggests that bonefish appear to be most susceptible to predation immediately after angling. Evading predators requires that bonefish are able to swim faster and for longer periods than predators. At present, there is very little known about swimming speeds of bonefish, or predators such as barracuda and sharks. Laboratory-based respirometry studies could be used to quantify and contrast swimming speeds.

Evaluate the performance of circle hooks, Shelton hooks, and J-style hooks for both bait and fly fisheries. As outlined above, recent advances in terminal tackle show promise for reducing injury and mortality of fish. There is virtually no information on how terminal tackle affects injury and mortality in bonefish. There is wide latitude for basic hooking mortality studies that vary with bait type (e.g., lure, fly, bait), hook type ("J" style, circle, Shelton), presence of a barb, and size of hooks. To generate appropriate sample sizes, volunteer anglers and guides could be recruited to participate with logbook/angler diary programs.

Evaluate the effects of different gear types (i.e., fly-fishing and spinning) on bonefish. The majority of anglers that target bonefish use fly-fishing gear. Fly-fishing gear provides the angler with incredible control over the fight relative to spinning gear, but only when the rod and line strength are appropriately matched to the fish. Based on the knowledge that fighting bonefish to exhaustion is detrimental to their condition and survival, there is a need for studies that contrast the use of different angling gears, for example, light-fly equipment, heavy-fly equipment, light-spinning tackle, heavy-spinning tackle.

Evaluate abiotic tolerances and responses to multiple stressors for adult bonefish. Basic information on environmental tolerances and preferences is required to provide a framework for interpreting field-derived data. Assessments of bonefish mortality and physiological disturbance at higher water temperatures are needed to understand how seasonal influences may need to be considered in catch-and-release angling for bonefish. This type of research would also benefit our understanding of the basic environmental biology and habitat relations of bonefish.

Assess the sublethal effects of angling-related behavior on growth and other fitness-related variables. Growth and other fitness-related indices can also be affected by catch-and-release angling either directly through reduced food intake or indirectly through sublethal acute or chronic stress (see Cooke et al., 2002a). Development of a bioenergetics model for bonefish would permit assessments of the costs associated with different angling practices and provide a framework for contrasting the effects of using different angling gear and practices. Physiological telemetry devices such as tilt tags (to indicate foraging), opercular tags, or locomotory activity tags could all be used to construct field energy budgets. When the reproductive ecology of bonefish is better understood, it will also be possible to evaluate how different stressors associated with angling can affect factors such as quality and quantity of gametes, reproductive behavior, viability of offspring, etc (see Cooke et al., 2002a, for comprehensive list of possible fitness alterations).

CONCLUSIONS

Catch-and-release angling assumes that fish released following capture have a reasonable chance of survival. It is clear that bonefish mortality rates can be rather high, potentially resulting in issues of sustainability. At present, there is insufficient information to assess whether bonefish recreational fisheries may be compatible with no-take marine reserves (Bartholomew and Bohnsack, 2005). Theoretically, this would require fishing mortality to be zero, a number that is likely unattainable. We contend that the education of anglers and guides on the strategies to minimize the sublethal effects of catch-and-release angling and to maximize survival are key to generating sustainable recreational fisheries for bonefish. This information should be based on data collected from empirical studies designed to develop and test hypotheses of basic and applied interest. As this information becomes available, it can be used to supplement or refine the bonefish catch-and-release guidelines that we present here. It is our hope that species-specific guidelines for catch-and-release can be developed based on sound science, instead of relying on inferences made from information obtained from studies of unrelated species (see Cooke and Suski, 2005).

ACKNOWLEDGMENTS

The Fisheries Society of the British Isles provided travel support to present these findings at the First International Tarpon and Bonefish Symposium. Financial support for the writing of this chapter was provided by the University of British Columbia, Illinois Natural History Survey, Natural Sciences and Engineering Research Council, the Killam Trust, the Canadian Foundation for Innovation, the Charles A. and Anne Morrow Lindbergh Foundation, and the University of Illinois. We also thank the many guides and anglers who contributed their ideas and expertise to this paper as well as Andy Danylchuk, Sascha Danylchuk, Tony Goldberg, Chris Maxey, and Jeff Koppelman. This is a contribution from the Flats Ecology Research Program, a partnership with the Cape Eleuthera Institute, the University of Illinois, and Carleton University.

REFERENCES

Anderson, W. G., R. Booth, T. A. Beddow, R. S. McKinley, B. Finstad, F. Økland, and D. Scruton. 1998. Remote monitoring of heart rate as a measure of recovery in angled Atlantic salmon, *Salmo salar* (L.). Hydrobiologia 371/372: 233–240.

Anon. 1999. Bonefish preservation. Meeting summary. Below Decks, Abaco Beach Hotel, 25 November 1998. Sponsored by the Bahamian Ministry of Tourism.

Anon. 2001. Marine conservation and research workshop. Proceedings Summary, August 16, 2000. College of the Bahamas. The Bahamas Environment, Science, and Technology Commission, Nassau, Bahamas.

Ault, J. S., R. Humston, M. F. Larkin, and J. Luo. 2002. Development of a bonefish conservation program in South Florida. Final report to National Fish and Wildlife Foundation on Grant No. 20010078000-SC.

Ault, J. S., M. F. Larkin, J. Luo, N. Zurcher, and D. Debrot. 2005. Bonefish-tarpon conservation research program. Final Report to the Friends of the Florida Keys National Marine Sanctuary. 91p.

Barthel, B. L., S. J. Cooke, C. D. Suski, and D. P. Philipp. 2003. Effects of landing net mesh type on injury and mortality in a freshwater recreational fishery. Fish. Res. 63: 275–282.

Bartholomew, A. and J. A. Bohnsack. 2005. A review of catch-and-release angling mortality with implications for no-take reserves. Rev. Fish Biol. Fish. 15: 129–154.

Black, E. C. 1958. Hyperactivity as a lethal factor in fish. J. Fish. Res. Bd. Can. 15: 573–586.

Bruger, G. E. and K. D. Haddad. 1986. Management of tarpon, bonefish, and snook in Florida, pp. 53–57, *In* R. H. Stroud (Ed.), Multi-jurisdictional management of marine fisheries, ed., National Coalition for Marine Conservation, Savannah, GA.

Burdick, B. and R. Wydoski. 1989. Effects of hooking mortality on a bluegill fishery in a western reservoir, pp. 187–196, In R. A. Barnhart and T. D. Roelofs (Eds.), Catch-and-release fishing—a decade of experience. Humboldt State University, California Cooperative Fisheries Research Unit, Arcata, CA.

Colborn, J., R. E. Crabtree, J. B. Shaklee, E. Pfeiler, and B. W. Bowen. 2001. The evolutionary enigma of bonefishes (*Albula* spp.): cryptic species and ancient separations in a globally distributed shorefish. Evolution. 55: 807–820.

Colton, D. E. and W. S. Alevizon. 1983a. Movement patterns of the bonefish (*Albula vulpes*) in Bahamian waters. Fish. Bull. 81: 148–154.

Colton, D. E. and W. S. Alevizon. 1983b. Feeding ecology of bonefish in Bahamian waters. Trans. Am. Fish. Soc. 112: 178–184.

Cooke, S. J. and D. P. Philipp. 2004. Behavior and mortality of caught-and-released bonefish (Albula spp.) in Bahamian waters with implications for a sustainable recreational fishery. Biol. Conserv. 118: 599–607.

Cooke, S. J. and C. D. Suski. 2004. Are circle hooks effective tools for conserving freshwater and marine recreational catch-and-release fisheries? Aquatic Conserv.: Mar. Freshwater Ecosyst. 14: 299–326.

Cooke, S. J. and C. D. Suski. 2005. Do we need species specific recreational angling catch-and-release guidelines to effectively conserve diverse fisheries resources? Biodiversity Conserv. 14: 1195–1209.

Cooke, S. J., K. M. Dunmall, J. F. Schreer, and D. P. Philipp. 2001. The influence of terminal tackle on injury, handling time, and cardiac disturbance of rock bass. N. Am. J. Fish. Manage. 21: 333–342.

Cooke, S. J., J. F. Schreer, K. M. Dunmall, and D. P. Philipp. 2002a. Strategies for quantifying sublethal effects of marine catch-and-release angling: insights from novel freshwater applications. Am. Fish. Soc. Symp. 30: 121–134.

Cooke, S. J., J. F. Schreer, D. H. Wahl, and D. P. Philipp. 2002b. Physiological impacts of catch-and-release angling practices on largemouth bass and smallmouth bass. Am. Fish. Soc. Symp. 31: 489–512.

Crabtree, R. E., C. Stevens, D. Snodgrass, and F. J. Stengard. 1998a. Feeding habits of bonefish, *Albula vulpes*, from the waters of the Florida Keys. Fish. Bull. 96: 754–766.

Crabtree, R. E., D. Snodgrass, and C. Harnden. 1998b. Survival rates of bonefish, *Albula vulpes*, caught on hook-and-line gear and released based on capture and release of captive bonefish in a pond in the Florida Keys, pp. 252–254, In Florida Marine Research Institute, Five year Technical Report to the US Fish and Wildlife Service. Sport Fish Restoration Project F-59. Investigation into nearshore and estuarine gamefish abundance, ecology, and life history in Florida, St. Petersburg, FL.

Crabtree, R. E., C. W. Harnden, D. Snodgrass, and C. Stevens. 1996. Age, growth, and mortality of bonefish, *Albula vulpes*, from the waters of the Florida Keys. Fish. Bull. 94: 442–451.

Davis, J. C. 1975. Minimal dissolved oxygen requirements of aquatic life with emphasis on Canadian species: a review. J. Fish. Res. Bd. Can. 32: 2295–2332.

Diaz, R. J. 2001. Overview of hypoxia around the world. J. Environ. Qual. 30: 275–281.

Diggles, B. K. and I. Ernst. 1997. Hooking mortality of two species of shallow water reef fish caught using recreational angling methods. Mar. Freshwater Res. 48: 479–483.

Diodati, P. J. 1991. Estimating mortality of hooked and released striped bass. AFC-22. National Marine Fisheries Service, Bethesda, MD.

Edwards, R. E. 1998. Survival and movement patterns of released tarpon (*Megalops atlanticus*). Gulf of Mexico Sci. 16: 1–7.

Falk, M. R., D. V. Gillman, and L. W. Dahlke. 1974. Comparison of mortality between barbed and barbless hooked lake trout. Technical Report Series CEN/T-74-1. Canada Department of Environmental Fisheries and Marine Service, Winnipeg, Manitoba.

Farrell, A. P. 2002. Cardiorespiratory performance in salmonids during exercise at high temperature: insights into cardiovascular design limitations in fishes. Comp. Biochem. Physiol. A. 132: 797–810.

Farrell, A. P., A. K. Gamperl, J. M. T. Hicks, H. A. Shiels, and K. E. Jain. 1996. Maximum cardiac performance of rainbow trout, Oncorhynchus mykiss, at temperatures approaching their upper lethal limit. J. Exp. Biol. 199: 663–672.

Farrell, A. P., P. E. Gallaugher, J. Fraser, D. Pike, P. Bowering, A. K. M. Hadwin, W. Parkhouse, and R. Routledge. 2001. Successful recovery of the physiological status of coho salmon on-board a commercial gillnet vessel by means of a newly designed revival box. Can. J. Fish. Aquatic Sci. 58: 1932–1946.

Ferguson, R. A. and B. L. Tufts. 1992. Physiological effects of brief air exposure in exhaustively exercised rainbow trout (*Oncorhynchus mykiss*): implications for "catch and release" fisheries. Can. J. Fish. Aquatic Sci. 49: 1157–1162.

Gallman, E. A., J. J. Isely, J. R. Tomasso, and T. I. J. Smith. 1999. Short-term physiological responses of wild and hatchery-produced red drum during angling. N. Am. J. Fish. Manage. 19: 833–836.

Gustaveson, A. W., R. S. Wydowski, and G. A. Wedemeyer. 1991. Physiological response of largemouth bass to angling stress. Trans. Am. Fish. Soc. 120: 629–636.

Humston, R. 2001. Development of movement models to assess the spatial dynamics of fish populations. Doctoral dissertation. University of Miami, Rosenstiel School of Marine and Atmospheric Science. Coral Gables, FL. 155p.

Humston, R., J. S. Ault, M. F. Larkin, and J. Luo. 2005. Movement and site fidelity of the bonefish (*Albula vulpes*) in the northern Florida Keys determined by acoustic telemetry. Mar. Ecol. Prog. Ser. 291: 237–248.

Jenkins, T. M. 2003. Evaluating recent innovations in bait fishing tackle and technique for catch and release of rainbow trout. N. Am. J. Fish. Manage. 23: 1098–1107.

Jolley, J. E., Jr. and E. W. Irby, Jr. 1979. Survival of tagged and released Atlantic sailfish (*Istiophorus platypterus*: Istiophoridae). Bull. Mar. Sci. 29: 155–169.

Kaufmann, R. 2000. Bonefishing. Western Fisherman's Press, Moose, WY. 415p.

Kieffer, J. D. 2000. Limits to exhaustive exercise in fish. Comp. Biochem. Physiol. A. 126: 161–179.

Kieffer, J. D., M. R. Kubacki, F. J. S. Phelan, D. P. Philipp, and B. L. Tufts. 1995. Effects of catch-and-release angling on nesting male smallmouth bass. Trans. Am. Fish. Soc. 124: 70–76.

Mann, K. and J. Lazier. 1996. Dynamics of marine ecosystems: biological-physical interactions in the oceans. Blackwell Science, Oxford, UK. 475p.

Masuda, R. and D. A. Ziemann. 2003. Vulnerability of Pacific threadfin juveniles to predation by blue trevally and hammerhead shark: size dependent mortality and handling stress. Aquaculture. 217: 249–257.

McIntosh, G. S. 1983. An assessment of marine recreational fisheries in the Caribbean, pp. 141–143, *In* J. B. Higman (Ed.), Proceedings of the Gulf and Caribbean Fisheries Institute, 35th Annual Session. Nassau, Bahamas.

McKenzie, D. J., G. Serrini, G. Piraccini, P. Broni, and C. L. Bolis. 1996. Effects of diet on responses to exhaustive exercise in Nile tilapia (*Oreochromis nilotica*) acclimated to three different temperatures. Comp. Biochem. Physiol. A. 114: 43–50.

Milligan, C. L., B. Hooke, and C. Johnson. 2000. Sustained swimming at low velocity following a bout of exhaustive exercise enhances metabolic recovery in rainbow trout. J. Exp. Biol. 203: 921–926.

Muoneke, M. I. and W. M. Childress. 1994. Hooking mortality: a review for recreational fisheries. Rev. Fish. Sci. 2: 123–156.

Newman, D. L. and T. W. Storck. 1986. Angler catch, growth, and hooking mortality of tiger muskellunge in small centrarchid-dominated impoundments. Am. Fish. Soc. Spec. Pub. 15: 346–351.

Payer, R. D., R. B. Pierce, and D. L. Pereira. 1989. Hooking mortality of walleyes caught on live and artificial baits. N. Am. J. Fish. Manage. 9: 188–192.

Pelzman, R. J. 1978. Hooking mortality of juvenile largemouth bass, *Micropterus salmoides*. Calif. Fish Game. 64: 185–188.

Pfeiler, E. 2001. Changes in hypoxia tolerance during metamorphosis of bonefish leptocephali. J. Fish. Biol. 59: 1677–1681.

Policansky, D. 2002. Catch-and-release recreational fishing: a historical perspective, pp. 74–94, *In* T. Pitcher and C. Hollingworth (Eds.), Recreational fisheries: ecological, economic, and social evaluation. Blackwell Science, Oxford, UK.

Prosser, C. L. 1991. Environmental and metabolic animal physiology. Wiley-Liss, Inc., New York.

Samson, J. 2001. The Orvis pocket guide to fly fishing for bonefish and permit. Lyons Press, New York. 150p.

Schaeffer, J. S. and E. M. Hoffman. 2002. Performance of barbed and barbless hooks in a marine recreational fishery. N. Am. J. Fish. Manage. 22: 229–235.

Schreer, J. F., S. J. Cooke, and R. S. McKinley. 2001. Cardiac response to variable forced exercise at different temperatures—an angling simulation for smallmouth bass. Trans. Am. Fish. Soc. 130: 783–795.

Siewert, H. F. and J. B. Cave. 1990. Survival of released bluegill, *Lepomis macrochirus*, caught on artificial flies, worms, and spinner lures. J. Freshwater Ecol. 5: 407–411.

Skomal, G. B. and B. C. Chase. 2002. The physiological effects of angling on post-release survivorship in tunas, sharks, and marlin. Am. Fish. Soc. Symp. 30: 135–138.

Skomal, G. B., B. C. Chase, and E. D. Prince. 2002. A comparison of circle hook and straight hook performance in recreational fisheries for juvenile Atlantic bluefin tuna. Am. Fish. Soc. Symp. 30: 57–65.

Suski, C. D., S. S. Killen, S. J. Cooke, J. D. Kieffer, D. P. Philipp, and B. L. Tufts. 2004. Physiological significance of the different components of live-release angling tournaments or largemouth bass. Trans. Am. Fish. Soc. 133: 1291–1303.

Taylor, M. J. and K. R. White. 1992. A meta-analysis of hooking mortality of nonanadromous trout. N. Am. J. Fish. Manage. 12: 760–767.

Taylor, R. G., J. A. Whittington, and D. E. Haymans. 2001. Catch-and-release mortality rates of common snook in Florida. N. Am. J. Fish. Manage. 21: 70–75.

Thorstad, E. B., T. F. Naesje, P. Fisker, and B. Finstad. 2003. Effects of hook and release on Atlantic salmon in the River Alta, northern Norway. Fish. Res. 60: 293–307.

Thompson, J. A., S. G. Hughes, E. B. May, and R. M. Harrell. 2002. Effects of catch and release on physiological responses and acute mortality of striped bass. Am. Fish. Soc. Symp. 30: 139–143.

Wilde, G. R. 1998. Tournament-associated mortality in black bass. Fisheries 23(10): 12–22.

Wilkie, M. P., K. Davidson, M. A. Brobbel, J. D. Kieffer, R. K. Booth, A. T. Bielak, and B. L. Tufts. 1996. Physiology and survival of wild Atlantic salmon following angling in warm summer waters. Trans. Am. Fish. Soc. 125: 572–580.

Wilkie, M. P., M. A. Brobbel, K. Davidson, L. Forsyth, and B. L. Tufts. 1997. Influences of temperature upon the post-exercise physiology of Atlantic salmon (*Salmo salar*). Can. J. Fish. Aquatic Sci. 54: 503–511.

Wood, C. M., J. D. Turner, and M. S. Graham. 1983. Why do fish die after severe exercise? J. Fish Biol. 22: 189–201.

Wu, R. S. S. 2002. Hypoxia: from molecular responses to ecosystem responses. Mar. Pollut. Bull. 45: 35–45.

Wydoski, R. S. 1977. Relation of hooking mortality and sublethal hooking stress to quality fishery management, pp. 43–87, *In* R. A. Barnhart and T. D. Roelofs (Eds.), Catch-and-release fishing as a management tool. Humbolt State University, Arcata, CA.

26 Florida Keys Bonefish Population Census

Jerald S. Ault, Sandy Moret, Jiangang Luo, Michael F. Larkin, Natalia Zurcher, and Steven G. Smith

CONTENTS

Introduction 383
Materials and Methods 384
    Study Location and Sampling Domain 384
    Field Census and Data Collection 385
    Statistical Sampling Design 386
Results 391
Discussion 396
Acknowledgments 397
References 397

Bonefish, the gamest fish that swims.

—**Zane Grey (1919)**, *Tales of Fishes*

INTRODUCTION

Bonefish (*Albula vulpes*) are spectacular, highly prized saltwater game fish known for their elusive behavior, burst swimming speeds, and tremendous fighting power. Bonefish commonly prowl the nearshore shallow "flats" where they are angled by sight fishers, but recent acoustic telemetry and conventional tagging studies in the Florida Keys have shown that bonefish also occur periodically on the deeper offshore coral reefs (Humston et al., 2004, 2005; Ault et al., 2005a; Larkin et al., Chapter 19, this volume). The Florida Keys is considered the birthplace of bonefishing (Grey, 1919; Oppel and Meisel, 1987; Davidson, 2004). Since the early 1900s, anglers have traversed the legendary islands and flats fished by luminaries such as the likes of Zane Grey, Ernest Hemingway, Joe Brooks, Bill Curtis, and Curt Gowdy. Bonefish present quite a challenge to both novice and expert anglers alike. Because of the unique setting and unparalled opportunities to catch large bonefish, the Florida Keys fishery is world renowned (Fernandez, 2004; Ault et al., 2005a). The regional fishery has produced more than two thirds of the current bonefish world records (IGFA, 2006).

The great interest in bonefish angling over the years has helped fuel the explosive growth of recreational fishing in south Florida and the Florida Keys. Over the past few decades, high-profile anglers and novices have increased their efforts in pursuit of the elusive bonefish. In Florida, the substantial overall economic impact of recreational fisheries greatly surpasses the current commercial fishery revenues, which, in fact, now exceed the historically dominant Florida citrus industry (FFWCC, 2005). Today, the Keys' recreational bonefish fishery contributes about $1 billion per annum to Florida's economy.

To protect this valuable Florida fishery resource, bonefish are regulated as a restricted species, available only to recreational anglers with a one-fish daily limit (www.myfwc.com), but are typically pursued as a catch-and-release species. However, Florida's coastal marine environment is undergoing extensive changes due to rapid regional growth of human populations, recreational fishing fleets, exploitation effects, and environmental changes (Bohnsack and Ault, 1996; Ault et al., 1998, 2005c), a situation magnified by the fact that overfishing is decimating other popular fisheries in the Florida Keys (Ault et al., 1998, 2001, 2002, 2005a,b). Although commercial harvests for bonefish have been prohibited in Florida, experienced professional fishing guides and others knowledgeable with the fishery believe that bonefish have declined in abundance (Larkin et al., in review). Bill Curtis, a legendary professional bonefish guide and IGFA Hall of Fame inductee, estimates that by the early 2000s, the visible bonefish population had declined 90% compared to the late 1940s (Curtis, 2004).

Because of the ecological and economic significance of bonefish, a principal question of anglers, scientists and managers is, "How many bonefish are there in the principal fishing grounds of the Florida Keys?" This information is critical to resource management because there are few data available to determine stock status relative to sustainability benchmarks. In this chapter, we develop and implement a statistically rigorous census methodology to establish a precise quantitative baseline estimate of bonefish population size for the entire Florida Keys, an objective metric, which will allow precise determination of resource trends for this incredible gamefish.

MATERIALS AND METHODS

STUDY LOCATION AND SAMPLING DOMAIN

The Florida Keys bonefish sport fishery, centered at Islamorada, Florida (24.8647°N, 80.7213°W), ranges throughout the coral reef ecosystem that extends about 400 km from northern Biscayne Bay at Key Biscayne near Miami, down south through Islamorada, on to the Seven Mile Bridge, Bahia Honda, the Contents Keys, past Key West and then out west to the Marquesas (Figure 26.1). Because we lacked any type of initial estimates of population size or stock spatial distribution, we employed the help and expert knowledge of several key professional guides and experienced anglers to design the initial pilot census. This process delineated a census survey domain composed of 19 geographical zones that corresponded to areas of high, moderate, and low mean density of bonefish (Figure 26.1). For statistical reporting and analysis, these zones were grouped into four regions or statistical strata (i.e., Biscayne, Upper Keys, Middle Keys, and Lower Keys) circumscribing geological and hydrographic features of the seascape (Marszalek et al., 1977; Shinn et al., 1977; Wolanski, 2000; Porter and Porter, 2001).

Florida Keys Bonefish Population Census

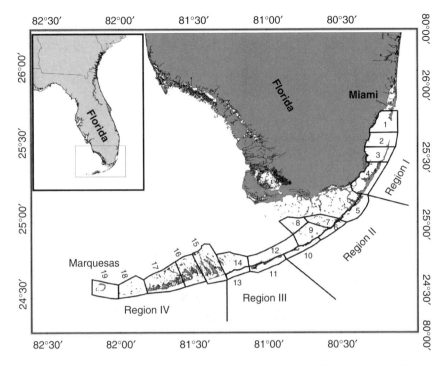

FIGURE 26.1 Map of the Florida Keys showing the 19 geographical zones from Miami to the Marquesas for the annual bonefish population census, covering 1575 mi^2. Also shown are the four regions (strata) used for statistical reporting and analysis.

FIELD CENSUS AND DATA COLLECTION

The annual 1-day Florida Keys bonefish population census was implemented using the volunteered services of professional guides and expert anglers provided by the Florida Keys Fishing Guides Association in collaboration with the 503.1 C nonprofit corporation Bonefish & Tarpon Unlimited (www.tarbone.org). The census surveys were conducted in early fall (September and October) for three consecutive years (2003–2005). This time period was chosen because it is the prime season for large bonefish throughout the Florida Keys and there are very few tourists on hand.

The survey assessment methodology was based on a standard line transect visual estimate, in this case, bonefish counted by two observers along a poled trackline of known distance (e.g., Gunderson, 1993). Generally, two-person teams of fishers–observers were in each flats boat: one was typically a professional bonefish guide who poled the craft at the rear of the vessel positioned atop a platform located approximately 5 ft above the water line, and the other an experienced angler stationed at the bow (Figure 26.2). Each observer wore polarized sunglasses to facilitate bonefish sightings. The boat generally moved in the direction of the prevailing tides and winds, and to the extent possible with the sun behind the observers, to further highlight the bonefish against the coral sand and sea grass–covered bottoms. To ensure adequate coverage of the fishery area during a census, 2–8 boats were assigned to each of the 19 geographic zones. The well-lit portion of the entire day (0800–1600) was

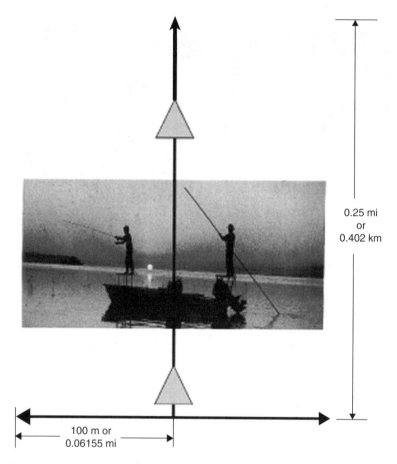

FIGURE 26.2 Pairs of guides and anglers in a flats boat were assigned specific zones where they sighted and counted all the bonefish within 100 m on either side of the vessel as they poled along the flats.

spent looking and fishing for bonefish. Survey participants recorded where they fished, the distance poled, how many bonefish were seen, and the weight and length of landed fish. Many captured bonefish were fitted with plastic anchor tags and then released (see Larkin et al., Chapter 19, this volume). Captains also provided their assessments of weather- and water-related sighting conditions during the day. Information obtained from the participants in the census was subsequently assimilated into a digital database.

STATISTICAL SAMPLING DESIGN

Each 1-day census employed a stratified random sampling design to estimate bonefish population abundance in the Florida Keys. Statistical sampling design procedures are described by Cochran (1977) and Ault et al. (1999, 2002, 2006). A glossary of sampling design statistics is given in Table 26.1. The total area A of the bonefish census–sampling domain was divided into 19 geographical zones to ensure that

TABLE 26.1
Glossary of Sampling Design Statistical Symbols Used in the Florida Keys Bonefish Census

Symbol	Description	Units
h	Stratum subscript	
j	Sample unit (vessel) subscript	
n_h	Number of samples in stratum h	
n	Number of samples in all strata combined	
n^*	Number of samples required to achieve a specified variance	
s_h^2	Sample variance of density in stratum h	
w_h	Stratum weighting factor	
A_h	Stratum area	mi^2
A	Area of all strata combined	mi^2
D_{hj}	Density observation by vessel j in stratum h	Bonefish/mi^2
$\bar{D}_h$	Mean density in stratum h	Bonefish/mi^2
$\bar{D}_{str}$	Domain-wide mean density for a stratified random survey	Bonefish/mi^2
N_h	Number of total possible sample units in stratum h	
N	Number of total possible sample units in all strata combined	
P_h	Population abundance in stratum h	Number of bonefish
P	Domain-wide population abundance	Number of bonefish
$\bar{T}_h$	Mean fished (searched) area in stratum h	mi^2
$V(\bar{D}_{str})$	Target variance of density in a future survey	
var[ ]	Variance of an estimate	
SE [ ]	Standard error of an estimate	
CV [ ]	Coefficient of variation of an estimate	

fishing effort was distributed across the ecosystem. A geographical information system (GIS) was used to compute the area of each zone. These 19 zones were subsequently grouped into 4 regions ($h = 4$) or statistical strata of area A_h for reporting and analysis (Figure 26.1; Table 26.2).

The basic statistical observation of density was the number of bonefish observed per area searched for a single vessel day. Mean density $\bar{D}$ and associated sample variance s^2 of bonefish in stratum h were computed, respectively, as

$$\bar{D}_h = \frac{1}{n_h} \sum_{j=1}^{n_h} D_{hj}, \tag{26.1}$$

and

$$s_h^2 = \frac{\sum_j [D_{hj} - \bar{D}_h]^2}{n_h - 1}, \tag{26.2}$$

TABLE 26.2
Geographic Coverage, Area (A_h, mi²), Sample Unit Characteristics ($\bar{T}_h$, N_h), and Number of Samples (i.e., Boats Participating n_h) for 19 Fishing Zones within Four Geographical Regions (h, Strata) of the Florida Keys for the Bonefish Population Census during 2003–2005

Region (h)	Zone	Geographic Coverage	A_h(mi²)	$\bar{T}_h$(mi²)	N_h	n_h 2003	2004	2005
	1	Key Biscayne to Stiltsville to Gables-by-the-Sea	94.68			1	2	1
	2	Stiltsville to Sands Cut	81.60			3	2	5
	3	Sands Cut to Old Rhodes Key	75.60			3	1	4
	4	Old Rhodes Key to Garden Cove	123.21			4	6	4
1		**Biscayne**	**375.29**	**0.173**	**2169**	**11**	**11**	**14**
	5	Garden Cove to Tavernier Creek (oceanside)	68.41			2	3	4
	6	Between zones 5 and 6 (bayside)	32.24			4	3	3
	7	Tavernier Creek to Indian Key (oceanside)	56.54			1	1	0
	8	Tavernier Key (bayside) to Whipray to Panhandle	54.02			4	2	2
	9	Panhandle to 9 mi bank to zig-zag area	39.26			1	3	1
	10	Indian Key to Duck Key (oceanside)	86.65			3	6	4

Florida Keys Bonefish Population Census

II		**Upper Keys**	**337.12**	**0.288**	**1171**	**15**	**18**	**14**
	11	Duck Key (bayside)	51.51			1	1	1
	12	Duck Key to 7 mi bridge (oceanside)	113.84			0	2	0
	13	7 mi bridge to Bahia Honda (oceanside)	25.47			1	2	1
	14	West Content Key to Bullfrog to West Bahia to US 1	90.84			2	3	1
III		**Middle Keys**	**281.66**	**0.176**	**1600**	**4**	**8**	**3**
	15	West Bahia Honda to Big Pine Key to Contents	148.14			1	7	5
	16	Contents	111.36			4	5	6
	17	Snipes (bayside) to Mud Keys	153.72			3	3	1
	18	Key West to Boca Grande	89.36			0	2	6
	19	Boca Grande to Marquesas	77.40			2	1	1
IV		**Lower Keys**	**579.98**	**0.236**	**2458**	**10**	**18**	**19**
		Florida Keys total	**1574.05**		**7398**	**40**	**55**	**50**

Note: Symbols are defined in Table 26.1.

where n_h was the number of samples obtained in stratum h, and D_{hj}, the observed density by vessel j in stratum h. Mean bonefish density for the Florida Keys survey domain, that is, all strata combined, was estimated by

$$\bar{D}_{str} = \sum_h w_h \bar{D}_h, \qquad (26.3)$$

with stratum weighting factor w_h defined as

$$w_h = \frac{N_h}{N}, \qquad (26.4)$$

where N_h was the total possible sample units in a stratum and N, the total possible sample units in all strata combined. We estimated N_h by dividing stratum area A_h by average searched (fished) area $\bar{T}_h$ of vessels in stratum h:

$$N_h = \frac{A_h}{\bar{T}_h}. \qquad (26.5)$$

Variance of $\bar{D}_{str}$ was estimated by

$$\text{var}[\bar{D}_{str}] = \sum_h w_h^2 \left(1 - \frac{n_h}{N_h}\right)\left(\frac{s_h^2}{n_h}\right). \qquad (26.6)$$

Bonefish population abundance P_h (numbers of fish) in stratum h was obtained by multiplying stratum mean density with stratum area

$$P_h = \bar{D}_h A_h. \qquad (26.7)$$

Variance of P_h was estimated as

$$\text{var}[P_h] = A_h^2 \left(1 - \frac{n_h}{N_h}\right)\left(\frac{s_h^2}{n_h}\right). \qquad (26.8)$$

Total bonefish population abundance, P, in the survey domain was obtained by summing Equation 26.8 over all strata:

$$P = \sum_h P_h. \qquad (26.9)$$

The associated variance, var[P], was obtained in a similar manner:

$$\text{var}[P] = \sum_h \text{var}[P_h]. \qquad (26.10)$$

The standard error of P was computed as

$$SE[P] = \sqrt{\text{var}[P]}. \qquad (26.11)$$

Florida Keys Bonefish Population Census

To evaluate a measure of relative precision of the sampling designs for census years, we computed the coefficient of variation (CV) of population abundance as

$$\mathrm{CV}[P] = \frac{SE[P]}{P}. \tag{26.12}$$

The 95% confidence interval of the population abundance estimate was calculated as

$$95\% CI[P] = P \pm t_{\alpha, n-1} SE[P], \tag{26.13}$$

where t is the critical value of Student's t-distribution with $\alpha = 0.05$. Samples needed in a future survey to obtain a specific precision were calculated using

$$n^* = \frac{\left(\sum_h w_h s_h\right)^2}{V(\bar{D}_{\mathrm{str}}) + \frac{1}{N}\sum_h w_h s_h^2}, \tag{26.14}$$

where $s_h = \sqrt{s_h^2}$ and $V(\bar{D}_{\mathrm{str}})$ is the desired variance of stratified mean density. Equation 26.14 presumes that samples will be optimally (Neyman) allocated among individual strata based on estimates of stratum size (area) and stratum variance of bonefish density (Cochran, 1977):

$$n_h = n^* \frac{w_h s_h}{\sum_h w_h s_h}. \tag{26.15}$$

RESULTS

Annual 1-day Florida Keys bonefish population censuses were conducted in early fall in consecutive years (i.e., October 2003 and 2004, and September 2005). The survey sampling domain of each annual census covered 1574 mi². The number of sampling vessels participating in the survey was 40, 55, and 50, in 2003, 2004, and 2005, respectively (Table 26.2). The distribution of survey sampling effort for the 19 geographic zones and 4 regional strata over the three census years is shown in Table 26.2. The most general results for the three survey years were: (1) in 2003, survey teams spotted 1899 and caught 23 bonefish during the 8-h day; (2) 1598 sighted and 36 caught in 2004; and (3) 1601 sighted and 35 caught in 2005. The distribution of bonefish count observations seen by teams each year was roughly equivalent (Figure 26.3). Overcast skies and windy weather, both of which hinder the sighting of bonefish, may have contributed to the greater number of zero counts recorded in 2004. The frequency distributions of density observations (number of bonefish per square mile) were also roughly equivalent among years.

On average, the majority of bonefish were estimated to occur in the Upper Keys and Biscayne Bay regions over the 3 years sampling period, although there was some fluctuation in regional abundance estimates among years (Table 26.3; Figure 26.4). Florida Keys–wide abundance estimates ranged from 259,395 to 340,552

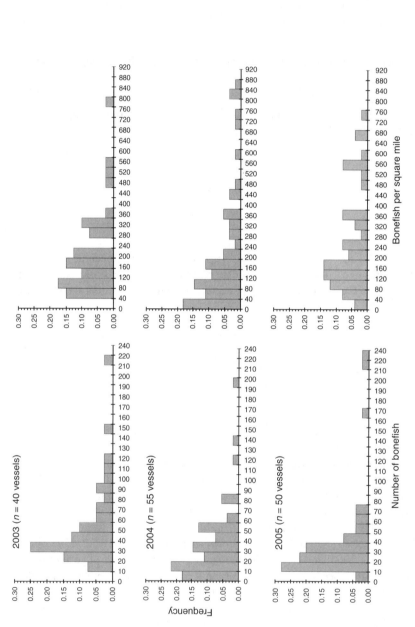

FIGURE 26.3 Frequency distributions of bonefish seen per boat (left column) and seen per square mile (right column) for the Florida Keys bonefish census: 2003 (upper panels) to 2005 (lower panels).

TABLE 26.3
Survey Statistics by Region Strata for the Florida Keys Bonefish Population Census in 2003, 2004, and 2005

Region	n_h	Mean Density $\bar{D}_h$ (fish/mi²)	s_h^2	Abundance P_h (number)	var$[P_h]$
October 7, 2003					
Biscayne	11	166.98	8183.10	62,665	104.244
Upper Keys	15	228.70	46107.86	77,100	344.868
Middle Keys	4	202.82	26321.72	57,127	520.739
Lower Keys	10	107.77	12951.19	62,503	433.873
October 26, 2004					
Biscayne	11	255.86	55147.52	96,023	702.519
Upper Keys	18	295.44	94699.75	99,598	588.731
Middle Keys	8	68.61	9383.81	19,323	92.591
Lower Keys	18	101.33	10258.72	58,768	190.305
September 29, 2005					
Biscayne	14	278.95	51625.54	104,685	516.008
Upper Keys	14	223.07	46072.66	75,201	369.539
Middle Keys	3	137.98	30691.35	38,864	810.087
Lower Keys	19	210.01	28734.84	121,801	504.785

Note: Symbols are defined in Table 26.1. Units for var $[P_h]$ are 1×10^6.

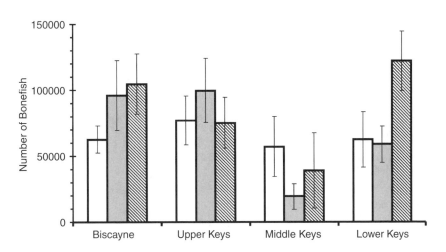

FIGURE 26.4 Comparison of regional estimates of bonefish abundance by census year: (2003) open; (2004) gray; and (2005) hatched. Bars denote one standard error.

for 2003 to 2005, with associated coefficient of variation ranging from 13.8 to 14.5% (Table 26.4). The 2005 population estimate was a slight increase from estimates for 2003 and 2004, but not statistically different (Figure 26.5). The 3 years average population abundance estimate was 291,220 bonefish in the Florida Keys. The census population estimate represents bonefish of the size classes (i.e., generally >14 in total length (TL)) in the Florida Keys flats fishery and does not include juveniles.

Aspects of domain stratification and sample allocation were investigated with respect to improving the performance of future bonefish surveys. Simple random sampling (SRS) design estimates of abundance were generally higher and CVs were more variable compared to abundance estimates for the stratified random design (Table 26.4). SRS estimates were made considering the entire domain as a single

TABLE 26.4
Florida Keys–Wide Estimates for the Stratified Random Survey of Bonefish Population Abundance in 2003, 2004, and 2005, and Comparison with Estimates from a Simple Random Survey Design

		Stratified Random			Simple Random		
Year	n	Abundance P	SE[P]	CV[P] (%)	Abundance P	SE[P]	CV[P] (%)
2003	40	259,395	37,466	14.44	281,608	40,023	14.21
2004	55	273,712	39,676	14.50	300,646	49,060	16.32
2005	50	340,552	46,909	13.77	359,901	43,854	12.19

Note: Symbols are defined in Table 26.1.

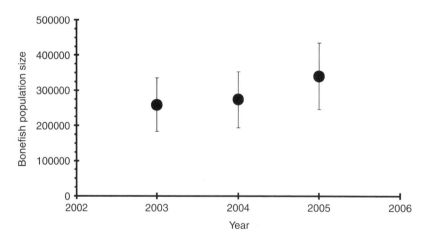

FIGURE 26.5 Comparison of total population abundance estimates by years. Bars denote 95% confidence intervals.

stratum. A plot of standard deviation against mean density for the 19 geographical zones for the combined 2003–2005 surveys (Figure 26.6) indicates that bonefish density and variance are heterogeneous throughout the Florida Keys sampling domain, in contrast to what is expected for a simple random design. Several alternative sampling designs were compared using poststratification analysis, a procedure that entailed reanalysis of survey data for stratification schemes that differed from the one actually implemented. Estimates of the performance measure n^*, the sample size required to achieve a specified precision (i.e., CV), suggested that stratifying the domain by the 19 geographical zones may perform better than the 4-strata regional scheme (Table 26.5; Figure 26.7). Estimates of n^* presume that samples will be allocated among strata according to both stratum size and stratum variance, i.e., more samples will be allocated to larger and more variable strata. As illustrated in Figure 26.7, the CVs actually achieved in the surveys were generally higher than would be predicted by the CV-n curve for the 4-strata (actual) design, an indication that past sample allocations were suboptimal to some degree. Thus, performance of future surveys would likely be improved via more refined stratification and allocation.

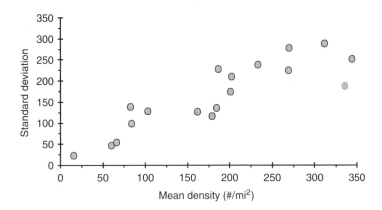

FIGURE 26.6 Results from the 2003–2005 Florida Keys bonefish censuses showing standard deviation of mean bonefish density (number of fish per square mile) dependent on mean density for survey vessels in the 19 statistical strata.

TABLE 26.5
Poststratification Analysis Results

Stratification Variable	Number of Strata	Mean Density $\bar{D}_{str}$	CV[$\bar{D}_{str}$] (%)	n^* (10%)
None (simple random)	1	199.52	14.30	98
Region (actual design)	4	185.27	14.36	87
Fishing zone	19	177.66	14.62	74

Note: Symbols are defined in Table 26.1.

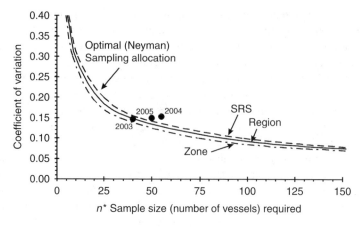

FIGURE 26.7 Optimal stratified random sample allocation using stratum (zone) areas and bonefish density estimates as a means of achieving high precision population estimates. Note that the 2004 sampling allocation of 55 guided flats boats achieved a coefficient of variation (CV) of about 14.5%, and with optimal allocation this could be lowered to less than 10% CV.

DISCUSSION

Bonefish would appear to be excellent candidates for assessing ecological changes in the Florida Keys ecosystem. Because of their large size, bonefish rely on and are sensitive to changes in prey whose populations are not as easy to assess. Bonefish also inhabit shallow, clear inshore waters on grass and sand flats that are highly susceptible to human impacts. These attributes also make them ideal candidates for a visual census in that, to the trained eye, they are easy to spot and count.

Lacking former knowledge of the bonefish population abundance distribution, we hypothesized that the spatial distribution of bonefish was proportional to the distribution of guided-fishing effort. This concept was borne out by the first 3 years of census results, which also showed that bonefish population abundance remained relatively constant. This generally indicates that there have been no dramatic changes in the environment or exploitation effects in the Keys over the past 3 years as they relate to bonefish. While our results suggest that there is room for design improvements in future surveys, we obtained relatively precise estimates of population abundance during this initial phase of the bonefish census. These findings give researchers and participants confidence in the statistical and field methodologies employed and help validate the "census" concept.

The census population estimates are important benchmarks of resource status and a quantitative measure for assessing future change. Long-time bonefish anglers have often remarked on the dramatic decreases they have observed in this popular sport fish. For the current bonefish population, there could be cause for considerable concern if the population were to drop from 300,000 to 200,000.

In addition to being good indicators of the ecosystem's health, bonefish are valuable for economic reasons. As a premier sport fish in a world-class destination, bonefish bring in a significant amount of tourism. Bonefish contribute approximately

$1 billion annually to the Florida economy. To arrive at this number, one considers indices provided by the National Oceanic and Atmospheric Administration, guided fishing trip bookings, surveyed hotel bookings, vessel costs, and license and tackle purchases. So just how much is this small, but formidable gamefish worth to the Florida Keys? An approximate figure can be determined from census and tourism statistics. Dividing the average census bonefish population estimate into annual industry value points out the remarkable fact that each bonefish represented in the census is worth about $3500 per year to the industry. Since bonefish live up to more than 20 years, prorated over its lifetime in the fishery, the lifetime value of each bonefish may be worth in excess of $75,000 to the fishing industry, making the fishery per fish one of the most valuable in the world.

Finally, the importance of bonefish, not only in Florida but at many locations throughout the world, highlights the potential for broader application of the census methodology developed here. This approach would facilitate regional assessments of bonefish populations and provide a quantitative basis for comparing ecosystems that support bonefish fisheries. The Florida Keys bonefish census has now become an annual phenomenon through the support of the Bonefish & Tarpon Unlimited, Florida Keys Fishing Guides Association, and the University of Miami. The annual bonefish census links experienced anglers and professional guides with research scientists in an effort to build a sustainable fishery. In the near term, integrating census results with ongoing tagging studies could strengthen the reliability of the both approaches to assess bonefish resource status, in turn providing a sensitive indicator of coastal ocean health and regional economic vitality. If census results are any indication of the importance of this small and formidable silver gamefish, then Florida should consider bonefish as "bars of silver" and ensure their protection and sustainability accordingly.

ACKNOWLEDGMENTS

This research was supported by funds from Bonefish & Tarpon Unlimited (BTU), Sanctuary Friends Foundation of the Florida Keys, the National Fish and Wildlife Foundation (Grant No. 2004-0000-000), and Florida Fish and Wildlife Conservation Commission. Professional guides from the Florida Keys Fishing Guides Association, the Lower Keys Fishing Guides Association, and the Key Largo Fishing Guides Association played an integral role in the census' success by providing boats and manpower. We especially thank Tad Burke, Rick Bouley, and Steve Venini for help with organizing the guide's efforts. We also greatly appreciate the technical assistance by Ivy Kupec of the University of Miami RSMAS, and essential fiscal support and intellectual guidance provided by Tom Davidson and Russ Fisher of BTU.

REFERENCES

Ault, J.S., J.A. Bohnsack, and G.A. Meester. 1998. A retrospective (1979–1996) multispecies assessment of coral reef fish stocks in the Florida Keys. Fishery Bulletin 96(3): 395–414.

Ault, J.S., G.A. Diaz, S.G. Smith, J. Luo, and J.E. Serafy. 1999. An efficient sampling survey design to estimate pink shrimp population abundance in Biscayne Bay, Florida. North American Journal of Fisheries Management 19(3): 696–712.

Ault, J.S., S.G. Smith, G.A. Meester, J. Luo, and J.A. Bohnsack. 2001. Site characterization for Biscayne National Park: assessment of fisheries resources and habitats. NOAA Technical Memorandum NMFS-SEFSC-468. 185 p.

Ault, J.S., S.G. Smith, J. Luo, G.A. Meester, J.A. Bohnsack, and S.L. Miller. 2002. Baseline multispecies coral reef fish stock assessment for the Dry Tortugas. NOAA Technical Memorandum NMFS-SEFSC-487. 117 p.

Ault, J.S., M.F. Larkin, J. Luo, N. Zurcher, and D. Debrot. 2005a. Bonefish-tarpon conservation research program. Final Report to the Sanctuary Friends Foundation of the Florida Keys. 91 p.

Ault, J.S., S.G. Smith, and J.A. Bohnsack. 2005b. Evaluation of average length as an estimator of exploitation status for the Florida coral reef fish community. ICES Journal of Marine Science 62: 417–423.

Ault, J.S., J.A. Bohnsack, S.G. Smith, and J. Luo. 2005c. Towards sustainable multispecies fisheries in the Florida USA coral reef ecosystem. Bulletin of Marine Science 76(2): 595–622.

Ault, J.S., S.G. Smith, J.A. Bohnsack, J. Luo, D.E. Harper, and D.B. McClellan. 2006. Building sustainable fisheries in Florida's coral reef ecosystem: positive signs in the Dry Tortugas. Bulletin of Marine Science 78(3): 633–654.

Bohnsack, J.A. and J.S. Ault. 1996. Management strategies to conserve marine biodiversity. Oceanography 9(1): 73–82.

Cochran, W.G. 1977. Sampling Techniques. 3rd Ed. John Wiley & Sons. New York. 428 p.

Curtis, B. 2004. Not exactly fishing (alligator fishing). Chapter 11, pp. 167–171 in Bonefish B.S. and Other Good Fish Stories, T.N. Davidson (ed.). Hudson Books. Whitby, Ontario, Canada. 204 p.

Davidson, T.N. 2004. Bonefish B.S. and Other Good Fish Stories. Hudson Books. Whitby, Ontario, Canada. 204 p.

Fernandez, "Chico" J.M. 2004. Fly-fishing for bonefish. Stackpole Books. Mechanicsburg, PA. 192 p.

Grey, Z. 1919. Tales of Fishes. The Derrydale Press. Lanham, MD. 267 p.

Gunderson, D.R. 1993. Surveys of Fisheries Resources. John Wiley & Sons. New York. 248 p.

Humston, R., D.B. Olson, and J.S. Ault. 2004. Behavioral assumptions in models of fish movement and their influence on population dynamics. Transactions of the American Fisheries Society 133: 1304–1328.

Humston, R., J.S. Ault, M.F. Larkin, and J. Luo. 2005. Movements and site fidelity of bonefish (*Albula vulpes*) in the northern Florida Keys determined by acoustic telemetry. Marine Ecology Progress Series 291: 237–248.

IGFA (International Game Fish Association). 2006. World Record Game Fishes: freshwater, saltwater, and flyfishing. Dania Beach, FL. 384 p.

Larkin, M.F., J.S. Ault, and R. Humston. In review. Mail survey of south Florida's bonefish charter fleet. Transactions of the American Fisheries Society.

Marszalek, D.S., G. Babashoff, M.R. Noel, and D.R. Worley. 1977. Reef distribution in south Florida. Proceedings of the 3rd International Coral Reef Symposium 2: 223–229.

Oppel, F. and T. Meisel. 1987. Tales of Old Florida. Castle Press. Seacaucus, NJ. 477 p.

Porter, J.W. and K.G. Porter. 2001. The Everglades, Florida Bay, and Coral Reefs of the Florida Keys: an ecosystem sourcebook. CRC Press. Boca Raton, FL. 1000 p.

Shinn, E.A., J.H. Hudson, R.B. Halley, and B. Lidz. 1977. Topographic control and accumulation rate of some Holocene coral reefs: south Florida and Dry Tortugas. Proceedings of the 3rd International Coral Reef Symposium 2: 1–7.

Wolanski, E. 2000. Oceanographic processes of coral reefs: physical and biological links in the Great Barrier Reef. CRC Press. Boca Raton, FL.

27 Science in Support of Management Decision Making for Bonefish and Tarpon Conservation in Florida

Luiz R. Barbieri, Jerald S. Ault, and Roy E. Crabtree

CONTENTS

Introduction 399
Current Status of Fisheries Management in Florida 400
    Tarpon 400
    Bonefish 400
Resource Impacts and Future Directions 400
    Fishery: Addressing Catch-and-Release Mortality 400
    Environmental: Protection of Nursery Habitats 402
Conclusions 402
References 403

INTRODUCTION

Atlantic tarpon (*Megalops atlanticus*) and bonefish (*Albula vulpes*) support very popular and economically important recreational fisheries in Florida, which has a long and rich history of angling for these species (Oppel and Meisel, 1987). The fishery for bonefish is restricted almost exclusively to south Florida, especially the Florida Keys and Biscayne Bay, where mature bonefish are the target of a specialized and highly directed sport fishery (Crabtree et al., 1998; Ault et al., Chapters 16 and 19, this volume). The majority of anglers pursuing bonefish do so with professional fishing guides on a daily charter basis. The Florida fishery for tarpon is triggered by seasonal availability and is less geographically restricted, occurring in coastal waters statewide. Although the Florida Keys and Boca Grande Pass (southwest Florida) represent two of the best-known and most-celebrated tarpon fishing spots in the world; other locations such as Jupiter and Sebastian Inlets on Florida's east coast, and Tampa Bay and Homosassa Springs on the west coast have rapidly expanding and increasingly popular tarpon fisheries.

The Florida tarpon fishery is also somewhat seasonally dependent on tarpon long-range migrations (Ault et al., Chapter 16, this volume; Luo et al., Chapter 18, this volume). The Florida fishery appears to be seasonally connected to North Carolina in the north, Texas, Louisiana and Mexico in the west, and Cuba and perhaps Belize in the south.

CURRENT STATUS OF FISHERIES MANAGEMENT IN FLORIDA

Tarpon

The Florida tarpon fishery has been intensely regulated for some time. Dating back to 1953, the Florida Legislature took the first steps in managing the fishery by granting tarpon game fish status (i.e., harvest or possession of tarpon for commercial purposes was prohibited by law). However, because tarpon were never considered a food fish in Florida (Bruger and Haddad, 1986), and the fact that a large number of the fish harvested by anglers were being used for trophy taxidermy mounts, game fish status has afforded little protection to tarpon populations. As a result, the Florida State Legislature in 1989 established a new regulation that required anglers wishing to harvest (or possess) a tarpon to purchase an individual permit costing $50. Each permit allows the harvest of one tarpon a year and requires anglers to attach a numbered tag to the tarpon's lower jaw (for enforcement purposes). Establishment of the tarpon permit system has resulted in substantial reduction of tarpon harvests in the State of Florida. Today the Florida fishery is almost exclusively catch-and-release.

Bonefish

The current bonefish fishery in Florida is also believed to be predominantly catch-and-release (Crabtree et al., 1998; Ault et al., Chapter 26, this volume). Bonefish were granted gamefish status in 1972, which prohibited their commercial sale in Florida. Recreational regulations developed at that time and still in effect today are a bag limit of one bonefish per angler per day, with a minimum size limit of 18 inches total length (TL). Bonefish are thought to be lightly exploited with minimal harvests for food in the state, so that concerns over take have never been viewed as sufficient to warrant development of a "kill tag program" similar to that established for tarpon.

RESOURCE IMPACTS AND FUTURE DIRECTIONS

Fishery: Addressing Catch-and-Release Mortality

Establishment of the Florida Tarpon Tag Program and the relatively nonconsumptive nature of the tarpon and bonefish fisheries in Florida have led to the assumption of limited directed harvest (i.e., intentional take of the species) and very low exploitation levels. While catches are difficult to quantify, the largest source of fishing mortality for these species is presumed to be postrelease mortality due to recreational angling, that is, fish that are released after hook-and-line capture but do not survive (Crabtree et al., 1995, 1996; Cooke and Philipp, Chapter 25, this volume). Studies have shown that postrelease mortality due to hooking and handling trauma or postrelease predation of tarpon and bonefish can be highly variable, and in some cases may be significant. Edwards (1988) evaluated survival of 27 tarpon caught and released by anglers at Boca Grande Pass, Florida. In this study, ultrasonic transmitters were attached to tarpon after capture by hook-and-line gears, then fish movements were monitored for up to 12-h postrelease. Edwards (1988) found that only one tarpon died during the tracking period for <4% postrelease mortality—the one fish that died had been lifted from the water for a photograph prior to release. Guindon et al. (2004) conducted a more intensive follow-up study in the Boca Grande Pass area using similar methods. They reported higher postrelease mortality on the order of 10–20%, and further, identified predation of tarpon by sharks as the predominant source of postrelease mortality, with all predation events occurring shortly after and nearby release locations.

Catch-and-release research on bonefish has been limited in comparison to other species (e.g., Bartholomew and Bohnsack, 2005). However, existing studies indicate that for fish that are improperly handled during angling and capture, postrelease predation may represent the largest source of fishing mortality. Crabtree et al. (1998) used hook-and-line gears to repeatedly catch-and-release bonefish held in a large experimental pond in the Florida Keys. They reported a 96% survival rate, and a few of the bonefish that ultimately died had been caught 5–10 times each. These findings suggest that wild bonefish that are properly hooked, handled, and released may have relatively high survival rates. In other recent studies, Cooke and Philipp (2004; see also Chapter 25, this volume) assessed bonefish postrelease behavior and mortality at two regions of differing shark abundance in the Bahamas. They reported that injuries to the fish from a variety of fly, lure, and bait hook types were minimal; bleeding was rare; and that most fish were hooked in the upper or lower lip and did not swallow the hook. All observed mortalities were stated to have occurred within 30 min of release and were directly attributed to shark predation. They found that all bonefish survived in areas of low shark abundance, whereas in high shark abundance areas, 39% postrelease mortality was observed. Furthermore, Cooke and Philipp (2004) reported that bonefish angled to exhaustion and exposed to air had significant problems maintaining equilibrium following release. Typically, these fish moved little for the first 30 min following release. The Cooke and Philipp (2004) study touted the benefits of short-duration fights when angling, minimizing handling and air exposure, and quickly and gently releasing the fish.

Marine recreational fishing effort and angling intensity in Florida has increased dramatically over the past few decades (Harper et al., 2000; Ault et al., 2005). This trend is accelerating and given this scenario and the popularity of tarpon and bonefish angling in Florida, it is not unreasonable to expect that fishing pressures directed at both species will not only remain high, but will likely increase significantly over time. The large numbers of tarpon and bonefish being caught and presumably released annually by anglers statewide could contribute to substantial postrelease mortality and have deleterious population-level impacts on these species. Nelson (2002) showed that the high angling effort directed at common snook (*Centropomus undecimalis*) in Florida caused catch and release mortality to account for >30% of total fishing mortality for this species. As a point of comparison, these signs are harbingers of problems that are likely to escalate over the next few decades for tarpon and bonefish populations in Florida.

ENVIRONMENTAL: PROTECTION OF NURSERY HABITATS

Juvenile tarpon (<100 cm) utilize a wide range of estuarine and tidal areas such as the upper reaches of salt marshes, mangrove forests and shorelines, and tidal creeks as nursery habitats (Zale and Merrifield, 1989; Crabtree et al., 1995; Shenker et al., 1995). Man-made mosquito control impoundments and storm-water management drainage systems also provide important habitats for young-of-the-year tarpon. It has been known for some time that activities associated with human development—including dredging, filling, channelization, and shoreline hardening—have resulted in widespread direct physical impacts and the loss of many critical nursery areas (e.g., Robins, 1977). In addition, land-based hydrologic modifications associated with regional transportation, agriculture, flood protection, storm-water conveyance, water supply, and wastewater treatment systems have altered the quantity, quality, location, and timing of freshwater inflows to the tidal reaches of many Florida rivers, streams, estuaries, and coastal lagoons (Porter and Porter, 2001). There is a great concern that these activities have negatively impacted the condition, productivity, and dynamics of Florida's coastal ecosystems. Preservation and management of these critical natal and juvenile habitats is of paramount importance for the long-term health, sustainability, and economic viability of Florida's tarpon fisheries.

With bonefish we face even bigger challenges. Nursery habitats, prey dynamics, and sources of recruits to the adult fishery in the Florida Keys are virtually unknown (Adams et al., Chapter 15, this volume; Dahlgren et al., Chapter 12, this volume). These critical information gaps prevent us from identifying potential conservation threats and may represent the greatest challenge to effective management and conservation of bonefish in Florida.

CONCLUSIONS

The catch-and-release nature of Florida's tarpon and bonefish fisheries emphasizes the need for a comprehensive, ecosystem-based fishery management strategy that integrates habitat protection, water quality, and trophic interactions (e.g., predator–prey dynamics) with more traditional fisheries management approaches. Quantifying fishery impacts is difficult because, with virtually no fish being landed (i.e., brought back to the dock), access to fisheries-dependent biological samples is very limited. These samples are critical for assessing the status of stocks because the most powerful and reliable stock assessment models integrate demographic and biological information (e.g., age composition of the catch, size and age at sexual maturity, and sex ratios) with fishery catch and effort data over time and space. A potential solution would be to develop directed data-collection programs focused on obtaining information on fishing effort, and the number and size composition of tarpon and bonefish that are caught and released in Florida. These data could then be used to estimate total catch and effort, and to develop catch-at-age information for stock-assessment purposes.

Efforts to control fishing mortality, that is, preventing it from increasing beyond a reasonable level, should be focused on maximizing catch-and-release survival. Angler education programs promoting fishing styles and techniques that improve survival—use appropriate tackle, do not angle your fish to exhaustion, avoid pulling the fish out of the water, etc.—should be expanded and intensified. We need to recognize that the power to reduce postrelease mortality belongs to the anglers, not fishery managers. Real gains will not be achieved until we succeed in engaging anglers to the point that they see themselves as part of the solution.

Finally, we need to develop an integrated and well-coordinated research effort to address the most critical knowledge gaps in tarpon and bonefish life history and population dynamics:

- *Reproductive biology.* More detailed information is needed for both species. Specific topics include spawning locations, basic reproductive parameters (e.g., batch fecundity, spawning frequency, size, and age at maturity), and spawning-site fidelity.
- *Recruitment.* Identification of bonefish nursery habitats and sources of recruits to the Florida fishery is needed. A better understanding of the role of tarpon and bonefish larval transport dynamics on recruitment variability and population connectivity is required.
- *Stock structure, movements, and migration.* Determine tarpon and bonefish "unit stocks" and the connectivity of populations in the western Atlantic, Gulf of Mexico, and the Caribbean. Evaluate the role of long-range migrations on tarpon population dynamics.

This list is by no means exhaustive. However, addressing these knowledge gaps would give us the basic information needed for development of management strategies aimed at the long-term sustainability and continuing a legacy of world-class tarpon and bonefish fisheries in Florida.

REFERENCES

Ault, J.S., J.A. Bohnsack, S.G. Smith, and J. Luo. 2005. Towards sustainable multispecies fisheries in the Florida USA coral reef ecosystem. Bulletin of Marine Science 76(2): 595–622.

Bartholomew, A. and J.A. Bohnsack. 2005. A review of catch-and-release angling mortality with implications for no-take reserves. Reviews in Fish Biology and Fisheries 15: 129–54.

Bruger, G.E. and K.D. Haddad. 1986. Management of tarpon, bonefish, and snook in Florida. Pp. 53–57 in Stroud, R.H. (ed.), *Multi-Jurisdictional Management of Marine Fishes*. National Coalition for Marine Conservation, Inc. Savannah, GA.

Cooke, S. J. and D. P. Philipp. 2004. Behavior and mortality of caught-and-released bonefish (*Albula* spp.) in Bahamian waters with implications for a sustainable recreational fishery. Biological Conservation 118: 599–607.

Crabtree, R.E., E.C. Cyr, and J.M. Dean. 1995. Age and growth of tarpon, Megalops atlanticus, from South Florida waters. Fishery Bulletin 93: 619–28.

Crabtree, R.E., C.W. Harnden, and D. Snodgrass. 1996. Age, growth, and mortality of bonefish, *Albula vulpes*, from the waters of the Florida Keys. Fishery Bulletin 94(3): 442–52.

Crabtree, R.E., D. Snodgrass, and C. Harnden. 1998. Survival rates of bonefish, *Albula vulpes*, caught on hook-and-line gear and released based on capture and release of captive fish in a pond in the Florida Keys. In *Investigations into Nearshore and Estuarine Gamefish Abundance, Ecology, and Life History in Florida*. Five-year Technical Report to the U.S. Fish and Wildlife Service, Sport Fish Restoration Project F-59. Florida Department of Environmental Protection, Division of Marine Resources, Florida Marine Research Institute. 566 p.

Edwards, R. 1988. Survival and movement of released tarpon. Gulf of Mexico Science 1998(1): 1–7.
Guindon, K.Y., L.R. Barbieri, and C. Powell. 2004. Summary report on the tarpon catch-and-release mortality study in Boca Grande Pass, 2002–2004. Florida Fish and Wildlife Research Institute Report.
Harper, D.E., J.A. Bohnsack, and B.R. Lockwood. 2000. Recreational fisheries in Biscayne National Park, Florida, 1976–1991. Marine Fisheries Review 62(1): 8–26.
Nelson, R.S. 2002. Catch-and-release: A management tool for Florida. Pp. 11–14 in Lucy, J.A. and A.L. Studholme (eds.), *Catch-and-Release in Marine Recreational Fisheries*. American Fisheries Society, Symposium 30. Bethesda, MD.
Oppel, F. and T. Meisel. 1987. Tales of Old Florida. Castle Press. Seacaucus, NJ. 477 p.
Porter, J.W. and K.G. Porter. 2001. The Everglades, Florida Bay, and Coral Reefs of the Florida Keys: An Ecosystem Sourcebook. CRC Press. Boca Raton, FL. 1000 p.
Robins, C.R. 1977. The tarpon—unusual biology and man's impact determine its future. Pp. 105–112 in Clepper, H. (ed.), *Marine Recreational Fisheries*, Volume 2. Sport Fishing Institute, Washington, DC.
Shenker, J.M., R. Crabtree, and G. Zarillo. 1995. Recruitment of larval tarpon and other fishes into the Indian River Lagoon (abstract). Bulletin of Marine Science 57: 284.
Zale, A.V. and S.G. Merrifield. 1989. Species profiles: life histories and environmental requirements of coastal fishes and invertebrates (South Florida)—ladyfish and tarpon. U.S. Fish and Wildlife Service Biological Report 82(11.104). U.S. Corps of Engineers, TR EL-82-4. 17 p.

28 How to Get the Support of Recreational Anglers

Doug Kelly

CONTENTS

Introduction 405
Problems of Perception—and Reality 408
    "You Don't Know What You're Talking About" 410
Steps to Take—and Not to Take 410

> *Doug Kelly lives in Tampa, Florida, and serves as executive director of the Florida Outdoor Writers Association. He is a representative for the International Game Fish Association and a member of the Coastal Conservation Association, Recreational Fishing Alliance, Southeast Outdoor Press Association, and Society of American Travel Writers.*

INTRODUCTION

Just as a cart gets pushed along better when more shoulders are behind it, so too is the outcome with projects and initiatives involving fisheries management. With fewer shoulders pushing—or worse yet, shoulders pushing in opposing directions—management tools become far less effective. True, there are unimaginative people who feel opposition should be ignored or stepped on, but many others recognize the power of collective action.

Collective action should not be confused with total support. Believe me, I know all too well that 100% unanimity is a pipe dream. It does not take long for a neophyte journalist to learn that a nice article about the Cherry Pie Association will beget an angry letter from the Apple Pie Association. It goes with the territory, and the same with fisheries management. While you do not expect unanimity to every initiative or rule proposal, it is also not wise to face an organized opposition with considerable resources and, if necessary, ballot-box power.

Speaking of journalists, I have been involved with the media most of my life. At various times I have served as a freelance writer; photographer; newspaper columnist; magazine editor; public speaker; web site publisher; producer of TV, video, and radio shows; and so on. While I have covered topics as diverse as golf and football, the centerpiece of my stories has always been fishing—particularly saltwater fishing.

As a result, I have had the great fortune to have traveled to many places around the world in pursuit of fish, and to transform those experiences into stories for readers. Each trip widens my circle of friends and contacts, including local anglers, offshore charter boat captains and backcountry guides, regional outdoor journalists, and fisheries employees of all levels. Via the keyboard, I stay in touch with many of them to keep abreast of what is going on in their neck of the woods. This allows me to stay up-to-date about fisheries elsewhere and the management of them, and to gain a more realistic picture of the true circumstances rather than relying on representations by others.

The foregoing is not an exercise in self-aggrandizement. Rather, I offer it in the hope you will perceive that I have a credible and objective basis for some of the viewpoints I am about to raise. You may disagree or remain unconvinced, but at least you will have a clearer understanding of these positions. While I do not purport to represent anyone's position other than my own, I have remained involved in most aspects of recreational fishing.

Without doubt, my experience in the trenches of a statewide constitutional amendment was particularly eye opening. In the early 1990s, lawmakers in Florida refused to listen to recreational interests about concerns involving the effects of gillnetting. Gill nets were depleting mullet from Florida's waters. The nets catch fish as they try to swim through the mesh openings, trapping, and thereby strangling them. While the nets were indiscriminate in killing snook, bonefish, redfish, and the like, the targeted mullet were fast disappearing from bays and rivers. Mullet represent an important forage species upon which many game fish depend, particularly tarpon and snook. But as if losing the mullet themselves was not bad enough, mullet roe was being ripped from the fish's bellies and sold to Japanese wholesalers.

Working it out with the commercial industry did not work. And they, along with Florida's governor and legislators at the time, disregarded our threat to make it a constitutional amendment to ban gillnetting. "It will never happen," we were told. "You're bluffing," they laughed. But it was no bluff, and we made it happen. When the amendment reached the ballot box, 72% of Florida's citizens voted it in, putting an end to gillnetting and restricting other types of nets as well. To be sure, it was not a simple accomplishment, beginning with collecting hundreds of thousands of signed petitions and followed by a long and bitterly contested campaign.

This is not meant as a condemnation of commercial fishermen—it is their gear and methods we deplore. Many of these people are hardworking, honest, and mean well. And like anglers, they have had to live with myriad rules, regulations, and restrictions, whether by acquiescence or fighting them tooth and nail.

Support for restrictions on fishing practices is usually overwhelmingly accepted and supported by the recreational fishing sector. This may be because we consider fishing a sport rather than a livelihood, but we are no less passionate about the resources. In Florida, my home state, most such regulations have come about only in the past 20 years. Considering the increase in anglers and boats, it is totally understood that changes must be made when necessary—which is exactly the purpose of fisheries management. These relatively modern rules and restrictions have included fishing licenses, additional tag fees for keeping certain species, size and bag limits on virtually every type of fish, and off-seasons on some species. In addition, a catch-and-release ethic has emerged on the part of more and more anglers due to the heightened interest in conservation of our resources.

Recreational anglers are willing to accept more restrictions *if*—and I accentuate the word if—such restrictions are fair and seem likely to be helpful to our resources. If instead, such an initiative is draconian and appears to be pushed for a particular agenda, resistance will be pronounced and ongoing. And if measures are jammed down our throats or lawmakers will not remedy the situation, we will take our case directly to the public for relief. We have successfully done so before and will again if necessary.

Waging battles and obtaining constitutional remedies are not what the recreational industry wants. Conflict is too energy sapping. We seek cooperation, and we want to sit down with cool heads with mutual respect and a level playing field. Stack the committee, cite flawed data or play heavy-handed, and any semblance of cooperation evaporates.

The number of fisheries management entities to deal with is bewildering. A multitude of federal agencies may have cojurisdiction in the Department of Commerce's National Oceanic and Atmospheric Administration (NOAA), such as the National Marine Fisheries Service (NMFS), the Fish & Wildlife Service, the National Ocean Service, and regional fishery management councils. The Department of Interior's National Park Service also has a say-so in areas where the park takes in estuaries and saltwater coastal areas, such as Everglades National Park. And then you have got state agencies and local city and county governments. Each has their own flow chart of decision makers, researchers, support staffs, advisory councils, and media specialists pumping out press releases on policies and issues.

Add the surrounding cast of interest groups, user groups, individuals, and organizations dependent on those management decisions, and you have quite a diverse menagerie of characters lobbying their points of view. It therefore comes as no surprise that disagreements and personality conflicts often arise. Consequently, I have witnessed encouraging examples of cooperation and maddening examples of near-fisticuffs.

And yet setting aside differences for the cause of conservation is a human capability that does not exist in the rest of the animal world. A pride of lions will never convince a herd of wildebeest to agree to be eaten to save the rest of the herd; a boa will not be swayed to pass up a rabbit because it is undersized; and a grouper will eat every grunt (Haemulidae) it can catch today and worry about tomorrow, tomorrow.

But altruism and the promise of sound conservation is usually a lesser sway than money and on the winning side. People do things for their reasons, not someone else's—a maxim all diplomats recognize when tailoring the most appealing approaches. Explaining to a fisherman that an increase in license fees will inevitably help the resource is so nebulous that it is guaranteed not to enlist support; tell him part of the license money will go toward fixing the ramp where he likes to fish, and he will listen.

Unfortunately, one side often cooperates and later regrets it. If the ramp never gets repaired, you have lost that fisherman's trust for years to come, if not forever. Help an official by accurately reporting your fish catches only to have the data used to shut down your fishing area, well, who would not feel betrayed? It is one thing to bury me, but having me dig my own grave ends any hope of future cooperation, even if my doing so would save the snail darter. These types of flip-flops—whether unintentional or not—create very passionate "antis" who will lobby their peers to also not cooperate.

But the situation is not hopeless. Leaders of opposing countries at times deal with other leaders they do not personally like, because reaching an agreement might save lives. Stockholders of companies often merge or negotiate buyouts of major competitors even though the CEOs have been anything but friendly to one another in the past. When elected officials on opposite sides of the political spectrum work together on issues for the good of the nation, that is statesmanship. When people from different nations pool resources to build a bridge that benefits a community, that is cooperation. When people with extreme differences in approaches to fisheries management can join forces for the betterment of the resources, that is collaboration.

Politics makes strange bedfellows, and it does not have to lead to marriage. Indeed, such a rare window of opportunity arose during the aforementioned Net Ban Amendment in Florida when all environmental groups (except what is now the Ocean Conservancy) rallied to the side of banning gill nets. They did so not because they loved the recreational fishing sector, but because it fit into their sense of conservation.

PROBLEMS OF PERCEPTION—AND REALITY

Look at any poll asking citizens about their faith in government, and what do you see? A majority harbor a very low opinion. It is the same with views on politicians and politics in general.

Why do so many U.S. citizens distrust their government? It is not because people are unpatriotic. Without a doubt, very few of those who say they distrust the government would live anywhere else.

I believe there are two underlying reasons for this attitude:

1. In everyday life, we encounter indifferent bureaucrats at the driver's license bureau; in the school system; at the courthouse; at the water department, etc. We encounter long lines, long waits, and red tape, yet have to pay fees and obtain licenses for just everything except breathing.
2. People are fed up hearing the same old reason for more regulations and more restrictions. A governmental official pronounces that a major problem exists and issues the assurance that "the government is here to fix it." Far from knowing what is good for us, it is often the government itself to blame for resource problems. Just look at the damage wrought by the U.S. Army Corps of Engineers from years ago installing drainage canals, locks, levees, interstate highways, and other nature-altering projects. Years later we realize that all that redirection of natural water flow threw ecosystems completely out of balance—including saltwater estuaries.

Take the Everglades, for example. How many billions are now being spent (by the government) to restore water flow that was redirected in the past (by the government)?

It should be little wonder, then, that anglers along with most other citizens do not place a great deal of faith in the competency of government agencies. These agencies blame "resource users" for lack of fish, and then admit pollution and effluents have not been controlled on land, coastal development continues unabated, and they even ram their own research vessels into reefs, as NOAA did years ago off the Keys.

According to recent reports (Dr. Ault), as much as 50% of the coral reefs off the Florida Keys have died in the past 10 years—under the federal stewardship of the Florida Keys National Marine Sanctuary (FKNMS).

But now let us look at the other side of the coin. The FKNMS cannot stop coral bleaching due to circumstances beyond its control. And the Army Corps of Engineers simply does what it is instructed to do by higher government authorities. The point is that the finger of blame can point in many directions other than just fishing impacts. But on balance I happen to like my government as do most anglers, and I believe in the value of many good programs such as our National Park system.

I also think that marine sanctuaries such as the FKNMS can render many benefits to the resources, even though without question all the division and derision due to the push for MPAs there and around all of our coastlines is causing more harm than any other issue to the relationship between the recreational industry and government regulatory bodies. The payoff, if any, simply is not significant enough in the opinion of most anglers to warrant the incredibly unhealthy and contentious battles that are also causing battle lines to be drawn between user groups. That is a whole different debate, but it is germane to this chapter about garnering support and obtaining active participation by recreational anglers.

As we know, the U.S. government can make things happen both good and bad, depending on whose grant is being funded or whose privileges are being curtailed. As mentioned previously, the Department of Commerce's NOAA controls an immense flow chart of services, councils, and departments. They have legions of researchers on staff working full time to attain the goals they want. This vast bureaucracy with thousands of employees issues grants totaling many millions of dollars to hundreds of entities that, not surprisingly, support NOAA's programs. NOAA has major political clout and many allies.

Private enterprise can make things happen too. The recreational fishing industry is composed of boat companies, tackle manufacturers, associated gear such as clothing and electronics companies, and thousands of retailers and fishing guides. Anglers themselves number over 61 million nationwide and represent a multibillion dollar impact in terms of state and federal tax revenues (sales taxes on goods, fuel taxes for boats, license fees, etc.). Traveling anglers spend money on airfares and taxis, hire fishing guides, rent hotel rooms, eat in restaurants, etc., which has an enormous tourism impact. This economic power is many times that of, say, commercial fishing or the diving industry. Therefore, the recreational angling industry has clout too, and many allies.

By combining these two powerful forces, we could together push so many carts that lawmakers would clamor to provide support. Fisheries initiatives would go forward quickly, enthusiastically, and with a semblance of teamwork. Right now, that sense of camaraderie seldom exists between the recreational sector and federal fisheries managers.

"YOU DON'T KNOW WHAT YOU'RE TALKING ABOUT"

It is said that if two partners always agree on everything, it means one of them is not necessary. In fisheries management, finding disagreement on virtually every issue is a guarantee. One group's ox gets fed, the other group's ox is gored. That in itself is not problematic as long as cool heads prevail and issues are negotiated or handled in a fashion that is not oppressive or outright unjust. The "partners" just need to cooperate more often than not.

That is easier said than done, to be sure. And at times no matter how tactfully an issue is handled, the losing side reacts immaturely. Schisms occur when opposition or differences become one-sided or personal. Without going into all the parameters, I can say that such a schism has evolved between divers and anglers due to the push for more and more marine-protected areas. Commercial fishing interests have historically clashed with recreational interests to the point of schisms.

But the one that is the most disappointing to me over the past 20 years or so is the lack of mutual respect between marine scientists and anglers. Scientists often disregard observations from fishing guides and experienced anglers as mere "anecdotal evidence" worthy of little or no credence. Likewise, anglers have come to consider most fisheries scientists as "bald-and-bearded eggheads" who sit in chrome-and-leather swivel chairs in office buildings making recommendations on fisheries regulations and yet do not know which side of a fishing rod to pick up.

Why have these schisms developed, and what lessons can be learned from them going forward? While it is tough to show much love for researchers recommending measures that limit our vocation or avocation, distrust by recreational anglers—particularly in Florida, California and other coastal states—should not reach a level where it stunts the progress of doing what is really needed for our fisheries resources.

Specifically, the fishing industry in Florida generally supports the state's Fish and Wildlife Conservation Commission. They are looked upon as more of a partnership with the angling community. That is generally not the same view reserved (pun intended) for NMFS at this point in time, and anyone who does not recognize that fact is either naïve or disingenuous.

The good news: anglers and scientists can work in harmony, and it is occurring more and more often with bona fide research scientists working with nonprofit entities interested in funding solid research.

If this can be accomplished in private enterprise, it can be done elsewhere. But it will only happen when anglers and scientists appreciate what each brings to the table.

STEPS TO TAKE—AND NOT TO TAKE

Just as a commitment obtained by force will be broken at the first opportunity, a commitment obtained willingly will be kept a long time—sometimes forever. True, circumstances change, but generally speaking, people who are treated respectfully and whose viewpoints are given genuine credence are more apt to accept a compromise.

Much depends on the baggage of preconceived attitudes. As a member of various advisory panels, councils, and the like over the years, I have noted how the organizers do things and I reflect on their planning regimen. Without exception, the groups that encourage discourse among those with divergent opinions accomplish far more. You can see it even during the lunch break. Researchers sit together and if not ignoring those on the panel with whom they disagree, they certainly make no effort to socialize. At the cocktail party at day's end, you do not feel welcome entering certain circles of conversation. When the opposite occurs—as being invited to join the "opposition" for lunch or cocktails—it is amazing how common ground is actually sought between the parties. The apprehension melts away in almost every case.

Considering the foregoing, I offer the following pet peeves followed by the solutions for fisheries managers genuinely interested in winning broader support for their initiatives. You may not always capture the hearts of those opposing you, but playing fair and refraining from ostracizing people with a difference of opinion will pay heavy dividends in the short and long term.

1. It is a given fact that we can all accomplish a lot more by finding ways to work together. This does not require liking each other, mind you, just a genuine desire to be conciliatory even if the other party never agrees with your position. That civility pays off when gridlock does occur because the losing side is more apt to let it fall off their shoulders. If a measure is instead tactlessly jammed down the throat, the heavy-handed offenders will suffer accordingly when the political pendulum swings in the other direction. And it does swing in one fashion or another over the course of time.

    Solution: Act with class, even if you are holding all the cards.

2. I cannot tell you how many "public hearings" I have attended and left realizing that it was staged only because agency policy required holding them. The panelists' eyes would often glaze over, they would glance at each other with that "I really wish I didn't have to be here" look, and generally appear disinterested. It is a hearing, but they are not really listening. It is a subterfuge. They have made up their minds already and anything logical that may be said by a responsible and thoughtful party is missed or not absorbed.

    Solution: Listen, really listen, to opposing arguments. It is boring at times, to be sure, and much will be redundant, but some may include salient points that can lead to conciliation if not compromise.

3. Do not stack committees, councils, and panels in a heavy-handed manner and then wonder why true recreational interests quit showing up. You do not have to be politically savvy to recognize that a voting majority of those sitting around the table are there because the organizer already knows how they are going to vote on issues of importance. This in turn usually renders those with the audacity to vote the "other way" to be made to feel like uncompromising hotheads for not capitulating.

    Solution: While I am not naïve enough to believe that organizers will not stack committees—they do not want to derail their own train—at least do it moderately, so true debate can take place. If nothing else, reasons for opposition will be more clearly understood and even a workable compromise could emerge.

4. Regarding the above point, it is particularly insulting—and about as subtle a finesse as being hit over the head with a tire iron—when someone is appointed to serve on a council who purportedly represents recreational fishing interests when in fact on major issues they vote with the antifishers. Such people were obviously chosen either with the perception he or she is wishy-washy and can be easily won over or their positions are already known to the organizer. This is how "unanimous" votes often occur supporting an issue that unquestionably most anglers would oppose. It is not a coincidence that such shills who vote for measures that few in the recreational angling industry would support soon get appointed to other advisory groups to "represent" recreational interests. What is sad is that often these people mean well and think the appointments are due to their shrewd input, and do not even realize they are being used.

   Solution: Instead of appointing a "yes-man" for your cause, ask a legitimate group such as the Coastal Conservation Association, Recreational Fishing Alliance or International Game Fish Association to send or select a representative. You might not always get their vote, but at the least you will be including their voice. Doing so may negate the public relations ploy of a unanimous vote, but you will not be insulting the intelligence of the recreational fishing side and misleading the public.

5. Anyone with a financial interest on an issue should recuse himself or herself on that vote. To do otherwise is not any different than, say, a county commissioner voting to repaint the courthouse and then his paint company bids for the job. In other words, someone representing an organization should not vote for a fisheries action and then his organization applies for, or receives, money regarding that action. Can anyone not recognize this as a conflict of interest? Even so, it is allowed on every fisheries advisory group I have seen.

   Solution: Do not allow people in an advisory capacity to vote their pocketbooks.

6. Quit using the stick and carrot that a measure will be tried and if it does not work, it will be removed in 5 years. It is an argument convincing to only the most gullible of listeners. Call me cynical if you will, but show me anything that the government has taken and then given back for open use by the public.

   Solution: Come up with a stick and carrot that is genuine.

7. If you want to offer scientific evidence to support your initiative, do not try to adopt and adapt old studies that in some cases do not support your case. For example, I read proposals for a marine protected area (MPA) site that included data from a study on gyres and currents off the Florida Keys. But in reading the referenced study (yep, I went to the trouble), I noted that the shallowest reading taken was in water twice as deep as the deepest portion of the proposed MPA—and not anywhere close to it. How can a bona fide scientist argue that larval transport will occur via current propagation by citing data in a study that did not even take place at the MPA site? That kind of deception destroys credibility and stiffens opposition.

Remedy: Do your homework first. Identify, say, 25 potential sites that might be successful MPAs. Do fish censuses in each for 5 years. Conduct grid and cross-shelf mapping to determine optimum nursery environments. Design dye and floating buoy tests to document gyres and currents. Observe the intensity of resource use by anglers, commercial fishermen, divers, and others to gauge capacity levels. Photograph and measure sea grasses, corals, and other substrates. After 5 years of carefully collecting and analyzing you announce, "Of the 25 potential sites, we've pinpointed three that have a high degree of potential success due to an overall declining fish population, desirable currents and nursery conditions, and evidence of deleterious impacts from resource users. We propose making those three sites MPAs, and all user groups will sacrifice equally by staying out and the follow-up studies will be conducted by an independent team." If that were to take place, recreational anglers would help install the boundary buoys. And do not tell us NOAA does not have enough money to do it that way.

8. Follow-up studies are themselves worthy of separate mention. One reason major corporations hire an independent firm to conduct their own financial audits is because doing so themselves lacks objectivity. Why, then, are the very people doing follow-up studies on a project the same who vigorously proposed the project to begin with? While it would be nice to say all marine researchers are totally above reproach, human beings are human beings. They do not let you grade your own papers in school, but NOAA allows people to grade their own papers (the studies). Guess what? It will not surprise you to know that I have to yet read a follow-up study that said, "Our project fell short of the scientific model we designed to measure its success. Therefore, we won't be requesting further funding for it."

    Remedy: Allow others to do the periodic studies, especially those known to be objective (not the so-called "peer groups" that merely rubber stamp each other's work). If that were to be done, the recreational community—and truly objective researchers—would fully recognize such results as scientifically valid.

9. Do not push for MPAs up and down our coastlines that allow for-profit dive operators unfettered access while banning all forms of fishing. How can anyone wonder why the overwhelming preponderance of the recreational industry does not support MPAs. (And if you do not believe that is the case, name any major recreational fishing group or sportfishing publication that does support them.) You cannot crow about "erring on the side of conservation" on the one hand and on the other allow tens of thousands of mostly inexperienced divers interact with the very resources you are trying to protect. Studies show irrefutably that diving activities cause significant impacts and their presence biases fish behavior. Even logic should be enough to figure that out.

    Remedy: Although other reasons exist that call into question the value of MPAs off U.S. shorelines, making the rules the same for all user groups would at least reduce the ire of sport fishers. Just as anglers can fish elsewhere, divers can dive elsewhere. But do not say you want to set aside an area free from human disturbances and then allow entry by one group and ban another, no matter who causes more or less damage. Doing so will *never* lead to acceptance of MPAs when anglers cannot stop their boats in an MPA as they pass 12 dive boats sitting atop the reefs supposedly under protection.

Much of the resolution to these nine points boils down to good old fashion manners. A fellow I knew for years always came across in a difficult, prickly manner. Nothing pleased him. One day I tactfully said that he needs to lighten up a bit. He looked at me with an amused smile and said, "My father told me that there are two kinds of people in this world: those who give ulcers, and those who get them." I shook my head. "No, you're wrong," I replied. "There's a third kind of person: someone who neither gives ulcers nor gets them." He was dumbfounded. It never occurred to him that his dad could be wrong until he heard it from someone else.

If only life were always that easy! But if I could likewise be convincing, I would advise federal fisheries managers to also lighten up. Rather than quashing dissent, open the doors and drop any strident posturing. By doing so, our differences of opinion do not cross the line of being personal. And that goes for those on the recreational fishing side as well.

A fine example of cooperation involves my relationship with Dr. Jim Bohnsack, who until a promotion in 2005 was the main NMFS lightning rod pushing MPAs. Although I have been one of his most vocal critics, Jim never hesitated when I requested pertinent studies or materials—even though I often used them against him. He even let me rifle through his files one day completely unsupervised. We have gone head to head in magazine articles and debated before audiences. But afterward, we go to lunch, share a few beers, and although we might still debate MPAs one on one until we run out of spittle, after all was said and done we did not let it become personal. I have had a similar relationship with Jerry Ault, as we have collaborated closely over the years on matters relating to research efforts.

Speaking of Jerry, I told him that my only condition for contributing this chapter is that it not be censored. Although he does not work for NOAA/NMFS—my main punching bag throughout this chapter—Jerry's viewpoints and mine on several pressing conservation issues clash 180°. I can therefore assure you that he had to bite his tongue and maybe gnaw a few fingernails as well when reviewing what you just read. But to sanitize my thoughts would be tantamount to boiling the flavor out of the broth. That he agreed to allow my unabridged concerns and unsanitized remedies to appear in these pages is itself a testament to cooperation.

Without honest dialogue, what use is dialogue?

29 Sustaining Tarpon and Bonefish Fisheries: Scientists and Anglers Working Together

Mark Sosin

CONTENTS

Introduction 415
The Writer's View 416

An award-winning writer, photographer, radio personality, and television producer, Mark Sosin has an impressive list of credits that span virtually all phases of outdoor communications. He is the executive producer and on-camera host of Mark Sosin's Saltwater Journal, *broadcast to all 50 states and several foreign countries on The Outdoor Channel. More than 3000 of his articles have been published in major magazines and he is currently writing his 30th book.*

Considered a leading educator and one of America's most knowledgeable fishing authorities, Mark teaches outdoor techniques through a series of seminars and clinics, serves as a consultant to national companies, is an advisor to the IGFA, and shares his expertise with government agencies and conservation groups. He is a director emeritus of The Billfish Foundation and a former trustee of the University of Florida's Whitney Laboratory.

A past president of the Outdoor Writers Association of America and recipient of its coveted Excellence In Craft Award as well as its prestigious Ham Brown Award, Mark is a member of the American Society of Journalists and Authors, Society of Professional Journalists, Southeast Outdoor Press Association, and Metropolitan Outdoor Press Association. He has been enshrined in both the IGFA Fishing Hall of Fame and the Freshwater Fishing Hall of Fame.

INTRODUCTION

An unknown fisherman once opined that if we could accurately predict what a fish would do, catching that fish would not be nearly as much fun. I can tell you that I caught my first bonefish and tarpon more than 60 years ago, and I will safely add that I still do not understand very much about them. Struggle as I may over all those years, I still have not broken the code held sacred by these species.

Someone needs to tell me why a flat is loaded with bonefish one day, but it fails to hold even a single fish the next day on the same stage of the tide, identical water temperature, and no apparent change in the weather. While you are sharing knowledge and insights with me, explain why tarpon suddenly show up in the backcountry in January or February after 2 or 3 days of warm, calm weather, and then take off the instant the wind clocks around to the west. How do they know conditions are right for them, and how do they know conditions are about to change?

You can travel the flats over 100 miles of backcountry when the water temperature is 74°F and see a handful of tarpon. The next day, the water temperature is 75°F and the flats are loaded with hundreds and hundreds of the silver king. Where were they? How did they know?

I have seen bonefish roll like tarpon, and tarpon that would spook at the cranking of an engine get used to go-fast boats with three and four engines flying over their heads. Why do tarpon refuse flies much more frequently today than they did 30 years ago, or a school of bonefish totally ignore a well-placed fly that used to be considered a well-placed tidbit? The list of questions is endless and most attempts at answers rely on speculation and heresay rather than scientific fact. We probably know more about the moon (having placed men on it) than we do about the tarpon and bonefish native to our waters.

THE WRITER'S VIEW

More than 35 years ago, I wrote a book titled *Through the Fish's Eye* (Sosin, 1973) with information supplied by fisheries scientist John Clark. It was the first attempt to relate fisheries science to recreational angling and the sport fisherman. The book was considered so revolutionary at that time that the publisher held the manuscript for a year and a half before deciding to publish it. The abridged book was finally published, using only 40% of the original manuscript.

Through my work in researching magazine stories at marine laboratories back then, I had learned that scientists frowned upon any colleague who shared information with the lay public. Researchers talked only to other researchers. Few, if any, considered the rewards of sharing their findings with people who could benefit from the results of that research, while ignoring input and observations from those who fished for those species on a regular basis. And, in those days, an unsophisticated public never demanded it or seemed to care. You have to remember that early research focused on species that had commercial value in the marketplace, because the federal agency was the Bureau of Commercial Fisheries of the Department of Commerce. That later became the National Marine Fisheries Service (NMFS), although some of the long-time employees of the agency never made the transition to consider the recreational side.

When I started listening to dialogue about lateral lines, feeding behaviors, body shapes, the senses, and so forth, I asked where researchers got that information. "It's Ichthyology 101," they would reply, grinning over the fact that I did not know the answer. "Any college freshman in a fisheries science program knows those things," they would add. "Why don't fishermen know those basics?" I asked. You already know the answer. Scientists talk to scientists.

All this happened 35–40 years ago. Today, a considerable amount has been written on the basics of ichthyology 101, but little is shared beyond that. Research papers are circulated among the scientific community, but outsiders rarely ever get even a brief peek at an abstract. Scientists have to be more forthcoming with information and also develop research aimed at helping the recreational angler. Much of fisheries science is now funded through recreational licensing and excise taxes on fishing tackle as well as through organizations such as Bonefish & Tarpon Unlimited (BTU).

If it were not for the recreational angler, some scientific budgets would be severely limited or cancelled altogether. I have been asked on occasion to write a magazine article about a scientific project to help generate funding so that researchers could demonstrate that the research is in the public interest and benefits the public. Even on television every week, I try to share as much science as I can with my audience.

In recent years, a growing number of conservation and watchdog organizations have scientists on their staff. The International Game Fish Association (IGFA) has taken a leadership role in that area, with Glenda Kelley first and now Jason Schratwieser. Rob Kramer, the president of IGFA, and his wife, Lara, both have scientific backgrounds. The Billfish Foundation retains scientists to help them come up with hard facts about the species of concern to them. The Recreational Fishing Alliance relies on science as they lobby for legislation favorable to recreational anglers. The ongoing argument centers on the fact that fisheries management and legislation must be based on hard science and not emotion.

The challenge I hold out to you today is one of scientists and anglers working together to share knowledge and information. The time has come to push aside the mistaken notion that scientists should only talk to scientists or else they lose professional credibility. With more demands on our fisheries and more sophisticated boats and electronics to find fish, along with increased loss of critical habitats, many species are going to need all the help we can give them to maintain viable populations and even survive. The observations of fishermen in the field can often lead scientists in a specific direction, while scientific data can produce increased benefits for anglers on the water.

I can point with pride to the Florida Fish and Wildlife Conservation Commission and their education and outreach programs. For the past 5 years, we have been making major videos for them showcasing the work of researchers in the St. Petersburg lab and the satellite labs around the state. The purpose of these videos is to educate the public on the work these scientists are doing and how important it is to ensuring that Florida remains the *fishing capital of the world*. Much of the research funding comes from revenues generated through fishing license sales in Florida along with our share of the federal funds from the excise tax on fishing tackle.

That is a beginning. BTU and the Fisheries Conservation Foundation (FCF) have formed a strategic alliance, and together through a unique relationship have begun to draw information and guidance from both anglers and scientists. The purpose lies in establishing targets and goals that will prove beneficial to both. It is the proverbial single step that begins the journey of a thousand miles. There will always be unanswered questions and new challenges to undertake, but surmounting the waves and troughs of tomorrow will become easier if all of us pull on the oars together today.

REFERENCES

Sosin, M., and J. Clark. 1973. *Through the Fishes Eye: An Angler's Guide to Gamefish. An Outdoor Life Book*. Harper & Row, New York.

30 Incorporating User-Group Expertise in Bonefish and Tarpon Fishery Research to Support Science-Based Management Decision Making

Robert Humston, Jerald S. Ault, Jason Schratwieser, Michael F. Larkin, and Jiangang Luo

CONTENTS

Introduction 419
Unique Fisheries, Close Communities 421
Examples and Notable Results 422
    Bonefish Charter Captain Survey 422
    Bonefish Mark-Recapture Research 422
    South Florida Bonefish Census 423
    Tarpon Pop-Up Archival Transmitting (PAT) Tags 424
    Angler-Based Conservation Organizations 425
Summary 427
References 427

INTRODUCTION

Fisheries management planning and decision making have traditionally considered substantial amounts of input and involvement of user groups (Mikalsen and Jentoft, 2001). Providing opportunities for stakeholder input can increase their support for policies and procedures, and ensure compliance with regulations (Born and Stairs, 2003; Peterson and Evans, 2003; Schratwieser, 2006). Increasingly, environmental resource management and conservation agencies incorporate inputs from the stakeholders and public interest groups during the planning and implementation phases of policy development (Sharp and Lach, 2003; Pitcher, 2005). In particular, the idea of "comanagement" of fisheries has become more prevalent, despite the inherent difficulties of balancing competing interests and variable representation (Hernes and Mikalsen, 1999; Mikalsen and Jentoft, 2001; Sutinen and Johnston, 2003; Jentoft, 2005).

Cooperative research, by contrast, enlists fisheries stakeholders to work side by side with scientists and contribute their knowledge and expertise to the research process (NRC, 2004; Conway and Pomeroy, 2006). In this way, stakeholders can be actively involved at grassroots levels in developing the science from which final management decisions are based. Properly implemented cooperative research can generate stakeholder buy-in, improve compliance with management decisions and serve to perpetuate working relationships between stakeholders and fisheries management decision makers (Conway and Pomeroy, 2006; Schratwieser, 2006). However, the stakeholder's active voice in management discussions gives way to direct evidence provided by research data. That is, anglers may be reticent to assist in research for fear that the data or findings may be "used against them" (NRC, 2004). For example, data may justify stricter harvest regulations, area closures, or may simply reveal otherwise "closely guarded" locations of high stock abundance (Conway and Pomeroy, 2006). These disincentives create natural tensions in the tenuous relationships between scientists and anglers, and at times can impede the development and maintenance of cooperative research programs. Additionally, caution must be exercised when considering how anglers should participate in cooperative research. Initiatives need to be objective driven and should not be undertaken solely as a political gesture to demonstrate cooperation. Stakeholders should be selected because they have assets or special skills that will improve the quality of the research. A review of these and other sociological aspects of cooperative research in commercial fisheries was provided by the National Research Council (NRC, 2004), and more recently by Conway and Pomeroy (2006).

Management of recreational or "sport" fisheries, in particular, can benefit from the effort and experience of user groups (NRC, 2006). Lack of adequate catch-and-effort data from recreational fisheries presents unique challenges for studying stock dynamics, and many target species, especially in catch-and-release fisheries, have received only limited attention from fisheries biologists. Worldwide, sportfisheries for tarpon and bonefish embody these characteristics. Through the end of the twentieth century, most studies of the species' biology were focused on early life history and processes regulating larval development and metamorphosis (see Ault et al., Chapter 16, this volume). Research compiled and presented in this book demonstrates the increasing relevance of these species and their regional economic importance for charter fleets, sportfishing, and tourism-related industries throughout their range. A review of the author's credentials reveals a remarkable variety of experience and perspective being brought to bear in the current understanding of these types of fisheries. Several aspects of the recreational angler and charter captain communities plying bonefish and tarpon fisheries make them uniquely suited for contributing to research and management on these species. This chapter reviews some contributions these angling communities have made to the study of bonefish and tarpon—and their attendant benefits to research and management—with a focus on south Florida.

UNIQUE FISHERIES, CLOSE COMMUNITIES

The storied history of bonefish and tarpon fisheries in the southeastern United States is chronicled in other contributions to this volume (see Moret, Chapter 20; Apte, Chapter 21; Stearns, Chapter 22, and Sosin, Chapter 23). In North America, these species are regarded highly as sport fish, but generally disdained as table fare. An early and seminal conservation ethic has made these fisheries primarily catch-and-release with only the occasional trophy or record fish brought to the docks. The pioneers of these fisheries were also their earliest custodians, and they engendered a sense of ownership and responsibility for future sustainability of the resource among growing angler and charter captain communities. As the popularity of these fisheries—and of south Florida as a premier destination for international traveling anglers—increased, so grew the economic benefits and the regional pride for local angling opportunities and expertise. The result was a knowledgeable and tight-knit community of charter captains, outfitters, and anglers that recognized their stake in the present and future state of these resources.

The recreational and charter angling community in south Florida has been largely responsible for many citizen-based initiatives that have greatly influenced fisheries management decisions at local, state, and federal levels. Most notable were strong public involvement in the stringent regulation of gillnetting in Florida coastal waters, and designation of marine-protected areas (MPAs) throughout the Florida Keys coral reef tract. The University of Miami's bonefish and tarpon conservation research program was initiated when a group of anglers and captains approached scientists with their concerns about perceived declines and potential threats to local stocks. Since then, this and other programs have benefited significantly from the input and efforts of experienced anglers and captains.

The strong conservation ethic among these communities *generally* fosters an interest in supporting research whose primary objective is conservation and sustainability of the fishery. However, this has not necessarily been the rule and interactions between scientists, managers, and stakeholders in these fisheries have been subject to the same difficulties that have historically dogged these types of relationships (NRC, 2004; Conway and Pomeroy, 2006). The protective nature of these communities for their resource extends to all perceived threats; this may be characterized by an initial reticence to support or involve themselves in research activities. This group may be a majority or a vocal minority, but regardless, their impact is pervasive in the community. It is a mistake on the part of scientists to dismiss this early on as a widespread sign of apathy. Persistent and directed outreach efforts can generally assuage and overcome early resistance, if conservation goals are coincident between scientists and stakeholders (cf. Helvey, 2004; Conway and Pomeroy, 2006; Schratwieser, 2006). This was the case for University of Miami bonefish research in south Florida. Despite being initiated in collaboration with concerned anglers and captains, the program saw scattered, though staunch, support with relatively slow growth for the first 3–4 years of its existence. Today the support from charter captains and anglers is widespread and represents a cornerstone of the program's efforts and accomplishments.

The synergistic relationship in this example has hinged on communication: stakeholders are given a strong role and voice in research activities, and progress and results are relayed back in an efficient, interesting, and timely manner. The knowledge amassed among the more experienced anglers and captains—some of whom have specialized on targeting these species for over 50 years—represents a remarkable repository of information about individual, population, and community levels of organization in the resource ecology and fishery dynamics of bonefish and tarpon. They have built their angling successes upon an intimate understanding of behaviors, population cycles, spatial dynamics, and the trophic ecology of these fish as both predator and prey. When we first framed the goals and objectives of bonefish and tarpon research at the University of Miami, there was only a very limited scientific knowledge base on which to build and to guide early efforts. As a result, we sought out the input of anglers and captains as a matter of necessity; there was simply no alternative for developing a cost-effective research program that could ensure early productivity and generate acceptable progress. As the program developed and research took shape, data and analyses were reported to the support community for their reactions and responses, with the goal of achieving an efficient feedback loop of information transfer across all levels. In turn, support grew quickly as individuals realized the potential for making their voices heard in directing research and addressing vexing questions. With this support came an enormous volunteer effort that now forms the backbone for research activities.

EXAMPLES AND NOTABLE RESULTS

BONEFISH CHARTER CAPTAIN SURVEY

As discussed previously, the earliest efforts in bonefish research at the University of Miami were shaped by information collected from experienced charter captains. This included the distribution of mail surveys in 1997 (Humston, 2001) and 2002 (Larkin et al., submitted) designed to gather data on the charter fishery (e.g., fleet size, fishing effort, and catch rates) and captains' opinions on aspects of population biology and ecology of bonefish.

These surveys revealed remarkable consistency among individual opinions, and provided new information for designing experiments and directing early efforts. Researchers gained unique insights into seasonal variations in bonefish abundance on fishing grounds, general habitat preferences, and suspected patterns of movement and migrations. This information was instrumental in streamlining fieldwork and developing testable hypotheses. Surveys also served to disseminate news about the research efforts, and helped recruit vocal support and volunteer effort across the survey's geographic distribution. This approach was essential for gathering enough support for angler-based mark-recapture efforts (see below). Finally, data from these surveys have been instrumental in initial assessments of cryptic (release) fishing mortality and evaluating the quality of available data on mortality rates in the stock (Ault et al., Chapter 16, this volume).

BONEFISH MARK-RECAPTURE RESEARCH

The volunteer-based tagging program in the Florida Keys initially recruited highly skilled charter captains for assistance rather than recreational anglers. Annual fishing effort (in days fished) and catch rates of charter captains generally far exceed capabilities of sport anglers, therefore this recruiting strategy maximized the tagging output per kit distributed. Support among charter captains was limited in the first 3 years of the program, but we capitalized on the dedicated efforts of a core minority and worked to involve them in the scientific process. The early days were a difficult learning curve for both parties, and when communication waned, so did the interest of volunteers. However, when this was realized we took steps to become more consistent and proactive in our written and oral communications. Through the feedback and advice of patient captains, we refined the program to meet the needs of volunteers as well as scientists. The result was a productive relationship that has achieved groundbreaking results that challenged angler wisdom and scientific hypotheses alike. For example, early recaptures indicated remarkable long-distance (>200 km) movements by some bonefish along the longitudinal axis of the Florida Keys. This contradicted many angler opinions on movement patterns and was instrumental in unifying charter captain support across regions in the fishery and charter fleet. More recently, a bonefish tagged in Florida was recovered in the waters of the Bahamas—an incredible result that flies in the face of scientific hypotheses concerning stock structure, and suggests a wealth of new research questions. Data such as these that provide inferences on fish behaviors are greatly appreciated by anglers and charter captains, as individual and stock behavior are the areas in which they have the greatest personal and vested interests.

The volunteer base for the tagging program has increased steadily from 10 dedicated captains in 1998 to more than 75 captains in 2006 (see Larkin et al., Chapter 19, this volume). In addition, select and highly experienced recreational anglers (~30) have been furnished with tagging kits; these individuals were recruited based on their high angling effort and profile in the community. Limiting volunteers in this way maximizes tagging output per unit cost of equipment. Information and updates on the program are communicated via the program's website (www.bonefishresearch.com), where volunteers and supporters can enter and access information on tagging activity, recaptures, and other components of research effort. As participation among charter captains has increased, tagging has been increasingly incorporated into competitive angling events (specialized fishing tournaments) as well. In some cases, tournament organizers have furnished tagging kits to participating anglers and captains at their own cost, and more recently are awarding bonus points and cash awards for fish tagged during tournaments.

To date, these volunteer men and women have tagged over 4600 bonefish, with approximately 162 (3.5%) of these fish subsequently recaptured. This represents the largest data set available anywhere on bonefish catch-and-release activity. These data can be used to examine size structures and selectivity of the recreational catch as well as spatiotemporal variation in catch characteristics (catch rates, sizes, etc.). Individual size and growth data (of recaptured fish) have provided some initial evidence challenging previously described aspects of stock biology (e.g., maximum size and age). In the long run, these catch data will also be used to monitor status of the fishery and identify trends in stock abundance.

SOUTH FLORIDA BONEFISH CENSUS

While logbooks, surveys, and tagging programs can provide catch-and-effort rates in recreational fisheries, the inferences these data may provide on stock abundance are not without bias (Pollock et al., 1994; NRC, 2006). As in all stock assessments, a fishery-independent estimate of stock abundance is preferred when possible. Bonefish abundance is particularly difficult to assess with traditional sampling methods. Their high swimming speeds and wary nature increases escapement from sampling gear such as haul seines, trawls, or purse seines. Their preferred habitats likewise pose unique challenges for effective use of these gears for sampling adult abundance, or for in-water visual surveys.

Anglers intent on catching bonefish often employ "sight fishing" techniques, where the shallow expanses of flats are visually monitored for signs of bonefish feeding activity. Using shallow draft boats fitted with raised platforms, anglers can locate feeding bonefish or moving schools of fish on these flats from significant distances (sometimes >100 m) by cueing on water disturbance, localized turbidity (caused by benthic feeding), or even the appearance of bonefish fins (particularly caudal fins) above the water's surface. When bonefish are closely approached in this manner, individual fish can easily be distinguished under average conditions of visibility.

This method of angling provides a suitable foundation for a visual survey of bonefish abundance. However, two potentially confounding factors need to be considered: (1) the potentially high movement rates and large-scale ranging of bonefish could affect spatial point estimates of abundance (Colton and Alevizon, 1983; Humston, 2001; Humston et al., 2005; Larkin et al., Chapter 19, this volume); and (2) variation in climatic conditions can affect bonefish foraging behavior and presence on shallow flats (Humston, 2001; Humston et al., 2005; Larkin et al., in review), as well as the observer's ability to detect their presence. To compensate, an ideal design would include simultaneous, system-wide spatial coverage of sampling. This seemingly would require a veritable "army" of observers, mobilized in a structured and coordinated fashion.

In 2003–2006, the recreational and charter angling community of south Florida provided the volunteer workforce to accomplish just that. Led by Dr. Jerald Ault and outfitter Sandy Moret (Florida Keys Outfitters), a volunteer base of captains and anglers aboard 45–60 fishing vessels completed visual surveys over >1500 mi^2 of bonefish habitat. The results of these surveys provide the first fishery-independent estimate of bonefish stock size in south Florida (~300,000 individuals). The survey methods and results are detailed in a separate contribution (Ault et al., Chapter 26, this volume), but the method appears robust, consistent, and repeatable. This remarkable effort requires a significant commitment on the part of the angling community, not only in sheer effort but also in adhering to the detailed "rules" of sampling to provide accurate results. This underscores the enthusiasm and dedication this community has for the conservation of this valuable resource.

Tarpon Pop-Up Archival Transmitting (PAT) Tags

This component of the University of Miami's bonefish–tarpon conservation research program has benefited greatly from the support, direction, and assistance of captains and anglers. The high unit cost of archival satellite-transmitting tags inherently limits large sample sizes, and puts the onus on effective experimental design to generate useful data from a limited number of deployments. However, the advanced technology of this equipment appeals to nonscientists and therefore can generate a great deal of support from the angling community. Tags deployed in this research program have been donated almost exclusively by recreational anglers and more recently by state resource managers. Moreover, charter captains and anglers have likewise donated their time, boats, fishing equipment, and expertise to assist in the capture of fish and deployment of PAT tags. This has greatly reduced the operational costs associated with this type of research, and enhanced the capabilities of putting tags in the water in a variety of locations around Florida, southeastern United States, Gulf of Mexico, and the wider Caribbean Sea (see Luo et al., Chapter 18, this volume).

The importance of organized fishing tournaments and groups of concerned anglers and guides working in concert with scientists to capture fish and make them available for tagging with this space-age technology cannot be overemphasized. The aspect of local knowledge being integrated into these efforts concerning temporal availability and location of the fish, and the appropriate gear, baits, and lures to make catches has been an instrumental component of our regional successes. This has also greatly reduced the amount of time required on the water to accomplish program goals. An effective feedback mechanism has been to provide focused scientific presentations at captains meetings prior to tournaments to convey the somewhat complex findings in terms that are compatible with the angler's interests. This type of feedback loop has also provided an opportunity to discuss the issues with anglers, or for these same fishermen to offer future logistical or critical financial assistance to ensure the viability of the research program.

Angler-Based Conservation Organizations

Nonprofit conservation organizations have gained increased visibility, relevance, and respect in the fisheries and wildlife management community over the past few decades, following the path of notably successful organizations such as Ducks Unlimited (www.ducks.org), Trout Unlimited (www.tu.org), and the Coastal Conservation Association (www.joincca.org). Many of these organizations began as small, relatively localized grassroots efforts, spurred on by the concerns of citizens who felt management policy and actions had not met their interests or expectations for various conservation targets. Their influence is now felt throughout state capitals, U.S. House of Representatives and Senate and, most importantly, in having impact on the conservation and restoration of our coastal marine resources. These unique partnerships with scientists include provision of funding for research, sponsoring public and scientific conferences, and assisting with dissemination of relevant new findings. Their most influential role is as high-profile advocates for active conservation of natural resources, at both the public outreach and political lobbyist levels.

One of the most exciting, long-term developments in bonefish and tarpon management and conservation research has been the formation of nonprofit organizations dedicated solely to funding new research and advocating sound management and conservation of these flagship fisheries. Bonefish & Tarpon Unlimited (BTU; www.tarbone.org) and Tarpon Tomorrow (www.tarpontomorrow.org) have been able to bridge anglers, scientists, and managers in a coordinated effort to increase the state of understanding of these species. BTU was created by a group of Florida anglers and scientists that wished to see more directed, organized, and focused research and management of these species worldwide. With an objective of both stabilizing and enhancing the quality of shallow water fisheries, BTU is proactively defining and supporting research projects to unlock the puzzle of the factors that limit and control the number of adult fish in our shallow water fisheries. The organization has provided a venue for concerned anglers to become advocates for their favorite flats species. Moreover, BTU has increased the influence of angler expertise in fisheries science and management by bringing scientists, managers, and experienced anglers to the same table time and time again. This is accomplished through regular meetings of the scientific advisory board (which has representation from all three groups), and by subsidizing international research conferences on a semiannual basis. These conferences have increased communication between scientists and managers representing almost all the world fisheries for these species, as evidenced by this volume. Panel discussions at these meetings challenge all stakeholders to voice and debate their visions and concerns for the future of these fisheries: their study, management, and ultimate conservation goals. Education and outreach within angling communities are facilitated by sponsoring fishing tournaments (including on-site research presentations) and publishing newsletters and updates in mailing and web-based formats.

In addition to the contributions from species-specific organizations, research on bonefish and tarpon fisheries has benefited from the support of broader-based recreational angling organizations. The International Game Fish Association (IGFA; www.igfa.org) was founded in 1939 and is a not-for-profit organization committed to the conservation of game fish and the promotion of responsible, ethical angling practices through science, education, rule making, and record keeping. Throughout its history, IGFA has always had strong ties with fisheries research, and has endeavored to promote recreational angling not only as sport, but also as source of scientific data. In addition to compiling decades of catch information from around the world, IGFA staff, trustees, and international representatives have participated in cooperative research and management efforts worldwide. IGFA represents recreational anglers on numerous regional, national, and international fisheries management panels, and also funds research relating to game fish and their habitats.

Whether participating in cooperative research or management, IGFA's core purpose is to provide a link to and facilitate interaction and information exchange between recreational anglers, fisheries scientists, and managers. In addition, IGFA's historical and international status allows it to function as an "umbrella organization" for recreational angling, which in turn facilitates productive working relationships with other angling organizations as well. Unlike organizations that focus on a single or suite of game fish species (e.g., BTU, The Billfish Foundation, Bass Anglers Sportsman's Society, Tarpon Tomorrow), IGFA chronicles angling history and record catches, promotes ethical angling practices, and participates in research and conservation for all game fish. While this holistic approach allows IGFA broad visibility in the global recreational angling community, logistical constraints limit the number of long-term or intensive species-specific initiatives it can undertake as an individual organization. By partnering with species-focused organizations, IGFA can participate in more initiatives and also help improve the quality of research by providing unique and complementary assets to those brought by the organization spearheading the research.

IGFA's International Committee of Representatives has been particularly influential in facilitating research by fisheries scientists traveling around the world.

Representatives are well respected, influential members of the recreational fishing community and provide a primary mechanism by which IGFA functions in the international fishing arena. IGFA representatives have helped assist the University of Miami and BTU in several tarpon-tagging initiatives in Mexico by providing logistical support, purchasing tags, and coordinating anglers to participate in tagging efforts. Future cooperative tarpon-tagging initiatives are also being planned for Angola, Africa. IGFA headquarters also hosted and partially funded the International Tarpon and Bonefish Symposia held in 2003 and 2006, partnering with BTU and other conservation organizations to coordinate these landmark research and management workshops.

There is a common perception that recreational angler organizations are in direct competition for support and resources. However, the long-term conservation of valuable recreational fisheries such as bonefish and tarpon requires that these organizations work together to coordinate efforts with fisheries research and management. Angling organizations ultimately bear a larger conservation impact collectively rather than independently. Together organizations like BTU and IGFA have increased awareness of conservation issues for bonefish and tarpon among managers, scientists, and anglers alike. This likely represents the strongest example—and greatest contribution—of the user-group involvement in the science and management of these fisheries.

SUMMARY

Cooperative research studies on recreational fisheries—particularly volunteer-based tagging and logbook programs—have a longstanding history and important place in the study and management of sport fisheries (Cooke et al., 2000; Arlinghaus, 2006; Conway and Pomeroy, 2006; Schratwieser, 2006; Cooke and Philipp, Chapter 25, this volume). As the only potential source of data on catch rates in release fisheries, such studies can fill a persistent gap in the knowledge of bonefish and tarpon fisheries in particular. The cooperative activities of the bonefish and tarpon conservation research programs described here are by no means unique in their structure or achievements. Like all other cooperative fishery research programs, their success is based on establishing trust and consistent communication between participating anglers and scientists (Conway and Pomeroy, 2006). Indeed, whenever communication and connections between anglers and scientists are enhanced, the likelihood of meaningful progress in the management of recreational fisheries is practically guaranteed (Dumont and Long, 2003; Kelly, Chapter 28, this volume; Sosin, Chapter 29, this volume). We reviewed these examples to highlight their contributions to the scientific understanding of these species, and expect them to shape the management of their fisheries in the not-so-distant future.

REFERENCES

Arlinghaus, R. (2006) Overcoming human obstacles to conservation of recreational fishery resources, with emphasis on central Europe. *Environmental Conservation* 33: 46–59.

Born, S.M. and G.S. Stairs (2003) An overview of salmonid fisheries planning by state agencies in the United States. *Fisheries* 28(11): 15–25.

Colton, D.E. and W.S. Alevizon (1983) Movement patterns of bonefish, *Albula Vulpes*, in Bahamian waters. *Fishery Bulletin* 81(1): 148–154.

Conway, F.D.L. and C. Pomeroy (2006) Evaluating the human—as well as the biological—objectives of cooperative fisheries research. *Fisheries* 31(9): 447–454.

Cooke, S.J., W.I. Dunlop, D. MacClennan, and G. Power (2000) Applications and characteristics of angler diary programmes in Ontario, Canada. *Fisheries Management and Ecology* 7: 473–487.

Dumont, S. and C. Long (2003) Why fisheries management professional should go fishing. *Fisheries* 28(9): 32–33.

Helvey, M. (2004) Seeking consensus on designing marine protected areas: keeping the fishing community engaged. *Coastal Management* 32: 173–190.

Hernes, H.K. and K.H. Mikalsen (1999) From Protest to Participation: Environmental Groups and the Management of Marine Fisheries. Paper prepared for the 27th ECPR Joint Sessions of Workshops, Mannheim.

Humston, R. (2001) Development of movement models to assess the spatial dynamics of marine fish populations. Doctoral dissertation. University of Miami, Miami, FL.

Humston, R., J.S. Ault, M.F. Larkin, and J. Luo (2005) Movement of bonefish (*Albula vulpes*) in Biscayne Bay determined using ultrasonic transmitters and a passive acoustic receiver array. *Marine Ecology Progress Series* 291: 237–248.

Jentoft, S. (2005) Social science in fisheries management: a risk assessment. Pp. 177–184. In: Pitcher, T.J., J.B. Hart, and D. Pauly (eds). *Reinventing Fisheries Management*. Kluwer Academic Publishers, Dordrecht, the Netherlands.

Larkin, M.F., J.S. Ault, and R. Humston (In review) Mail survey of south Florida's bonefish charter fleet. *Transactions of the American Fisheries Society*.

Mikalsen, K.H. and S. Jentoft (2001) From user-groups to stakeholders? The public interest in fisheries management. *Marine Policy* 25: 281–292.

NRC (National Research Council) (2004) *Cooperative Research in the National Marine Fisheries Service*. National Academies Press, Washington, DC.

NRC (National Research Council) (2006) *Review of Recreational Fisheries Survey Methods*. National Academies Press, Washington, DC.

Peterson, J.T. and J.W. Evans (2003) Quantitative decision analysis for sport fisheries management. *Fisheries* 28(1): 10–21.

Pitcher, T.J. (2005) Back-to-the-future: a fresh policy initiative for fisheries and a restoration ecology for ocean ecosystems. *Philosophical Transactions of the Royal Society B: Biological Science* 360(1453): 107–121.

Pollock, K.H., C.M. Jones, and T.L. Brown (1994) Angler survey methods and their applications in fisheries management. American Fisheries Society Special Publication 25, Bethesda, MD.

Schratwieser, J.E. (2006) Integrating cooperative research and management: perspectives from a recreational fishing organization. Pp. 223–225. In: Read, A.N. and T.W. Hartley (eds). *Partnerships for a Common Purpose: Cooperative Fisheries Research and Management*. American Fisheries Society, Symposium 52, Bethesda, MD.

Sharp, S.B. and D. Lach. (2003) Integrating social values into fisheries management: a Pacific Northwest study. *Fisheries* 28(4): 10–15.

Sutinen, J.G. and R.J. Johnston (2003) Angling management organizations: integrating the recreational sector into fishery management. *Marine Policy* 27: 471–487.

Index

A

acoustic tag surgery, 42–3
acoustic telemetry (AT), 88, 302–6, 308, 310–4
acoustic transmitters, 43
Africa tarpon
    characteristics of, 132, 219
    map showing range of, 116
air breathing, 9, 15, 23
air exposure, 366–8, 374
air temperature, significance of, 117
Albua sp. A, 151
Albua sp. B, 151–2
Albua sp. C, 152
Albua sp. D, 152
Albua sp. E, 152
Albua spp., *see* bonefish
Albuliformes, 218
alleles, 135, 137, 141–2
Alligator Rivers, Indo-Pacific tarpon studies, 13
allometric relationships, 13, 227, 229, 232
allozymes, 132, 141
alongshore winds, 168, 171
Ambassidae, 17, 21–2
ambient turbidity, 277
Ambloplites rupestris, *see* rock bass
American alligator, nesting sites, 101
among-population genetic diversity, 136–7, 139
Amphipods, consumption of, 49
analysis of variance (ANOVA), 15, 33
Anchoa sp., 108
anchor-tagged bonefish, 302, 306–10, 317–9
anchovies, MDS analysis, 22
Andros Island, 83–5
angling activities, effects on bonefish
    air exposure, 366–8, 374
    degree of exhaustion, 364–5, 371, 374
    handling, 365–71, 374–5
    schematic illustration, 362–3
    visualization of, 367
Anguilla japonica, *see* Japanese eel
annelids, 189
annuli, 213
anoxia, 187
apex predators, 52
Applied Biosystems (ABI), 207–8
Apte, Stu, 58
Archosargus probatocephalus, *see* sheepshead
ARGOS satellite network, 279, 285–6, 295
Ariidae, 20–2
Ariid catfishes, 10
ARLEQUIN version 2.0, 134–5

artisanal fisherman, 122, 127
artisanal fishers, 5, 112
assignment test, 140–1
Atlantic bonefish
    essential habitats for, 250–1
    feeding habits, 251
    fisheries exploitation and human impacts, 247–9
    life cycle, 236–8
    population dynamics
        age and growth, 242–6, 251–2
        fecundity, 247
        lifetime survivorship, 246, 251
        maturity, 246–7, 251
        maximum age, 246, 251
        maximum size, 246
    resource ecology
        diet, 239–41
        regional movements and migration, 238–9, 242, 250
        spawning, 250
Atlantic bonito, 64
Atlantic Ocean, 203
Atlantic salmon
    air exposure, effects of, 366–7
    conservation management strategies, 363–4
    habitat, 271
Atlantic tarpon
    age and growth, 224–5, 227–32
    fecundity, 233–4
    fisheries exploitation and human impacts, 234–5
    life cycle, 219–22
    lifetime survivorship, 233
    maturity, 233–4
    maximum age, 232–3
    maximum size, 230, 233–5
    migration
        diadromous, *see* diadromous migration of Atlantic tarpon
        influential factors, 221–4, 252
        seasonal pattern and vertical habitat utilization from satellite PAT tags, *see* seasonal pattern and vertical habitat utilization by Atlantic tarpon
    population dynamics
        age and growth, 224–5, 227–32
        fecundity, 233–4
        lifetime survivorship, 233
        maturity, 233–4
        maximum age, 232–3

population dynamics (*contd.*)
    maximum size, 230, 233–5
    research parameters, 226
  research challenges, 218
  research parameters, 226
  resource ecology
    diet, 222
    regional movement and migration, 115, 221–4, 252
Australia
  Great Barrier Reef (GBR), 7, 11
  Northeastern Queensland, Indo-Pacific tarpon, 11–24
Australian Institute of Marine Science, 13
Australian National Sportfishing Association (ANSA), as information resource, 6, 8
autoregressive integrated moving average (ARIMA) 162

B

Bahamian bonefish fisheries
  body size, 86
  conservation
    needs, 88–9
    program strategies, 87–8
  ecology
    distribution and abundance, 83
    feeding, 85
    general applicability, 81–83
    habitat use and movements, 83–5
    population dynamics, 85–6
  economic importance, 79–80
  environmental influences, 88–9
  genetic analysis, 82
  historical perspectives
    recreational fishery, 80–1
    subsistence fishery, 80
  management strategies, 87–8
  research needs, 88–9
Bahamian Archipelago, *see* Bahamian bonefish fisheries
Bahamas
  Atlantic bonefish, 236, 247–8
  bonefish biology research, 203–4, 210
  bonefish fisheries, *see* Bahamian bonefish fisheries
  catch-and-release strategies, 368, 370–1
  fly rod quarry, 325–7
  leptocephali, 180, 190
  recreational anglers, 169
  water temperature, 362–3
bait, 372
barracudas, 95, 122, 127, 368
barramundi, 5, 22, 368
bathymetry, 83
Batrachoidiformes, 111

beachrock shorelines, 209
beach seines, 122
beetles, consumption of, 34–5
Belize
  bonefish biology research, 203, 205, 210
  fisheries, *see* Turneffe Atoll, Belize, tarpon and bonefish fishery
benthic fishes, 85
Bertalanffy equation, 232
Big Cypress National Preserve, 353
BigDye® Terminator v1.1, 208
billabongs, Indo-Pacific tarpon, 5, 23
billfish, 224
Billfish Foundation, 417
Biscayne National Park, 354, 356–7
bivalves, consumption of, 49, 85, 107, 110, 240
black noddies, 28
blacktip reef sharks, 43, 52–3
blue crabs, 222
bluefin tuna, 364
bluegill, 371
blue shark, 364
Boca Grande World's Richest Tarpon Tournament, 59
Boca Grip, 366
Bohnsack, Jim, Dr., 414
bonefish, *see specific types of bonefish*
  Belize fishery, 99–102
  characterized, 52, 147–8
  coastal ecosystem management, 93–8
  current fishery management in Florida, 400
  economic value of, 396–7
  evolutionary lineages and taxonomy
    phylogenetics, 149–50
    taxonomy, 148, 150–2
  fisheries, *see* Palmyra Atoll National Wildlife Refuge, recreational bonefish fishery
  larval development, 196
  reproductive biology, 105–8
  social stigma of, 80
Bonefish & Tarpon Unlimited (BTU), 239, 397, 417
Bonferroni corrections, 134, 136
bonga, 122, 124
bony fishes, 239
bootstrapping, 134, 139
bottom trawls, 121
brackish estuaries, 60
brackish water, 126, 265
Bray-Curtis index of similarity, 13
Brazil, Atlantic tarpon, 220, 234
bridge trolling, 346
British Virgin Islands
  Atlantic tarpon PAT tag deployment, 283, 286
  bonefish biology research, 210

Index **431**

British West Indies
    Atlantic tarpon PAT tag deployment, 281
    bonefish biology research, 210, 212–3
Brody growth, 33
broodfish management, 133, 142–3
brown shrimp, 222
Buck Island Reef National Monument, 354
bull sharks, 232, 368
buoyancy, Indo-Pacific tarpon, 9

C

caging effects, 44
calcium levels, 189
California sheephead, 53
canals, man-made, *see* Man-made canals
capture fishery, 122–4
Carangidae, 21–2
Caranx hippos, see crevalle jack
Caranx latus, see horse-eye jack
Carcharhinus leucas, see bull sharks
Carcharhinus melanopterus, see blacktip reef sharks
Caribbean, bonefish biology research
    annuli formation, 213
    conservation strategies, 211
    future research directions, 213
    growth rates, 212
    juvenile *vs.* adult habitats, 211
    methodologies
        otolith samples, 208–10
        tissue sample and genetic analysis, 206–9, 213
    observed lengths, 211
    overview, 203–4
    sampling, 204–6, 209, 212, 213
    sandy beaches, 209, 211–2
    spawning, 212
cast nets, 122
catch per hour (CPUE), 12, 18, 41, 47–8, 58
catch-and-release angling/fishing
    development of, 52–3
    occurrence of, 29, 87
    promotion strategies, 352
    stress response in bonefish, 43–7, 52
catch-and-release bonefish fisheries, sustainability improvements
    angling
        activities, impact of, 364–8
        guidelines for, 374–6
    considerations, 361–2
    degree of exhaustion, 364–6
    emerging issues
        facilitating recovery, 373–4
        novel hook design, 372–3
    environmental factors
        oxygen concentrations, 363
        water temperature, 362–3
    hooking injury, 371–2
    mortality estimates, 369–1
    overview of, 360–1
    predation issues, 368–9
    research recommendations, 376–7
catch-and-release programs/strategies
    Atlantic bonefish, 248
    Atlantic tarpon, 232, 234–5
    benefits of, 95
    mortality rates, 252
    types of stress, 235
caudal fork length (CFL), Indo-Pacific tarpon, 8, 13
Cayman Islands, bonefish biology research, 210, 212–3
census, as information resource, 383–97, 423–4
Central America, Atlantic tarpon, 221
Centropomidae, 9, 20, 22
Centropomus undecimalis, see snook
Cephalochordates, 49
channel nets
    characteristics of, 35–6, 158–9
    sampling, 162–3, 165, 167, 169, 171–3
channels, Indo-Pacific tarpon, 23
charter guide fishing, 238
chi-square analysis, 168
chondroitin sulfate, 181
chord-distance matrix, 134–8
circle hooks, 372–3
Clark, John, 416
Clupcidae, 17, 22
Clupeidae, 21
cluster analysis, Indo-Pacific tarpon, 21
Clypeasteridae, 49
coastal bays, Indo-Pacific tarpon, 5
Coastal Conservation Association, 412
coastal marsh habitats, 270
coastal waters, Indo-Pacific tarpon, 8
coastal zone development, 252
cobia, 64
coefficient of variation (CV), 391, 394–5
Columbia, Atlantic tarpon, 220
commercial fish pellets, 126
commercial fishing
    gill-net, 11–12, 14
    in Belize, 101–2
    Indo-Pacific tarpon, 5
computer software applications
    ARLEQUIN version 2.0, 134–5
    GENECLASS, 135
    GENEPOP, 134
    GENETIX version 4.04, 134
    PROC MDS program, 135
    Tools for Population Genetic Analyses, version 1.3, 135
conch, 101
confidence interval, 13

Conger myriaster, 190
conservation genetics, of tarpon
   allozymes, 141
   assignment test, 140–1
   geographic differentiation, 142
   haplotype diversity, 132
   heterozygosity, 142
   microsatellite variation research
      allele frequencies, 135, 137, 141–2
      among-population diversity, 136–41
      chord-distance matrix, 134–5
      collection methodologies, 133
      isolation and characterization methodologies, 134
      significance of, 132–3
      statistical analysis, 134–5
      within-population diversity, 135–6
   multidimensional scaling (MDS), 135, 137, 139, 141
   resource monitoring, 133
conservation management
   in Florida
      decision-making, scientific support of catch-and-release mortality, 400–3
      current fishery management, 400
      nursery habitats, 402–3
      significance of, 399–400, 402–3
   goals of, 421
   in National Parks, 355–7
conservation programs, significance of, 50, 52, 64, 66
conservation strategies, 211
controlled experiments, 376
CoonPop®, 60–1, 63, 65
cooperatives, 252
coral reefs, 5, 35–6, 85, 317
coral rubble, consumption of, 34
correlation coefficient, 168
cortisol levels, implications of, 44–5, 190
Coryphaena hippurus, see dolphinfish
Costa Rican tarpon, 132, 141–2, 196, 219–21, 228, 233–4
crabs, 11, 34–5, 48–9, 85, 126, 239–40
crawling crustaceans, consumption of, 35
crayfish, 117, 122, 124, 126
creeks, Indo-Pacific tarpon, 23
crevalle jack, 64
crocodiles, 368
Crocodylus porosus, see crocodiles
cross-correlations, bonefish and, 162–3, 178
cross-shelf winds, 168, 171
crustaceans, 10–11, 34–5, 85, 108–9, 189
cubera snapper, 96
cues, environmental, 190
culture fishery, 125–6
cupleiformes, 107, 111

Curtis, Bill, 384
Cynoscion nebulosus, see spotted seatrout
Cynoscion regalis, see weakfish
cytochrome b, 152
cytochrome sequencing, 207–9

D

daisy-chaining fish, 199, 343
dart tags, 42–3
decapods, 107, 111
degree of exhaustion, 364–5, 371, 374
diadromous migration research, Atlantic tarpon
   habitats
      adult, 271–2
      nursery, 270–1
   methodologies, 260–4
   otolith strontium levels, 264–70
diet, as influential factor, 222, 239–41
dissolved oxygen, 126
DNA, isolation in genetic analysis, 206–7
dolphinfish, 64
Domecq, Max, 57
drums, MDS analysis, 22
Ducks Unlimited, 425

E

early life history of bonefishes, Leptoceaphalos larva
   characteristics of, 180–2
   development phases, 181–2
   drawings of, 182
Echinoderms, consumption of, 49
ecology of bonefish, transition from late larvae to early juveniles
   overview of, 155–6
   larva stage
      behavioral adaptations, 160–1
      influx, interannual variability in, 168–169
      onshore migrations and settlement, timing of, 157–60
      physical transport processes, 161–8
      planktonic, duration of, 156–7
      recruitment, spatial variability in, 156, 169–73
   juvenile stage, settlement and habitats of, 155–6, 173–4
   research priorities and management applications, 174–5
   spawning, 157
ecotourism, 126
educational materials, 373
eels, 156, 186–7, 190, 196, 218, 270
electron transport system (ETS), 183, 184
El Niño-Southern Oscillation (ENSO), 184, 190

Index

Elopiformes, 218
Elopomorpha, 180–1, 218
Empire-South Pass Tarpon Rodeo, 59
Engraulidae, 21–2
Epinephelus striatus, see Nassau grouper
epipelagic organisms, 187
Esox masquinongy, see muskellunge
estuaries
    hyposaline, 189
    Indo-Pacific tarpon, 5, 11, 14, 23
    nursery habitats, 197
    set gill net fisheries
        barramundi, 5
        Indo-Pacific tarpon, 5, 14, 16
Ethmalosa fimbriata, see bonga
eutrophication, 9
Everglades National Park, bonefish and tarpon conservation, 353–4, 356
evolutionary theory, 148
exploitation by fisheries, 234–5, 247–9
exploratory sampling, 205

F

feeding habits, 111
finfish species, 64
fire worms, 35
fish consumption, by other fish, 11
Fish and Wildlife Service (FWS), 29
FishBase, 8, 10
Fisheries Conservation Foundation (FCF), 417
fishery management, critical knowledge gaps
    population dynamics, 251–2, 403
    reproductive biology, 403
    resource ecology, 250–1
    types of, 252–3
fishing gear, selection factors, 377
fishing license, saltwater, 66
fishing records, tarpon, 331–3
fishing tournaments, 58–9, 63–6, 363, 372–3
flats fishing, 203, 355–7
floodplains, Indo-Pacific tarpon, 5, 9, 23
flood tides, 86, 160, 163
Florida
    Atlantic tarpon PAT tag deployment, 277, 280, 282–3, 285
    bonefish, 111
    Fish and Wildlife Conservation Commission, 417
    Fish and Wildlife Conservation (FWC) Saltwater Record Program, 233
    Fish and Wildlife Research Institute, 207
    tarpon, 63, 132, 141–2
Florida Keys
    Atlantic bonefish, 236, 240–2, 248
    Atlantic tarpon, 219, 221, 227–8, 233

bonefish
    biology research, 203–7, 210–2
    historical perspectives, 347
    patterns in, 81, 84, 86
    population census, 383–97
    recruitment, 158
    settlement, 173
    spawning, 157
    fishing industry, 183, 199
    fly rod quarry, 323–7
    tarpon
        1940s–1960s, 346–7
        record, 333–8
Florida Keys National Marine Sanctuary (FKNMS), 409
Florida Marine Research Institute, 369
Florida Tarpon Tag Progam, 400
fly fishing
    catch-and-release, 95
    hook design, 372
    lightweight, 80
    saltwater, 58
    techniques, 112
fly rod
    quarry, 325–7
    record tarpon on, 329–38
follow-up studies, 413
food consumption, 10. *See also* diet
Food and Agricultural Organization of the United Nations, 5
foraging behaviors, 239
fork length (FL), 196, 242–4, 246–7, 277
freshwater systems, 23–4, 368
F-tests, 231

G

Gambusia affinis, see mollies
gastropods, 49, 107, 111
GENECLASS, 135
GENEPOP, 134
genetic analysis
    bonefish biology
        cytochrome sequencing, 207–9
        environmental influences, 213
Genetic-species-identification (GSI), 207
GENETIX version 4.04, 134
genomic DNA, 134, 207
Georgia, Atlantic tarpon PAT tag deployment, 277, 284
ghost crabs, 34–5
giant African threadfin, 122, 127
Gibson, Tom, 63
gills, development phases, 187
glassfish, MDS analysis, 22
glycosaminoglycans (GAGs), 181
gobies, 35, 85

Gold Cup Invitational Tarpon Fly Tournament, 59
Golden Meadow Tarpon Rodeo, 59
gonadosamatic index (GSI), 32–3, 107–8
Great Barrier Reef (GBR), Indo-Pacific tarpon studies, 7, 11
growth
    angling-related behavior, 377
    seasonality patterns, 34
growth rate
    calculation methods, Gulland-Holt, 8
    influential factors, 8, 15, 31–6, 111, 212
grunts, MDS analysis, 22
guides, angling guidelines, 374–5
Guidry, John, 65
gulf toadfish, 108, 110
Gulf of Mexico (GOM)
    Atlantic bonefish, 195–6, 200, 223–4, 237
    strontium levels, 268
    tarpon, 271
gulper eels, 218

H

habitats
    National Parks, 351–7
    selection factors, 238
Haemulidae, 21–2
hammerhead sharks, 232, 369
handling trauma, 365–71, 374–5, 400
haplotype diversity, 132
Hardy-Weinberg equilibrium, 134, 136, 142
Harengula humeralis, 108
hatching, Atlantic tarpon, 219
Hatiguanico River, Cuba, 94, 96–7
Hawaiian Islands
    bonefish population, 37–8, 40, 148
    National Parks, 354
head bones, 189
herrings, MDS analysis, 22
heterozygosity, 135–6, 142
high-salinity waters, 220, 224
histograms
    bonefish anchor tagging, 306
    strontium levels, 264–5
holistic angling, 80
homogeneity, 137
hook(s)
    barbless, 29, 366, 373, 375
    construction of, 372
    design factors, 372–3
    self-releasing, 373
hook-and-line capture, 400
hooking
    depth of, 371–2
    injury from, 371–2

techniques, 343, 346
    trauma, 400
hormones, 190
horse-eye jack, 95–6
hyaluronan, 181
Hybaid® thermocyclers, 207
hydrophones, 238, 304–5
hydrostatic pressure, larval development, 182–3
hyperglycemia, 44
hypersaline environment, 265–6
hypothalamic-hypophyseal axis, 190
hypoxia, 9, 187–8, 362–3

I

Ilaje people, 117
image analysis software, applications of, 36
Indo-Pacific tarpon
    air-breathing ability, 9, 15, 23
    capture fisheries, time series of, 5
    catch-and-release practices, 6
    catch data, 5, 17
    common names for, 5
    ecology, 10
    food items consumed by, 10–11, 23
    growth rate, 8, 15
    habitats, 5–6, 11, 13, 23
    human consumption of, 5–6
    information resources, 6, 8
    Northeastern Queensland populations, ecological analysis
        biotic interactions, 17, 20–22
        catch, characterization by fishery-independent surveys, 13–17
        distribution patterns, abiotic factors, 15–20
        ecological summary, 23
        fish data and analysis, 11–13
        growth rates, 15
        length-frequency histogram, 14–15
        length-weight relationship, 13, 15
        life history summary, 22
        study area, 11–12, 14
        vulnerability assessment, 23–24
    occurrences of, 3–5
    photo of, 9
    recreational catches, 6, 8
    stages of
        adult, 8–9, 22
        larval and juvenile, 7–8, 23
        swim bladder function, 6–7, 9
        tagging, 8
    tidal traps, capture during, 6–7, 15
    worldwide distributional map, 4
insects, consumption of, 11, 23, 85
insulin growth-like factor (IGF) system, 44
International Game Fish Association (IGFA), 8, 228–9, 233, 243–4, 412, 417

Index

International Grand Isle Tarpon Rodeo (IGITR), 59, 63–6
islands, insular oceanic, 212. *See also specific islands*
isopods, consumption of, 35

J

Japanese eel, 270
John G. Shedd Aquarium, 232
J-style hooks, 372–3
juvenile bonefish, 7–8, 23, 47–8

K

keraton sulfate (KS), 181
Key West National Wildlife Refuge, 158
kill tag programs, 343, 400
king mackerel, 64
Kiritimati (Christmas) Atoll, bonefish research
    artisanal fishing activities, 49
    characterized, 31, 37–40, 49–50, 52
    Fisheries Division, 49
    habitat degradation, 52
    life history, 49
    "no-kill" areas, 49
    prespawning, 50–2
    recruitment, 157–8
Kolmogorov-Smirnov two-sample test, 106

L

lace net, 125
lactate levels, implications of, 44
ladyfish, 156, 196
lagoonal habitats, Palmyra and Pacific bonefish, 42–3
lagoons, ecosystem in, 94–5
Lake Nicaragua tarpon, migration research, 260–71
largemouth bass, 363
larva, bonefish
    migration, 190
    Palmyra and Pacific, 35–40
    physiological ecology
        environmental factors, 182–5
        feeding ecology and nutrition, 188–9
        osmoregulation, 186–7
        overview of, 179–80
        oxygen availability and respiration, 187–8
        salinity tolerance, 186–7
Lates calcarifer, see barramundi
Lee Stocking Islands, Bahamas, 169
Leiognathidae, 17, 21–2

lemon sharks, 88, 368–9
Lepomis macrochirus, see bluegill
Leptocephali
    biology research in Caribbean, 204–13
    head development, 197
    larval stage, 156, 220
    Phase I development
        depth distribution of, 182–3
        metabolic rate, 187
        nutritional sources, 188–9
        osmoregulation, 186
    Phase II development
        calcium and phosphorus balance, 189
        characteristics of, 184
        nonfeeding, endogenous nutrients, 189
        oxygen requirements, 187–8
        salinity tolerance, 186–7
light attenuation, 277–8
Line Islands, bonefish population, 37–40
lizardfishes, 35
Liza sp., *see* mullets
Lobotes surinamensis, see tripletail
lobster fishing, 101, 117
logbook program, Palmyra and Pacific bonefish, 40–1
log-likelihood estimates, 135
Los Roques Archipelago (LRA) National Park, Venezuela, bonefish recreational fishery
    age in sample, 111
    description of, 103, 105
    economic importance of, 105
    feeding habits, 108, 111
    growth in sample, 111
    management programs, 113
    map of, 104
    "no-take" marine-protected area, 105, 113
    pre-Hispanic and artisanal fishery, 111–2
    recreational fishery, 112–3
    reproductive biology, 105–8
Louisiana
    Atlantic tarpon PAT tag deployment, 277, 282–3
    recreational tarpon fishery
        conservation program, 64, 66
        current perspectives, 61–62
        distribution patterns, 61–2, 64
        historical perspectives, 59–61
        International Grand Isle Tarpon Rodeo (IGITR), 59, 63–6
        management practices, 58, 66
        seasonal migration, 57–8
        spawning period, 62
        top 10 landed and weighed in, 63
        tournaments, 58–9, 63–6
        200 lb tarpon club and record tarpon, 62–3
        water temperature, 61–2

lugworms, 35
lunar cycle, impact of, 86, 159, 168
lure(s)
    artificial, 344, 346
    CoonPop, 61
Lutjanus cyanopterus, see cubera snapper

M

Macrophthalmus spp., *see* ghost crabs
Majidae, 107
Man-made canals, Indo-Pacific tarpon, 5, 17
manatees, 101
mangrove creeks, Indo-Pacific tarpon, 10
mangroves habitats, 10, 15, 23, 95, 99
Mantel test, 137
mantis shrimp, 35
maps
    Los Roques Archipelago (LRA) National Park, 104
    Palmyra Atoll National Wildlife Refuge, 29
    Peninsula de Zapata National Park, 94
    strontium, 267–9
    tarpon
        African, 116
        Nigerian, 116, 119
        worldwide, 4
    Turneffe Atoll, Belize, fisheries, 100
    vertical and thermohabitat utilization, 293
marine biodiversity, 87
marine catfish, 222
marine phytoplankton, 10
marine protected areas (MPAs), 84, 89, 102, 105, 174, 412–4, 421
marine reserves, 87
Markov-chain randomization, 134
mass spectrometry, 271
measurement methods, 117, 122, 127
MEGA version 2.1, 208
Megalopidae, 20, 126
Megalops atlanticus, see Atlantic tarpon
Megalops cyprinoides, see Indo-Pacific tarpon
menhaden, 222
metamorphosis
    Atlantic bonefish, 236
    bonefish conservation strategies and, 363
    hypoxia and, 188
    Indo-Pacific tarpon, 7–8, 22
    inshore, 190–1
    larval development phases, 181–3
    onset of, 190–1
    rates of, 183, 184
Mexico
    Atlantic tarpon
        characteristics of, 219, 221, 224, 228, 233
        PAT tag deployment, 277, 280–2, 285
    bonefish biology research, 203

microchemistry applications, 270, 272
Micropterus dolomieu, see smallmouth bass
microsatellites, 132–3
migration, *see specific species*
    bonefish in South Florida, 303, 317
    decision-making factors, 403
    fishery management strategies, 252
    onshore, 159–60
migratory tarpon, 58
Mithracinae, 108
mitochondrial DNA (mtDNA), 36, 38–9, 132, 141, 148, 152, 208
MODIS satellite, 293
Mola mola, see ocean sunfish
molecular markers, 36–7, 142
mollies, 222
mollusks, 85, 108–10, 189, 239
Morone saxatilis, see striped bass
mortality rate
    catch-and-release programs, 400–3
    species-specific, 232, 248
mud crab, 35
Mugillidae, 20–2
Mugil spp., *see* mullets; striped mullet
mullet, 5, 22, 122, 353
multidimensional scaling (MDS), 13, 21–2, 135, 137, 139, 141
multivariate analyses, 12–13
muskellunge, 366
mutations, 39
mysid shrimp, 35

N

NaCl efflux, 186
Nassau grouper, 161
natal bay philopatry, 132
National Marine Fisheries Service (NMFS)
    functions of, 415–6
    Gamefish Tagging Program, 222
National Oceanic and Atmospheric Administration (NOAA)
    Fisheries, 66
    functions of, 58, 287, 292, 397, 409, 413
    Grand Isle Station (GDILI), 61–2, 64
National Park of American Samoa, 354
National Parks (NP)
    characteristics of, 351–2
    conservation management
        general management plans (GMPs), 356–7
        objectives of, 352–3
National Recreational and Indigenous Fishing Survey, 6
nearshore marine fisheries, 366
Negaprion brevirostris, see lemon sharks

Index

neighbor-joining (N-J)
  algorithm, 134, 136, 139
  phenogram, 134
Nematopaleamon hastatus, see crayfish
neuroendocrine stress response, 44
neuston nets, 156
New Caledonia, Indo-Pacific tarpon studies, 13
Nigerian Institute for Oceanography and Marine Research, 133
Nigerian tarpon
  adult, 126–7
  capture fishery, 122–4
  catch rate, 127
  climate, impact on, 118
  culture fishery, 125–6
  description of, 115–6
  distribution of, 118, 120, 126
  ecotourism, 126
  fishing festivals, 117
  historical perspectives, 116–7
  larva, 126–7
  maps
    showing range of, 116
    study area, 119
  meristic characters of, 121–2, 127
  morphology, 121–2
  recruitment problems, 116
  research methodologies, 117
  resource ecology, 118, 120–2
  seasonal abundance, 120–1
  spawning, 126
  stock depletion, 127
  tidal influence, 118
nitrogen levels, 20
North Carolina, Atlantic tarpon PAT tag deployment, 280, 282
Notacanthiformes, 218
no-take marine protected areas, 87, 113
nursery habitats, 155

O

Oahu, Hawaii, bonefish research
  catch-and-release fishery, 46–7
  feeding, 48–9
  juvenile recruitment, 47–8
  stock identification, 45
  types of, 31, 37, 45
ocean currents, 155
ocean sunfish, 277
Ogyrididae, 48–9
Oncorhynchus spp., see Pacific salmonids; rainbow trout
one-way analysis of similarity (ANOSIM), 17
open-water spawning, 132
Opheliidae, 48–9
Opsanus beta, see gulf toadfish

osmoregulation, 186
otoliths
  bonefish biology research, 208–9
  growth bands of, 36
  migration research, 264–70
outreach programs, 373
overfishing, 50, 74, 211
overharvesting, 28
overlapping distribution, 148
Overseas Highway, 345, 347
oxeye herring, see Indo-Pacific tarpon

P

Pacific Ocean
  bonefish, 355
  National Parks, 354
Pacific of Panama tarpon, 132
Pacific salmonids, 368
Pacific threadfin, catch-and-release programs, 370
Palmyra Atoll National Wildlife Refuge, recreational bonefish fishery
  acoustic receivers, illustration of, 29
  annual rainfall, 28
  climate, 28
  conservation program, 52
  daily catch rates, 41–2
  environmental benefits, 28–9, 53
  fish stock definition, 30–1
  genetic(s)
    analysis, 36
    diversity, 38–39
    haplotypes, 39
  geographic location, 28
  growth
    allometric, 31–2
    and mortality, 33–4, 52–3
  information provided by fisheries
    bonefish physiological responses to catch-and-release stress, 43–4, 52–3
    remote monitoring of bonefish movement, 42–3
    spatial and temporal trends in bonefish catch, 30–2, 40–1
    tagging program, 30, 41–2
  information provided by other islands
    Hawaii, 31, 37, 45–9
    Kiritimati (Christmas) Atoll, 31, 37–40, 49–50, 52
    overview, 43–4
    Tarawa Atoll, Kiribati, 31, 50–2
  larval biology, characterized, 35–6, 52
  life history, 49, 53
  map of, 29
  population
    age of, 40

population (*contd.*)
    effective size, 39–40
    isolation, 37
    recruitment, 37–8, 47–8, 158–9
    sex ratios and reproductive condition, 31–3, 52
    size of fish, 30–2, 52
    spawning, 50
    sportfishing, 29
    stomach contents, 34–5
Papua New Guinea, Sepik River, 10
Pasiphaeidae, 49
Pate, Billy, 58
PCR amplification, 134. *See also* Polymerase chain reaction
peanut worms, 34–5
pelagic fishes, 364
penacid shrimp, 11, 23
Peninsula de Zapata National Park, Cuba, bonefish and tarpon sportfishing
    carrying capacity, 97
    catch rates, 95–7
    coastal ecological management program, 97–8
    description of, 93–4, 97
    government involvement, 98
    Hatiguanico River, 96–7
    La Salinas, 94–6
    map of, 94
    natural conditions, 97
    seasonal fishing, 95–8
    tourism, 93–4
    vegetative ecosystem, 94
permit, 95
Philippines, Indo-Pacific tarpon, 5
phosphorus levels, 20, 189
phylogenetics, bonefish, 149–50
phytoplankton, 126
pilchards, 353
pineal gland, 190
pink shrimp, 222
Pisonia, 28
pituitary gland, 190
plankton, 197
planktonic larval duration (PLD), 36
plants
    as food item, 11
    terrestrial, 10
plugs, 343–4, 346
pocket nets, 112
Poecilia spp., *see* mollies
polychaete worms, 34, 48–9, 107, 110
Polydactylus sexfilis, see Pacific threadfin
Polydactylus quadrifilis, see giant African threadfin
Polymerase chain reaction (PCR), 207–8
Polynemidae, 21–2

ponyfish, MDS analysis, 22
pop-up archival transmitting (PAT) tags, seasonal patterns in Atlantic tarpon
    characteristics of, 275–8
    data retrieval, 279
    deployment
        regional, 278–9
        in U.S., Mexico, and Trinidad, 280–1
    implications of, 281–6, 424–5
    photograph of, 278
    pop-up locations, 279
    technological advances, impact of, 276, 293, 296–7
population bottlenecks, 39
population census of bonefish, in Florida Keys
    annual 1-day, 391
    benefits of, 396–7
    frequency distribution, 392
    historical, 393
    research methodologies
        data collection, 385–6
        geographic coverage, 388–9, 395
        statistical sampling design, 386–7, 391
population dynamics, *see specific species*
    bonefish
        historical perspectives, 339–44
        in South Florida, 301–19
    fishery management concerns, 251–2
Portunus spp., *see* crabs
predation
    pressure, 52
    rates, 88
    risk of, 53
predator(s)
    detection by, 159
    detection of, 42
    overharvesting, 28
    tackle selection and, 373, 375
    vulnerability to, 190
predatory fishes, 368–9
prey of opportunity, 246
PRIMER V6, 13
principal component (PC) analysis, 16–20
Prionace glauca, see blue shark
PROC MDS program, 135
protective cover, types of, 238
Pterothrissus gissu, 183
Puerto Rico
    Atlantic bonefish, 236–7, 241, 244
    Atlantic tarpon, 219
    bonefish biology research, 206, 208, 210, 212–3
    Department of Natural Resources, 133
    tarpon migration research, 271
Puregene® DNA Isolation Kit, 207
p value, 168

Index

Q

queen conch, 111
Quickstep2 kit, 208

R

Rachycentron canadum, see cobia
rainbow trout, 366
raptors, 9
recreational anglers, support strategies
 importance of, 405–8
 management–angler relationship guidelines, 411–4
 perception and reality problems, 408–9
 preconceived attitudes and, 410–1
 schisms and, 410
recreational fisheries, catch-and-release programs, 248–9
Recreational Fishing Alliance, 412
recruitment strategies, 37–8, 47–8, 116, 156, 158–9, 169–73, 403
red drum, 64, 364
red-footed boobies, 28
regional movement, *see specific species*
remote detection, 160–1
remote monitoring, 42–3, 88
reproductive biology
 Atlantic tarpon, 219–20
 bonefishes
  length-frequency distribution, 106–7
  in South Florida, 316
 tarpon
  larval history, 196–7
  observations of, 197–200
  overview of, 195–6
  synthesis, 200–1
  video documentation, 197
reservoirs, Indo-Pacific tarpon, 5
resource ecology, fishery management concerns, 250–1. *See also specific species*
restriction fragment length polymorphisms, 132
ribbonfish, 222
rivers, tarpon habitat, 5
rock bass, 366, 373
rolling pods, 60

S

Saccopharyngiformes, 218
salinity levels, significance of, 20, 23, 126–7, 227
Salmo solar, see Atlantic salmon
salt marshes, 94
Salt River Bay National Preserve, 354
saltwater ecosystem, sportfishing in, 94–5

Salvinia molesta, 10
Sander vitreus, see walleye
sandy beaches, 209, 211–2
Sarda sarda, see Atlantic bonito
sardines, 353
Sarotherodon melanotheron, see tilapia
scatter plot, tarpon history, 74
Schouest, Captain Lance, Sr., 58, 60–2, 65
Sciaenidae, 21–2
Sciaenops ocellatus, see red drum
science-based decision-making
 Florida Keys, 399–3
 user-group expertise
  angler-based conservation organizations, 425–7
  benefits of, 419–420
  bonefish charter captain survey, 422
  bonefish mark-recapture research, 422–3
  conservation management goals, 421–2
  South Florida bonefish census, 423–4
  tarpon pop-up archival transmitting (PAT) tags, 424–5
scientist-angler relationship, importance of, 415–6
Scomberomorus maculatus, see king mackerel
seabirds, nesting locations, 28
sea catfish, MDS analysis, 22
seagrass, 10
sea urchin, 239
seasonal fishing, impact of weather conditions, 95–8
seasonal pattern and vertical habitat utilization, from satellite PAT tags
 future research directions, 297
 implications of, 275–86, 296–7
 transmitting technology, 276, 293, 296–7
 vertical and thermohabitat utilization, 286–93, 296
seawater, artificial, 189
Secchi disk visibility, 126
seine nets, 112
seine sampling, 206–7
sensory reception, 190
settlement
 habitats, 155–6, 173–4
 timing of, 159–60
shallow flats, 383
shallow shoreline habitats, 212, 236
sharks, 64, 122, 127
sheepshead, 53, 64
sheltered embayments, 5
Shelton self-releasing hooks, 373
shrimp, consumption of, 34–5, 48–9
shrimps, 222, 239–40
sight fishing, 424
silicate, Indo-Pacific tarpon, 20, 23

silver king
    anglers, 58, 62–3
    as goal, 276
    decline of, 342–4
silversides, 222
similarity percentages (SIMPER) analysis, 17
similarity analysis (ANOSIM), 111
Sipuncula, 110
size of bonefish, degree of exhaustion and, 365
smallmouth bass, 364
snails, 240
snapping shrimp, 35
snook, 95–6, 363, 406
Snow, Capt. Harry, Sr., 339
South Carolina, Atlantic tarpon PAT tag deployment, 280
Southeast Florida and Caribbean Recruitment (SEFCAR), 196
South Florida bonefish, 111. See also Florida Keys
Southwest Florida, tarpon history and trends
    background to, 69–70
    decline in captures, 71–2
    justification, 70
    juvenile tarpon, 75
    length-frequency histogram, 73
    methodology and materials, 70–1
    mortality rates, 75
    population structure changes, 75–6
    seasonal tarpon landings, 71–2
    size and weight of tarpon, 71–2, 74
spatial variability, 156
spawning, see Reproductive biology
    activity, influential factors, 157, 212
    Atlantic bonefish, 236
    Atlantic tarpon, 220, 223
    bonefish
        characteristics of, 7–8, 22, 107, 126, 132, 157, 195, 200
        rituals, 343
Sphyraena spp., *see* barracudas
Sphyrna mokarran, *see* hammerhead sharks
Sphyrna sp., *see* sharks
spinning technique, 112
spiny eels, 218
splash fishing, 50
sportfishing
    characteristics of, 6, 80, 93–8
    clubs, 69
    Peninsula de Zapata National Park, Cuba, 93–8
spotted seatrout, 64
squids, 239
standard deviations, 17
standard length (SL), 197, 242
state fishing regulations, 58
statistical analysis
    analysis of variance (ANOVA), 15, 33
    coefficient of variation (CV), 391, 394–5

multivariate linear model, 231
multivariate analysis, 12–13
standard deviation, 395
t-tests, 12
stocked ecosystems, 133
Stomatopods, 49
stream habitats, 270
stress response, significance of, 43–4
striped bass, 364, 374
striped mullet, 317
Strombus giigas, *see* queen conch
strontium maps, 267–9
subsistence fisheries, 80
Suncoast Tarpon Roundup, 59
surrounding-net fisheries, Indo-Pacific tarpon, 5
surveys, as information resource, 422
survival times, 191
survivorship, influential factors, 43. See also specific species
sweepstake recruitment, 39
swimbladder, 187
swimming behavior, larvae development, 189
swimming crabs, 35
swimming shrimp, 35
swimming speed, 376–7

T

tackle, 373, 375, 377
tag-and-recapture studies, 43
tag-and-release programs, benefits of, 46, 63–4, 85
tagging program, 8, 41–3
tag shedding, 41–2
tagging programs
    Atlantic bonefish, 245–6, 248–9
    bonefish in South Florida research methodologies, 301–2
    size measurements, 310, 314–7
Taq polymerase, 208
Tarawa Atoll, Kiribati, bonefish research, 31, 50–2
tarpon
    Belize fishery, 99–102
    coastal ecosystem management, 93–8
    current fishery management in Florida, 400
    description of, 131–2
    larval stage, 156
    PAT
        T-03, 286–9
        T-24, 289–293
    rodeos, 58–9, 65–6
    scales, historical, 72–4
    spawning, 132
t-bar tags, 42
teleosts, 107
temporal variability, 156

Terrebonne Sportman's League Anuual Rodeo, 59
Tertrapturus albidus, see white marlin
Texas
    Atlantic tarpon PAT tag deployment, 277, 280–3
    recreational fisheries, 70, 73
Texas Parks and Wildlife Department (TPWD), 133
The Nature Conservancy (TNC), 28–9, 54
threadfin, MDS analysis, 22
Through the Fish's Eye (Ault), 416
thryoxine, 190
Thunnus albacares, see yellowfin tuna
Thunnus thynnus, see bluefin tuna
tidal currents, 83
tidal influences, 118, 161
tidal traps, 6–7
tide-dominated estuary systems, Indo-Pacific tarpon, 15, 17, 21–22
tilapia, 122, 126
Tools for Population Genetic Analyses, version 1.3, 135
Trachinotus falcatus, see permit
traps, 6–7, 122
trevally, MDS analysis, 22
triiodothyronine, 190
tripletail, 64
trophy tarpon, 62
tropical bay, tarpon habitat, 11
tropical fisheries, 232
tropical rivers, tarpon habitat, 11
t-tests, 12
tuna, 224
turbidity, 227
Turks and Caicos, *see* Bahamian bonefish fisheries
Turneffe Atoll, Belize, tarpon and bonefish fishery
    description of, 100–1
    environment, 99
    government opportunities with, 101–2
    map, 100
    recreational anglers, 101
Turneffe Island Coastal Advisory Committee, 102

U

underground hydrologic system, 94
unit stocks, 252, 403
United Nations Educational, Scientific and Cultural Organization (UNESCO), 93
United States
    Fish and Wildlife Service, 28–29
    government program funding, 409
urchin, 49

V

VEMCO Model V16, 303
VEMCO V8SC-IL pingers, 42–3
Venezuala
    Atlantic bonefish, 241
    bonefish biology research, 203
vessel grounding, 356
Virgin Islands Coral Reef National Monument, 354
Virgin Islands National Park, 354
visible implant alphanumeric tags (VI Alpha), 42
von Bertalanffy growth function (VBGF), 33, 111–2
vulnerability, types of, 23, 190

W

walleye, 371
walleye pollack, 271
War-in-the-Pacific National Historic Park (Guam), 354
warm waters, 218, 288–9
water-breathing fishes, anoxic conditions, 9
water temperature
    larval development and, 182–4, 190
    reproductive biology and, 197
    significance of, 18–20, 23, 33, 117, 126, 295, 343, 362–3, 371
wave-dominated estuary systems, Indo-Pacific tarpon, 15, 17, 21–3
weakfish, 271
websites, as information resource, 6
white marlin, 364
white milt, 199
white shrimp, 222
Wildlife Computers, 277
Williams, Ted, 58
wind, calculation methods, 162
within-population genetic diversity, 135–6
world record tarpon, 57–8, 62–3
World's Richest Tarpon Tournament, 59
worldwide distributional map, of tarpon, 4
World Wildlife Fund, 102
worms, consumption of, 11, 34–5, 239–40

Y

yellowfin tuna, 364
YOY bonefish, 237

Z

zooplankton, 160
Z-score, 135